Learning as Self-Organization

Learning as Self-Organization

Learning as Self-Organization

Edited By:
Karl H. Pribram
Stanford University &
Radford University
and
Joseph King
Radford University

We gratefully acknowledge the interest, affection, and sustenance that Samuel Leven has provided which have made all of these conferences possible. We also want to acknowledge the sponsorship of the International Neural Network Society and their first Neural Network Leadership Award presented to Karl H. Pribram in 1994.

Additionally, we would like to acknowledge the efforts of Terry Lynn Hayes and Shelli D. Meade in preparing the manuscript for publication.

LONDON AND NEW YORK

First published 1996 Lawrence Erlbaum Associates, Inc.

Published 2022 by Routlege
2 Park Square, Milton Park, Abingdon, Oxon OX14 4RN
605 Third Avenue, New York, NY 10017

Routledge is an imprint of the Taylor & Francis Group, an informa business

Chapter 12, Morphogenesis and Mental Process, by J. Brown, is reprinted from *Development and Psychopathology, 6*, 1994, 551-563, with the permission of Cambridge University Press.

Chapter 14, Brain Regions Associated with Retrieval of Structurally Coherent Visual Information, by D. L. Schacter et al., is reprinted from with permission from *Nature, vol. 376*, August 17, 1995, Macmillan Magazines Limited.

Publisher's Note

The publisher has gone to great lengths to ensure the quality of this reprint but points out that some imperfections in the original copies may be apparent.

Library of Congress Cataloging-in-Publication Data

Learning as self-organization / edited by Karl H. Pribram and Joseph
  King.
    p.   cm.
  Includes bibliographical references and indexes.
  ISBN 0-8058-2586-X (pbk. : alk.paper)
  1. Learning, Psychology of. 2. Self-organizing systems.
3. Behaviorism (Psychology) 4. Neuropsychology. I. Pribram, Karl
H., 1919-   . II. King, Joseph.
  BF318.L3857 1996
  153.1'5--dc20   96-31389

ISBN: 978-0-805-82586-2 (pbk)

TABLE OF CONTENTS

Foreword

Karl H. Pribram

Foreword

Learning as Self-Organization

Karl H. Pribram

Foreword

Learning as Self-Organization

by Karl H. Pribram

A year before his death B.F. Skinner wrote that "There are two unavoidable gaps in any behavioral account: one between the stimulating action of the environment and the response of the organism and one between consequences and the resulting change in behavior. Only brain science can fill those gaps. In doing so it completes the account; it does not give a different account of the same thing." This declaration ended the epoch of radical behaviorism to the extent that it was based on the doctrine of the "empty organism", the doctrine that a behavioral science must be constructed purely on its own level of investi- gation.

However, Skinner was not completely correct in his assessment. Brain science on its own can no more fill the gaps than can single level behavioral science. It is the relation between data and formulations developed in the brain and the behavioral sciences that is needed.

To this end, a series of Appalachian Conferences on Behavioral Neurodynamics was initiated at Radford University in 1992 under the auspices of the International Neural Network Society (INNS) with a contribution from the National Institutes of Mental Health. The first three of these conferences were aimed at filling the first of Skinner's gaps: that which exists between the stimulating action of the environment and the response of the organism. The conferences were organized to flesh out a reasonably complete account of configural vision presented by Pribram (1991) in "Brain and Perception: Holonomy and Structure in Figural Processing." This monograph details the contribution of each brain structure to the perceptual process, ranging from the retina, to the far frontal cortex, within the context of mathematical models for each contribution. The first conference emphasized the synaptodendritic and nanoneurological (cytoskeletal) levels of processing and their significance both for theory construction and in practical applications using massively parallel processing architectures. The title of the conference publication is "Rethinking Neural Networks: Quantum Fields and Biological Data."

The second conference on Behavioral Neurodynamics continued the themes of the first. Emphasis was placed on how synaptodendritic processing could be inferred from the analysis of spike-train recordings; on the self-organizing aspects of image processing; and on the contributions of the brain's various motor control systems to forging entities (objects) within images. The title of the publication of the proceedings of this conference is "Origins: Brain and Self-Organization."

The third of these conferences convened last September. Its title reflected the concerns of the series as a whole: "Scale in Conscious Experience: Is the Brain too Important to be Left to Specialists to Study?" Prominent mathematical physicists, neurophysiologists, neuropsychologists and philosophers presented their contributions and adequate time was given for in-depth discussion.

The time is ripe for planning the fourth conference and to take the series in a new direction. The aim of the fourth and subsequent conferences is to explore the second of the gaps in the behavioral account noted by Skinner, that between consequences and the resulting changes in behavior.

During the 1970s major advances were made in our understanding of the brain processes involved in perception. It was found that visual pattern processing is akin to that involved in auditory functioning, that the "eye" was more like the "ear" than had been suspected (see DeValois and DeValois 1988 and Pribram 1991 for review). This allowed general mathematical models of the perceptual process to be developed and to be tested. The first three Appalachian Conferences on Behavioral Neurodynamics were based on these developments.

During the 1990s a similar series of advances appear to be emerging with regard to our under- standing of the brain processes involved in "valuation". In the language of behavioral science the issue centers on the nature of the process of reinforcement. The fourth conference aims to bring together those who are making these advances on the behavioral level (mainly working in the tradition of operant conditioning) and those working with brains (mainly amygdala, hippocampus, and far frontal cortex).

Our reading of these advances is the following: during the heyday of stimulus-response psychology, reinforcing events were considered to be either drive inducing or drive reducing. This view foundered on neurobehavioral demon- strations that after lesions of the ventromedial hypothalamus, a rat would become obese if given food ad libitum but would starve if it had to overcome an obstacle or press a panel in order to obtain a reward. (For review, see Pribram 1970). How could a rat have both increased and decreased drive depending on the situation? How could reward be both drive inducing and drive reducing depending on the situation? The cognitive construct "effort" came closer to describing the results of the experiment than did the stimulus-response construct "drive". What then might effort be? Under what circum- stances would effort be expended? What is the relationship between effort and reinforcement? In a seminal study, David Premack (1965) provided the first steps toward and answer to this question. Premack showed that reinforcement occurs whenever a response with a lower independent rate coincides, coheres, with stimuli that govern the occurrence of a response with a higher independent rate. Thus, the organism tends to increase the response of the lower rate to approach the rate of the response of the higher independent rate. The organism expends effort. Premack used running in an activity wheel and licking a drinking tube to measure behavioral rates of response, and showed that the reinforcing relationship was reversible depending on deprivation circumstances.

All of this detail may seem irrelevant to practical concerns. That this is not so, can be illustrated with an example from research with monkeys. A monkey is rewarded with a peanut for making a correct response. He places the peanut in his food pouch. On the next trial he makes a mistake. No peanut is given. The monkey takes a peanut from his pouch and munches it with obvious relish. This behavior is repeated but the monkey learns the problem readily. On another occasion, the now experienced monkey continues testing for hundreds of trials without any reward -- the peanut delivery machine has gone awry and the experimenter failed to notice it.

We tell persons in psychotherapy to reward themselves when everything seems to be going wrong: to take control over their own well being. We note that some people remain motivated despite a paucity of rewards. What makes such highly motivated people tick? Can we help maintain motivation in the classroom and in the clinic?

To summarize: The aim of the conference was to explore the aphorism: <u>The motivation for learning is self organization</u>. In keeping with this aim and in the spirit of previous conferences, the mission of the conference was to acquaint scientists working in one discipline with the work going on in other disciplines that is relevant to both.

Keynote

Konrad Z. Lorenz

Innate Bases of Learning

by Konrad Z. Lorenz

Life as a Knowledge Process

The greatest wonder of organic life is that it develops, in seeming defiance of all the laws of probability, from the simple to the complicated, from systems of lower to systems of higher harmony. However, infractions of the second law of thermodynamics simply do not occur, and, to achieve what it does, life is dependent on a gradient in the general, all-pervading flow of energy dissipation. Life lives on negative entropy. Any living species is a system that, very much like a prairie fire, greedily gathers energy and, in a positive feedback cycle, becomes able to gather the more energy, and to do it the quicker, the more it has already acquired.

What distinguishes the organism from this and other inorganic systems that also gather energy with a positive feedback is its special structure, which is molded by evolution to make probable the gain of energy and to exploit highly specific sources of energy. The process of this molding is called adaptation, and, thanks to the old Darwinian theory of natural selection and to the recent findings of biochemistry, we have a pretty good idea of the mechanisms achieving it. These insights not only justify but demand our asking a question unknown to physics and chemistry, the question, "what for?" This is not the *teleological* question concerning the ultimate aims and reasons of creation; it is just shorthand for the Darwinian question: "which is the function whose survival value exerted the selection pressure *causing* the species to evolve the characteristics that our question concerns?" Colin Pittendrigh suggested the term *teleonomy* for this kind of approach, hoping, by analogy, to divorce the concept of teleonomy from that of teleology as strictly as the concepts of astronomy and astrology are divided.

Every process of adapting a species to a certain given in its environment creates a new correspondence between the properties of the organism and those of its ambience. By virtue of its causation by selection, this change of the organism is one that increases the chances of its gaining energy—in other words, the probability of its survival. As it is the organism that changes in the process and not (or not to an appreciable degree) the environment, it is the former that develops a mold, in fact an image, of its natural environment. The fin of a fish reflects the properties of water, much as a horse's hoof reflects those of the hard, even ground of a steppe, or as an eye reflects the properties of light emanating from the sun.

In the parlance of information theory, it is correct to say that the process of adaptation increases the transinformation between the living system and its environment. However, the concept of information, as defined by information theorists, is formed in intentional disregard of the semantic level, consciously abstracting from the *meaning* that information concerning the environment may have for the organism in the interest of its survival. In the parlance of information theorists, it is impossible to speak of information *about* something. If I should confine myself to their terminology, I should have to forgo the important teleonomic aspects by which the biological concept of adaptation is determined. I shall, therefore, use the term "information" as it is used in common parlance and as it has been exactly defined by Bernhard Hassenstein (27). To define this concept in the terms of information theory, one might say that it is that kind of transinformation, between an organism and its environment, that is effected by the adaptation of the former to the latter. "Information" in common parlance means relevant, teleonomically organized information that has a meaning for the organism receiving or possessing it. In other words, it means *knowledge*. After long and somewhat bitter experience in the attempt to avoid misunderstanding, I shall speak of "information" when discussing cognitive functions of lower organisms, in order to avoid the reproach of ascribing to them human, conscious knowledge processes; and I shall speak of "knowledge" when dealing with the latter, in order not to countenance the pernicious error of equating knowledge with information in the sense of information theory. However, *the terms "information" and "knowledge" will be used synonymously* in this paper, both meaning relevant, teleonomically meaningful information. This will facilitate our approach to the all-important question of how, by creating in itself a progressively detailed image of its environment, the organism improves its chances of gaining energy and thereby surviving.

Any relevant information that the living system acquires concerning its environment improves its chances of increasing its capital of energy—in other words, its rate of propagation. Otto Rössler (73) was the first to formulate clearly that this positive feedback of energy-gain (which in itself is paying compound interest!) is linked with another cycle of positive feedback—that of information-gain—because the increased rate of propagation made possible by newly acquired information, in its turn, opens new possibilities for further gain of information, since among an increased progeny there also are increased chances of successful new mutations and recombinations of genes.

This twin cycle of positive feedback is characteristic of all that is alive, even of the "borrowed life," as Weidel (93) calls it, of a virus. Though true, it is misleading to say that evolution proceeds solely by random change and by weeding out the unfit. What life really does is conduct a closely knit and active enterprise, in which research and gain of capital collaborate to each other's advantage. The way a modern commercial concern continuously spends a certain proportion of its financial gains on further research is not a model, but a special case of this essential life process.

Thus viewed, life itself is a knowledge process. Donald Campbell, in his essay on "Evolutionary Epistemology" (8), argues that "the natural selection paradigm for such knowledge increments can be generalized to other epistemic activities, such as learning, thought and science." This is entirely true; in fact it is more or less what I propose to do here, though keeping in mind that life is, on the other side, an economical (one is tempted to say a commercial) process. It is the economic advantages of having more knowledge about environment that, since the very beginning of life, have exerted the selection pressure responsible for the progressively higher evolution of all knowledge-acquiring organic structures and functions. The general trend to more complex, "higher" organization, discernible in most evolutionary processes, is explicable on this principle; in my opinion, there is no need to assume a "demiurgic intelligence," as J. G. Bennett does in his *Dramatic Universe* (4). In fact, the levels of what, in living creatures, we call "lower" and "higher" organization cannot be better or more objectively defined than by the amount of relevant, teleonomically organized information they possess—and this applies as well to information contained in the genome as to that which an individual or, for that matter, a human culture has acquired in its life span. In our emotional estimate of the lower and the higher, we are also swayed by the potential ability of an organism to acquire new information, new knowledge.

A condensed survey of the physiological functions that have been evolved to acquire information can serve, simultaneously, to furnish impressive illustrations of how evolution in general proceeds and also to give us an unbiased view of the position that *learning* holds in the frame of reference of all the other knowledge processes.

In describing evolution, we are forever hampered by the fact that our vocabulary was created by a culture not yet aware of phylogeny. All the existing terms (development, evolution, *Entwicklung,* etc.) imply the unfolding of something preexisting, wrapped closely into a tight bundle, as a flower is in its bud. They are wonderfully expressive of what they are made to express, the processes of ontogeny, but they fail miserably to do justice to what is the essence of evolution, the coming-into-existence of *something entirely new, which simply did not exist before.* Even the German word *Schöpfung* implies etymologically that some preexisting substance is being ladled out of a great reservoir. Some philosophers of evolution, feeling the inadequacy of these words and groping for a new one, have rather pathetically hit on the term "emergence," which, worse than any other, suggests that an entirely preexisting thing, like a surfacing walrus, puts in an appearance above the water, which previously, to a literally superficial view, had seemed empty. Some theistic philosophers have coined, for the act of creating something entirely new, the term *fulguratio* "lightning," which implies that a creative stroke of lightning emanates from an all-knowing and eternal god. By an etymological fluke of coincidence, this term is more descriptive of what really happens than are all those aforementioned. To us, the thunderbolt of Zeus is an electric spark like any other, and the first thing that comes to our mind on seeing a spark at an unexpected point in a system is a short circuit. When the beginning and the end of a one-way chain of causation establish a connection, so that the end effect influences the first cause, a feedback cycle is established; in other words, the previously linear chain is transformed into a system possessing entirely new systemic properties.

Cybernetics and general systems theory have relieved the coming-into-existence of entirely new properties and functions of the odium of being a miracle. It is no miracle if a preexistent linear chain of causes and effects becomes closed into a circle, though the consequences may be truly epoch-making. One of the events that must have happened at the origin of life is the "fulguration" of a feedback cycle containing, in its chain of causation, one link whose effect on the next one bore a negative sign. If all the effects in a feedback circle are positive, they inevitably result in a snowballing of events that must sooner or later spell destruction to the system. The better the prairie fire burns, the sooner it burns itself out; the avalanche is of still shorter duration. If the great twin cycle of positive feedback of which I have spoken has not, as yet, led to a dangerous disturbance of the equilibrium between life and its environment, this is exclusively due to the overpowering odds against which life has to contend in a pitiless, inorganic universe. In fact, with man's increasing mastery of our planet, the positive feedback cycle under discussion may still lead to destruction.

No permanent equilibrium of an open system can be established in a variable environment without the aid of the negative, or self-regulating, feedback cycle. It is all but impossible to construe even a thought model of life without reference to this principle. Very probably, the first regulating cycles, the principle of homeostasis, must have come into being simultaneously with, or even before, the molecular structure performing self-reduplication and the trial-and-success procedure of gaining information. Indeed, the regulating cycle itself can be regarded as the most primitive (or at least the simplest) form of a knowledge process, because, as I shall explain later, it is able, quite by itself, to feed into the organic system relevant information about the environment.

The establishment of a regulating cycle is perhaps the most important, but it is far from being the only possible, "fulguration" of new systemic properties coming into existence at the integration of several independently preexisting subsystems into a functional whole of higher order. Hassenstein's (27) electric model, represented in Figure 1, is sufficient to explain the principle. The cryptic-sounding statement of gestalt psychologists that the whole is more than its parts contains exactly that amount of truth.

Very many and very different scientists and philosophers have recognized the role that this type of integration has played in evolution. Goethe defined "development" as differentiation of the parts and as their subordination to the whole. W. H. Thorpe (92), in his book *Science, Man and Morals,* has given many examples of how "unity out of diversity" constitutes a creative principle in evolution; L. von Bertalanffy, in his book on theoretical biology (5), has discussed it with great exactitude; and Teilhard de Chardin (88) has expressed it in the shortest and most poetic way in saying, *"Créer, c'est unir."*

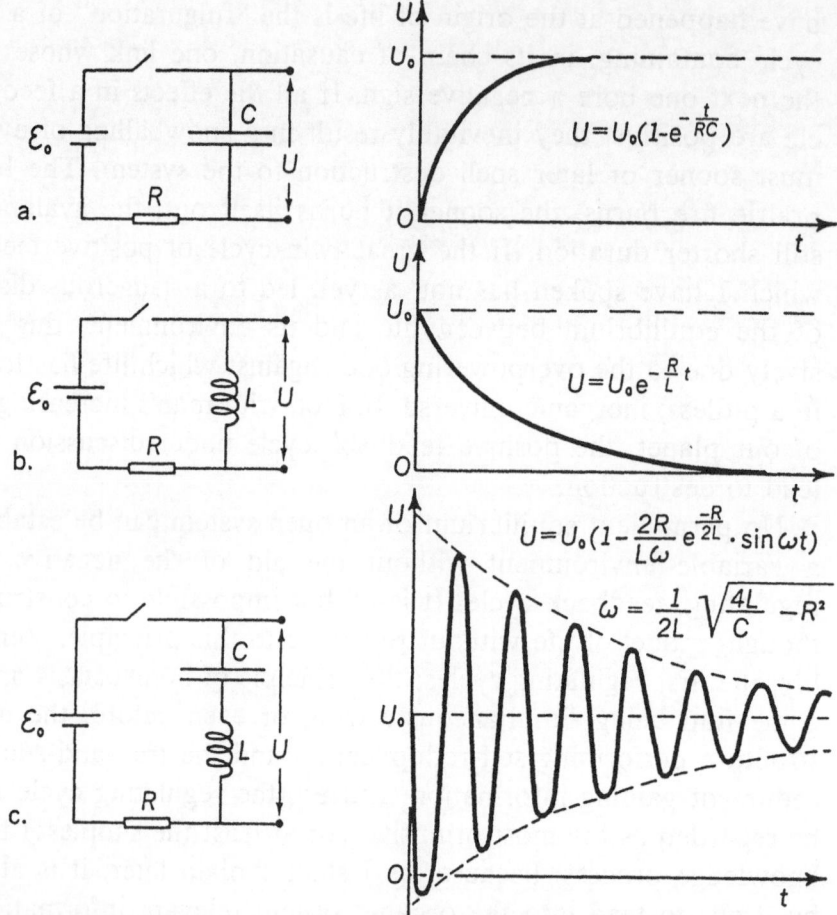

FIGURE 1 Three circuits, among them an oscillating cycle, illustrate the concept of "systemic property." The poles of a battery with the electromotoric power E_o and with the potential tension U_o are joined by a conduit. The resistance of the circuit, in ohms, is supposed to be concentrated in R. In circuit a, there is a condenser of the capacity C; in circuit b, an induction coil L; and in c, both condenser and coil together. At two terminals, the voltage U can be measured. The diagrams at the right represent the changes occurring in this tension after the circuit is shut. In a, the condenser is gradually charged up, until the tension U_o is reached. In b, the current, which at first had to overcome the resistance offered by the self-induction of the coil, increases until the current intensity determined by E_o and R is reached, the tension U then being theoretically zero—the resistance of the circuit being thought of as concentrated in R. In c, the closing of the switch causes an oscillation of decreasing amplitude. It is apparent that the systemic properties of c are not the result of an additive superposition of those characteristic of a and b. The diagram is valid for the following values: $c = 0, 7 \cdot 10^{-9}$ F; $L = 10^{-3}$ Hy; $R = 10^3$ Ω; $\lambda \approx 1, 2 \cdot 10^{-6}$ sec. The last value also defines the temporal axis, which is identical for all three curves. (SOURCE: B. Hassenstein, *Kybernetik und biologische Forschung.* Frankfurt: Akad. Verlagsgesellschaft Athenaion, 1966. Calculations by E. U. v. Weizsäcker.)

Although this unification, in itself, certainly implies an increase in the system's complication, it is in further evolution very often followed by processes of simplification. The progressive division of labor among the subsystems permits each of them to restrict its function to the part allotted to it in the context of the whole. Even our own brain cells, which together and as an integrated supracellular unit perform all the highest functions of the human mind, are vastly inferior to an amoeba or a paramecium in regard to the individual function performed by each of them and to the adaptive knowledge underlying these functions. This simplification of any primarily independent subsystem in the course of its integration into a superimposed greater unit is a phenomenon to be found on every level of evolution. In the psychosocial evolution of mankind in general and of science in particular, it poses a difficult problem. Progressive specialization has the unavoidable consequence that the individual "knows more and more about less and less"—as an old joke has it—until finally the specialist loses his bearings and ceases to know what place his special knowledge holds in the frame of reference of supraindividual, collectively human culture and science. I shall come back to the problem of the specialist's stultification in the chapter on pathology of knowledge and dehumanization.

A different type of simplification consists in arranging the relationship of the integrated parts in the most efficient manner. Anybody who has ever constructed some sort of apparatus for a certain purpose will have noticed that the moment one has finished the work, he is struck by the realization that there was a simpler way to do it. Causal concatenations as well as the exchange of information between the parts of a system may, in this manner, be simplified and, at the same time, rendered more efficient. When this type of improvement is achieved in social systems of human culture, we usually speak of "better organization."

It is, then, not too difficult to understand how, with the integration of preexisting subsystems into a functional whole, entirely new systemic properties come into existence. Also, it is easy to see that in consequence of this, there arises a one-sided stratification of lower and higher systems, a relationship that exists, not only between the subsystem and the whole, but between lower organisms and their more highly evolved descendants as well. This relationship is highly characteristic of the whole world of organisms. When examining a living system of a higher level, we must constantly keep in mind that it possesses, for the most part, all the systemic properties of the subsystems of which it is composed, and that none of the natural laws, down to those of chemistry and physics, which prevail in its components, suffer any infraction in the functioning of the whole. The one-sidedness of the relationship between the higher and the lower levels of integration lies in the fact that, conversely, the properties of the superimposed, integrated system are not contained in any of its components and are certainly not predictable on the basis of a knowledge that comprehends only the single subsystems and does not include the way in which they are put together. In other words, one must know the *structure* of the whole in order to understand its systemic properties.

In this sense, the higher system is not "reducible" to its elements, as Michael Polanyi (65, 66) has emphasized again and again. Living systems are not reducible to inorganic matter and processes; nor, for that matter, are man-made machines. However, this does not mean that the higher system cannot be *explained* and *understood* on the basis of a thorough level-by-level analysis of the components subsystems *and the structure* in which they are put together. Polanyi does not imply that supranatural vitalistic factors are necessary to explain the existence of living systems, and, to avoid this suspicion, I prefer to say that the higher system is not *deducible* from the knowledge of its subsystems. Though it is indeed *made* of them, this making consisted in a historically unique event of "fulguration," which is the evolutionary counterpart of what, in cultural development, we call an *invention*.

This stratified building of organisms has the important consequence that the living system, in the course of its evolutionary progress, *remains* what it was in respect to practically everything it was before, while simultaneously becoming something entirely different in respect to new, *additional* properties and faculties. Life processes are still physical and chemical processes and something very different besides; man is still an animal and a primate, while, additionally, he is everything he claims to be in his proudest moments. From their ontological aspects, all these facts have been recognized with superlative clarity by the German philosopher Nicolai Hartmann (24, 25, 26). A number of definitely nonvitalistic biologists—some of whom, I am quite sure, never have read Hartmann—have said the same, emphasizing that living beings comprise a whole series of levels, each of which we must study in terms of its own distinctive conceptions without for a moment forgetting its relationship to the levels below it.

An amazing number of scientists, otherwise biologically minded people among them, seem unable to grasp the fact that the stratified structure of the whole world of organisms absolutely forbids the conceptualization of living systems or life processes in terms of "disjunctive"—that is to say, mutually exclusive—concepts. It is nonsense to oppose to each other "animal" and "man," "nature" and "culture," "innate programming" and "learning," as if the old logical diagram of alpha and nonalpha were applicable to them. Man, as I have just said, is still an animal; human nature persists in and is the basis of culture; and all learning is very specifically innately programmed. The fallacy of disjunctive conceptualization is characteristic of philosophical anthropologists of a certain type and has given rise to endless and fruitless discussions of the question, "in what way is man different from other living creatures?" These authors fail to understand that *any* "fulguration" of a new systemic property, however small and unimportant, produces a difference that cannot by any effort of conceptualization be described as a difference only in degree. Our system *c* in Figure 1 is indubitably different in kind, not only in degree, from systems *a* and *b*. Failing to grasp the principle of stratified and integrated levels, and interested only in the properties by which man is distinctly set off from and raised above "the animal," philosophers of this type remain blissfully ignorant of the fact that differences at least as fundamental as those between chimpanzee and man are to be found anywhere, between any two steps in the evolutional ladder. Whenever one of these people used the disjunctive concepts of "man" and "the animal," my teacher Oskar Heinroth very kindly interrupted with the question: "Excuse me, please. When you speak of the animal, are you thinking of an amoeba or of a chimpanzee?"

An example of the fallacy under discussion being spun out to its last bitter consequences is furnished by Mortimer J. Adler's book *The Difference of Man and the Difference It Makes* (1). Understanding neither that differences "in kind," however great and trenchant, do not contradict the continuity of life, nor that speciation can be explained in a natural way, the author unavoidably arrives at the conclusion that man is radically different in kind from other animals, and that, besides possessing the well-known peculiarities of conceptual thought and syntactic language, he possesses "contra-causal freedom of choice by virtue of having a nonphysical or immaterial factor in his make-up, a factor that has a certain measure of autonomy and causal efficacy."

The analogous, but inverse, fallacy lies in ascribing to the lower and simpler subsystems or ancestral forms systemic properties that come into existence only at a specific, higher level of integration. A typical example of this error is the persistent attempt of psychologists and even ethologists to find adaptive modifiability through learning, not only in lower animals that simply do not learn, but worse, in those subsystems that, in higher organisms, not only are not programmed to learn but are specifically programmed to be resistant to any modification. To a psychologist, most of whose knowledge of behavior stems from the study of human beings or mammals, and who, furthermore, was brought up in the firm conviction that the conditioned response is a simple and elementary mechanism of behavior, it may seem natural to assume that at least oriments or phylogenetical precursors of conditioning must be discernible in protozoa and the lowest invertebrates. This erroneous assumption accounts for the almost pathetic persistence with which many scientists have tried to demonstrate true conditioning in these animals and also for the great number of self-deceptions that occurred in this research.

Cognitive Processes Not Involving Learning

Like all life processes, the mechanisms gathering and storing information are stratified, many layered, and closely interwoven. Even a preliminary attempt to analyze their relationship may be helpful in clearing away a number of current conceptual confusions. In the following survey of simpler and of more complicated examples of these mechanisms and of their functions, I want to emphasize three points that, in fact, already emerge from what has been said in the preceding pages.

First: A new and more complex function very often, if not always, arises from the integration of several preexisting simpler ones, which, so far from disappearing or losing their importance, persist as indispensable parts of the new systemic whole.

Second: Simpler systems are perfectly able to function on their own and indeed do so in lower organisms, which otherwise could not live or produce more highly organized descendants.

Third: It is entirely in vain to search, in the independent functioning of simple and primitive mechanisms, for traces or oriments of those systemic properties that come into existence only after these mechanisms have been integrated into a system of higher order.

THE TRIAL-AND-SUCCESS SYSTEM OF THE GENOME

The basic process of acquiring and storing information is that performed by the chain molecules of DNA and RNA in the genome. I need not bother to explain its mechanism. No limit is known to the quantity of information it can acquire and retain; it by far surpasses all that a human culture, with all its libraries, can achieve in that respect. No organic structure, including that of the human brain, has been devised by any other method than the trial-and-success procedure of the genome. It has not lost any of its importance on the way from the first beginnings of life to the greatest achievements of humanity.

Its enormous range and capability notwithstanding, the trial-and-success procedure employed by the genome to gather information does not seem able, by itself, to keep up a state of adaptedness that is continuous enough to guarantee survival. The obvious reason for this is that it cannot cope with *quick* changes in the environment. It cannot "know" about the success of one of its experiments until at least one generation has lived its life, and it can achieve adaptation only to such environmental givens as are present, with a sufficient statistical constancy, over comparatively long periods of time. In the parlance of biocybernetics, the duration of one generation is the "dead time" that has to elapse before the mechanism under discussion even begins to register an input.

THE REGULATING FEEDBACK CYCLE

All mechanisms that *bridge* this dead time by acquiring and exploiting information at shorter notice are dependent on the functioning of structures programmed by the trial-and-success method of the genome. This is true of the simple regulating feedback cycle as well as of the highest processes of knowledge. We are faced with the hen-and-egg problem: how could a genome come to exist without a negative feedback cycle, and vice versa?

An organism's recovery of its inner equilibrium after a disturbance caused by a change in its environment indicates that information concerning the quality and the quantity of the change has been fed into the living system. If an organism accelerates its breathing rate in a medium of lowered oxygen concentration, if it discontinues ingesting food in an ambience in which food abounds, this means that the organic system possesses information, not only concerning its own present and specific demands, but concerning the supply that the environment is offering at the moment.

Regulating feedback cycles exist in all degrees of complication ranging from simple homeostasis that functions on a "merely chemical" level up to the most complicated processes in which sensory and behavior functions play a part.

Like many other mechanisms by which the individual is enabled to get instant information about its environment, that of regulating cycles can function again and again without any change of its underlying genetically programmed machinery. In other words, the structural substratum of the function remains the same. The instant information is exploited at once *but not stored* in any way. In fact, it is essential to the survival value of these mechanisms that their function can be repeated an unlimited number of times, keeping the organism adapted to quick environmental changes that can be expected to be reversed almost at once.

IRRITABILITY

All higher processes by which instant information is fed into and acted upon by the organism are dependent on its ability to respond to stimuli. "Irritability," as it has been called, is closely associated with motility, at least in its primitive forms. It is only by the division of labor between the nerve cell and the muscle that, in higher metazoa, irritability and motility have become divorced. We do not know of any organism that is able to move and yet is devoid of irritability. In principle, this is not impossible, since locomotion alone, without any input of instant information, could well be able to increase the organism's chances of gaining energy. On the whole, however, the faculty to respond to stimuli is mostly combined with that of locomotion. One is tempted to think that the primary and still most important function of both is to let the organism run away from dangerous surroundings, unless we think of a motility still simpler than locomotion—that of the organism contracting to a minimum space, offering the smallest possible surface of thickened and corrugated skin to the external world. The third type of response to stimuli is, of course, secretion. In all these cases, some overt response of obvious survival value proves that the organism has received instant information about a relevant given in its environment. It is only in the case of organisms with a highly complex central nervous system that stimulation can be received and its information content exploited and stored, while no direct response is immediately observable.

Otto Rössler (73) has proposed to define as behavior any process in which the information about the biological success of an activity is fed back into the living system by any way other than the selective effect on the genome. In my opinion, this definition is too widely inclusive, as it would embrace all gaining of information and energy achieved, in the way just described, by simple cycles of negative feedback and also by modification, of which I shall speak later. I would propose to confine the term "behavior" to all those activities in which motility and irritability combine their functions in gaining information and thereby increase the probability of an immediate energy gain.

PSEUDOPOD RESPONSE

Curiously enough, the most primitive response to stimuli that we know, that of the amoeboid cell, is directed in space, either at the stimulus or away from it. A similar faculty to aim a response in one particular direction of three-dimensional space does not appear again in the realm of organisms until a much higher level is reached; functionally, the directional response of amoeboid cells is analogous to what Kühn (48) calls a topotaxis, which involves the functions of a central nervous system and complicated sense organs. Yet the response simply consists in the ectoplasm lowering its surface tension wherever a "beneficial" stimulus impinges and, conversely, tightening it wherever a noxious one does. Pressure then forces the protoplasm in the direction of lowest resistance, causing the extrusion of a pseudopod toward a localized positive stimulus or contracting and retracting the body surface away from a negative one. Both these processes have been successfully simulated in nonliving models. However, having familiarized myself with the behavior of an amoeba, I believe this explanation to be too simple. I think I can see in direct observation that the ectoplasm of an amoeba possesses the faculty of changing, at a moment's notice, from gel to sol and back again. Though I cannot find any reference,* I am sure that this is well known to cytologists. What appears on superficial observation to be a mere increase of surface tension really is a quick jelling immediately below the surface, which is always covered with a skin of gelated protoplasm except where it sticks to a solid body, such as food particles. In gelating, the plasm contracts slightly, thus producing effects similar to those of an increase in surface tension. Furthermore, I do not

* Herbert Fischer has kindly brought to my attention the chapter on amoeboid movement in L. V. Heilbrunn, *An Outline of General Physiology,* in which an article by Mast, *Journ. Morph. and Physiol.,* 41 (1926) 347, is cited. On the basis of thorough investigations, Mast arrives at identical conclusions.

believe that the streaming of protoplasm is wholly passive. Identical currents of plasm occur in plant cells that are enclosed in a cellulose capsule and in which no change of external pressure explains the movement.

When a comparatively simple motor pattern can be turned to many different uses, the main part of the innate information necessarily is residing in the releasing mechanism that determines where and when the pattern is used to the organism's advantage. The amoeboid cell's one motor pattern serves locomotion, feeding, and escaping and must, correspondingly, be releasable in many different ways, causing the organism to move into favorable conditions (in the case of free amoebae actually effecting habitat selection), making it retract and flee from noxious stimuli and "effusively" run toward promising ones. Whoever has watched an amoeba in near-natural surroundings will have been surprised at the seeming "intelligence" of its behavior—in other words, at the wealth of innate and environmental information on which it acts.

KINESIS

The Proteus-like ability of the amoeboid cell to sprout a functional front, in whatever direction of space the organism is required to move, makes it superfluous for the cell to possess any other spatial guiding mechanisms. The loss of this capability is the price to be paid by evolution for the differentiation of any solid body structure. Only a few radially symmetrical metazoa are able to move, at least in one plane, in whatever direction an impinging stimulus makes desirable. If the octopods seem to have conquered the limitations imposed by structure, their apparent freedom of moving, amoeba-like, in any direction they want to is based not on the absence of structure, but on a supreme mastery of it.

All organisms that developed, in the interest of locomotion, structures stiffening and streamlining the body along one longitudinal axis need mechanisms that collect instant information concerning the desirability of spatial directions and steer the animals accordingly. The acquisition of instant information is, in these cases too, the function of a phylogenetically evolved machinery that *does not undergo any modification* in the process. The reaction therefore remains the same after any number of repetitions. The simplest orienting mechanism of this type is what Fraenkel and Gunn (16) have termed a kinesis. It consists simply in speeding up a randomly moving organism whenever conditions are noxious and slowing it down when they are beneficial. The effect is comparable to that which, though in this case most undesirably, a bad stretch of road has on automobiles. The enforced slowing down causes the traveling cars to come close together, their distance being inversely proportional to their speed. Without influencing the direction of locomotion, kinesis thus causes the organisms to spend more time in circumstances promising a gain of energy than in those that do not. This effect is increased if the organisms, as many animals do, move not in a straight line but in a zig-zag and if, on reaching a favorable environment, they increase the angles of their wavering to and fro. This response, called "klino-kinesis" by Fraenkel and Gunn, is by no means confined to protozoa and lower invertebrates but occurs in higher animals as well, for instance in grazing mammals and in humans picking mushrooms. However, in the phylogeny of behavior, kinesis seems to have been the first mechanism by which a nonamoeboid organism became able to receive and exploit spatially orienting information.

PHOBIC RESPONSE

A further advance in the evolution of information-gaining organizations is represented by what Alfred Kühn (48) called the phobic response. It is characterized by its inability to take into account the direction in which the moving organism hits upon a noxious stimulus situation. If the organism moves along a gradient of "improving" environmental conditions, it shows no response—unless, as often happens, a kinesis is at work simultaneously. However, on meeting a gradient along which conditions deteriorate, or on running up against a strongly noxious stimulus, the animal stops, goes into reverse, and, after backing up a certain distance, resumes its forward movement in an altered direction. The alteration of its course is achieved by a turning behavior *that bears no direct relation to the direction from which the releasing stimulation impinged.* In some protozoa, on strong stimulation, the turn may chance to comprise exactly 360° so that the organism, on starting to move forward again, runs into the releasing stimulus a second time and has to repeat its performance. Thus, while imparting directional information lacking in the kineses, one phobic response can tell the organism only what direction *not* to take, leaving it to a process of elimination to determine which course is free of noxious stimuli. Of course, the information concerning which stimuli are noxious and which beneficial is phylogenetically acquired and may be coded in very simple terms. Some ciliates "prefer" a certain concentration of H-ions, giving a phobic response when getting out of it. Statistically, this acidity signals the presence of putrescent organic matter and bacteria on which the animals feed. These organisms do not possess any information that there are any acids other than CO_2 and get caught in a death trap if, for instance, a drop of oxalic acid is offered.

TAXES

On a much higher level, in respect to the amount of information gained as well as to the complication of physiological mechanisms involved, we find the type of orienting behavior that Kühn has termed *taxes,* or *tropic reactions.* Since the publication of his now classic book *Die Orientierung der Tiere im Raum* (48), a vast amount of research has been done on the subject, and the biocybernetical approach of H. Mittelstaedt (61, 37) and his school has helped us understand the mechanisms involved in the more subtle process of orientation in space. I need not attempt to give a survey of all this work; it is sufficient for my present purpose to say that all these mechanisms, from the simplest "tropotactic" response of a planarian that finds its prey by turning toward an olfactory stimulus until receptors on both sides of its head receive an equilibrated amount of stimulation, to the highly complicated feedback circles enabling a mantis to strike at a fly with great precision, one and all cause the organism to turn directly into that direction in space that promises the greatest probability of gaining and/or preserving energy. In other words, *the angle at which the animal turns is directly determined by the angle between the impinging stimulus and the longitudinal axis of the organism.* While the phobic response tells the organism only that one particular direction in space is "bad," the tropic response conveys the information as to which one, among all the innumerable directions in space that it could choose to take, is the most promising.

If one compares the efficiency of the three principles of spatial orientation—kinesis, phobic response, and taxis—not only direct observation, but also an assessment of the information imparted by each, reveals the fact that the phobic response conveys a multiple of the instant information furnished by kinesis and that taxes surpass both of these by another multiple.

RELEASING MECHANISMS AND FIXED MOTOR PATTERNS

In the preceding sections on the pseudopod response of the amoeboid cell, on kinesis, and on phobic responses, I have already emphasized the fact that a great amount of specific information concerning certain relevant conditions prevailing at the moment in the environment is

16

fed into the organism by the *selective* function of the releasing mechanisms. It is the *afferent* part of what Ivan Pavlov (63) called an unconditioned reflex that acts like a filter, permitting only very specific kinds of stimulation to take effect. This function has also been likened to that of a lock that can be opened only by a specifically structured key—hence the term "key stimulus."

In protozoa and lower metazoa, in which the inventory of possible behavior patterns is limited and consists mainly of going away from danger and toward situations promising energy gain, no very high demands are made on the selectivity of the releasing mechanism. Though the faculty of the amoeba to know "good" from "bad" is remarkable, it is less so in some ciliates, as in paramecium, which works on the hypothesis that all acids of a certain concentration are "good," poisonous ones included. Also, it is to be supposed that all releasing mechanisms in protozoa act on a chemical basis rather than on the principle of stimulus conduction found in organisms with a centralized nervous system.

Organisms of the latter type, particularly those with an articulated skeleton—arthropods and vertebrates—invariably possess elaborate and complicated motor patterns whose coordination is entirely blueprinted in the genome. Their coordination is, in most cases, achieved without proprioceptors, as E. von Holst (32) has shown in his classic papers on central coordination, while their orientation in space is accomplished by an elaborate apparatus of multiple taxes. Quite lately, E. Taub and A. J. Berman (87) have shown that even in primates the coordination of fixed motor patterns is largely independent of proprioceptors. The guidance by afferent controls is not by any means an indispensable prerequisite for the performance of fixed motor patterns, as is proved by the occurrence of so-called vacuum activities, in which a whole pattern is performed without the object toward which it is normally oriented. A fixed motor pattern is, in many cases, an extremely specialized tool that can be applied to advantage only in one particular situation. The wonderful coordination with which a quelea weaver attaches the first grass fibers of its nest to a branch can be used for absolutely nothing except this function. Since a species may possess quite a number of equally specialized motor patterns, it is of the greatest importance that the organism possess rather detailed innate information determining which pattern ought to be performed in what situation. Furthermore, quite a number of behavior patterns of very different functions can be released by the same sense organs, and the performance of a pattern in the wrong place may be disastrous. If a baby apistogramma were to approach a fish other than its mother in its specific schooling-and-following behavior, this would spell destruction just as surely as if it should respond to its mother with an all-out escape response.

In such cases, the selective filtering of stimuli must obviously be done by the central nervous system. We have little information as yet as to how it is done, though the work of Letvin and others on the retina of frogs and fishes has given us some idea of the mechanisms achieving this function. (The retina, incidentally *is* part of the central nervous system in vertebrates.) All evidence agrees with the assumption that it is *configurations* of stimuli that are responded to selectively. What has been termed "key stimulus" in ethological literature is, in many cases, such a specifically effective configuration, and this not quite precise expression has led to misunderstandings with stimulus physiologists. Much experimental work has been done by ethologists in investigating the releasing mechanisms, and in many cases the innate information contained in them has been analyzed with all desirable precision—for instance, in the Kuenzers' study (47) of the young of apistogramma and nannacara. For a general survey of all these investigations I recommend W. Schleidt's review in the *Fortschritte Zoologie,* 1964 (76).

Observing the function of releasing mechanisms under natural conditions, one is tempted to overassess the amount and the specificity of innate information contained in them. If one observes a tame hand-reared young raven, which up to that moment has never paid any attention to living prey, make a determined dash at the one sick jackdaw among dozens of healthy ones and kill it skillfully with one well-aimed blow at the back of its skull, one is amazed at the amount of innate knowledge underlying this behavior. The bird seems to know that it could not prey on healthy jackdaws; it seems to recognize the symptoms of illness, to conclude that they promise success to an attack, and to know exactly how and where to launch the attack. On close examination, the innate information underlying all this complicated sequence of behavior boils down to very few and simple, if important, data. The raven, like many birds of prey and also many carnivores, possesses a mechanism that releases motor patterns of prey-catching and responds to irregularities in the prey's locomotor actions. A slight stumbling, an irregularity of wing-beat, or the like elicits a predatory attack with the mechanical predictability of a reflex, as trainers of big carnivores have learned to their cost. An analogous response in hawks and falcons is used by falconers in the construction of the "lure." Furthermore, the raven's fixed motor pattern of grasping with both feet and delivering a fearful blow of the bill exactly between them is supplemented by a built-in orienting mechanism that directs it at the back of the prey's head. It is characteristic of all releasing mechanisms that the innate information contained in them is coded in a surprisingly simple and, at the same time, most effective way. For example, the stinging response of the common tick, *Ixodes rhizinus,* is released by any object having a temperature of roughly 37°C and smelling of butyric acid. Simple though these key stimuli are, it is difficult to visualize a natural situation in which the response could be elicited by anything except the animal's adequate host, a mammal. Similarly, the baby apistogramma will follow any object that shows black and yellow markings roughly corresponding to that of the mother and that performs jerking movements of a certain frequency and amplitude; the male stickleback will fight any simple dummy that has a red undersurface; the jackdaw will attack any living creature carrying something black and dangling; and so on.

In spite of the simplicity of its coding, the instant information that the releasing mechanism conveys about a present, relevant given in the animal's environment is reliable enough to permit a rigid linking of the receptor and the effector patterns, even in those cases in which the pattern consists in a very specific fixed motor pattern that would be not only useless but even detrimental in any but the adequate situation for which it is phylogenetically programmed.

The combination of a releasing mechanism with a fixed motor pattern released by it must develop a very great survival value if one is to judge by the frequency with which it is found in higher animals. Indeed, it is so common that it was regarded as the prototype of instinctive activity by O. and M. Heinroth (28). They avoided definitions on principle, but all their examples of what they called *angeborene Triebhandlungen* are combinations of unlearned releasing mechanisms and fixed motor patterns. A tremendous amount of phylogenetically acquired information can be packed into this type of program, because a few selective mechanisms acquiring instant information guarantee the motor patterns being performed at the right time and at the right place.

The realization of the fact that at least three intrinsically different sorts of mechanisms—taxes, releasing mechanisms, and fixed motor patterns—worked together as subsystems in one *Triebhandlung* led to further analysis. I discovered independently what Wallace Craig (10) had known for many years—namely, that the threshold of key stimuli is dependent on the discharge of the fixed pattern, being lowered whenever it had not been realized for some time and raised after it had. Craig also had clearly realized what I had not—that the internal buildup of the motor pattern does, in this manner, affect not only the thresholds of key stimuli but the organism as a whole, causing a general state of unrest and making the animal move about, thus increasing the probability of its meeting the releasing stimulus situation. In its simplest form, this "appetitive behavior," as Craig called it, is essentially the same as a kinesis.

Tinbergen and Baerends were the first to show that the three elementary subsystems—appetitive behavior, releasing mechanism, and fixed motor pattern—are very often linked together in other and more complicated sequences. An initiating appetitive behavior achieves a stimulus situation that, instead of immediately releasing the terminating motor pattern, causes a switch of appetitive behavior that now is directed at another equally specific stimulus situation. Such links may follow each other in considerable number, and the chain of events may branch out, because, on reaching a certain situation, the animal develops an equal readiness for several subsequent activities, and which of them is released depends on further key stimuli. G. P. Baerends, in his classic work on the digger wasp *ammophila* (2), has achieved a deeply searching analysis of this hierarchical organization of instinctive behavior.

Hierarchically organized chains of appetitive behavior, releasing mechanisms, and fixed motor patterns are far more adaptable to present environmental circumstances than are simpler arrangements, for the obvious reason that every one of the multiple releasing mechanisms and of the built-in orienting mechanisms feeds its own instant information into the organism. The hierarchical organization permits an enormous amount of innate information to be packed into a program without incurring the disadvantage of the rigidity characteristic of any single, specialized fixed motor pattern. The studies of P. Leyhausen (50) have shown that in higher mammals such as cats, this type of organization can account for an enormous plasticity of behavior, and also that it persists unchanged even if it is modified, overlaid, or even completely hidden by learned forms of behavior.

Although, as I shall explain in the next chapter, the epoch-making new feedback of success or error that underlies all true learning very probably originated in rather complicated and highly integrated systems involving appetitive behavior, releasing mechanisms, and fixed motor patterns, it would be a grave error to assume that their function is dependent on any form of adaptive modification by learning. In fact, we know that it is not; there are many animals in which such systems function only once in the individual's life, so that no opportunity of learning by experience ever arises. The work of the Peckhams (64) and of Jocelyn Crane (11) on spiders, and that of Ernest Reese (69, 70) on hermit and coconut crabs, provides many convincing proofs of this fact. Even in higher animals, in which these systems are typically improved upon by experience, they must be able to function without this help at least enough so that the success essential to successful conditioning is reached after not too many unsuccessful attempts.

Learning

All the procedures hitherto discussed, by which the organism acquires instant information about conditions prevailing in its environment, are the functions of physiological mechanisms that can repeat their performance any number of times *without undergoing any change, least of all any adaptive change, of their inner machinery*. The only way in which such a change can be wrought is the devious route of feeding back, to the genome of the species, whatever biological success a mutation or recombination of genes may have achieved by altering the machinery of behavior.

In spite of the fact that such machinery can be changed, as organs can, only by the procedures of evolution, its function can achieve an amazing adaptability to the circumstances prevailing, at the moment, in the organism's environment. A complicated behavior system consisting of several hierarchically organized appetences, releasing mechanisms, and fixed motor patterns (p. 32), guided in space by innumerable built-in orienting mechanisms including most complicated feedback organizations (p. 27), commands a wealth of cognitive processes that gather information and enable the organism to act on it instantaneously. As M. Konishi (44) has pointed out, it is a fundamental error to equate "unlearned" or "phylogenetically programmed" with "stereotyped." In fact, the fixed motor pattern is the only subsystem of phylogenetically programmed behavior to which the attribute "stereotyped" can be applied with any justification.

ADAPTIVE MODIFICATION

There is, however, one type of cognitive process that is entirely different, physiologically, from all those based on the repeatable function of phylogenetically programmed mechanisms and must, therefore, be strictly distinguished from the latter. This process is based on changes in the organism's structure caused, during its ontogeny, by environmental influences. Such a change is called a *modification*.

Modification in itself is a ubiquitous phenomenon; practically any permanent change in an organism's environment, provided it is not immediately lethal, will cause some modification or other, at least in young and growing individuals. But modifications of this kind are very far from being necessarily adaptive; indeed, their chance of being so is no greater than that of any random mutation or recombination of genes, and this, as geneticists assure us, is extremely small. If a modification is the regular response to a certain environmental change and is, at the same time, clearly *adaptive to* this particular influence, we are safe in assuming that this modifiability is "selected for"—in other words, that it is the function of a built-in mechanism, already programmed by the trial-and-success procedure of the genome. Such an "open program," as Ernst Mayr (60) called it, has for its prerequisite, not less, but much more genetically acquired information. This seeming paradox might well be illustrated by a parable. A prefabricated house can be erected only on a perfectly level and hard basis, such as the terrasses of "pa hoe hoe" lava in Hawaii, on which part of Hilo is built. The necessity to adapt an otherwise similar building to a site of irregular structure and composition demands, on the part of the architect, methods of acquiring much additional information concerning the ground, as well as the learned knowledge of how to deal with it adequately in adapting the understructure of the house to its substratum. Any form of regularly adaptive modifiability—like that of a plant stretching toward the light when there is too little of it, and thus obtaining sufficient illumination, or that of a mammal's fur growing longer in a cold climate—necessarily presupposes a genetical program so cleverly constructed by phylogeny that it can realize different adaptive possibilities according to different information coming from the environment. Thus, adaptive modification is a truly cognitive process.

This kind of open program, which is able to react adaptively to different environmental inputs, is familiar to the biologist from the results of experimental embryology. The embryo's ectoderm possesses the prospective potency (*prospektive Potenz,* Spemann) of forming a neural tube and a brain, or the lens of an eye, or simple skin; it possesses all the innate information necessary to build any of these very different structures. Which of them it realizes is dependent on the input it receives from its tissue environment. It builds a neural tube when there is a chorda dorsalis beneath it, a lens when there is an eye bladder, and simple skin when it is not thus specifically influenced. The process of calling forth, by a specific environmental influence, the realization of that part of an open program that is adaptive (or, in other words, phylogenetically constructed to fit that particular influence) is called *induction* by experimental embryologists. For our consideration of cognitive processes, it is immaterial whether the input determining the realization of such an adaptively modifiable open program comes from another subsystem pertaining to the same organism or from an environmental circumstance to whose regular occurrence the species is phylogenetically adapted. In other words, all adaptive modification is *essentially identical with induction*.

Adaptive modification is found in the lowest organisms. In some bacteria with organellae serving the intake of certain chemicals—for instance, phosphorus—are increased in number when the phosphorus concentration of the medium is lowered. It takes the cell some time to grow these contrivances; the compensatory response to the dearth of phosphorus is by no means instantaneous. Conversely, if the phosphorus content of the culture medium is suddenly brought back to normal, the organism will actually "overeat" until it has again dismantled the surplus organellae. The cognitive function of this adaptive modification in a low organism is closely akin to that of a simple feedback cycle, which, as I have explained (pp. 23–24), procures for the organism information about the availability (one might say the market situation) of certain substances.

In all adaptive modification, it is *the machinery* on which the organism's way of responding is dependent that is altered by what may be called loosely "individual experience." In respect to the time it requires to gather information, adaptive modification is intermediate between the cognitive procedure of the genome and all the processes gaining instant information. Unlike the latter, it is able to *store* knowledge over varying periods. Some processes of adaptive modifications, such as those of embryonic induction, are irreversible for the life span of the individual; some, such as those in our example of the bacterium, are reversible within a comparatively short time; some are between these two extremes. In any case, adaptive modification combines the extremely valuable faculties of *acquiring* information in a comparatively short time and of *storing* it during time lapses of varying size order. Neither the cognitive function of the genome, nor that of the mechanisms acquiring (but not storing!) instant information, is able to perform *both* of these functions!

The more complicated a biological system, the less likely it is that a random modification will effect anything but disintegration. In the

23

whole world, there is hardly a system more complicated than the central nervous organization underlying the behavior of a higher animal. One of the greatest achievements of phylogeny is to have constructed systems of this sort in such a way that they are still adaptively modifiable by an input occurring during the individual's life. There never was a greater error in the history of science than the empiricist philosophers' belief that the human mind, before any experience, was a *tabula rasa,* a blank, unless it is the reciprocal, but intrinsically identical, assumption of nonbiological psychologists that "learning" must, as a matter of course, "enter into" any physiological behavior processes whatever. The worst aspect of both these reciprocal errors is that they obscure the central problem of learning—the question: how does learning come to be adaptive?

TYPES OF LEARNING OTHER THAN CONDITIONING

In *Evolution and Modification of Behavior* (55), I gave an intentionally all-embracing definition of learning, equating it with any adaptive modification of behavior. Also, I have tried to give a survey of many different types of modifiability of behavior, laying more stress on those that do not involve conditioning than on those that do. To avoid repetition as far as possible, I shall do the opposite here, discussing only three examples of learning other than conditioning.

Motor Facilitation by Exercise.   The machinery of a new car is undoubtedly modified in an adaptive way by the process known as "breaking in." Something similar occurs with some motor patterns of young animals. Wells (97) has shown that in the young of the common squid, *Sepia officinalis,* the prey-catching patterns, though not formally different from those of the experienced animal, are noticeably slower and less sure of their aim. In young domestic chickens, E. Hess (30) has demonstrated a similar process in the development of food-pecking, which becomes considerably more exact with exercise. The pecking becomes surer of its aim, even if the aim is quite wrong. When Hess fitted newly hatched chicks with prism goggles shifting the image of the food particles sideways, the birds never learned to compensate for this misleading shift, but the straying of the hits round the point at which they erroneously aimed diminished rapidly with "exercise." This proves that actual success in the hitting and taking up of food plays no role in rendering this motor pattern more exact.

Habituation.   It is an extremely common phenomenon that stimulus situations that release a response at their first occurrence gradually cease to do so after a number of repetitions. This waning of the response is *not* influenced by whether or not the stimulus situation, to which the animal gets habituated, is followed by a reinforcing stimulus. Though superficially similar to fatigue (and perhaps evolved out of fatigue of specific receptor mechanisms), habituation has its survival value in *preventing* fatigue on the motor side of the response. This is accomplished by the *specificity* of habituation to a certain stimulus only. A hydra primarily contracts in response to a great number of stimuli—to a touch, to being passively moved in the water, to a concussion of the substratum, and so on. A hydra sitting in running water gets habituated to being moved to and fro by the current, while its threshold to all other contraction-eliciting stimuli is *not* changed. In other words, the motor side of the response is not affected; habituation is *stimulus-specific desensitization*.

In higher animals, this specificity can be so elaborate and so selective that it is impossible to explain without resorting to gestalt perception as an explanation. Wildfowl normally react to anything furry moving along the bank of their lake by a combination of escape and cautious mobbing, which is obviously a good defense against foxes and similar predators. Our birds in Seewiesen became specifically habituated not to chow dogs, which primarily represent supernormal key stimulation for the response, but to our individual chows; they mobbed those of a chance visitor as intensely as they would a fox. It is only in the context of the complex gestalt perception of the well-known individual dog that the key stimuli furry, red, and moving along the water's edge lose their effectiveness. Even when the known dog appears in an unaccustomed corner of the lake, there may be a recrudescence of the reaction.

In the process of habituation, an innate releasing mechanism becomes associated with the individually acquired perception of a gestalt in such a manner that the primarily releasing key stimuli cease to be effective, provided that they are being received together with and in the context of the complex configuration of very many stimuli characteristic of that particular acquired gestalt. In contexts that are different from this, the key stimuli retain their original effectiveness. Very often a slight change in the situation, scarcely perceptible to the human observer, causes the habituation of an animal to break down completely, as in the example of the *Anatidae* mobbing the well-known dog when it appears in an unaccustomed place.

This very specific form of habituation confronts us with one great riddle. We know quite a number of innate responses that, in spite of their obvious survival value, wane rapidly if released a number of times in quick succession, as W. Schleidt (75) has shown in the escape reaction released in turkeys by flying predators and as R. Hinde (31) has demonstrated in the owl-mobbing response of the chaffinch. In the latter case, the response did not regain its original intensity even after a rest period of several months, nor did the strongest possible reinforcement—letting the subject be chased by a real owl—counteract the desensitization. It is hard to believe that mechanisms so elaborately adapted by phylogeny are made to function only once or twice, losing most of their efficacy after that. There must be something we miss in our experiments—perhaps only that we ruin the response by impatiently releasing it too often in too short a time, or that we do not change in our laboratory experiments the concomitant stimulus situation often enough.

Sensitization of Avoidance Responses. There is a type of learning process generally subsumed under the conception of conditioning that I propose to treat as something entirely different, because it is not necessary to assume, for its explanation, the complicated feedback apparatus reporting success or failure of which I shall speak in Chapter 5.

A key stimulus eliciting escape reactions of the utmost intensity is associated, after very few repetitions (often after a single exposure), with the concomitant or immediately preceding stimulus situation. In low invertebrates, the effects of this process grade from mere sensitization by these "conditioned" stimuli to their complete releasing of the escape response without the help of the primarily eliciting unconditioned stimulus. All the "conditioning" hitherto demonstrated in animals not possessing an integrated central nervous system is of this type; to the best of my knowledge, no case of conditioning by reinforcement (which would involve the feedback cycle, to be discussed later) has been proved below the level of annelid worms.

In higher vertebrates, this kind of learned avoidance response involves the functioning of the most complicated perceptual mechanisms, much in the same way as habituation does, except that the effect is the very opposite of a desensitization. A dog that once had been squeezed in a revolving door retained the avoidance response thus acquired for years, possibly for its life, to the extent of insisting on crossing the street to pass, at a gallop and with its tail between its legs, the locality of the trauma on the opposite sidewalk. Horsemen know to their displeasure how stubborn analogous responses are in their mounts.

Constancy Functions and Gestalt Perception. A very peculiar and, in the highest organisms, extremely important cognitive function, which involves an equally peculiar form of sensory learning, is performed by those of our brain mechanisms that process the multiplicity of incoming sensory data and organize them so as to make meaningful *perceptions*. Punctiform, isolated data are meaningless in principle; the organism does not react specifically to single, absolute stimuli, like a note of determined pitch, or to light of a certain wavelength, but to *patterns* that are characteristics of *objects* or *situations*. The response is never dependent on the sameness of the single, absolute sensory data that make up the pattern and is therefore independent of constant conditions of perception. Differences in color of illumination, in distance or direction, and in absolute pitch do not make any difference to the response. A bee, as Wilhelm Ostwald (62) put it very graphically, is not in the least interested in recognizing a certain wavelength of colored light; what it must be able to do is to recognize a certain flower by what we call "its" color—that is to say, by its property of reflecting certain colors in preference to others. Also, the bee must be able to do so independently of the color of the illumination in which it happens to see the flower at the moment. Erich von Holst (32) has shown in what a comparatively simple manner the special mechanism of color constancy computes, from the color prevailing in the visual field, the probable color of the impinging illumination and, by relating this to the color actually reflected by an object, the color-reflecting properties of the latter—in other words, "its" color.

Analogous mechanisms ascertain the size of an object independently of the distance at which it is seen and of the size of its retinal image, which varies with the square of this distance. Others make sure that we perceive an object as remaining stationary in a certain direction, though with the movements of our body and our sense organs the stimuli emanating from it impinge from altogether different angles. Again, it is Erich von Holst to whom most of our knowledge of these fascinating processes is due.

The most amazing of the constancy functions is the one that permits us to perceive an object of complicated structure as having a constant and rigid form, even when we view it from different sides. We are too familiar with this achievement to realize the tremendous stereometrical computations that it must indubitably perform. If I turn the spectacles I hold in my hand to and fro in front of my eyes, their retinal image assumes very different forms; yet a built-in apparatus functioning without my awareness interprets all these changes correctly as movements of a rigid body and not as changes of form. If the spectacles should contract or wriggle during the process, I should perceive it immediately. This interpretation functions without the help of stereoscopic vision: if I shut one eye, it still works and does so even when I see only the shadow of a thing, except that, in this case, the sense in which it turns is ambiguous.

It must not be forgotten that the original connotation of the German word "gestalt" is simply "external form." Indeed, the most important of the criteria postulated for a typical gestalt by Christian von Ehrenfels are clearly given even in the simpler constancy functions: the perception of the object remains the same, though it can be "transposed" into different colors, sizes, and aspects, and it is certainly something very different from the simple sum of sensory data that are integrated into it. Gestalt perception, even in its highest form, is nothing but a constancy function, though it contains, as integral parts, most other constancy functions known to science.

The functions of constancy perception are "objectifying" in the most literal sense of the word. It is they that make it possible for us to perceive an object as an entity that remains identical in the course of time. In essence, all these functions are based on a process of abstracting, from a chaos of accidental and ever varying sensory data, those few patterns that are constantly characteristic of an object. The operations by which they achieve this abstraction are so closely analogous to those of rational thought that a great man like Helmholtz could actually mistake them for "unconscious conclusions" (*unbewusste Schlüsse*). However, they are performed by neural organizations of a level altogether different from that of ratiocination, and they are altogether inaccessible to our self-observation. Egon Brunswik (7) coined for them the term "ratiomorphic."

Ratiomorphic perceptual processes first developed in phylogeny in the service of objectification, in what gestalt psychologists called *Ding-Konstanz*. Their survival value first lay exclusively in enabling the organism to recognize certain relevant objects as remaining identical through time, the variability of conditions under which they are perceived notwithstanding. A superlatively important change of cognitive function took place when the mechanism attuned to do just this proved to be able to achieve, by virtually the same operations, something that came close to conceptualization. An apparatus made to abstract, from varying accidental sensory data, those that are characteristic of one object is also able to abstract, from many objects, those regularly recurring configurations of data that they have in common. The same sort of built-in computer that enables me to recognize my own, individual dog, in all kinds of illuminations, in all possible body postures, viewed from nearby or from a distance and from all possible angles, is able just as well to perceive the gestalt quality of "dogginess" in a St. Bernard, a Pekinese, a Dachshund, and so on. A small child, just able to call all kinds of dogs "Bowwow," has not abstracted the zoological diagnosis of the species but responds to just this super-individual gestalt quality, which is *not* a concept but is certainly the prerequisite and precursor of all conceptualization.

It is my conviction that what many thinkers call "intuition," and what Einstein called *Einfühlung,* is the function of the same ratio-morphic mechanism. In my paper on "Gestalt perception as fundamental to scientific knowledge" (57), I pointed out that very probably all scientific discoveries of laws of nature are primarily guided by the discoverer's gestalt perception. The first sign that the perception of a lawfulness is beginning to stand out against the background of the "white noise" of chaotic accidental data is that the phenomena in which it prevails begin to assume a quite particular, if as yet quite undefinable, and intriguingly attractive quality. The potential discoverer finds himself irresistibly compelled to occupy himself with the phenomena in question and quite automatically to gather more and more information about them until, often quite suddenly, the gestalt of the suspected lawfulness stands out from the background of accidental data with such convincing clarity that one wonders how one could have overlooked it for such a long time.

Quite obviously, the highly complicated perceptual processes here under discussion imply something like an unconscious memory—in other words, a *storage* of information that is inaccessible to our self-observation. Equally obviously, the final "clicking" of perception takes place at the moment when redundancy of the information gathered is sufficiently great to compensate for the "noisiness" of the channel through which it is received. Judging from the time intervals that often lie between subsequent discoveries made by the same discoverer on the same object, a very retentive storage mechanism must be at work.

Deductively minded scientists tend to underrate the importance of the perceptual process just described, or even deny its existence. Its subconscious nature makes it very easy to overlook or to "displace" (in the psychoanalytical sense), at least for anybody who has an emotional bias against anything as "unscientific" as intuition or *Einfühlung.* Such people honestly believe that all their scientific work on a successful discovery consisted in verifying, by deductive procedures, randomly chosen hypotheses, while all the time it is their gestalt perception to which their choice of object and hypothesis—and, therewith, their success—is due.

Karl Popper (67, 68) has cast serious doubt on the validity of the logic of induction, which proceeds from many singular statements to a universal one. He points out the important asymmetry between verifiability and falsifiability: a universal statement can never be derived from a singular one, while conversely a singular statement, verified beyond reasonable doubt, can falsify a universal statement, irrespective of the number of singular data from which it has been inferred. Though seemingly irrefutable logically, this statement is none the less erroneous. For one thing, there does not exist, in practice, any theory that fits absolutely, covering all the facts. When we attempt to explain a tolerably complicated process by a hypothesis, there always remains quite a number of facts that refuse to fit in snugly. Donald Campbell (9) has shown, most convincingly I think, that the procedure in verifying a hypothesis is nothing but a special case of the process of *pattern matching,* which is the basis of all indirect, objectivating knowledge, which Egon Brunswik (7) has termed *distal* knowledge. "Science," as Campbell says, "is the most distal form of knowledge. . . . The processes and entities posited by science . . . are all very distal objects very mediately known via processes involving highly presumptive pattern matchings at many stages." The attempt to match theory and data is never *entirely* successful. In the practice of research, we do not abandon a theory on the finding of one datum that apparently contradicts it. We do not discard the laws of gravity when we see a toy balloon seemingly fall upward. What does cause us to reject a theory is the finding of an alternative one that fits the facts more exactly, though still by no means to perfection—which no theory ever achieves.

The second consideration is that real contradictions—not seeming ones, like the toy balloon falling upwards—very rarely do occur once a tolerable degree of certainty is reached. Our present-day cosmology may still undergo some radical changes, but I simply refuse to consider the possibility that facts will ever come to light that give the lie to the statements that the earth is revolving round the sun and that man is descended from animals.

The third reason is that gestalt perception proceeds inductively with such very satisfying results, its admitted well-known errors notwithstanding. Even if the long-term storing of information, of which I have spoken, should have no other function than that of compensating for the noisiness of accidental data by redundancy of relevant information, gestalt perception would still be a highly important cognitive process. However, I suspect that it is much more. Without it, the procedure of scientific induction would be impossible. The abstraction of a general lawfulness out of a mass of singular data would indeed be as impossible as Popper holds it to be, if those data were just randomly collected information in the sense of information theory. However, they are not, having been preselected and predigested by a procedure that involves Campbell's principle of pattern-matching, sifts the relevant from the irrelevant and accidental, and accumulates, not an amorphous mass of data, but organized knowledge. The ratiomorphic process of gestalt perception, aided by rational induction, is the sole means at our disposal by which we are able to *detect unsuspected* lawfulnesses. Gestalt perception does so directly; there *is* no other way for human beings to "see," for the first time, a new regularity in the chaos of impinging sensory data. Rational induction performs a similar function only by supplying gestalt perception with that redundant information we call the basis of induction.

The most important cognitive function of gestalt perception is that function which, on the ratiomorphic level, is analogous to rational abstraction and is, indubitably, the indispensable prerequisite of conceptualization. Wolfgang Köhler (42) does not hesitate to call the result of perceptual abstraction a concept. His term *anschaulicher Begriff* is not easy to translate into English. *Anschaulich* is something that can be imagined, visually or acoustically; it is doubtful whether gestalt perception occurs in the realm of any other senses.

Nonverbal Thought. Cognitive functions closely akin to the formation of these perceptual concepts have been closely investigated by Otto Koehler and his pupils (41). Pigeons, shell parakeets, jackdaws, and ravens proved to be able to distinguish *simultaneously* presented quantities by the *number* of their parts, irrespective of the individual size of the latter. The stimulus objects were, in most cases, plasticine splotches, which were applied in different and widely varying forms and sizes (and differently in each of the experiments) to the covers of the little dishes containing the reward. The number is recognized as such, even if the objects to be "counted" change quite suddenly. A jackdaw trained to open the dish whose cover showed four plasticine spots and to leave alone one showing three was offered four and three mealworms lying openly on top of the dish covers. The bird at once ate the four mealworms and then turned away, ignoring the three others and not opening another dish.

Animals that had proved able to "see" numbers presented to them simultaneously could also be trained to *act* a number of times successively. Pigeons, magpies, and others could be trained without any punishment to open one dish after the other until they had obtained a number of rewards. Kinaesthesis and rhythm were excluded as clues by distributing the rewards in an ever varying way over a greater or smaller number of dishes, so that the bird one time got the prescribed number of rewards by opening one or two dishes, while another time it had to open a number of dishes—in some cases, many more than the learned number of rewards. In another series of experiments, pigeons were taught to take a certain number of grains falling out of a chute successively at quite irregular intervals. In one of Koehler's films, it is very impressive to see one of his birds peck up, in quick succession, six grains and then wait patiently for a period of more than 30 seconds (which is a long time for a delayed response in a bird) for the last grain "due" to it. Thus, the pigeons learned to "act to" seven rewards, irrespective of whether their distribution in ten identical dishes with unmarked covers was 2111110000, 4300000000, 0231010000, 0111400000, or 1111100020.

The most amazing feat achieved by this kind of perceptual conceptualization in animals was their choosing, by sample and by doing so, from a temporal sequence of "counted" events to a number of objects seen simultaneously and vice versa. A jackdaw learned to take as many of the mealworms arranged in a circle as was indicated by the number of plasticine spots shown in its center. A magpie was faced with the following arrangement: seven dishes, which the bird had learned to open one after the other, contained in random distribution a varying number of mealworms ranging from one to seven. Another row of seven dishes was marked one to seven by irregular plasticine spots on their covers. The bird learned to open that dish whose number corresponded to the number of mealworms it had found by opening the first row of seven dishes. An African grey parrot learned to take the number of hempseeds indicated by a number of whistles given in varying intervals and durations on a flute.

A different kind of perceptual abstraction was achieved by common domestic mice. The animals, which had been blinded, had mastered the task of learning a maze containing twenty T-shaped bifurcations at which they had to decide between a right and a left turn. All possible clues, such as olfactory traces on the path, orientation marks from outside the maze, and so on, had been carefully eliminated. After thorough training, the mice were confronted with new mazes differing from the original one in the following four points. In the first, the straight runs were of double length. In the second, the 90° angles were replaced by angles of 45°, and in the third by angles of

135°. The fourth represented a mirror image of the maze in which the mice had been trained. Surprisingly, all animals proved to be able to run through the new mazes without committing appreciably more errors than they had in the last stages of their training in the original maze.

Imprinting. A fixation of a response to a stimulus situation encountered only once or a very few times is effected by the process called "imprinting." If, at a particularly sensitive period in its ontogeny, the individual is exposed to a complex stimulus situation that contains certain key stimuli, a whole complicated system of behavior patterns may be fixated, in a more or less irreversible manner, on that complex situation. As Heinroth already knew and as E. Hess (29), F. Schutz (77, 78), M. Schein (74), C. Immelmann (35, 36), and others have conclusively demonstrated, all of the sexual behavior of a bird, for instance, may in hand-reared individuals thus be "imprinted" on humans or on another species used as foster parents in an experiment.

Imprinting differs from other learning processes in several important characteristics. It is confined to a particular, and often very short, sensitive period in the individual's life; it is never quite reversible; and so on. In all these respects it is closely akin to the embryogenetic process of inductive determination. What seems to distinguish it from true conditioning is that a strong positive response is fixated on a complex releasing stimulus situation without the function of those factors that usually act as reinforcements. In fact, the classic object of imprinting experiments, the sexual behavior of some birds, is fixated on its object at an ontogenetic phase at which the strongest reinforcement of this response, the consummatory act of copulation, is a long way from having matured. Nor can the effect of imprinting be undone in those cases in which the experimenter succeeds in the conclusive experiment of inducing his subjects to copulate with a species other than that on which their sexual responses had been imprinted. However, the fact that the usual reinforcements of sexual responses are not effective does not preclude the possibility that other reinforcements, as yet not recognized as such, are at work and that imprinting still may be a special case of true conditioning, though a very special one.

FEEDBACK OF SUCCESS AND FAILURE, OR
"TRUE" CONDITIONING

At a certain level of central nervous organization, an entirely new cognitive process has come into being, which derives information from the success as well as from the failure of a behavior pattern *just performed*. A temporal sequence of behavioral events becomes "circuited" in such a manner that a report on the biological success or failure of the physiological processes that terminate the chain is conducted back to its initiating links and effects their adaptive modification.

With the "fulguration" of this new feedback cycle, a new information-acquiring system has come into being, which conveys specific information about what to do and what not to do, not to the species' store of genetic information but directly to the physiological machinery that determines the individual's behavior and retains the message, if not with the pertinacity of the genome, nevertheless in the form of an adaptive modification that may last for an individual's life span. One single performance of a sequence of behavior patterns can thus bring, within minutes or even seconds, an adaptive change of behavior equal to that which the genome's primal method would need at least the time lapse of one generation to achieve. In fact, the gain is twice greater, because the genome gains information by its successes only, while the individual learns by its failures as well. These advantages are so obvious that it is easy to understand why the cognitive mechanism of conditioning has evolved in practically all metazoa possessing a sufficiently complicated and integrated central nervous system.

On the other hand, it is equally easy to understand why it could not develop in protozoa, or in metazoa with a more or less diffuse neural organization: obviously, a system able to feed back the success or failure of a behavior pattern into the machinery of another that has preceded it in time has, as a prerequisite of its function, a certain minimum number, as well as a considerable minimum complexity, of the subsystems integrated into it. The initiating mechanism, which sets the whole sequence of behavior going, must possess a special adaptive modifiability attuned to the message of success or failure coming from the activities terminating the whole action. The latter, on their side, must be furnished with proprio- and exteroceptor mechanisms that contain sufficient innate information to enable them to distinguish reliably between biological success and failure. *I do not see how one could construct a simpler thought model than this and yet be able to account for the biological function of true conditioning.*

The system of physiological mechanisms that undergoes an adaptive modification through conditioning never is "one reflex," as Pavlovian terminology implies. Even the classical salivating response is really only one part of a complicated system of feeding behavior, *all* of which is activated by the conditioned stimulus and which, in the experiment, is prevented from being performed only by the simple means of tying the dog to a frame. My late friend Howard Liddell told me about an unpublished experiment he did while working as a guest in Pavlov's laboratory (51). It consisted simply in freeing from its harness a dog that had been conditioned to salivate at the acceleration in the beat of a metronome. The dog at once ran to the machine, wagged its tail at it, tried to jump up to it, barked, and so on; in other words, it showed as clearly as possible the whole system of behavior patterns serving, in a number of *Canidae,* to beg food from a conspecific. It is, in fact, this whole system that is being conditioned in the classical experiment. It is very far from my mind to disparage the methods or the value of Pavlov's experiments. In order to make quantification possible, it is entirely legitimate to isolate, by artificial means, a response like salivation; my point is only that, while doing so, one must keep thoroughly aware of what he has done. What is completely forbidden by the laws of systems analysis is deceiving oneself into believing that the artificially isolated part is all that counts and is sufficient to explain the functional properties of the system.

If one surveys, from the viewpoint of a biologically minded systems theory, the essential facts known about conditioning by success and error, he is confirmed in the opinion that this type of adaptive modification occurs only in behavior systems with a rather elaborate genetical program. There is, in most cases, a releasing mechanism that gains in selectivity, and appetitive behavior that is modified, and even new motor patterns that are constructed on the basis of the individually gained, instant information furnished by the reports of success or failure. Practically all the simpler and phylogenetically older information-gaining mechanisms go, as integral parts, into the making of such an adaptively modifiable system, but they are not modified themselves.

Before I discuss the important question of where in such a system the innate information is situated that *teaches* the organism what to do and what not to do, I want to say a few words about the physiological nature of adaptive modifiability of behavior through conditioning, or through learning generally. The search for the engram, as Lashley (49) has called it, has not been too successful as far as its localization in the nervous system is concerned; nor do we know

35

much about the changes in the single nerve cells or synapses, on which it is based. This state of affairs explains why, when the coding and storing of phylogenetically acquired information was discovered, many serious scientists tended to the hypothesis that learned information was also coded and stored in the form of chain molecules. This hypothesis, however, makes it necessary to postulate the function of two physiological mechanisms, the existence of which is, to say the least, improbable. One of them would be needed to transpose into the form of chain molecules, not only sequences, but configurations of neural impulses, much as a Morse apparatus writes a series of electrical impulses on a paper tape, but multidimensionally. A second mechanism would have to perform the reciprocal task of retranslating the information thus coded into series and configurations of nervous impulses causing well-coordinated behavior patterns. Furthermore, this hypothesis fails to explain in the least why the faculty of learning is as directly correlated to the size and complication of the central nervous system as it obviously is. Lately, an eminent biochemist, Gierer (20), has shown by convincing arguments that the theory of individual experience being coded in chain molecules is untenable. Furthermore, many of the experimental results that seemed to support it have been proved to be irreproducible under more critical conditions. In view of all of this, it still seems most probable that the adaptive change wrought by learning in the machinery of behavior is, like all modification, a process closely related to, if not identical with, induction (in the sense of experimental embryology) and that the locus at which it takes place is in the synapses.

INNATE TEACHING MECHANISMS

The open program furnished, ready made by phylogeny, to each individual of a species of higher animals is always elaborately constructed so that the "open" parts concern variable, unpredictable parts of the environment, while other parts, which are necessarily genetically programmed in a practically invariable form, contain the information telling the individual what to do and what not to do. As I said on page 46, and as I explained more explicitly in my book *Evolution and Modification of Behavior* (55), any physiological mechanism effecting reinforcement or the opposite must contain sufficient innate information not to confound biological success and failure in the report that it feeds back to the machinery of precedent behavior. The existence of "innate teaching mechanisms" must be postulated, unless one prefers to explain the obvious adaptive function of learning by the assumption of a prestabilized harmony between the organism and its environment. Paraphrasing Kant's definition of the a priori, one can say that the innate is what is there before all learning and must be there in order to make learning possible. In order to achieve its programmed modifying effect, it must be resistant to modification.

It is a fascinating endeavor to ascertain, in such a modifiable system of behavior patterns, which of its subsystems are being adaptively modified by learning and which are the ones that do the teaching. In the type of behavior system that was called an "aversion" by Wallace Craig (10) and an "appetite for quiescence" by Monika Meyer-Holzapfel (33), an appetitive behavior, which may be represented by all possible gradations from a simple kinesis to learned motor skills, is continued until the organism has reached a specific stimulus situation in which it comes to rest. The most frequent function of this simple system is to guide the organism in selecting habitat conditions favorable to its survival—optimal temperature, humidity, or illumination; the right kind of cover; and so forth. The innate information resides, in this case, exclusively in a receptor organization that "knows" the right conditions and reinforces all precedent behavior that has successfully led to establish them. As in many other instances, the immediate reinforcement here consists in a relief of tension. This is the classic case of Hullian conditioning.

In systems comprising specialized and complicated fixed motor patterns, the teaching information often is contained in the motor coordination itself. When performed in the adequate situation, and only then, the fixed motor pattern produces a specific extero- and proprioceptor feedback, which is wired back, as a report of biological success, to the physiological mechanisms initiating the activity. This feedback has the double function of bringing the activity to an end for the time being and of strongly reinforcing all behavior patterns precedent to the achievement of this end. A good example of a simple behavior system built on this principle is found in the nest-building behavior of the jackdaw and other corvides. Standing on the potential nest locality, with nesting material held in its beak, the bird performs a downward and sideward sweeping movement, which brings the material into forceful contact with the substratum or, later on in the process, with other nesting material already accumulated. The moment the twig or branch carried by the bird meets a resistance, the sideways shoving becomes stronger and, at the same time, saccaded into a series of quick, trembling thrusts, similar to those a man performs with a pipe-cleaner in order to get it through an obstructed part of the pipe stem. When these thrusts succeed in wedging the twig in, so that it offers an increased resistance to the movement, the latter gains in intensity, to end in an orgiastic maximum once the twig really sticks fast. After this consummatory act the bird loses interest for the time being. Unlike many other songbirds, the jackdaw possesses no highly specified releasing mechanism containing innate information as to what the nesting material should be like. When, for the first time in its life, the naive bird is aroused to a nest-building mood, it will grab, carry, and tremble-shove practically all objects small enough to be handled, the most unlikely on my records being pieces of ice and settings of small electric bulbs. None of these things ever gets so firmly lodged by being tremble-shoved as to procure the drive-assuaging stimulus situation that spells biological success. This failure very quickly extinguishes the individual's response to inadequate objects, while an equally quick positive conditioning is effected to adequate ones. In fact, the birds become "connoisseurs" of that kind of twig that is just flexible enough to be shoved into crevices, just twisted enough to stick well, and so forth. Hence, very often most of the material used in all the nests of a jackdaw colony comes from one species of tree.

There are other and more complicated systems of behavior that make use of the extremely specific information forthcoming from the feedback of fixed motor patterns performed in their several adequate situations. The brown rat, as Eibl-Eibesfeldt (14) has shown, possesses three motor patterns achieving the collecting and general arrangement of nesting material. The first is running out (from a potential nest site that has to be determined by precedent learning), grabbing nest material, carrying it back, and dropping it at the point of departure. (Inexperienced rats, deprived of material, did exactly this with their own tails, so that the experiment had to be repeated with tail-less rats.) The second motor pattern consists in the rat sitting in the nest center, turning to and fro and heaping up, with its forepaws, a more or less circular wall of nesting material. The third is patting the inside of this wall with the forepaws so as to tamp down and smooth the inner surface of the nest cavity. A naive rat, offered paper strips or other soft material for the first time, will get into a frenzy of all three of these activities, each of which is performed to complete perfection, not differing even on analysis by slow motion pictures from those of an experienced rat. However, the naive rat does something the experienced one never does: after having carried two or three paper strips, which are lying flat on the ground, it will perform the heaping-up movements in the empty air above them and even do the patting movements, tamping down a nest wall not yet in existence. It is the failure to get the "rewarding" reaffirmation that teaches the rat not to do the heaping-up movements before enough material has been carried in, or the patting movements before a sufficiently high nest wall has been heaped up.

Fascinating illustrations of the necessity to ascertain the localization of innate information have come to light through the work of Mazakazu Konishi (43, 45). In songbirds, the sound utterances denoting simple signals—such as warning calls, flight calls, and the like —are inherited as simple fixed motor patterns, just as are the calls in gallinaceous birds, anatidae, and many others. The song of some passerines, however, is not based upon any inherited motor patterns, even in species in which a bird reared in the isolation of a sound proof room develops a recognizable species-specific song. Konishi demonstrated that birds that were deafened before a certain age developed nothing but an absolutely amorphous twittering, which was more like a noise than a note. The innate information about how the specific song ought to sound is situated in a template that lies *exclusively on the receptor side*. The young bird, which, in the so-called subsong, utters a wide range of sound combinations, much as the human baby does, matches fortuitously produced utterances with its auditory template and retains those that match best. In this respect, subsong plays the role of exploratory play.

Something similar is true of fixed motor patterns as well in all those cases in which they are not linked, in a closed program, to highly specific innate releasing mechanisms. The spontaneity of the fixed pattern then plays an important role in the teaching process. Progressive lowering of the threshold of the unreleased pattern urges its provisional discharge at substitute objects or in not quite adequate situations, and recurrent appetitive behavior forces the organisms to try again and again, even in spite of otherwise strongly extinguishing and discouraging experiences, until at last the fully rewarding consummatory situation is hit upon.

There are, of course, many other and very different "teaching mechanisms," of which I shall discuss only one more because its importance was recognized long ago when the actual problem, the nature of the innate information, was not fully realized. The widely accepted assertion that it is the fulfillment of tissue needs that acts as a reinforcement is certainly quite correct, as far as it goes; but there remains the question of how the modifiable parts of behavior—those that achieve the choice and intake of the necessary substances—get the information indispensable for doing so correctly. That this information is very detailed indeed is shown by Curt Richter's experiments (71, 72) in which rats successfully put together a perfectly balanced diet even when the nutriments furnished to them were split up into the simplest possible ingredients, the component amino acids of protein being offered in separate dishes. Since the rat as a species cannot possibly have innate information about these components of its food, much less about the correct proportions of the ingredients, the most economical explanation of this amazing feat seems to be that the organism has "feelers" (in the cybernetical sense), in all the homeostatic cycles of its metabolism, that report any deviation from the biologically correct values of reference. In agreement with this assumption, the rat eats tentatively very little of any foodstuff as yet unknown to it, thus gaining the opportunity to record "how it feels afterwards"—or, more objectively expressed, to let the food intake affect its metabolism. This kind of feedback, far from creating only avoidance responses, causes all the specific food preferences described by Richter, as the results of John Garcia and his colleagues (18) demonstrate quite conclusively.

These examples are sufficient to show two facts. First, the innate information underlying the adaptiveness of learning may be localized in quite unexpected parts of the behavior system that, as a whole, is modified in the process. Second, the learning process itself cannot, in principle, be understood without understanding the whole system that is being modified. This is true of all other teaching mechanisms as well, a discussion of which would lead us too far afield.

PROCESSES MODIFIABLE BY TEACHING

In *Evolution and Modification of Behavior* (55), I have shown, I hope conclusively, that, unless one believes, again, in the miracle of a prestabilized harmony, it is quite impossible to assume a general adaptive modifiability of *all* physiological mechanisms of behavior. All known observational and experimental facts support what common sense tells us anyway: the information relevant for one specifically adaptive process must be gathered by an equally specific cognitive mechanism, and there can be, in each species, only a limited number of such mechanisms.

If one surveys, from the obligatory viewpoint of systems analysis, the physiological mechanisms underlying the behavior of a species of higher animal, one encounters a limited number, not only of those physiological processes that are able to acquire and to feed back to precedent behavior mechanisms the information spelling success or failure, but also of those mechanisms that are adaptively modified by this report. To the best of my knowledge it was Otto Storch (86) who first called attention to the fact that an adaptive modification of receptor patterns, termed *Erwerbs-Rezeptorik* by him, occurs so much more frequently and at much lower levels of evolution than does *Erwerbs-Motorik,* the analogous improvement of motor patterns.

I need not enlarge on the fact that the most frequent form of learning consists in feeding new information into the releasing mechanism, making it more selective and, at the same time, susceptible to stimuli that regularly precede in time the unlearned key stimuli, thus enabling the organism to prepare for action. In vertebrates, there is hardly one releasing mechanism known that is not thus adaptively modified by conditioning, as shown by W. Schleidt (76), to whose paper I refer the reader for detailed facts.

As regards the adaptive modification of motor activities, the simplest and most primitive effect seems to be the coupling of two or more preexistent fixed motor patterns, as exemplified by Eibl-Eibesfeldt's rats. I doubt whether the way in which the acquisition of apparently new motor patterns of learned motor skills is achieved is physiologically different. As I have explained in more detail in *Evolution and Modification of Behavior,* that which is called a path habit in mice and many other small mammals can be regarded as "one" motor skill, because the motor units contained in it are welded into one coherent sequence. Each single unit, however, consists of a mechanism of fixed motor patterns and taxes that is also encountered in other combinations or sequences. In other words, the new skill is achieved by stringing, end to end, in a particular and specifically adapted sequence, single motor patterns that, as such, are phylogenetically ready-made possessions of the species. In man and in those mammals able to acquire highly differentiated new motor skills, this faculty is dependent on the existence of extremely small elementary motor patterns, each of which is independently releasable and at the disposal of the will—hence the term "voluntary movements." They can be welded together in practically any coordination, which, once established, functions with the expediency of a fixed motor pattern.

SELECTION PRESSURE EXERTED BY LEARNING

Once, somewhere back in the evolutionary process, the "fulguration" of the great feedback cycle of conditioning had integrated, into a new functional whole, a number of physiological mechanisms that previously had performed their species-preserving functions separately or in the framework of smaller, preexisting systems, each of them was allotted, in the context of the integrated organization, a function that was in some ways different from the one it had hitherto performed. By this change of function, a new selection pressure was brought to bear on the mechanisms of appetitive behavior, releasing mechanisms, fixed motor patterns, consummatory actions, and so on. All processes that had to perform the function of innate teaching mechanisms became specialized in the function of feeding back, into the initiating links of the chain, as unambiguous and energetic reports of success or failure as possible, while the initiating processes themselves evolved an ever growing degree of adaptive modifiability.

The structure and functional properties of the consummatory act, in particular, were never fully understood before their teaching function was realized. In a nonconditionable system of behavior patterns that functions only once in the individual's life, such as that of copulation in many arthropods, the only feedback emanating from the consummatory act is the one that switches off appetitive behavior as well as the receptiveness of the releasing mechanism. For obvious neurophysiological reasons, a much greater quantity of nervous impulses, as well as a much more complicated configuration of their message, is necessary to perform the new function of a conditioning feedback of success or failure. A comparison between the consummatory act in nonconditionable and conditionable systems seems to agree with this deduction. After one has observed the elaborate and excited courtship dance of a salticid spider, one is astonished to watch the quiet way in which it proceeds to copulate. Conversely, the enormous general excitation pervading the whole organism during the consummatory act of copulation in a mammal, for instance in a stallion, makes it very probable that these fireworks are not mere epiphenomena but are essential for creating sufficient nervous impulses to impress and modify the machinery of antecedent behavior, effecting a strong reinforcement.

The innate information underlying the function of all teaching mechanisms must, for obvious reasons, be shielded against any random change by individual modification. For this reason the terminating links of any conditionable chain of behavioral systems invariably are those that are most rigidly phylogenetically programmed: "the end of the chain is always instinctive," as Wallace Craig (10) put it. It is the initiating processes of a behavior system on which the selection pressure making for greater modifiability takes effect. Thus, appetitive behavior and releasing mechanisms have been selected for the greatest possible degree of adaptive modifiability.

It is in the framework of appetitive behavior that all motor learning, all of the *Erwerbs-Motorik,* has evolved. The random activity that, in the primitive and unconditionable forms of appetitive behavior, had been the only effect of the internal buildup of an unassuaged motor pattern—or, in the case of an appetite for quiescence, of a disturbing, biologically threatening stimulus situation—constituted the matrix that could be organized to form the open program of adaptively modifiable behavior. There are many examples of primitive unconditionable appetitive behavior, but the opposite is not true: there does not seem to exist any known case in which conditioning by reinforcement could be demonstrated in a behavioral system *not* including typical appetitive behavior.

The releasing mechanism, as the afferent part of a conditionable system of behavior, is under a selection pressure directly opposed to the one exerted upon it when learning does not enter into the picture. In the latter case, the releasing mechanism itself contains all the innate information concerning when and where the particular behavior is to be discharged with the best chances of gain, and therefore a maximum selectivity is of survival value. If, on the other hand, all or most of this information is obtained from a teaching apparatus that records success or failure at the ultimate end of a sequence of behavior patterns, high selectivity of the releasing mechanisms ceases to be an asset. Particularly if a well-developed appetitive behavior impels the animal to try repeatedly, so that repeated failure does not do serious damage by postponing ultimate success all too long, a rather sketchy, unselective releasing mechanism may increase the scope for trials, thus increasing the adaptability of the whole behavioral system. The method of trial and error really gains more essential information: in the case of the nest-building of the jackdaw, it really extracts from the object the information about those properties on which its use as nesting material really depends. As is to be expected, trial-and-error behavior is very definitely correlated with an "intentional" (that is, a "selected for") unselectivity of releasing mechanisms.

EXPLORATORY BEHAVIOR

Out of the extremely common trial-and-error behavior that practically all the young of higher vertebrates show to some degree, a new type of cognitive function has evolved. While all young and inexperienced animals "explore" to the extent of trying their several behavior patterns in this situation and that, typical exploratory behavior is characterized by a strong and autonomous appetitive motivation directed at stimulus situations *new* to the individual. A young corvide bird, confronted with an object it has never seen, runs through practically all the inventory of its behavior patterns, except social and sexual ones. It treats the object first as a predator to be mobbed, then as a dangerous prey to be killed, then as a dead prey to be pulled to pieces, then as food to be tasted and hidden, and finally as indifferent material that can be used to hide food under, to perch upon, and so on. All these behavior patterns, the initial cautious mobbing excepted, are identical with those serving the experienced bird to obtain food. However, it would be a great mistake to assume that hunger is the motivating force behind this behavior. Quite the contrary; if the bird were really hungry, it would not indulge in exploratory behavior but resort to a method it already knows as leading to satiation—for instance, begging from the human foster parent. Exploratory behavior can function only in what Gustav Bally (3), following Kurt Lewin, terms *das entspannte Feld*—"the field free of tension." In exploratory behavior, motor patterns that are clearly adapted to very definite functions are thus performed under a motivation entirely different from the one that activates them in the biological situation in which they are "seriously" applied to achieve the survival value under whose selection pressure they have evolved. In this respect, exploratory behavior is closely akin to what is generally described as *play*. The appetite for new situations, which we usually call curiosity, supplies a motivation as strong as that of any other appetitive behavior, and the only situation that assuages it is the ultimately established familiarity with the new object—in other words, new knowledge. The young raven experimenting with feeding behavior on a new object does not want to eat; he wants to know whether it is edible *in principle*. The information acquired by exploratory behavior is *objective* in the most literal sense of the word.

As I have discussed exploratory behavior more explicitly in other places, I need to say very little more here. Exploratory behavior furnishes an excellent example of how a new cognitive function—and an extremely efficient one—can come into existence by just "wiring" a number of common, preexistent subsystems in a slightly different way. Fundamentally, the new "invention" consists in connecting a strongly motivated appetitive behavior, not essentially different from that which brings about the stimulus situation releasing a consummatory act, with a stimulus situation in which a conditionable system of the type already described (p. 46) can acquire a maximum of new and relevant information. All the physiological mechanisms thus integrated into a new cognitive process are those we already know, but the new systemic properties coming into existence with this new "fulguration" make a tremendous difference. In fact, *most of the difference between man and all the other organisms is founded on the new possibilities of cognition that are opened by exploratory behavior.*

INSIGHT LEARNING

When the totally inexperienced young animal performs, for the first time, a conditionable sequence of behavior patterns, this trial is by no means as blindly random as one of the genome's experiments in mutation and recombination of genes. Even in animals in which, as I have explained, a maximum development of exploratory behavior makes it desirable for releasing mechanisms not to be too selective, they still supply the inexperienced animal with a tolerably good "hypothesis" concerning when and where first to try a certain behavior pattern. Furthermore, every one of these first trials is guided, in space and time, by all the wealth of instant information supplied by the phylogenetically adapted mechanisms described in Chapter 3 and by others that have not been mentioned.

What is generally termed "insight" is usually defined by negative attributes only. Behavior is regarded as intelligent or guided by insight when the organism confronted with a new situation proves able to cope with it at once, although there are neither any phylogenetically programmed releasing mechanisms or motor patterns nor any learned perceptual responses or motor skills fitting the requirements of this special problem. One is tempted to add to this negative definition the exclusion of all the phylogenetically adapted mechanisms acquiring instant information. On close examination, however, this proves to be impossible. If a fish succeeds in circumnavigating an obstacle by the simple means of two simultaneously effective taxes, one that directs it toward the prey behind a little semidiaphanous glass screen and another that causes it to avoid the screen, this appears, at first sight, to be similar to the movement of a projectile in the resultant of the two forces of inertia and gravity. Yet there are all possible gradations between this simplest way of solving a detour problem on one hand and the highest achievements of the human intellect on the other. In another paper (58) I have tried to show how the faculty of complex insight has evolved with the necessity of more and more detailed information about the spatial structure of an organism's environment and how tree-climbing animals with prehensile hands stand in particular need of such information. Man's intellectual faculties have evolved "hand in hand with the hand," and even our present terms for our highest cognitive activities, such as insight, method, concept, object, and so on, still bear all the earmarks of their provenience from processes of spatial orientation.

In the present context, it is sufficient to say that insight is nothing but the function of complex systems of phylogenetically programmed mechanisms gaining instant information. Among these, it is mainly a multiplicity of taxes and of perceptual functions ensuring object-constancy that are integrated in the complex intellectual function.

Though the physiological processes of insight and of learning are essentially different from each other, their functions are practically always united in a joint act of cognition. Even in the most primitive kind of trial-and-error learning, the animal does not run, scratch, or peck indiscriminately in all directions but, by virtue of some taxis or other, possesses a measure of "insight" that quite considerably improves the chances of success. From this, there are all gradations to a type of behavior in which an intelligent animal gains an almost complete insight into the problem with which it is confronted, so that it is able to find, with high probability, an immediate solution and to retain it, after one single reinforcement, well enough never to err again. Insight actually is contained, to some degree, in all trial-and-error learning; what we choose to call "insight learning" is a question of arbitrarily defined quantitative differences.

If it is correct to say that insight enters into most of the more highly differentiated learning processes, it is equally correct to state that learning enters into all of the more complex achievements of insight. Even in solving, by insight, a comparatively simple detour problem, the animal will look around, taking in the structural details of the obstacle one after the other; and, though each of the mechanisms supplying the instant information on which the ultimate solution is based belongs to the type discussed in Chapter 3, the animal must *remember,* having looked to the left and having perceived an opening in the wire fence, that this exit is more easily accessible than the one it discovers a second later on its right. In the classical insight experiments conducted by W. Köhler (42) on chimpanzees, the successive memorizing of all the single details of the problem situation was clearly observable. This learning very probably was of the perceptual type discussed on pages 43–45.

An important relation between insight and learning was found and duly emphasized by Köhler. A problem-solution that was demonstrably achieved by insight tends to degenerate, after very few repetitions, into a routine performance learned by heart, so that a subsequent slight change in the problem, which certainly would not have prevented the animal from finding the solution at the first attempt, does so now, because the silly beast persists in the once successful procedure. This "degeneration" of insight into rote procedure has both its advantages and its dangers. It relieves insight from unnecessary strain—as, for instance, when we use mathematical formulas or logarithms without consciously realizing what we are doing and what they really are. On the other hand, the process under discussion may prevent us from finding obvious solutions because we are blocked by procedures and thought habits that force us to miss them by a hair's breadth only, but miss them nevertheless.

SELF-EXPLORATION

Most of the cognitive mechanisms discussed here take part in the highest achievements of the human mind. Some of them, like the "objectivating" abstraction performed by constancy perception, like the equally objectivating function of exploratory learning, and like insight, have been erroneously regarded as being specifically human. So was tradition of which I shall speak in the next chapter.

There is only one cognitive function that came into existence with the origin of man or that, to be more exact, constituted humanity by coming into existence. Here, once again, the fulguration of an entirely new principle came about, joining into a circle what up to then had been a linear process. With the extreme differentiation of man as a "specialist for nonspecialization" and with the concomitant development of his urge to explore, it seems to have been unavoidable that he discovered himself as an object that is very rewarding to explore. Though as simple as a snake biting its own tail, this process has given to the human community new systemic properties that are entirely absent in the animal world.

In another place, I have recently (54) discussed the consequences that the discovery of the self had for human social behavior. This paper, whose title is "The Innate Bases of Culture," would indeed form a logical sequel to what has been said here. In order not to exceed the frame set by the title of this lecture, I shall confine myself to summarizing what is relevant from the viewpoint of cognitive processes.

Seeing, for the first time, one's own reflection in the mirror of self-exploration need not necessarily have been accompanied by that wonderment at that which was hitherto treated as a matter of course, with that amazement that is the birth of philosophy. The simple matter-of-fact knowledge that the own subjective self is inhabiting a

47

creature essentially similar to any fellow member of the species is sufficient to set off a new feedback cycle of knowledge processes that is epoch-making. Very probably the reflection by which the subject becomes aware of its own subjectivity has been the origin of all specifically human achievements and first of all of conceptual thinking, the prerequisite of verbal language, without which cultural tradition could never have come into being; it is the basis of conscience, which is primarily the simple consciousness of being one member of a society and which is the basis of rational morality. Of course, there are functions analogous to all these in animals. There is phylogenetically programmed social behavior that is analogous to rational morality—and is often confounded with it by anthropomorphically minded observers. There is true tradition in some social animals, of which I shall speak anon. There is something akin to the asking of questions and to the understanding of answers; there is true insight in the faculty of solving spatial problems in apes and some other mammals; there is a function closely resembling conceptualization in the objectivating abstraction achieved by constancy perception; and there are, last though not least, the phenomena of preverbal thinking described by O. Koehler (41) and discussed on pages 43–45. None of these functions has become unimportant, for all of them represent integral parts of conceptual thought, but this unique cognitive faculty did not come into being until the crucial moment when man perceived simultaneously, in the world he was exploring, not only the object he grasped, but his own grasping hand and his own act of grasping. It was then that the whole sensory and nervous process of "grasping" blossomed into conceptualization and that the central nervous image of whatever was being grasped became a concept.

If one observes a bored chimpanzee playing with his own hand, bending and extending the fingers slowly while intensely watching the process, one is tempted to believe that self-exploration of this kind may be at the root of reflection, and also that the action of the hand right in the center of the ape's visual field may have played an important role. On the other hand, a very social creature with an extremely strong urge to explore might well have directed its exploratory behavior at a fellow member of the species rather than at its own body. The play of asking questions and receiving answers, typical of all exploration, may have been played mutually by two individuals. The first mirror in which each of them saw himself may well have been the other. The close relationship between conceptual thinking and verbal communication make it probable that both can have evolved only in a species with a well-developed social organization. The relationship between these two faculties is so close that the question of which came first is yet another hen-egg problem. True, conceptual thinking can demonstrably function independently of verbalization, but it never would have reached the heights it did had it not been the indispensable prerequisite of verbal speech, thus developing under the selection pressure of the necessity to communicate. No system of concepts that are common to the members of a social group and communicable between them could ever have developed without verbal speech.

I do not overrate the function of speech as a help to truly cognitive functions. Formulating a thought in verbal speech has a function similar to that of jotting down on paper a sequence of mathematical operations—in other words, it is not much more than a memory aid. This function of storing knowledge becomes of superlative importance only when tradition develops to the point at which it is able to pass on, from one generation to the next, knowledge coded in the symbols of the spoken and the written word.

REFERENCES

1. Adler, M. J. *The difference of man and the difference it makes*. New York: Holt, Rinehart and Winston, 1967.
2. Baerends, G. P. "Fortpflanzungsverhalten und Orientierung der Grabwespe, *Ammophila campestris*." *Tijdsch. Ent.*, 84 (1941) 68–275.
3. Bally, G. *Vom Ursprung und von den Grenzen der Freiheit: eine Deutung des Spieles bei Tier und Mensch*. Basel: Birkhäuser, 1945.
4. Bennett, J. G. *The dramatic universe*. Mystic, Conn.: Verry, 1967.
5. Bertalanffy, L. von. *Theoretische Biologie*. Berlin: Bornträger, 1933.
6. Born, M. *Von der Verantwortung des Naturwissenschaftlers*. München: Nymphenburger Verlagshandlung, 1965.
7. Brunswik, E. "Scope and aspects of the cognitive problem," in Bruner, J. S., et al. (eds.), *Contemporary approaches to cognition*. Cambridge: Harvard Univ. Press, 1957.
8. Campbell, D. T. "Evolutionary epistemology," in Schilpp, P. A., *The philosophy of Karl R. Popper*. La Salle: Open Court Publishing Co., 1966.
9. Campbell, D. T. "Pattern matching as an essential in distal knowing," in Hammond, K. R. (ed.), *The psychology of Egon Brunswik*. New York: Holt, Rinehart and Winston, 1966.
10. Craig, W. "Appetites and aversions as constituents of instincts." *Biol. Bull.*, 34 (1918) 91–107.
11. Crane, J. "Comparative biology of salticid spiders at Rancho Grande, Venezuela, IV: An analysis of display." *Zoologica*, 34 (1949) 159–214.
14. Eibl-Eibesfeldt, I. "Angeborenes und Erworbenes im Verhalten einiger Säuger." *Z. Tierpsychol.*, 20 (1963) 705–54.
16. Fraenkel, G. S., and Gunn, S. D. *The orientation of animals*. Oxford: Clarendon Press, 1961.
18. Garcia, J., and Ervin, F. R. "Gustatory-visceral and telereceptor-cutaneous conditioning: Adaptation in internal and external milieus." *Comm. in Beh. Biol.* In press.
20. Gierer, A. "Uber die Funktion von Desoxyribonukleinsäuren und die Theorie der Regulation der Genwirkung." *Naturwiss.*, 54 (1967) 389–96.
24. Hartmann, N. *Die philosophischen Grundlagen der Naturwissenschaften*. Jena: Fischer, 1948.
25. Hartmann, N. *Der Aufbau der realen Welt*. Berlin: W. de Gruyter, 1964.
26. Hartmann, N. *Grundzüge einer Metaphysik der Erkenntnis*. Berlin: W. de Gruyter, 1949.
27. Hassenstein, B. *Kybernetik und biologische Forschung*. Frankfurt: Akademie Verlagsgesellschaft Athenaion, 1966.
28. Heinroth, O., and Heinroth, M. *Die Vögel Mitteleuropas*. Berlin: Behrmühler, 1924–28.
29. Hess, E. H. "Imprinting, an effect of early experience." *Science*, 130 (1959) 133–41.
30. Hess, E. H. "Space perception in the chick." *Sci. Am.*, 195, I (1956) 71–80.
31. Hinde, R. A. "Factors governing the changes in strength of a partially inborn response, as shown by the mobbing behavior of the chaffinch (*Fringilla coelebs*)." *Proc. Royal Soc. B*, 753 (1960) 398–420.
32. Holst, E. von. "Regelvorgänge in der optischen Wahrnehmung." *Pflüg. Arch.*, 236 (1935) 149–58.
33. Holzapfel, M. "Triebbedingte Ruhezustände als Ziel von Appetenzhandlungen." *Naturwiss.*, 28 (1940) 273–80.

35. Immelmann, K. "Prägungserscheinungen in der Gesangsentwicklung junger Zebrafinken." *Naturwiss.*, 52 (1965) 169–70.

36. Immelmann, K. "Zur Irreversibilität der Prägung." *Naturwiss.*, 53 (1966) 209.

37. Jander, R. "Die optische Richtungsorientierung der roten Wald-ameisen (*Formica rufa L.*)." *Z. vergl. Physiol.*, 40 (1957) 162–238.

41. Koehler, O. "Vom unbenannten Denken," in Friedrich, H. (ed.), *Lebendiges Wissen 99.* Wiesbaden: Dietrich, 1953, pp. 271–79.

42. Köhler, W. *Intelligenzprüfungen an Menschenaffen.* Berlin: Springer, 1964.

43. Konishi, M. "Effects of deafening on song development in two species of juncos." *Condor,* 66 (1964) 85–102.

44. Konishi, M. "The attributes of instinct." *Behaviour,* 27 (1966) 316–28.

45. Konishi, M. "The role of auditory feedback in the control of vocalisation in the white-crowned sparrow." *Z. Tierpsychol.,* 22 (1965) 770–83.

47. Kuenzer, E., and Kuenzer, P. "Untersuchungen zur Brutpflege der Zwergcichliden Apistogramma reitzigi und A. borelli." *Z. Tierpsychol.,* 19 (1962) 56–83.

48. Kühn, A. *Die Orientierung der Tiere im Raum.* Jena: Fischer, 1919.

49. Lashley, K. S. *In search of the engram. Symposia of the society for experimental biology 4: Physiological mechanisms in animal behaviour.* London: Cambridge Univ. Press, 1950.

50. Leyhausen, P. "Uber die Funktion der relativen Stimmungshierarchie." *Z. Tierpsychol.,* 22 (1965) 412–94.

51. Liddell, H. Personal communication (1951).

54. Lorenz, K. "Die instinktiven Grundlagen menschlicher Kultur." *Naturwiss.,* 54 (1967) 377–88.

55. Lorenz, K. *Evolution and modification of behavior.* Chicago: Univ. of Chicago Press, 1965.

57. Lorenz, K. "Gestaltwahrnehmung als Quelle wissenschaftlicher Erkenntnis." *Z. f. experimentl. u. angewandte Psychol.,* 6 (1959) 118–65. Reprinted in *Über tierisches und menschliches Verhalten.* München: Piper, 1965.

58. Lorenz, K. "Psychologie und Stammesgeschichte," in Heberer, G. (ed.), *Die Evolution der Organismen,* 2nd ed. Jena: Fisher, 1954, pp. 131–72.

60. Mayr, E. *Animal species and evolution.* Cambridge: Harvard Univ. Press, 1963.

61. Mittelstaedt, H. "Die Regelungstheorie als methodisches Werkzeug der Verhaltensanalyse." *Naturwiss.,* 8 (1961) 246–54.

62. Ostwald, W. *Mathetische Farbenlehre.* Leipzig: Unesma, 1930.

63. Pavlov, I. P. *Conditioned reflexes.* New York: Oxford Univ. Press, 1927.

64. Peckham, G. W., and Peckham, E. G. "Observations on sexual selection in spiders of the family *Attidae.*" Milwaukee: Occasional papers of the National History Society of Wisconsin, 1889.

65. Polanyi, M. "Life transcending physics and chemistry." *Chemical and Engineering News* (1967).

66. Polanyi, M. *Personal knowledge towards a post-critical philosophy.* Chicago: Univ. of Chicago Press, 1958.

67. Popper, K. R. *The open society and its enemies.* New York: Harper & Row, 1962.

68. Popper, K. R. *The logic of scientific discovery.* New York: Harper & Row, 1962.

69. Reese, E. S. "A mechanism underlying selection or choice behavior which is not based on previous experience." *Am. Zool.,* 3 (1963) 508.

70. Reese, E. S. "The behavioral mechanisms underlying shell selection by hermit crabs." *Behaviour,* 21 (1963) 78–126.

71. Richter, C. P. "The self-selection of diets." *Essays in biology.* Berkeley, Calif.: Univ. of California Press, 1943.

72. Richter, C. P. "Total self-regulatory functions in animals and human beings." *Harvey Lectures,* 38 (1942–43) 63–103.

73. Rössler, O. E. "Theoretische Biologie." Lecture delivered at Max-Planck-Institut für Verhaltensphysiologie, 1966.

74. Schein, W. M. "On the irreversibility of imprinting." *Z. Tierpsychol.,* 20 (1963) 462–67.

75. Schleidt, W. M. "Reaktionen von Truthühnern auf fliegende Raubvögel und Versuche zur Analyse ihrer AAM's." *Z. Tierpsychol.,* 18 (1961) 534–60.

76. Schleidt, W. M. "Wirkungen äußerer Faktoren auf das Verhalten." *Fortschr. Zool.,* 16 (1964) 469–99.

77. Schutz, F. "Homosexualität bei Tieren." *Stud. Gen.,* 5 (1966) 273–85.

78. Schutz, F. "Sexuelle Prägung bei Anatiden." *Z. Tierpsychol.,* 22 (1965) 50–103.

86. Storch, O. "Erbmotorik und Erwerbmotorik." *Anz. Mat. Nat. Kl. Österr. Akad. Wiss.,* 1 (1949) 1–23.

87. Taub, E., and Berman, A. J. "Movement and learning in the absence of sensory feedback," in Freedman, S. J. (ed.), *The neurophysiology of spatially oriented behavior.* Homewood, Ill.: Dorsey, 1968.

88. Teilhard de Chardin, P. *La vision du passé.* Paris: Editions du Seuil, 1957.

92. Thorpe, W. H. *Science, man and morals.* London: Methuen, 1965.

93. Weidel, W. *Virus, die Geschichte vom geborgten Leben.* Berlin: Springer, 1957.

97. Wells, M. J. *Brain and behaviour in cephalopods.* London: Heinemann, 1962.

The Behavioral Level: Learning

Theoretical Level Learning

1.
Respondents, Operants, and Emergents

Duane M. Rumbaugh

Respondents, Operants, and <u>Emergents</u>:
Toward an Integrated Perspective on Behavior

Duane M. Rumbaugh, David A. Washburn, and William A. Hillix
Georgia State University and San Diego State University

Key Words: respondents, operants, emergents, competence, comparative

Abstract

A triarchic organization of behavior, building on Skinner's description of respondents and operants, is proposed by introducing a third class of behavior called "emergents." Emergents are new responses, never specifically reinforced, that require operations more complex than association. Some of these operations occur naturally only in animals above a minimum level of brain complexity, and are developed in an interaction between treatment and organismic variables. (Here <u>complexity</u> is defined in terms of relative levels of hierarchical integration made possible both by the <u>amount</u> of brain, afforded both by brain-body allometric relationships and by encephalization, and, also, the elaboration of dendritic and synaptic connections within the cortex and connections between various parts/regions of the brain.) Examples of emergents are discussed to advance this triarchic view of behavior--the prime example is language. This triarchic view reflects both the common goals and the cumulative nature of psychological science.

Respondents, Operants, and <u>Emergents</u>: Toward an Integrated View of Behavior

Scientific psychology has been accused of failure to grow theoretically. Its critics claim that we do not integrate prior findings and explanations into contemporary perspectives (see, for example, the discussion and rebuttal by Posner, 1982). A goal of good science is progress, whether reflected in cumulative theoretical development, or through Kuhn's (1962) paradigmatic revolutions, cyclic and dramatic changes that are likely to exclude many central tenets of the previous theoretical regime in favor of "more enlightened" or "more accurate" approaches.

Science may have moved beyond the phase Kuhn described, in which paradigmatic development and rejection were the primary modes of change. Kuhn's unflattering claim that exponents of different paradigms could not communicate may have been a self– subverting law; scientists who knew about it may have tried harder to eliminate their intellectual provincialism. Technological advances like the "information highway" have countered most of the contribution that geographical distance made to intellectual distance. In any case, the present article is an attempt to circumvent revolution to achieve cumulative progress.

Psychology was (R. I. Watson, 1967), and perhaps still is, in a preparadigmatic stage characterized by a failure to agree sufficiently on the fundamentals to qualify for a Kuhnian paradigm. If we are right about the progressive substitution of cumulative science for paradigmatic revolutions, psychology may move smoothly from pre–paradigmatic to post–paradigmatic status without ever clearly having a Kuhnian paradigm.

Whether or not it is philosophically justifiable, it is trendy to discuss the "cognitive revolution" kindled in the 1950's and 1960's and evident in the current popularity of cognitive science. Behaviorism may not have been a true paradigm, but in any event cognitivists tended to challenge, discount, or ignore five decades of research in the behaviorist tradition. Conditioning, schedules of reinforcement, and similar topics once esteemed by behaviorists are rarely discussed in treatments of human cognition; rather, they receive limited attention in introductory and animal learning texts. Ironically, if behaviorism did have a kingly paradigmatic head that cognitivism has chopped off, its crown of objective methodology remains firmly in place.

It is true that behaviorism's metatheoretical commitment to associationism (Marx & Hillix, 1987) has been challenged by the camp of cognitivists most closely related to traditional computer science and artificial intelligence. However, the parallel distributed processing camp, technologically advanced and sophisticated though it is, relies on a connectionism that is fundamentally the same as that of Edward Thorndike (1898) or John B. Watson (1919). (Connectionists frequently do try to identify within hidden layers the rule-like patterns that mediate stimulus-response associations--patterns that are consistent with the thesis advanced here.) The historical roots of the connectionistic movement are often overlooked; even the very direct ancestors of parallel distributed processing (Selfridge, 1955, 1959; Rosenblatt 1958, 1962) are seldom cited.

Although there is thus a recidivistic/modern side of cognitive psychology, the present thesis is that the rise of cognitive psychology represents substantial progress—not just change. As one way of recognizing this progress, we suggest a trichotomous classification of behavior that recognizes and adds to Skinner's (1938) distinction between respondent and operant conditioning, while continuing to acknowledge the importance of antecedents, behavior, and consequences in psychological research. At the same time, we assert that there exist complex processes and determinants of behavior that go beyond those involved in operant or respondent behaviors. These emergent processes should not be confused with species–typical behaviors (instincts) that are fundamentally unlearned adaptations, such as imprinting, taste aversion (i.e., bait shyness), and courtship and migration patterns (see Alcock, 1979).

Consequently, we propose that a third category of behavior, emergents, be defined to extend the domain of inquiry for those who espouse an experimental analysis of behavior. The recognition of emergents will provide a unifying link connecting the several camps (e.g., behaviorist and cognitivist) that try to understand behavior through empirical, systematic research that identifies the antecedents and consequences responsible for the appearance, morphology, and disappearance of responses. This "new" class of behaviors is particularly likely to appear in organisms possessing cerebral complexity (see earlier definition) and encephalization (i.e., the extraordinary elaboration of the cortex relative to the rest of the brain; see Stephan, Bauchot, & Andy, 1970), as within the order Primates.

Emergents include alterations in the nature of the learning process (e.g., in the ability to learn relationally as well as associatively, to form both natural and arbitrary concepts, to recognize equivalence relations between stimuli that are not specifically trained/reinforced, and to develop the ability to solve novel problems in a single trial). Emergent abilities also enable an organism to learn to use symbols as representations of things and events not necessarily present, to comprehend and to use language, to speak and sing, to be able to learn vicariously from secondary records (e.g., written materials and other records), and to reflect upon past experiences and events projected in the future—to mention a few of the salient ones.

From a behavioral perspective, these alterations can be properly viewed as emergent response modes; from a cognitive perspective, they can be viewed as cognitive operations and structures. Either way, however, these alterations have properties that reflect the neuroarchitecture, neurophysiology, and neuropsychology of specific organisms as affected by specific experiences, treatments, or rearing conditions.

Precedents in the history of thought have led the present authors to label this third category of learning "emergent." In the 19th century, John Stuart Mill postulated a "mental chemistry" that coalesced simple ideas into complex ideas (see Heidbreder, 1933). Emergent complex ideas had their own distinguishing structures and properties and, hence, were more than just a composite of the simpler ideas on which they were based. In the 20th century, Nissen's (1954) discussions of possibly new and qualitatively different processes emerging as products of quantitative elaborations of the primate brain directed the senior author of the present report into research regarding their etiologies.

An interesting question that arises in this connection is whether phylogeny to some extent recapitulates the ontogeny of human development with respect to emergent behaviors. These behaviors, like all behaviors, depend on an interaction of organismic and experiential factors; thus, the full complement of emergents is available only to normal adult humans. It may be that some animals never get beyond the first stage of human development--according to Piaget, the sensory-motor stage. Higher stages may emerge in more complex animals. An argument can be made that linguistically trained chimpanzees, orangutans, and gorillas have manifested in rare cases some properties of Piaget's highest stage, the formal operational stage. Some aspects of the intermediate stages are almost certainly seen in nonhuman primates.

Another fascinating question is how precisely the fundamental elements of behavior should be described. It is well accepted that the formation of associations is one basic mental capacity. This involves one type of memory. The ability to compare stimuli with respect to various properties--size, color, shape, and desirability, for example--seems to be an emergent capacity. Several researchers, from Krechevsky (1932) to Levine (1971) have presented evidence that animals from rats to humans are able to generate and test hypotheses about the relationships between stimuli and reinforcers. These are only two of many possible emergent capabilities that might be suggested.

Before distinguishing emergents from Skinner's respondents and operants, consider their important dimensions of commonality. First, they are all forms of behavior. Second, the behaviors are observable and measurable. Third, all three are taxonomic groups of behaviors. As such, they categorize behaviors so that they can be better understood and studied with tactics appropriate to their defining features. It is important to note, however, that, as categories of adaptive behaviors, they are not to be confused with scientific explanations. Fourth, each category has antecedents and consequences that must be defined as parameters of behavior if valid scientific descriptions and explanations of the form and continuance of behavior are to be obtained. Fifth, none of the three categories can be accounted for satisfactorily by, or reduced to, the operations of any two of the other categories. Generally, respondents and operants provide the foundation for emergents; stimulus equivalence relationships, or expectancies (Tolman, 1959) may also be considered part of this foundation; alternatively, means-end readinesses and the expectancies on which they are based can themselves be regarded as emergents.

Brief definitions of each category of the behavioral trichotomy are as follows:

I. Respondents

Respondents are responses that are elicited, without prior training, by the presentation of specific stimuli, called "unconditional stimuli" (UCS) or their conditional associates. It is reasonable to view respondents as being basically unlearned, reflexive responses elicited by specific stimuli that organisms encounter in the natural world. All other things being equal, one can predict with considerable confidence the form and continuance of a respondent upon its initial elicitation given the identity of the subject's species, its state and context, and the specifics of the UCS. For a given species, set of circumstances, and UCS, a respondent is very likely to recur time after time in the same form. Generally, a respondent requires only the impact of the UCS upon a given specimen, not upon that specimen's history of reinforcement with the UCS. Pavlovian conditioning involves respondents; the reinforcer is a stimulus, the UCS, that is correlated with an initially neutral stimulus, the conditional stimulus (CS). The UCS both elicits the respondent to be conditioned and serves as the reinforcer. After repeated presentations of the CS–UCS pair, the CS will tend to elicit a response similar to, though not in detail identical to, the response elicited initially by the UCS.

II. Operants

In contrast to respondents, operants are responses that are emitted by the organism and that are modified by their consequences. There is no readily definable UCS that elicits the operant to be conditioned. Rather, the response is initially emitted with apparent spontaneity by the subject and is not directly produced by specific operations of the experimenter. The operant can come to be occasioned by an initially neutral stimulus—a discriminative stimulus (S^D)—that functions somewhat analogously to the CS in respondent conditioning. Operants function by operating upon the environment and are selected by the reinforcing properties of the environment (e.g., the locations of nourishment, contrasted with sources of pain and trauma). Reinforcers for operants can be any external stimuli that increase the probability that the operant will be emitted. Consequently, by contrast to respondent conditioning, where the reinforcer is a rather specific UCS, in operant conditioning any of a number of consequences (e.g., things and events) might sustain the acquisition and continuance of an operant.

In the case of both respondent and operant conditioning, the presentation of the antecedent stimulus may provide a necessary context for the conditioned response to be manifested (i.e, for a discriminated operant, or for a respondent). Their learning entails reinforcers as consequences. There are several different types of procedures for both respondent and operant conditioning, and for schedule–of–reinforcement effects, that are beyond the scope and purpose of this paper.

III. Emergents

Emergents are <u>new competencies and/or new patterns of responding</u> that were never specifically reinforced by operations of the experimenter. They are <u>not</u> relatively simple, unitary responses (e.g., salivating, eye blinking, jumping over a hurdle, pecking a target, pressing a bar, or even chains of such behaviors) as in the case of respondents and operants. Several good examples of what we call emergents are presented by Sidman (1994; see Rumbaugh, 1995, for a review) stimulus–equivalence paradigm, in which as a result of a few specifically reinforced responses to relationships between specific stimuli, a substantially larger number of unreinforced relations can be obtained that, in turn, demonstrate "stimulus equivalence," defined by the properties of reflexivity, symmetry, and transitivity. These associations have been described by Sidman as having "emerged"; hence, their classification here as emergents is congruent with Sidman's view of them.

Emergents occur in a variety of contexts, in addition to that of Sidman's stimulus equivalence paradigm. These examples of emergents will be discussed subsequently, but each of them has in common the following attributes: (1) All emergents are forms of silent learning—by which it is meant that learning or acquisition of new response patterns or the cultivation of new competencies (i.e., the emergents) might progress with no obvious manifestation. (In reference to various aspects of inhibition, excitation, second order conditioning, and so on, Flaherty, 1985, pages 126–127, uses the term "silent" in his discussions of kinds of learning that go unnoticed unless special tests are instituted.) Emergent behaviors/competencies may go unmeasured, if not anticipated, unless the subject is tested in unique/altered contexts for transfer of learning and novel patterns of behavioral adaptation. However, subjects may spontaneously manifest emergents if, during training, they markedly alter their responses in a way that is both novel and extraordinarily adaptive. (2) The emergent behaviors/skills were never intentionally or systematically reinforced as part of the experimenter's treatment procedures. (3) The emergent behaviors/skills are established through induction, so it would appear, by the organism. Again it should be noted that emergents sometimes surprise the observer when they first appear—a consequence of the fact that they were not specifically reinforced or trained by the experimenter. (4) Emergents are noted for their apparent appropriateness to new situations. Emergents can make their appearance in new contexts which only in principle are similar to those in which they formed. They generalize between contexts not on the basis of the of specific stimulus dimension, as in stimulus generalization, but rather on the basis of relationships between stimuli and/or rules. The relationships and/or rules referenced here can be between any kind or number of elements (stimuli, responses, reinforcers, etc.) that are shared by two or more contexts.

Interim Summary. Although emergents, like operants and respondents, provide for adaptation and generally gain in strength with time and experience, only emergents are characterized by their complexity (e.g., heirarchical integration and creativity) and by their adaptive value in highly novel contexts. These contexts must be novel enough that, as posited above, generalization on traditional stimulus and response dimensions cannot provide a sufficient account for the response. Additionally, whereas operants, respondents, and emergents all depend on antecedents and consequences, and are sensitive to contingencies, emergents are not as readily accessible to the experimenter for specific shaping by consequences as are operants. Hence, emergents are distinguished from respondents and operants in that they can appear in novel, unanticipated forms that frequently appear to be clever, creative, and, indeed, smart.

Emergents differ from respondents and operants in still other important characteristics: Whereas both respondents and operants are relatively specific responses that can become conditioned to initially neutral stimuli, emergents are modes of responding or solving problems that are not "forced" by specific antecedents/stimuli, such as a UCS or S^D. Also, the overt motoric response entailed in the conditioning of respondents and emergents is fundamentally the same as the resultant conditioned response, whereas an emergent response might be strikingly different from the behavior manifested by the subject during the training experiences that generated the emergent response. Whereas overt motor responses are generally required by the subject for the conditioning of respondents and operants (sensory preconditioning is a notable exception), emergent responses can be learned silently by an apparent passive subject through observation. Finally, the learning of respondents or operants can be easily charted, for example by a cumulative recorder, whereas the formation of emergent response modes may not be discernible, because neither their formation nor their probability of later emission necessarily are indexed by concomitant behaviors.

These distinctions between respondents, operants, and emergents are summarized in Table 1. Most important, however, is that emergents are much more likely to be revealed in treatment X organismic interactions, where "organismic" refers to both between– and within–species variables, than are either respondents or operants. Some species are able to benefit from treatment conditions that hinder others, or to which the latter species are oblivious. For example, although stimulus equivalence training can generate reflexive, symmetric, and transitive relations in normal 4–year– old children, it did not in rhesus macaques (Macaca mulatta; Sidman, Rauzin, Lazar, & Cunningham, 1982)—though that is not to conclude that macaques are incapable of stimulus–equivalence relations. After appropriate training on other pairs of numerals, Rhesus macaques can choose the larger of two numerals, never before encountered as a pair, and, thereby, obtain the greater number of reinforcers (Washburn & Rumbaugh, 1991, p. 191; see details below). This behavioral skill, like the acquisition of symmetric relationships, requires an advanced brain, but not one so advanced as that of the human child.

See Table 1.

Similarly, individuals within a given species benefit differently from treatment conditions because of parameters such as age, level of maturation, state of health, and so on. Emergents can be particularly sensitive to differences in early rearing conditions. Examples of emergents from areas of psychology in which treatment X organism interactions are more likely to be sought and defined—such as comparative, developmental, and stimulus–equivalence research—will be discussed to help distinguish emergents from respondents and operants. The examples listed in Table 2 do not exhaust those available from the literature, and future research will surely define additional ones.

See Table 2.

Examples of emergents

Learning set, defined by Harlow in his classic paper of 1949, operationalized procedures which resulted in the transformation of rhesus subjects from trial–and–error associative learners to one–trial, seemingly insightful, problem solvers. Complexity of the brain across species and integrity of the brain within species, along with levels of maturation, were demonstrated to be powerful organismic variables which, in interaction with the treatment of learning–set training, affected the probability that one–trial learning capabilities would emerge. The ability to choose the correct (reinforced) one of a pair of novel stimuli at nearly the 100% level after a single "testing" trial was the terminal point of learning set formation. From the cognitivist perspective, the organisms capable of learning set formation had learned an emergent strategy, "win– stay, lose–shift," that they applied to each new pair of stimuli. From a connectionist perspective, they had learned to strengthen or reduce associative strength to stimulus cues enough in a single trial so that they could choose the stronger association at near 100% levels after that trial. Part of the reason for that might be that all increments or decrements in associative strength were attached to the cues offered by the discriminanda rather than to other "error factor" cues like right vs left position.

Transfer of learning research has a long and rich history. Transfer of learning is quantified on a continuum that extends from strongly negative (e.g., transfer slows learning), through null (e.g., no transfer), to strongly positive (e.g., transfer facilitates new learning). Brain complexity, as represented within the array of species that comprise the order, Primates, is also a continuum that extends across several levels. When one examines transfer–of–training effects in reversal learning as a function of amount learned prior to the test for transfer, one finds a remarkable effect—transfer for prosimians with their relatively smooth, small brains becomes increasingly negative as pre–test trials increase, whereas the more encephalized, large-brained primates' transfer can become increasingly positive. The interactive effect between treatment (i.e., amount learned prior to transfer) and the organismic variable of brain complexity qualitatively alters the essence of the transfer effect (Rumbaugh & Pate, 1984). This phenomenon may be related to, and certainly confirms, the connection between brain complexity and the ability to form learning sets. In both cases, organisms with more complex brains are better able to "escape the bonds" formed by previous learning in order to form new associations quickly. Cognitively speaking, more complex organisms learn to identify "relevant" and discount "irrelevant" cues better than less complex organisms.

Learning processes also vary in relation to levels of brain complexity within the order, Primates. Primates with relatively smaller and simpler brains learn in accordance with the traditional stimulus–response associative models that apply best to the establishment of habits of responding to reinforced stimulus choices and of not responding to unreinforced stimulus choices in a multiple-problem, two-choice, discrimination- learning situation. Whereas some primates with relatively larger brains and cortical elaborations apparently learn as stimulus– response learners, others can learn in accordance with a mediational or relational model which enables the subject to take, for example, discrimination–reversal test trials seemingly as a continuance of the initial discrimination task (Rumbaugh and Pate, 1984). In other words, they discount the fact that the cue values of the discriminanda have been exchanged and continue to improve in the execution of choices. These emergent response modes alter transfer–of–learning effects, and the essence of the discrimination learning process itself. They are not a consequence of procedures used by the experimenter to establish such modes. Rather, they emerge as a consequence of how brains of greater and lesser degrees of complexity respond to the same treatments (e.g., the discrimination tasks and tests for transfer).

Of course, organisms that learn relationally do not cease to learn associatively. In fact, it seems likely that, for species with the capacity for relational learning, the propensity for relational versus associative learning improves both with phylogeny and ontogeny. Thus, rhesus monkeys have demonstrated the capacity for relational learning, but have also failed to extract rule-like relations from other tasks, responding stubbornly (but generally successfully) according to stimulus-response associations (Filion, Washburn, & Fragaszy, 1995).

Stimulus–equivalence training experiences result in differential outcomes depending upon the species (humans are markedly superior to nonhuman primates that, at best, have less ability to manifest equivalence, symmetry, and transitive relations) and, within humans, upon whether or not language is operative (Sidman, 1994).

Concept learning for both natural and arbitrary things and events varies markedly as a function of species and age level, when treatment variables (e.g., tasks) are held constant. Emergent behaviors may include generalized identity matching-to-sample, symbolic matching, and sameness-difference concepts.

The representational use of symbols, as an ability, is strongly controlled by brain complexity and the age at which such training/learning experiences are given to the subject. Chimpanzees are clearly capable of using symbols to represent things not present, as indicated by their ability to classify symbols into appropriate categories (for example, whether the symbol represents a tool or food; see Savage–Rumbaugh, 1986, for a review of relevant research).

Speech comprehension and the invention of proto–grammar appear to be strongly related both to the variables of brain complexity (e.g., monkeys, chimpanzees, and children) and rearing (i.e., treatment) conditions for the subject. Kanzi, a bonobo (Pan paniscus) has manifested the ability to understand novel requests, conveyed to him via sentences spoken by humans, at a level that compares favorably with a child whose mental age was 2½ years. He also has employed what would be termed grammar, if he were a human of 1-1/2 years, in the productive combinations of gestures and symbols that he uses to communicate complex messages/requests to his caretakers (for details see Savage-Rumbaugh, Murphy, Sevcik, Brakke, Williams, & Rumbaugh, 1993; Savage-Rumbaugh & Lewin, 1994; Greenfield and Savage–Rumbaugh, 1991). It is also significant that Kanzi did not develop his language skills as a result of specific, discrete–trial, reinforced training. Rather, his skills were acquired quite indirectly—through observation of efforts to teach his mother, Matata, to learn the appropriate use of word–lexigrams (i.e., geometric symbols) and use a "talking" lexigram board. Matata, who was then more than 15 years old, failed to learn any language skills, quite possibly because she was a feral animal until the age of about 6 years. For her, the years for the optimal learning of language had long passed. For Kanzi, however, they had not, for he played about in the context within which Matata received her scheduled language training from soon after birth to the age of 2½ years. Here we have, then, a prime example of the organismic variable of age (Matata was too old to learn language skills, while Kanzi was precisely the right age, as it turned out) interacting with the treatment condition that consisted not of language training, but, rather, of <u>exposure</u> to language usage.

 <u>Numerical cognition</u> by nonhuman animals provides an additional example of emergents. As mentioned earlier, Washburn and Rumbaugh (1991) reported that rhesus monkeys learn substantially more than which of two numerals is the one that pays off the most in food pellets. Two monkeys were trained with all but seven combinations of pairs of numerals 0 through 9; seven pairs were chosen to be used later as novel test pairs to determine whether, during training, the monkeys had learned only to pick one of each specific pair of numerals, or whether they learned something about the "value" of each numeral. If, for example, on a given training trial they were presented with a 5 paired with a 3, the selection of the 3 would result in the automatic delivery of 3 food pellets, whereas the selection of the 5 would result in the delivery of 5 pellets. During test trials on the seven new pairs in which the numerals 6, 7, and 9 were each used <u>twice</u> (i.e., 6:4, 6:5, 7:5, 7:6, 8:5, 9:7, and 9:8), one monkey made no errors on their first presentation, and the other made only two errors. If they had learned only which numeral to choose in the context of each training pair, they would not have been able to perform above chance on the novel pairings. Thus the monkeys performed significantly above chance—they may have learned something like a matrix of relative values.

 Alternatively, the animals could have learned a comparison strategy: they could have attached a value to each numeral as a result of the original training, and learned that they profited most by comparing each pair of numerals and choosing the one with the larger value. In Hull's theory, these "values" for each stimulus would be called "reaction potential." In contemporary cognitive terms, these would be representations of quantities corresponding to the meaning of the numerals. In either case, this type of comparison is a different process from immediately responding to any stimulus that has been reinforced, or even to the stimulus that had the greatest habit strength.

 Such an altered response mode was not specifically trained— nor could it have been demanded. It may, however, have been prepared through evolutionary selection for animals that try to obtain better nutrients, rather than selecting whatever food is available. Notwithstanding, the training, in interaction with the brain/learning capacity of the rhesus subjects, allowed the ability to execute ordinal judgments accurately to emerge, as reflected in their choice of the larger numeral in novel pairings presented for test.

Other examples of phenomena from the history of psychology that exemplify emergent response modes include latent learning (Blodgett, 1929; Tolman, 1948) and the effects of early rearing environments (Riesen, 1982; Bryan & Riesen, 1989; Stell & Riesen, 1987) upon patterns of brain development and complex learning skills, and still others that are listed in Table 2. In these experiments, treatment effects interacted with developmental, hence organismic, variables to determine whether or not learning was manifested subsequent to explorations of mazes without specific reinforcement, or whether learning, language, and speech were compromised as a consequence of deprivation of appropriate stimulation or of the opportunity to learn at appropriate levels of maturation.

Even Epstein's (Epstein, 1985; Epstein, Kirshnit, Lanza, & Rubin, 1984) simulation of "insight" in the pigeon illustrates what we call an emergent response mode in this paper. Epstein's pigeon, in a final test, moved a box into position, then stood on it in order to access a target that was otherwise out of reach. His account detailed the antecedents, but it was the pigeon's brain that processed the prior training and blended it to allow for a chimpanzee–like solution to a classic problem (Ellen & Pate, 1986). The importance of experiences relevant to task demands has been recognized by researchers with chimpanzees from the days of Köhler's (1925) classic studies.

Notwithstanding, it is the subject, be it pigeon or chimpanzee, whose brain operations generate a new response mode, an emergent, that allows for problem solution. That individual and specific prior conditioning of operants is part of the subject's training history is certainly relevant, indeed critical, to the emergent response mode; but it is the subject's brain's processes, contingent as they are upon the organization and complexity of the brain, that generate the new, emergent, response modes. The most salient attribute of those modes is that, in novel tests/contexts, they provide for adaptive novel behaviors that are substantially extended in form and organization beyond those manifested during "training."

Summary

Do operants and respondents operate in the manifestation of emergents? Most certainly they do, but it is the novel blending of them, their varied orchestration and patterning, their immediate manifestation, that reveals the emergents present in the brain's operations; it is not specific reflections of antecedents and contingencies provided by the environment or the experimenter.

Are emergents reducible to either operants or respondents? It is the argument of this paper that they are not, though, as stated above, operants and respondents surely are the behavioral elements and indicants of emergents. Indeed it is through behaviors that by tradition might be termed respondent or operant that emergents are manifested. Notwithstanding, it is precisely the non–respondent, non–operant nature that makes certain behavior an emergent. Emergents make their appearance as novel patterns of responding or choosing between alternatives, and they do so with some element of surprise to the observer. By contrast, both respondents and operants make their appearance as improved forms of what they were at the very beginning of training or conditioning. Their basic forms are not altered. Again, and by contrast, emergents do not have specific training histories. There is no reason to assert that they were there in some miniscule form that either became stronger or was shaped across time, as is the case of operants. This is not to contradict the argument, however, that emergents have their etiology in the experiences whereby organisms, particularly those with complex brains, acquired respondents and operants. Emergents are new competencies, new patterns of behavior, based in experience, that are produced by novel generative operations of the subject's brain—a brain whose operations depend on age, absence of trauma, experience, tasks, and species.

The category "emergents" encourages the behavioral researcher to use time–tested tactics that emphasize antecedents and consequences to study behaviors that are new patterns and demonstrate competence for adapting. Alternatively, one may study the same behavioral modes using a cognitivist point of view, but that is neither necessary nor necessarily advantageous compared to use of the framework herein advanced.

Science moves with the times and new findings. We here argue that it is timely for behaviorally oriented psychologists to evaluate the merits of extending Skinner's "respondent and operant" dichotomy to a trichotomy that includes the new category of <u>emergent</u>. The category "emergent" can facilitate the integration of large corpuses of comparative, developmental, and brain research into the behavioral framework, and thereby substantively enhance the science generated by the rich tradition of psychological research.

Author Notes

Preparation of this paper was supported by HD–06016 from the National Institute of Child Health and Human Development to Georgia State University. Additional support was provided by the College of Arts and Sciences of Georgia State University, and by a grant (NAG2–438) from the National Aeronautics and Space Administration. Dr. Shelly Williams and Dr. Daniel Cerutti gave helpful comments on early drafts of this paper, which is based on a presentation by the first author at the Association for Behavior Analysis, Atlanta, Georgia, 1991.

References

Alcock, J. (1979). <u>Animal behavior: an evolutionary approach</u> (2nd ed). Massachusetts: Sinauer Associates, Inc.

Blodgett, H. C. (1929). The effect of introduction of reward upon maze performance of rats. <u>University of California Publications in Psychology</u>, <u>4</u>, 113-134.

Bryan, G. K. & Riesen, A. H. (1989). Deprived somatosensory-motor experience in stumptailed monkey neocortex: Dendritic spine density and dendritic branching of layer IIIB pyramidal cells. <u>Journal of Comparative Neurology</u>, <u>286</u>, 208-217.

Epstein, R. (1985). The spontaneous interconnection of three repertoires. <u>The Psychological Record</u>, <u>35</u>, 131-141.

Epstein, R., Kirshnit, C., Lanza, R., & Rubin, L. (1984). "Insight" in the pigeon: Antecedents and determinants of an intelligent performance. <u>Nature</u>, <u>308</u>, 1 March, 61-62.

Ellen, P. & Pate, J. L. Is insight merely response chaining?: A reply to Epstein. <u>Psychological Record</u>, 1986, <u>36</u>, 155-160.

Ellen, P., Sotores, B. J., & Wages, C. (1984). Problem solving in the rat: Piecemeal acquisition of cognitive maps. <u>Learning and Motivation</u>, <u>12</u>, 2, 232-237.

Filion, C., Washburn, D. A., & Fragaszy, D. M. (1995, June). <u>Trajectory estimation by rhesus macaques</u>. Poster presented at the annual meeting of the American Psychological Society, New York, NY.

Flaherty, C. F. (1985). <u>Animal learning and cognition</u>. New York: Alfred A. Knopf.

Gallup, G. G. (1987). Self awareness. In J. Erwin (Ed.), <u>Comparative primate biology: Volume 2B</u>. Behavior, cognition and motivation, (pp. 3-16). New York: Alan R. Liss.

Greenfield, P. M. & Savage-Rumbaugh, E. S. (1991). Imitation, grammatical development, and the invention of protogrammar by an ape (<u>Pan paniscus</u>). In N. Krasnegor, D. M. Rumbaugh, M. Studdert-Kennedy, & R. L. Schiefelbusch (Eds.), <u>Biobehavioral Foundations of Language Development</u>. Hillsdale, NH: Erlbaum.

Harlow, H. F. (1949). The formation of learning sets. <u>Psychological Review</u>, <u>56</u>, 51-65.

Heidbreder, E. (1933). <u>Seven psychologies</u>. New York: Appleton-Century-Crofts, Inc.

Koehler, W. (1925). <u>The mentality of apes</u>. New York: Routledge & Kegan Paul.

Krechevsky, I. "Hypotheses" in rats. <u>Psychological Review, 39</u>, 516-532.

Kuhn, T. S. (1962). <u>The structure of scientific revolutions</u>. Chicago: University of Chicago Press.

Levine, M. (1971). Hypothesis theory and nonlearning despite ideal S-R-reinforcement contingencies. <u>Psychological Review, 78</u>, 130-140.

Nissen, H. W. (1951). Phylogenetic comparison. In S. S. Stevens (Ed.), <u>Handbook of</u>

experimental psychology (pp. 347-386). New York: Wiley.

Posner, M. I. (1982). Cumulative development of attentional theory, American Psychologist, 37, 168-179.

Riesen, A. H. (1982). Effects of environments on development in sensory systems. In W. D. Neff (Ed.), Contributions to sensory physiology, (Vol. 6, pp. 45-77). New York: Academic Press.

Rosenblatt, F. (1958). The perceptron: A probabilistic model for information storage and organization in the brain.
Psychological Review, 65, 386-407.

Rosenblatt, F. (1962). Principles of neurodynamics. New York: Spartan.

Rumbaugh, D. M. (1995). Emergence of relations and the essence of learning: A review of Sidman's Equivalence Relations and Behavior: A Research Story. The Behavior Analyst, 18, 367-375.

Rumbaugh, D. M., Hopkins, W. D., Washburn, D. A., & Savage-Rumbaugh, E. S. (1989). Lana chimpanzee learns to count by "Numath": A summary of a videotaped experimental report. The Psychological Record, 39, 459-470.

Rumbaugh, D. M., & Pate, J. L. (1984). The evolution of cognition in primates: A comparative perspective. In H. L. Roitblat, T. G. Bever, & H. S. Terrace (Eds.), Animal cognition (pp.403-420). Hillsdale, N. J.: Lawrence Erlbaum Associates.

Savage-Rumbaugh, E. S. (1986). Ape language: From conditioned responses to symbols. New York: Columbia University Press.

Savage-Rumbaugh, E. S. & Lewin, R. Kanzi: At the brink of the human mind. New York: John Wiley.

Savage-Rumbaugh, E. S, Murphy, J., Sevcik, R. A., Rumbaugh, D., Brakke, K. E., & Williams, S. (1993). Language comprehension in ape and child. Monographs of the Society for Research in Child Development, Serial No. 233, Vol. 58, Nos. 3-4, pp. 1 - 242.

Schrier, A. M., Harlow, H. F., & Stollnitz, F. (1965). Behavior of nonhuman primates. New York: Academic Press.

Selfridge, O. G. (1955). Pattern recognition by a machine. Proceedings of the Western Joint Computer Conference.

Selfridge, O. G. (1959), Pandemonium: A paradigm for learning. In D. V. Blake & A. M. Uttley (Eds.), The Mechanisation of Thought Processes. London: H. M. Stationary Office.

Sidman, M. (1994). Equivalence relations and behavior: A research story. (pp.606) Boston:Authors Cooperative, Inc.

Sidman, M., Rauzin, R., Lazar, R., & Cunningham, S. (1982). A search for symmetry in the conditional discrimination of Rhesus monkeys, baboons, and children. Journal of the Experimental Analysis of Behavior, 37, 23-44.

Skinner, B. F. (1938). The behavior of organisms: An experimental analysis. New York: Appleton-Century-Crofts.

Stell, M., & Riesen, A. (1987). Effects of early environments on motor cortex neuroanatomical changes following somatosensory experience: Effects of Layer III pyramidal cells in monkey cortex. Behavioral Neuroscience, 101, 341-346.

Stephen, H., Bauchot, R., & Andy, O. J. (1970). Data on size of the brain and of various brain parts in insectivores and primates. In C. R. Noback & W. Montagna (Eds.), The primate brain (pp.289-297). New York: Appleton-Century-Crofts.

Tolman, E. C. (1948). Cognitive maps in rats and men. Psychological Review, 55, 189-208.

Tolman, E. C. (1959). Principles of purposive behavior. In S. Koch (Ed.), Psychology: A study of a science. Vol.2. pgs. 92-157. New York: McGraw-Hill.

Washburn, D. A. & Rumbaugh, D. M. (1991). Ordinal judgments of numerical symbols by macaques (Macaca mulatta). Psychological Science, 2, (3), 1991.

Table 1.   A Summary of Similarities and Differences between Respondents, Operants, and Emergents

Parameter	Respondents	Operants	Emergents
A. well-defined CS or antecedent	yes	yes	no
B. acquisition depends upon experience with specific and limited antecedents and consequences	yes	yes	no
C. overt response required and recordable during acquisition	yes*	yes	No--their formation may be **SILENT**
D. conditionable to CS/S^D	yes	yes	no
E. based on histories that emphasize generalized classes of experiences	no	no	yes
F. repetition of trials or events important	yes	yes	yes?
G. new response mode form and provide for novel adaptations	no	no?	yes
H. appears in novel contexts/problems and transfer tests	no	no	yes
I. entails syntheses of individually acquired responses	no	no	yes
J. particularly sensitive to **Early Rearing** variables	no	no	yes
K. interactive products of **Task X Organismic** variables (e.g., brain complexity as per maturation and species)	no	no	yes

* sensory preconditioning is a notable exception

Table 2. Research Areas that Produce Emergents

Emergent	Investigators	Characteristics
Learning set	Harlow, 1949; see Schrier, Harlow, & Stollitz, 1965, for a review	Primates and children's learning changed from trial and error to 1-trial learning as a function of number of problems.
Transfer Index	Rumbaugh & Pate, 1984	As an interaction between increased brain complexity across taxa and increased learning prior to test, primates' transfer of learning changed from negative to positive
Mediational learning	Rumbaugh & Pate, 1984	In association with increased brain complexity across taxa, learning shifted from associative to mediational or relational
Ape-language research	Savage-Rumbaugh, 1986, Savage-Rumbaugh & Lewin,1994; Savage-Rumbaugh, Murphy, Sevcik, Brakke, Williams, & Rumbaugh, 1993.	Chimpanzees learned to use arbitrary symbols to represent items, to categorize them symbolically, and to communicate about them in their absence. Also, learned symbols by observation and came to comprehend syntax of human speech.
Stimulus equivalence	Sidman, 1994	Reinforced choices of specific stimuli in discrimination learning generated many other relations between stimuli.
Latent learning	Blodgett, 1929; Tolman, 1948	Subsequent to exploration of mazes, rats demonstrated learning had taken place and to obtain incentives in accordance with privation states.
Mapping	Menzel, 1978	Chimpanzees, carried and shown locations of foods in an open field, subsequently obtained them by travelling a route that required minimal effort.
Recognition of self in mirror	Gallup, 1983	Chimpanzees, if reared in social groups, come to recognize their images in mirrors, but do not do so if reared alone.
Counting by a chimpanzee	Rumbaugh, Hopkins, Washburn & Savage-Rumbaugh, 1989	Lana, chimpanzee, learned to count in that she could remove 1, 2, or 3 boxes from a video screen in accordance with the value of each trial's target number, 1, 2, or 3, with only her memory of intra-trial events to guide her choice.
Ordinal judgments of numerals by macaques	Washburn and Rumbaugh, 1991	In transfer tests, rhesus monkeys were able to choose the larger of two numerals, never before paired, as a consequence of learning the relative pellet-values of experience with other pairs of numerals, 0-9, during training. They had acquired a matrix of relationships between all numerals.
Integration of temporally-separated explorations of maze segments	Ellen, Sotores, & Wages, 1984	Rats learned a three-table "reasoning-type" problem via unreinforced exploration of separate segments on separate days.

2.

Analysis of Behavioral Selection by Consequences and Its Potential: Contributions to Understanding Brain-Behavior Relations

William J. McIlvane

Analysis of Behavioral Selection by Consequences and
Its Potential Contributions to Understanding Brain-Behavior Relations

William J. McIlvane, William V. Dube, & Richard W. Serna

E. K. Shriver Center, Massachusetts General Hospital,
Harvard Medical School, & Northeastern University

Correspondence to:

W. J. McIlvane
Shriver Mental Retardation Research Center
200 Trapelo Road
Waltham, MA 02254
(617) 642-0153

Throughout much of its history, behavioral science can be fairly characterized as an ongoing series of skirmishes and organized battles among individuals and groups espousing a variety of theoretical and methodological perspectives. Given the field's subject matter, the stakes are no less than arriving at a coherent, empirically justifiable understanding of such basic phenomena as perception, learning, consciousness, emotion, and so forth. The secondary gain includes faculty positions, research grants, and other benefits of academic life.

Among the best known of these campaigns occurred during the middle part of this century, when learning theorists like B. F. Skinner were competing with those involved in the "cognitive revolution," notably Noam Chomsky. As in most such competitions, each group concerned itself with some fundamental, essential truth about behavior, and thus each could justify its continued existence and relative prosperity. The 4th Annual Behavioral Neurodynamics Conference, in fact, can be seen as a celebration of the success of the many subdisciplines of behavioral science. In Pribram's expansive vision, each has a critical role to play in understanding the activities and operational processes of the brain. The program included not only learning theorists, but also accomplished representatives of the cognitive, biological, and engineering sciences.

Concerning the subdisciplines of behavior science, our hope is that the individuals involved have begun to see at least the rough outline of their future and that this future does not keep alive the old battles. Rather, an essential component will be an alliance with the brain sciences that uses the many advances in science and technology to help unravel the relationship between brain and behavior. An important part of forging this alliance will be articulating what each subdiscipline has learned that might contribute to this grand enterprise. This chapter will endeavor to elucidate some of the contributions of our field, behavior analysis -- a field that is based on principles articulated by Skinner and one that has contributed a large body of relevant theoretical and empirical work.

Behavior Analysis and Brain Science

A common misperception is that behavior analysis does not concern itself with the biological processes that underlie behavior. This misperception reflects a fundamental misunderstanding of the behavior-analytic position on the relationship between behavioral science and brain science. Perhaps the clearest expression of that position was offered in Skinner's 1989 *American Psychologist* paper: "There are two unavoidable gaps in any behavioral account: one between the stimulating action of the environment and the response of the organism and one between consequences and the resulting change in behavior. *Only brain science can fill those gaps.* In doing so, it completes the account; it does not give a different account of the same thing" (italics ours, p. 18). From its early days, the literature of behavior analysis has included studies of relationships between behavior and brain processes (e.g., see Mogenson & Cioe´, 1977). Why then have behavior analysts been labeled "black box" psychologists who were uninterested in the operations of the nervous system?

Perhaps behavior analysis has been misunderstood in part because of Skinner's frequent admonitions about studies of the "conceptual nervous system." Skinner was much concerned about efforts that used very limited data sets to guess how the biological nervous system worked. Central also to Skinner's thinking was the possibility of developing an independent science of behavior. Even if little progress were made in the brain sciences, he reasoned, one could still develop a science of behavior with its own body of data, principles, and theories. Moreover, Skinner understood that a well developed science of behavior would be needed even if every important operation of the brain was discovered. In order to relate brain and behavior, it would be essential to have a cohesive, integrated, intellectually rigorous account of behavior.

In essence, Skinner and his followers proposed a division of labor. Behavioral scientists would concern themselves mainly with delineating precise, quantitative relations between behavior and the controlling variables that could be directly observed or reasonably inferred from empirical analysis. Such relations would emerge in part from methodologies with increasing capabilities to predict and influence the behavior of both humans and nonhumans. Brain scientists would concern themselves with increasing understanding of the operations of the nervous systems. And as each discipline developed its knowledge base, it would become increasingly possible to define secure empirical relationships between brain and behavior.

As things have turned out, this division of labor has been rejected by many behavioral scientists, most notably those who have turned to cognitive psychology. Cognitivists were inspired, of course, by the development of the serial digital computer. One may characterize their goal as attempting to discern the mind/brain's computer program and its symbol system. This enterprise, while productive, has not been entirely satisfactory, and that dissatisfaction has given rise to connectionism and neural network models that do not rely on symbolic processes. Connectionists seek to account for complex behavior via simpler processes that are not dissimilar from those that concerned Skinner and other behavior analysts (cf. Donahoe, Burgos, & Palmer, 1993). This is not to say that connectionists have espoused behavior analysis, but merely that there has been some realization that behavior analysis does touch nature in some fundamental way.

Many behavior analysts, in turn, have begun to take on problems that have been the traditional subject matter of cognitive psychology. For example, the burgeoning field of stimulus equivalence research is now addressing the phenomena of stimulus-stimulus relations, the *sine qua non* of cognitive analyses. Moreover, there is growing appreciation within behavior analysis that advances in the brain sciences -- for example, imaging technologies like PET, MRI, EEG (Posner & Raichle, 1994; Tucker, this volume) -- have made it possible to observe directly events within the skin. This growing capability is rapidly removing barriers to experimental analysis of private events, a longstanding interest of behavior analysis (e.g., Skinner, 1974). For the remainder of this chapter, we shall briefly review some noteworthy accomplishments of behavior analysis over the last thirty years and endeavor to articulate a rationale for its participation in interdisciplinary studies of brain-behavior relations.

Overview of Behavior Analysis

From the outset, behavior analysis made a radical departure from methodological behaviorism and the S-R psychologies that dominated the first half of this century. Perhaps one of the most important distinctions is the behavior-analytic acceptance of private events (e.g., Posner's [1980] covert shifts of attention) as perfectly acceptable subject matter (cf. McIlvane, Dube, & Callahan, 1996). Also, behavior analysts tend to agree with cognitively oriented scientists that what is learned is contingencies -- relations or regularities among environmental events -- and not specific stimulus-response connections.

The fundamental analytical method is termed contingency analysis, which at its most basic level concerns itself with three types of events: antecedent events outside and within the skin that occasion behaving; behaviors such as listening, talking, moving, thinking, and so forth; and consequences, the events that influence the future probability of the behaviors occasioned by antecedents (i.e., positive and negative reinforcers; see below). Important variables of contingency analysis include: (1) antecedent stimulus variables (complexity, duration, intensity, modality, salience, etc.); (2) behavioral variables (type, latency, force, duration, etc.); and (3) consequential stimulus variables (type, magnitude, schedule, etc.). Potentially modulating the effects of one or more of these classes of variables are: (4) subject variables (age, sex, clinical diagnosis, behavioral history, etc.); and (5) state variables (biological establishing operations, disease, drug, etc.). The main goals of behavior analysis are to describe qualitative and ultimately quantitative functional relations among these variables.

It is important to note here that contingency analysis is not merely applied to study nonhuman animal behavior in "Skinner boxes." There is a large contingent who do experimental analyses of human behavior -- and not just its simple forms. Behavior analysis takes on subjects of varying complexity -- from individual responses to larger behavioral episodes (see McIlvane et al., 1996, for further illustrations). Among the more illustrative examples of this scope is the research program of Allan Neuringer and his students (e.g., Neuringer & Voss, 1993). This group studies reinforcement procedures that teach both humans and nonhumans to behave in a highly variable fashion, that is, to continually emit new behaviors that have not previously occurred in the experimental context. Humans' performance becomes statistically indistinguishable from that which would be predicted from a random number generator, and nonhumans approach that level.

The Riddle of Reinforcement

The problem of specifying the nature of the reinforcement process is significant enough that many behavioral scientists -- both within behavior analysis and outside it -- deal with it mainly by ignoring the complexities. Within behavior analysis, the most common approach has been to define reinforcers through an empirical analysis of their effects. A stimulus is reinforcing if its delivery contingent upon a given behavior is followed by a reliable change in the probability of that behavior.

This functional definition has long been recognized as circular, most notably by Paul Meehl (1950). Continued reliance on it is often justified on practical grounds; both experimental and applied behavior analysis have made useful contributions to basic science and clinical practice by defining reinforcers in this functional manner. Nevertheless, many behavior analysts have endeavored to define reinforcement in a more informative fashion. For example, John Donahoe (Donahoe & Palmer, 1994) has recently defined a reinforcer as a stimulus that elicits reflexive behavior that differs from the behavior currently ongoing, which in turn alters the probability of behavior that precedes the interruption.

This "unified reinforcement principle" is a very attractive definition in certain respects. It not only defines reinforcement functionally, but also begins to suggest a way out of the circularity problem and allows one to make reinforcement a central feature of both respondent and operant conditioning. Although a comprehensive presentation of Donahoe's thinking is beyond the scope of this chapter, it seems particularly noteworthy in that he offers this account as part of a broader effort to connect behavior analysis not only with connectionism but also with facts of developmental neurobiology.

Many years ago, there were other efforts to eliminate the circularity in the definition of reinforcement. Reinforcement was conceptualized in terms of drive states, drive reduction, and so on. Subsequent work proved devastating for these formulations. It was found, for example, that lesions of the ventral medial hypothalamus would cause rats to eat voraciously, suggesting a high drive state. If food was made contingent upon an operant response, however, the food would not function as a reinforcer, suggesting a low drive state. How could animals simultaneously exist in a high and low drive state? (See Pribram 1995, for additional discussion of this issue).

David Premack (1965) helped to circumvent this problem, when he articulated his now-famous Premack Principle. Briefly, he suggested that organisms tended to prefer to engage in certain behaviors over others, and that at any given moment there was a hierarchy of valued activities. Those behaviors with higher preference values, like eating if food had not been recently available, could serve as reinforcers for lower probability behaviors, like pressing a bar in an operant chamber. Notice that in Premack's formulation there is no reference to basic biological needs or drives that impel behavior. In their place, one talks about events that have preference or value for the organism, thus implicitly expanding the range of events that might serve as reinforcers.

Theoretical and empirical work subsequent to Premack's original contribution has refined and modified the value-based account. One such account, for example, suggests that what is critical to establish a given behavior as reinforcing is that access to that behavior is restricted, so that its ordinarily preferred (or baseline) level is not achievable (Timberlake & Allison, 1974). A related one suggests that animals have certain levels of the behaviors that they characteristically engage in -- a preferred behavioral organization -- if you will (Allison, Miller, & Wozny, 1979). Experimental operations that alter the preferred levels create organismic efforts to restore those levels, thus rendering opportunities to do so reinforcing. Each of these accounts may be seen as similar in that behaviors are reinforcers not because they reduce a drive or restore biological homeostasis but rather because behaviors or behavioral organizations have established value for the organism.

These new ways to conceptualize reinforcement were further enriched when Jack Michael (1982) helped to redefine operations that would establish events as reinforcers. He defined the "establishing operation" as an environmental event, operation, or stimulus condition that affects the organism by momentarily altering (a) the reinforcing effectiveness of other events and (b) the frequency of occurrence of that part of the organism's repertoire relevant to those events as consequences. The notion that operations can momentarily render events reinforcing is attractive for many reasons, but perhaps especially because it gives us a more plausible framework for talking about reinforcement processes, especially in humans. For example, imagine that you are the audio-visual specialist responsible at the 4th Annual Behavioral Neurodynamics conference, and it is your job to make sure that the speakers' slides can be read by the audience. Imagine further that the projector bulb fails. Failure of the old bulb is the establishing operation that makes finding a new bulb strongly reinforcing -- although that same spare bulb would have been at best a weak reinforcer or even a neutral stimulus just a moment before. Note that the establishing operation formulation applies to all three conceptions of the reinforcer described above.

These newer conceptions of reinforcement and reinforcement operations have been extremely useful for behavior analysis. They help the field to shed criticisms that have been leveled at reinforcement-based accounts of behavior. In the 1950s, the drive-based accounts of reinforcement stressed "primary" reinforcers. According to such accounts, the ongoing stream of behavior was largely motivated by a search for biologically significant reinforcers like food, sex, warmth, and secondary factors related to those reinforcers, such as status within the group. Thus, the drive reductionists were offering a sort of Rabelaisian view of existence that was not widely seen as an attractive picture of the human condition. The more recent formulations allow us to articulate the principles of reinforcement in new terms -- of values, preferred activities, or reducing behavioral discrepancies. By contrast with drive theorists, these modern formulations suggest a broader view that also outlines a research agenda: What are the circumstances under which activities become preferred?

The Nature of Selection by Consequences

Two longstanding, somewhat controversial issues in the analysis of behavior are (1) under what circumstances does selection by consequences occur and (2) what is selected? With respect to the former, behavior analysis has typically asserted that reinforcement does not require awareness or other intervening processes to operate. Among the most compelling demonstrations of this phenomenon was one reported by Hefferline and colleagues at Columbia (e.g., Hefferline, Keenan, & Harford, 1959). They showed that appropriate reinforcement contingencies could increase the frequency of minute, imperceptible muscle contractions -- events clearly out of the awareness of the humans who were their subjects. Thus, selection by consequences has been assumed to operate in a fairly broad fashion that emphasizes environment-behavior contingency, not awareness, attention, or some obvious selective process.

More recently, there have emerged apparent challenges to this assumption. Peter Killeen (1994), for example, has offered a mathematical model of reinforcement that suggested an important role for short-term memory processes. Based on certain characteristic delay functions obtained from empirical data, he argues that reinforcement selects not behavior directly but rather its representation in memory. It seems difficult to reconcile such analyses with findings like Hefferline's that suggest broader operation of reinforcement processes, and this seems to be a fertile area for future research and theoretical development.

With respect to what is selected by consequences, both early (Ray & Sidman, 1970) and recent (e.g., McIlvane & Dube, 1992) work from our group suggests that reinforcement selects not merely individual behaviors but rather environment-behavior relations as a unit (see Sidman [1986] and Donahoe and Palmer [1994] for further development of this issue). Central to this argument is the question of whether every behavior or behavioral episode has a potentially identifiable antecedent and conversely whether behavior can be emitted in the absence of an occasioning environmental event (see Shull, 1995, for a comprehensive discussion).

Outcomes of Selection by Consequences

The process of selection by consequences leads to a number of outcomes that seem directly relevant to the emerging field of behavioral neurodynamics. Three aspects of our current research program seem particularly relevant, and we shall describe them to exemplify research in this area.

Stimulus equivalence. If reinforcement is about value for the organism, the antecedent or discriminative stimuli that set the occasion for the reinforcer should have a detectable value. Further, stimuli discriminative for the same reinforcer should have equal values, that is, they should be equivalent to one another. From a great many examples available to us, we shall briefly describe an illustrative, recently reported study that sought to define a minimal experimental history for relating stimuli on the basis of common relations to reinforcers (Dube & McIlvane, 1995).

Figure 1 illustrates the procedures (*see fig. 1*). During baseline training, four different nonrepresentational forms served as samples and comparisons on identity-matching trials. Correct selections of two of the forms were always followed by a specific reinforcer, SR1 (e.g., a sip of juice), and those of the other two forms were followed by a second reinforcer, SR2 (e.g., a pretzel). Unreinforced test trials then asked whether this history was sufficient to establish the arbitrary-matching performances shown in the lower portion of Figure 1. That is, were stimuli that were antecedents for the same reinforcer equivalent to each other? If so, then they should be substitutable for one another in the matching task. Such performances, in fact, did emerge in a number of individuals with mental retardation, thus suggesting that the stimuli matched to each other had the same value (cf. Edwards, Jagielo, Zentall, & Hogan, 1982; Schenk, 1995).

Based in part on research such as this, Murray Sidman (1994) has argued that stimulus equivalence is not a derived phenomenon, as has been widely suggested (e.g., Horne & Lowe, in press). Rather, it may be a direct product of the reinforcement process. If so, then equivalence relations may turn out to be fundamental building blocks of cognition rather than its products. Although the equivalence work thus far has been mainly descriptive, there are now the beginnings of a structural and quantitative analysis of equivalence relations, including such variables as nodal distance -- how the number of connecting relations interacts with a subject's ability to demonstrate equivalence relations (Fields, Adams, Verhave, & Newman, 1990). There have also been recent efforts to develop neural network models of stimulus equivalence (e.g., Donahoe & Palmer, 1994), which may ultimately prove relevant to behavioral neurodynamics.

Studies of choice. If reinforcement establishes value, one should be also able to allocate behavior according to value -- to choose among available reinforcement sources. Behavior-analytic researchers have been studying choice situations for over 30 years (for reviews, see Baum, 1974, 1979; de Villiers, 1977; McDowell, 1982; Williams, 1988). Research in this area is extensive, quantitatively rigorous, and seems to compel the attention of those interested in behavioral neurodynamics. Because even cursory coverage of this area would consume many pages, we shall merely introduce it.

Richard Herrnstein (1970) has offered formal mathematical statements of the relation between reinforcement and behavior. According to this matching law, behavior among concurrently available response alternatives is distributed in the same proportion as the reinforcements are distributed among the alternatives. Matching has been investigated most widely with concurrent schedules paradigms, that is, those in which two independent reinforcement schedules are available. The following equation describes results of quantitative analyses in two-response situations:

$$B_1 / (B_1 + B_2) = r_1 / (r_1 + r_2) \tag{1}$$

where B_1 and B_2 are response rates or durations for behavioral alternatives 1 and 2, and r_1 and r_2 are the rates of reinforcement obtained from alternatives 1 and 2. An alternate form of the equation describes response-reinforcer relations in a single-response situation, where the choice is between the measured response and any other behavior. In a later section, we will suggest that the matching analysis may be applicable to an even broader range of problems than it has been heretofore.

Behavioral momentum. Not only does reinforcement establish value and permit choice, but we are now learning that it establishes what Tony Nevin has termed behavioral momentum (reviewed in Nevin, 1992). The momentum analysis makes analogies between the relationships described in the physics of motion and the psychology of behavioral persistence. The momentum of a moving body is defined in classical mechanics as the product of mass and velocity. The degree to which an outside force can perturb the motion depends upon momentum; increasing mass while holding velocity constant increases the resistance to change.

As a starting point for Nevin's analogy, Newton's Second Law (Equation 2) states that a change in the velocity of a body (ΔV) will be the product of the force applied (f) and the reciprocal of the body's mass (m).

$$\Delta V = f / m \tag{2}$$

Thus, given equal force applied to two bodies, the ratio of changes in their velocities will be inversely proportional to the ratio of their masses (Equation 3).

$$\Delta V_1 / \Delta V_2 = m_2 / m_1 \tag{3}$$

Nevin suggests a direct parallel in the domain of behavior. He argues that rate of responding is analogous to velocity, and the resistance of that rate to change by a perturbing operation (prefeeding, alternative reinforcement, punishment, etc.) can be used to index the analogue of mass. As expressed in Equation 4, if the same perturbing operation is applied to two responses, the ratio of changes in response rates (ΔR) will be inversely proportional to the ratio of the behavioral masses (b).

$$\Delta R_1 / \Delta R_2 = b_2 / b_1 \qquad (4)$$

Nevin and others have shown that behavioral mass is determined by the rate of reinforcement (Nevin, Mandell, & Atak, 1983; Nevin, Tota, Torquato, & Shull, 1990). Higher rates of reinforcement produce greater behavioral mass, and this occurs independently of the response rates. As suggested in Figure 2 (*see fig.2*), if two responses are maintained at different reinforcement rates, then a perturbing force will have less effect on the response with the higher rate of reinforcement (filled points) than on the response with the lower rate of reinforcement (open points).

Figure 3 presents some findings from our laboratory that illustrate the momentum analysis. A young woman with mental retardation (chronological age 17 years, mental age 2.8 years[1]) was given two discrimination tasks that she could perform accurately. Sessions alternated between blocks of trials with each task, that is, she was not asked to choose between the two, but rather to work on each at different times. On one task, every correct response produced a reinforcer (a small bit of cookie), and on the other task, every fourth response on average did so. Figure 3 shows the rate of trial completion, which was under her control. Baseline response rates for the two tasks are shown in Condition A, and they were roughly comparable. According to the momentum analysis, the richer reinforcement schedule (filled points in Figure 3) would have established greater behavioral mass, and thus rendered the behavior less vulnerable to perturbation (*see fig. 3*).

Condition B in Figure 3 shows the effect of our perturbing force, a concurrently available distraction, videotapes shown on a television next to the discrimination apparatus. Note that rate of behavior fell proportionally more with the leaner schedule. After a return to baseline, in an effort to increase the perturbing force, we added prefeeding to the test sessions (two cookies just before the session), which led to the more pronounced momentum effect shown in Condition B'.

Although still in an early stage, the analysis of behavioral momentum addresses a previously under-appreciated dynamic characteristic of the reinforcement process. In particular, the finding that behavioral persistence is largely independent of response rate, but rather related to reinforcement rate, may have important implications for the design of behavioral training and testing procedures. For example, one somewhat counterintuitive implication of momentum theory is that under some conditions a <u>decrease</u> in reinforcement rate may facilitate the transfer of behavioral control from instructional supports to the stimuli that are the training goal.

Integrating Behavior Analysis and Brain Science: Potential Benefits
Benefits for Behavior Analysis

As noted earlier, behavior analysis has the challenge of developing a coherent, broadly applicable theory of reinforcement. One possible benefit of integrating behavior analysis and brain science has been articulated by respondents to Killeen's (1994) target article on mathematical modeling of reinforcement in *Behavioral and Brain Sciences*. C. M. Bradshaw, for example,

---

[1]Peabody Picture Vocabulary Test - Revised mental age equivalent score.

mentioned a problem that plagues mathematical modeling: Behavioral data can frequently be modeled by more than one, and sometimes many, equations. Suggesting a solution, he wrote, "Skinner's eschewal of neurobiological explanations of behavior ... was in part pragmatic; neuroscience was not ready ... to provide a satisfactory account of ... environment-behavior interactions. Such accounts are now conceivable, and it may behoove us to place increasing reliance on physiological plausibility as a criterion for selecting equations to model operant behavior." (p. 137).

Bradshaw's suggestion also appears pertinent to behavior analysts such as those in our research group at the Shriver Center, who are interested in studying antecedent stimulus control of behavior and the nature of environment-behavior relations selected by the reinforcement process. (For those more comfortable with the older language of experimental psychology, the problem of stimulus selection.) Skinner was not much interested in this very difficult problem, perhaps because he felt that descriptive and quantitative analysis of response-reinforcer relations was more tractable.

Our group's efforts, however, have been directed toward defining and perhaps ameliorating problems of individuals with intellectual disabilities like mental retardation, autism, and other problems of clinical significance. With individuals who have clinical problems, merely establishing reinforcer control is typically not challenging. What is challenging, however, is establishing and maintaining the kinds of environment-behavior relations that comprise an adaptive behavioral repertoire. In responding to this challenge, we have become less concerned with operant response topographies. Rather, we have been drawn to the analysis of variations in stimulus control, which we have termed "stimulus control topographies" (see Dube & McIlvane [in press] for a comprehensive presentation of this concept). For the purposes of this chapter, suffice it to say that stimulus control topography corresponds roughly to "representation" or "the stimulus as encoded."

Other behavior analysts have begun to confront the same problems. For example, Mark Rilling, who authored the chapter on "Stimulus Control" in the widely read *Handbook of Operant Behavior* (1977), recently wrote, "When the stimulus is a complex event (e.g., a color slide of a scene in the real world) ... the experimenter is forced to confront two problems: perception and representation ... A theory of discrimination learning that cannot specify precisely the nature of the event about which [the subject] has learned will be unable to predict the events that control behavior in the future" (Rilling, 1992). To redress this potential problem, Rilling was attracted to J. J. Gibson's (1979) perspective, which seems to offer much to behavior analysis (cf. Green, Mackay, McIlvane, Saunders, & Soraci, 1990).

In addition to ecological perspectives, behavior analysis will likely benefit from the brain sciences in the analysis of stimulus variables. For example, consider the extensive analysis of visual perception that Pribram (1991) has accomplished in *Brain and Perception* and the preceding three volumes of this series. This analysis addresses the gap between the stimulating action of the environment and those aspects of it that would be available to guide behavior. By understanding how the brain decomposes complex sensory information, we may advance a step toward solving what Skinner called "the problem of the first instance" (Skinner, 1957). He referred to the logical problem of identifying the variables responsible for the first instance of new behavior, before that behavior is reinforced.[2] A related question asks, what are the variables responsible for noticing, for example, a new feature of a complex stimulus for the first time? We believe that the answer must be some interaction between our sensory abilities and capacities (products of phylogenic [survival] contingencies), and the cumulative experience that builds and tunes them (i.e., ontogenic [reinforcement] contingencies) (cf. Skinner, 1969).

There are two other areas in which behavior analysis might benefit from closing the gap between itself and the brain sciences. Above, we discussed several ways in which reinforcement processes have been defined in behavior analysis. Harkening back to Bradshaw's point, connecting with the neurology and neurobiology of the brain may give us clues that will help to further refine our understanding of reinforcement from the psychological perspective. Such work may also help to illuminate the nature of behavioral selection by consequences. Perhaps one reason that there remains some issue about what reinforcement selects -- responses or environment-behavior relations -- is that the immediate antecedents to some behavior are within the skin and thus difficult to study directly. Such antecedent events may be studied only via methods that render them detectable (e.g., appropriately targeted EEG recordings).

In addition, we think it unlikely that behavior analysis can, by itself, do an adequate job of coming to terms with how subject variables influence environment-behavior interactions. While psychological and ethological analyses have been reasonably successful in revealing species-specific behaviors selected by phylogenic contingencies, there are limits to what unguided studies can accomplish in this area. For example, our own studies of stimulus control and reinforcement processes in people with intellectual and other developmental disabilities require us to make daily contact with individuals who have an appalling variety of neurological problems, many of which are at best poorly understood. In our view, such problems cannot be fully understood without looking at the brain operations, defects, or dysfunctions that subsume them.

Benefits for Brain Science

While behavior analysis needs the brain sciences to complete the account, it offers a number of worthwhile contributions to brain science, two of which merit special attention here. The first is in developing and/or applying methodologies that will allow the brain scientist to study research subjects who do not have well-developed language skills and do not share a common frame of reference with the experimenter. To make this point, we will relay something of the history of our research group. The Shriver Center Behavioral Sciences Division exists today because Raymond D. Adams, Chief of Neurology at the Massachusetts General Hospital and the Harvard Medical School, realized in the 1950s the many limits of behavioral testing to evaluate neurological dysfunction and disease. He looked to the behavioral sciences for help, and this led him to recruit several behavior analysts, most notably Murray Sidman, to the Massachusetts General Hospital. In collaboration with several others, Sidman initiated a program that sought to apply the methods of behavior analysis to the problem of conducting neurological evaluations of individuals who had limited language abilities -- particularly those who had suffered strokes or were severely mentally retarded.

---

[2] With respect to new responses, behavior analysis has made substantial progress. For example, Galbicka's (1994) percentile schedules (see Killeen's chapter, this volume) may help us learn of the processes involved in shaping new responses.

Stimulus control shaping. Sidman and his colleagues chose to begin with visual perception, a sensible choice given that so much of the brain's activities are devoted to acquiring and integrating visual information. They needed a procedure that could evaluate visual perception in people with very limited language. The product of their effort is shown in Figure 4 (*see fig. 4*). To communicate without using words, Sidman and Stoddard adapted a methodology called "fading" which had been suggested by Skinner and had recently been demonstrated by Terrace (1963) at Columbia in an animal behavior model. They sought to use fading to establish a discrimination of a circle (S+) from ellipses of various sizes (S-) (Sidman & Stoddard, 1966).

Their fading program proceeded as follows: Initially, the circle was displayed on one of eight response keys. The key was lit, and the remaining keys were blank and dark (Figure 4, a). After typically rapid acquisition of this very easy circle vs. dark key discrimination, the blank keys were illuminated gradually (b-c). When the S+ and S- keys were illuminated equally, the subject was required to make a form vs. no form discrimination (d). When this discrimination was mastered, flat ellipses were faded onto the S- keys (e-g) until the subject was ultimately required to make a circle-ellipse discrimination (h). Finally, the ellipses were made gradually rounder (i-l), thus requiring a progressively finer circle-ellipse discrimination and providing a means for determining a circle-ellipse discrimination threshold.

The circle-ellipse program was a truly remarkable achievement. With it, they were able to evaluate visual perception in a variety of individuals with profound neurological problems. The test was successful, for example, with Cosmo, a microcephalic man with profound retardation, who became quite famous as a result of his participation in these studies. A similar program was used with another famous neurological patient, H.M., and the methods helped clarify the nature of his behavioral deficits (Sidman, Stoddard, & Mohr, 1968). These successes exemplify many more that have been achieved when behavior analysts and brain scientists collaborate.

Research since Sidman and Stoddard's seminal work has greatly expanded the scope and complexity of performances that can be taught by stimulus control shaping methods. The work has advanced to the point where one can reasonably entertain the possibility of rendering shaping less of an art and more of a science, hopefully a quantitative science. One use of stimulus control shaping, for example, is to establish matching performances based on arbitrary stimulus-stimulus relations, where the stimuli that are matched are not physically identical (e.g., matching "2" and "II"). Subjects begin by selecting a comparison stimulus that is identical to a sample stimulus. Then, the features of the comparison stimulus gradually change, over a series of graded steps, and ultimately become those of the new stimulus that is the training goal. We envision methodology whereby a computer program might be able to determine in advance what teaching steps are needed, by using the learner's judgments of stimulus similarity in the intended teaching sequence (see below). More generally, we believe that many barriers to effective stimulus control shaping could be removed by understanding better the determinants of stimulus control transfer. For example, what accounts for the smooth transition between steps in a successful stimulus control transfer program?

Recently, we have been pursuing a stimulus-class analysis of stimulus control transfer. Our analysis predicts the following: The optimal series of stimuli for a successful stimulus shaping program is one in which the learner regards the stimuli from adjacent steps as members of the same feature class. A feature class is one in which stimuli are grouped on the basis of similar, though not necessarily identical, features. If a pretest revealed, for example, that an experimental subject regarded the stimuli in the first three steps of a shaping program series as similar, we would predict that the second step would be superfluous. Conversely, if the subject did not regard the stimuli from fourth and fifth steps as similar, we would predict a breakdown in performance at that point in a shaping program. Most importantly, we would expect that specific feature-class judgments would vary from individual to individual.

Thus far, our studies have focused on methods for assessing the feature-class boundaries within graded series of stimuli like those used in stimulus-control transfer programs (examples of such series are shown in the rows in Figure 5) (*see fig. 5*). In one study, three individuals with mental retardation were trained to make "yes/no" similarity judgments (Serna & Wilkinson, 1995; cf. McIlvane, Kledaras, Lowry, & Stoddard, 1992). After extensive training, the following baseline performance was established with nonrepresentational form stimuli: Delayed matching-to-sample trials presented a sample stimulus, one comparison stimulus, and a "blank" comparison stimulus (a black mask large enough to cover any stimulus). If the sample and displayed comparison were very similar, but not identical, the subject made a "yes" response by selecting the displayed stimulus; if not, he/she made a "no" response by selecting the blank comparison. When this yes/no performance baseline was firmly established, stimuli from two new series of graded forms were presented on unreinforced test trials that were interspersed among the baseline trials.

Results of the test trials can be illustrated with individual "feature-class profiles" for each subject, shown in Figure 5. Feature classes within the two tested stimulus series are indicated by enclosing boundaries. A noteworthy characteristic of the three profiles is that no two are identical; judgments about stimulus similarity varied across individual subjects. It is also noteworthy that the graded steps within the series, constructed according to the experimenter's best guess as to equal-sized steps, did not yield tidy judgments by the subjects. Instead, each profile shows classes of varying sizes, as well as adjacent steps in which the participants did not regard the stimuli as similar.

If these series were subsequently used in programs designed to transfer control from the first stimulus in the series to the last, these profiles would lead to clear predictions about performance. For example, for Subject TFR (upper profiles in Figure 5), we would expect a breakdown in performance between E-4 and E-5 because the feature classes do not overlap at that point. No such breakdowns would be predicted for the G-H series. Further, we would predict that stimuli E-2 and E-3 are superfluous because E-1, E-2, E-3, and E-4 were all classed together. Similar, though idiosyncratic, predictions could be made for the other two participants.

The goal of making shaping less of an art and more of a science seems within our reach. Many of the building blocks for a quantitative stimulus control transfer technology already exist; for example, potential stimulus series of various gradations could be generated by applying algorithms like those that produce visual "morphing" effects in computer graphics. Most importantly, the feature-class analysis described above can provide a theoretical framework to help organize technology so that it will make effective contact with the behavior of the individual subject.

High-probability environment-behavior relations. A second, related contribution is the production of behavioral baselines of unusual purity, such that brain-behavior interactions might be more clearly revealed in individuals with neurological disease. For example, assume that one has the goal of using our ever increasing ability to image brain activity to trace the way in which the brain performs some simple discrimination task -- like detecting a single salient feature that distinguished two otherwise identical visual stimuli. Ideally, one would prefer to see crisp, high-probability stimulus control baselines, so that the relevant stimuli and corresponding brain events can be related. Such baselines are assumed when one tests adults with normal capabilities, but the same assumption does not hold when testing a young child or someone with neurological disease. Their baselines are often very messy, and high-probability stimulus control is a rarity. These and related observations, in fact, have led to gating theories of neurological deficit and dysfunction -- that the behavioral deficits result because the individual is unable to screen out irrelevant information (Hasher & Zacks, 1988).

Behavior analytic methods allow one to show, for example, that such gating or screening problems are due at least in part to contingencies of reinforcement that encourage too wide a gate. As such, contingencies can also narrow it. For example, Figure 6 illustrates a task that we have used for several years to study variables responsible for intermediate accuracy (*see fig. 6*). The left half illustrates our method for analyzing imperfect accuracy scores. In Condition A, every trial-initiation response is followed by a discrimination between visual stimuli that are identical except for one feature. The difference is obvious to individuals without developmental limitations; the positive form flashes, alternates with a black field, or appears on a colored background. Selecting the positive stimulus is followed by a reinforcer and selecting negative stimuli are not. In individuals with developmental limitations accuracy scores may stabilize at levels above chance but short of perfection, as shown for Condition A in the right portion of Figure 6.

In Condition B, the contingencies are altered such that each trial proceeds through two "trial states." Every trial begins with trial state 1 (TS1) by presenting negative stimuli on all keys. The appropriate response is to wait a few seconds until TS2, where one stimulus begins to flash, for example, and thus becomes S+. Any failures to wait merely extend TS1. Note that the stimulus display in TS2 is identical to that in Condition A. Characteristic data for Condition B are shown in the rightmost portion of Figure 6. When the subject waits, selections in TS2 may be always correct. Over sessions, competing control -- reflected in the TS1 responses -- gradually declines as it is extinguished.

The upper portion of Figure 7 shows data from four individuals with severe mental retardation (*see fig. 7*). An essentially flat, near "chance-level" function in Condition A is followed immediately or nearly so by close to 100% TS2 accuracy scores and declining TS1 responding in Condition B. Sometimes, however, we obtain stable, intermediate accuracy scores in Condition B like those shown for three other subjects in the lower portion of Figure 7. These scores are far enough above chance levels to indicate that the relevant stimulus control occurs with some frequency, but they do not improve with continued training.

Herrnstein's equation (Equation 1, above) and quantitative analyses of behavioral choice may offer one way to understand intermediate accuracy scores like these. When one views discrimination procedures from a perspective that permits multiple stimulus control topographies, they can be seen to present the subject with choices between the discriminative stimuli for concurrently available response options. One option is the relevant form of stimulus control that the experimenter is trying to establish -- in our case, select the flashing stimulus. If the subject's performance reflected only this type of stimulus control, the resulting accuracy score would be 100%, as illustrated by the shaded portion of the circle in the top left of Figure 8 (*see fig. 8*). The probability of reinforcement for behavior under the relevant stimulus control, B_1, is 1.0. The other options are irrelevant forms of stimulus control, and a performance consisting only of them would produce 50% accuracy scores in a two-choice task, as shown in the top right of Figure 8. The probability of reinforcement for behavior under irrelevant forms of stimulus control, B_2, is 0.5.

According to Herrnstein's equation, and as illustrated in Figure 8, if the proportions of B_1 responding matched the proportion of reinforcers obtainable by the R_1 reinforcement contingency, then B_1 would occur on two-thirds of the trials. If so, then the resulting accuracy score is shown at the bottom of Figure 8, 67% for perfect accuracy on two-thirds of the trials, plus 17% for chance accuracy on one-third of the trials, resulting in an overall accuracy score of 84%. If our analysis is correct, the problem of stable, intermediate accuracy may not be the result of faulty perception, poor gating of attention, or other such problems. Rather, such scores may be the ordinary outcome of known reinforcement processes and may be correctable, for example, by teaching subjects to maximize reinforcers rather than to match reinforcement probabilities in choice situations.

Summary of contributions of behavior analysis to brain science. To summarize our basic points, behavior analysis can contribute to brain science by helping to solve problems in communicating with and testing individuals with neurological disease. The methodology may also be helpful with nonhuman populations that participate in more invasive aspects of brain research. This methodological contribution, we believe, is one that behavior analysis can uniquely make. Yet another contribution is that behavior analysts are studying important subject matter that is not currently under study by other branches of the behavioral sciences. The focus on reinforcement processes such as equivalence, choice, and momentum, has produced impressive, quantitative analyses of a critical dimension of behavior. Data reviewed in the preceding section, for example, suggest that what have traditionally been interpreted as problems of attention may be due in part to failure to appreciate important aspects of reinforcement processes. If so, much unproductive and/or misdirected research might be avoidable by increased understanding of such processes.

Perspectives on Discipline Integration

From certain perspectives, the arguments that we have made thus far may seem curious. Some will ask whether behavior analysis-brain science disciplinary integration is not already a *fait accompli*. Those interested in the neurobiology and neuropsychology of learning and memory, for example, make routine use of behavioral testing methods in their work and have done so for years. Consider neurobiological studies of learning processes, which often use very simple organisms and respondent and/or operant methods. Consider also the long tradition in neuropsychology of studying the impact of naturally occurring or artificially created lesions on simple performances such as delayed matching (or nonmatching) to sample, which are extensively studied in behavior analysis. This tradition notwithstanding, we believe it fair to say that work thus far has only begun to scratch the surface of much deeper possibilities (see Donohoe, et al. [1993] and Schull [1995] for additional relevant commentary and illustrative analyses).

Discipline isolation seems particularly persistent, for example, in the experimental analysis of human behavior, and doubly so in the growing field of neural imaging. Imaging work is dominated by cognitive neuroscientists, who are interested in such topics as tracking the information flow that occurs in complex behavioral repertoires like reading (Posner & Raichle, 1994). Work in this area has not been much concerned with looking at reinforcement processes, that is, issues of stimulus equivalence, choice, and behavioral persistence, as exemplified by the momentum analysis.

In our view, there are at least two reasons for behavior analysts and brain scientists to join together to take on such problems. First, the problems are scientifically interesting, and their respective sciences have advanced to the point that such studies now appear to be feasible. Second, practical and humanitarian concerns dictate this path. It is being increasingly recognized that disabling conditions like autism and attention deficit hyperactivity disorder are in part related to reinforcement processes (Barkley, 1996). In sum, it seems that behavior analysis and brain science have developed common cause, and this in the final analysis is what sustains effective interdisciplinary efforts.

Learning as Self-Organization

To conclude, we shall speak to the overarching theme of this conference: Learning as self organization rather than self gratification. As an example supporting this interpretation, Pribram often relates his observation of a monkey who was performing a learning task apparently for peanut reinforcers. As the task progressed, the animal tended to store earned peanuts in his pouch, rather than eating them immediately. In addition, when erroneous responses occurred, the animal often ate stored peanuts shortly afterwards, apparently "self-reinforcing" the errors. Finally, Pribram observed that the animal sometimes responded incorrectly for long series of trials, apparently indifferent to the absence of peanut deliveries. Each of these phenomena seem inconsistent on their face with the suggestion that the animal was performing the task merely for the sake of gratifying its desires for peanuts. Indeed, observations such as these are not unusual, particularly with humans, and made remarkable only by taking an impoverished view of reinforcement and its role in supporting and organizing behavioral activity.

To develop our argument, we shall relate another observation from our own research program. Some time ago, we studied a severely autistic, mentally retarded boy on the circle-ellipse threshold task (Figure 4, h-l). As the boy proceeded through the task, he initially responded slowly, carefully scanning the display matrix to find the single circle among the many similar ellipses. Each success, like all others before it, resulted in an M&M, which the boy ate avidly. The M&Ms and auditory stimuli accompanying their delivery served as the only instructions. No one else was physically present during the testing, which was accomplished in an automated teaching environment (see McIlvane, Kledaras, Dube, & Stoddard [1989] and Stoddard [1982] for descriptions).

As the discrimination became more difficult, the boy encountered circle-ellipse differences that were beyond his ability to discriminate. The result was a characteristic shift in stimulus control; if his first response was not correct, he selected adjacent keys in rapid succession, "circling" the display matrix in search of the correct key. Because of a correction procedure, incorrect responses had no effect and every trial ended with a correct response. One consequence was that the boy came to earn chocolates even faster than before, because circling can be accomplished rapidly, while careful scanning typically requires more time. Another feature of the test procedure -- a backup following errors -- ultimately returned the boy to circle-ellipse differences that he could discriminate. Under such circumstances, the careful scanning and slower responding returned, despite the fact that the slower rate led to fewer chocolates.

On their face, such observations are not easily reconciled with a view of reinforcement as self-gratification. That is, one must explain why the boy abandoned the more lucrative circling in favor of more careful responding. Clearly, the chocolates *per se* were not the only factors influencing performance. This seems doubly interesting, given that this was a boy with autism working alone in an automated test environment. What other variables might account for his behavior? Perhaps one might argue that the prior, more extensive programmed training had established greater momentum in the more careful approach. Also, the careful approach led to a reduction in unreinforced responding (i.e., errors) which past research has shown may be aversive (Stoddard & Sidman, 1967; Terrace, 1971). Going beyond known effects of reinforcement procedures, we speculate that well-organized behavioral sequences may be inherently reinforcing, perhaps through the operation of phylogenic rather ontogenic contingencies (Skinner, 1969). One characteristic, in fact, of behaviors that are reinforcing is that they are inherently well-organized (e.g., eating, drinking, sex, etc.; cf. Pribram, 1995). Taking this line, one may think of learning as extending and elaborating biologically determined patterns of organized activity. Consistent with our overall argument, it seems likely that determining whether our speculations have merit will require not only comprehensive understanding of behavior but also of the brain processes that underlie it.

Acknowledgments

Our program has been supported mainly by the National Institutes of Child Health and Human Development. Work discussed in this chapter has been supported by NICHD grants HD 25995, HD 25488, HD 28141, and HD 32049. We also acknowledge support from the Department of Mental Retardation of the Commonwealth of Massachusetts. This chapter is dedicated to Dr. Raymond D. Adams, who understood earlier than most the necessity for interdisciplinary efforts to fully understand the relationship between brain and behavior. Address correspondence to W. J. McIlvane, Shriver Mental Retardation/Developmental Disabilities Center, 200 Trapelo Road, Waltham, MA 02254.

Literature Cited

Allison, J., Miller, M, & Wozny, M. (1979). Conservation in behavior. Journal of Experimental Psychology: General, 108, 4-34.

Barkley, R. A. (1996). Critical issues in research on attention. In G. R. Lyon & N. A. Krasnegor (Eds.), Attention, memory, & executive function (pp. 45-56). Baltimore: Brookes.

Baum, W. M. (1974). On two types of deviation from the matching law: Bias and undermatching. Journal of the Experimental Analysis of Behavior, 22, 231-242.

Baum, W. M. (1979). Matching, undermatching, and overmatching in studies of choice. Journal of the Experimental Analysis of Behavior, 32, 231-242.

Bradshaw, C. M. (1994). Variations in behavioral equations: Can neurobiology help? Behavioral and Brain Sciences, 17, 136-137.

de Villiers, P. A. (1977). Choice in concurrent schedules and a quantitative formulation of the law of effect. In W. K. Honig & J. E. R. Staddon (Eds.), Handbook of operant behavior (pp. 233-287). Englewood Cliffs, NJ: Prentice-Hall.

Donahoe, J. W. & Palmer, D. C. (1994). Learning and complex behavior. Boston, MA. Allyn & Bacon.

Donahoe, J. W. Burgos, J. E., & Palmer, D. C. (1993). A selectionist approach to reinforcement. Journal of the Experimental Analysis of Behavior, 60, 17-40.

Dube, W. V. & McIlvane, W. J. (1995). Stimulus-reinforcer relations and emergent matching to sample. The Psychological Record, 45, 591-612.

Dube, W. V. & McIlvane, W. J. (in press). Some implications of a stimulus control topography analysis for emergent stimulus classes. In T. R. Zentall & P. M. Smeets (Eds.), Stimulus class formation in humans and animals. North Holland.

Edwards, C. A., Jagielo, J. A., Zentall, T. R., & Hogan, D. E. (1982). Acquired equivalence and distinctiveness in matching to sample by pigeons: Mediation by reinforcer-specific expectancies. Journal of Experimental Psychology: Animal Behavior Processes, 8, 244-259.

Fields, L., Adams, B. J., Verhave, T., & Newman, S. (1990). The effects of nodality on the formation of equivalence classes. Journal of the Experimental Analysis of Behavior, 53, 345-358.

Galbicka, G. (1994). Shaping in the 21st century: Moving percentile schedules into applied settings. Journal of Applied Behavior Analysis, 27, 739-760.

Gibson, J. J. (1979). The ecological approach to visual perception. Boston: Houghton-Mifflin.

Green, G., Mackay, H. A., McIlvane, W. J., Saunders, R. R., & Soraci, S. A. (1990). Perspectives on relational learning in mental retardation. American Journal on Mental Retardation, 95, 249-259.

Hasher L. & Zacks, R. T. (1988). Working memory, comprehension, and aging: A review and a new view. In G. H. Bower (Ed.), The psychology of learning and motivation (Vol. 22, pp. 193-225). San Diego, CA: Academic Press.

Hefferline, R. F., Keenan, B., & Harford, R. A. (1959). Escape and avoidance conditioning in human subjects without their observation of the response. Science, 130, 1338-1339.

Herrnstein, R. J. (1970). On the law of effect. Journal of the Experimental Analysis of Behavior, 13, 243-266.

Horne, F. & Lowe, P. (in press). On the origins of naming and other symbolic behavior. Journal of the Experimental Analysis of Behavior.

Killeen, P. R. (1994). Mathematical principles of reinforcement. Behavioral and Brain Sciences, 17, 105-135.

McDowell, J. J. (1982). The importance of Herrnstein's mathematical statement of the Law of Effect for behavior therapy. American Psychologist, 37, 771-779.

McIlvane, W. J., & Dube, W. V. (1992). Stimulus control shaping and stimulus control topographies. The Behavior Analyst, 15, 89-94.

McIlvane, W. J., Dube, W. V., & Callahan, T. D. (1996). Attention: A behavior analytic perspective. In G. R. Lyon & N. A. Krasnegor (Eds.), Attention, memory, & executive function (pp. 97-117). Baltimore: Brookes.

McIlvane, W. J., Kledaras, J. B., Dube, W. V., & Stoddard, L. T. (1989). Automated instruction of severely and profoundly retarded individuals. In J. Mulick and R. Antonak (Eds.), Transitions in Mental Retardation (Vol. 4, pp. 15-76). Norwood, NJ: Ablex.

McIlvane, W. J., Kledaras, J. B., Lowry, M. J. & Stoddard, L. T. (1992). Studies of exclusion in individuals with severe mental retardation. Research in Developmental Disabilities, 13, 509-532.

Meehl, P. E. (1950). On the circularity of the law of effect. Psychological Bulletin, 47, 52-75.

Michael, J. (1982). Distinguishing between discriminative and motivational functions of stimuli. Journal of the Experimental Analysis of Behavior, 37, 149-155.

Mogenson, G. & Cioe´, J. (1977). Central reinforcement. In W. K. Honig & J. E. R. Staddon (Eds.), Handbook of operant behavior (pp. 570-595). Englewood Cliffs, NJ: Prentice-Hall.

Neuringer, A., & Voss, C. Approximating chaotic behavior. Psychological Science, 4, 113-119.

Nevin, J. A. (1992). An integrative model for the study of behavioral momentum. Journal of the Experimental Analysis of Behavior, 57, 301-316.

Nevin, J. A., Mandell, C., & Atak, J. R. (1983). The analysis of behavioral momentum. Journal of the Experimental Analysis of Behavior, 39, 49-59.

Nevin, J. A., Tota, M. E., Torquato, R. D., & Shull, R. L. (1990). Alternative reinforcement increases resistance to change: Pavlovian or operant contingencies? Journal of the Experimental Analysis of Behavior, 53, 359-379.

Posner, M. I. (1980). Orienting of attention. The VIIth Sir Frederic Bartlett Lecture. The Quarterly Journal of Experimental Psychology, 32, 3-25.

Posner, M. I., & Raichle, M. E. (1994). Images of mind. New York: Scientific American Library.

Premack, D. (1965). Reinforcement theory. In D. Levine (ed.), Nebraska Symposium on Motivation. Lincoln: University of Nebraska Press.

Pribram, K. H. (1991). Brain and perception: Holonomy and structure in figural processing. Hillsdale, NJ: Lawrence Erlbaum Associates.

Pribram, K.H. (1995) The Enigma of Reinforcement. In Neurobehavioral Plasticity: Learning, Development and Response to Brain Insults (pp. 381-403). Proceedings of the Bob Isaacson Symposium in Clearwater, FL. NJ: Lawrence Erlbaum Associates.

Ray, B. A., & Sidman, M. (1970). Reinforcement schedules and stimulus control. In W. N. Schoenfeld (Ed.), The theory of reinforcement schedules (pp. 187-214). New York: Appleton-Century-Crofts.

Rilling, M. (1977). Stimulus control and inhibitory processes. In W. K. Honig & J. E. R. Staddon (Eds.), Handbook of operant behavior (pp. 432-480). Englewood Cliffs, NJ: Prentice-Hall.

Rilling, M. (1992). An ecological approach to stimulus control and tracking. In W.K. Honig & J.G. Fetterman (Eds.), Cognitive aspects of stimulus control (pp. 347-366). New York: Erlbaum.

Schenk, J. J. (1995). Emergent relations of equivalence by outcome-specific consequences in conditional discrimination. The Psychological Record, 44, 537-558.

Serna, R. W., & Wilkinson, K. M. (1995). Methods for assessing feature-class membership in series of graded stimuli. Proceedings of the 28th Annual Gatlinburg Conference on Research and Theory in Mental Retardation and Developmental Disabilities, 78.

Shull, R. L. (1995). Interpreting cognitive phenomena: Review of Donahoe and Palmer's Learning and Complex Behavior. Journal of the Experimental Analysis of Behavior, 63, 347-358.

Sidman, M. (1986). Functional analysis of emergent verbal classes. In T. Thompson & M. D. Zeiler (Eds.), Analysis and integration of behavioral units (pp. 213-245). Hillsdale, NJ: Erlbaum.

Sidman (1994). Equivalence relations and behavior: A research story. Boston: Authors Cooperative.

Sidman, M., & Stoddard, L. T. (1966). Programming perception and learning for retarded children. In N. R. Ellis (Ed.), International Review of Research in Mental Retardation, Vol. 2 (pp. 151-208). New York: Academic Press.

Sidman, M., Stoddard, L. T., & Mohr, J. P. (1968). Some additional quantitative observations of immediate memory in a patient with bilateral hippocampal lesions. Neuropsychologia, 6, 245-254.

Skinner, B. F. (1957). Verbal behavior. New York: Appleton-Century-Crofts.

Skinner, B. F. (1969). Contingencies of reinforcement. New York: Appleton-Century-Crofts.

Skinner, B. F. (1974). About behaviorism. New York: Knopf.

Skinner, B. F. (1989). The origins of cognitive thought. American Psychologist, 44, 13-18.

Stoddard, L. T. (1982). An investigation of automated methods for teaching severely retarded individuals. In N. R. Ellis (Ed.), International review of research in mental retardation (pp. 163-207). New York: Academic Press.

Stoddard, L. T., & Sidman, M. (1967). The effects of errors on children's performance on a circle-ellipse discrimination. Journal of the Experimental Analysis of Behavior, 10, 261-270.

Terrace, H. S. (1963). Discrimination learning with and without "errors". Journal of the Experimental Analysis of Behavior, 6, 1-27.

Terrace, H. S. (1971). Escape from S-. Learning and Motivation, 2, 148-163.

Timberlake, W. & Allison, J. (1974). Response deprivation: An empirical approach to instrumental performance. Psychological Review, 81, 146-164.

Tucker, D. (this volume). **<<Need reference when available>>**

Williams, B. A. (1988). Reinforcement, choice, and response strength. In R. C. Atkinson, R. J. Herrnstein, G. Lindzey, & R. D. Luce (Eds.), Stevens' handbook of experimental psychology, Volume 2: Perception and motivation (pp. 167-244). New York: J. Wiley & Sons.

IDENTITY-MATCHING BASELINE

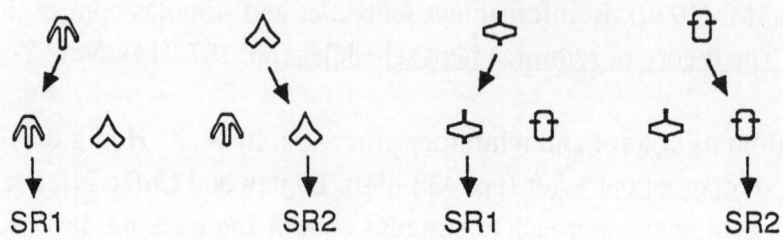

TEST TRIALS

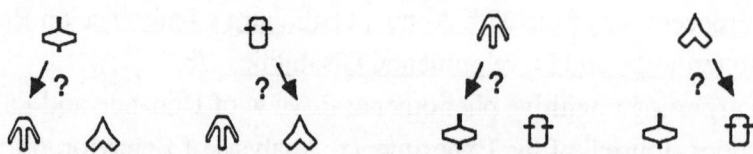

Figure 1. Identity-matching baseline and arbitrary-matching test trials for an investigation of the minimal experimental history necessary for emergent matching based on stimulus-reinforcer relations.

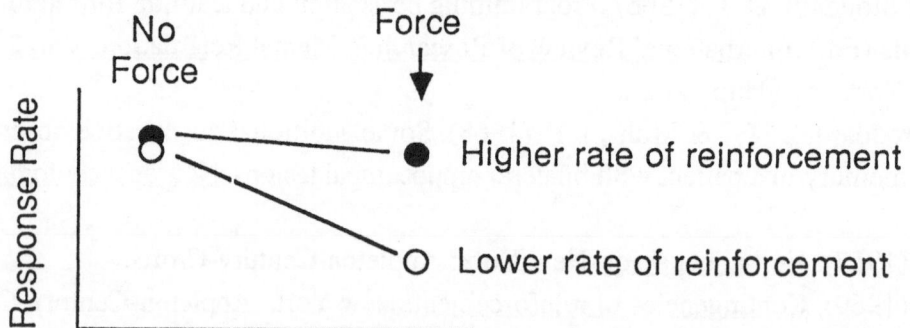

Figure 2. Hypothetical data illustrating effects of different reinforcement schedules on behavioral resistance to change (momentum).

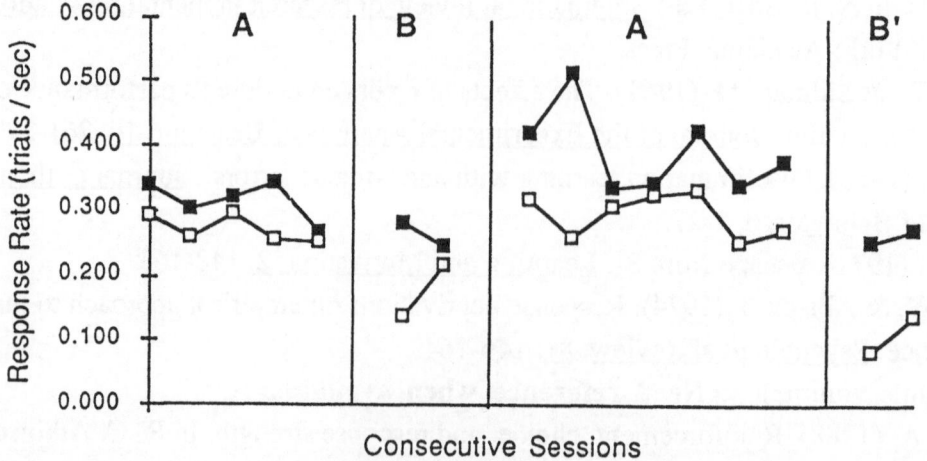

Figure 3. Actual data illustrating effects of different reinforcement schedules on behavioral momentum in an individual with severe mental retardation.

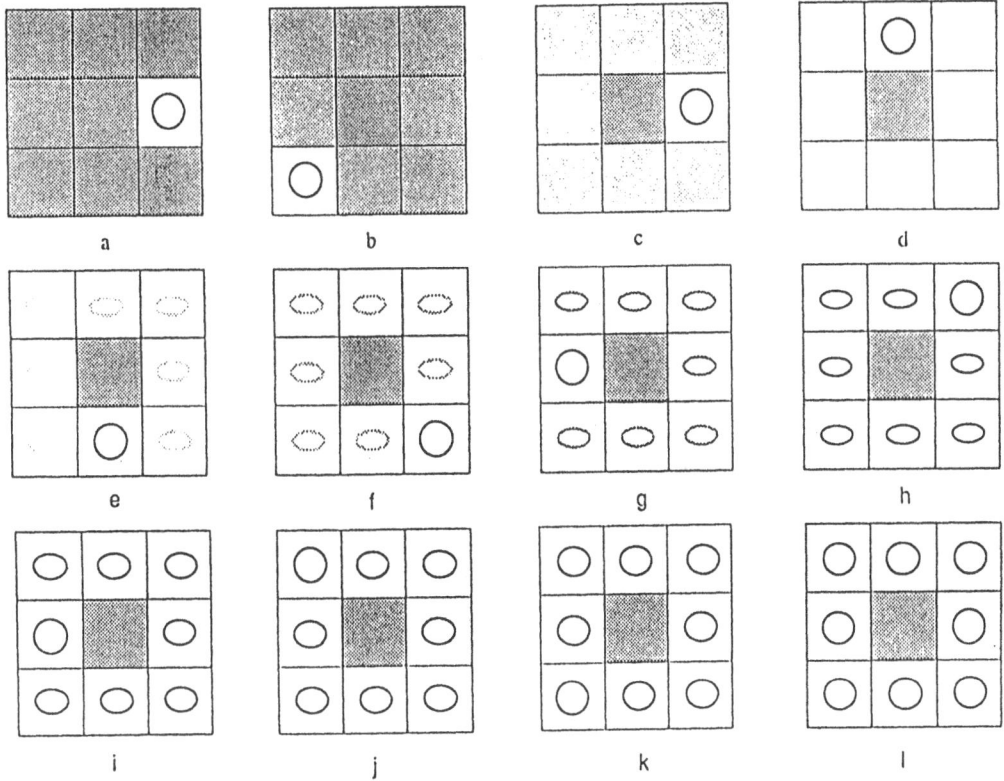

Figure 4. Sidman and Stoddard's (1966) circle-ellipse fading and threshold assessment program.

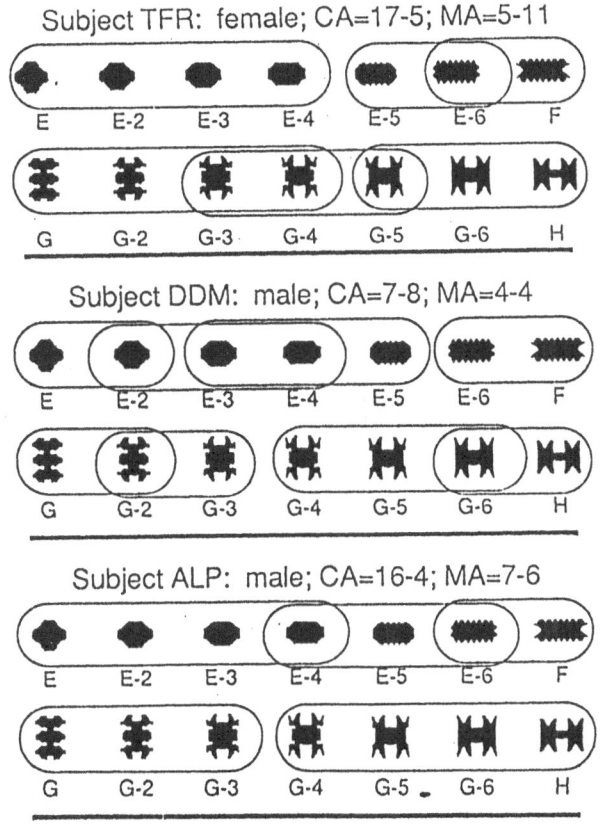

Figure 5. Feature-class profiles for three individuals with severe mental retardation.

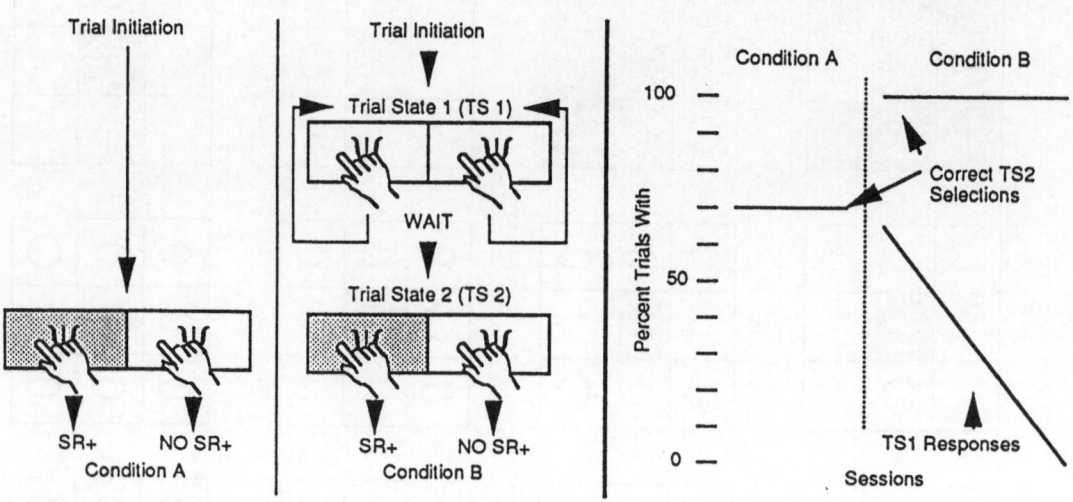

<u>Figure 6</u>. Schematic diagram and frequently obtained effects of imposing delayed S+ procedure on a baseline of low-probability discrimination performances.

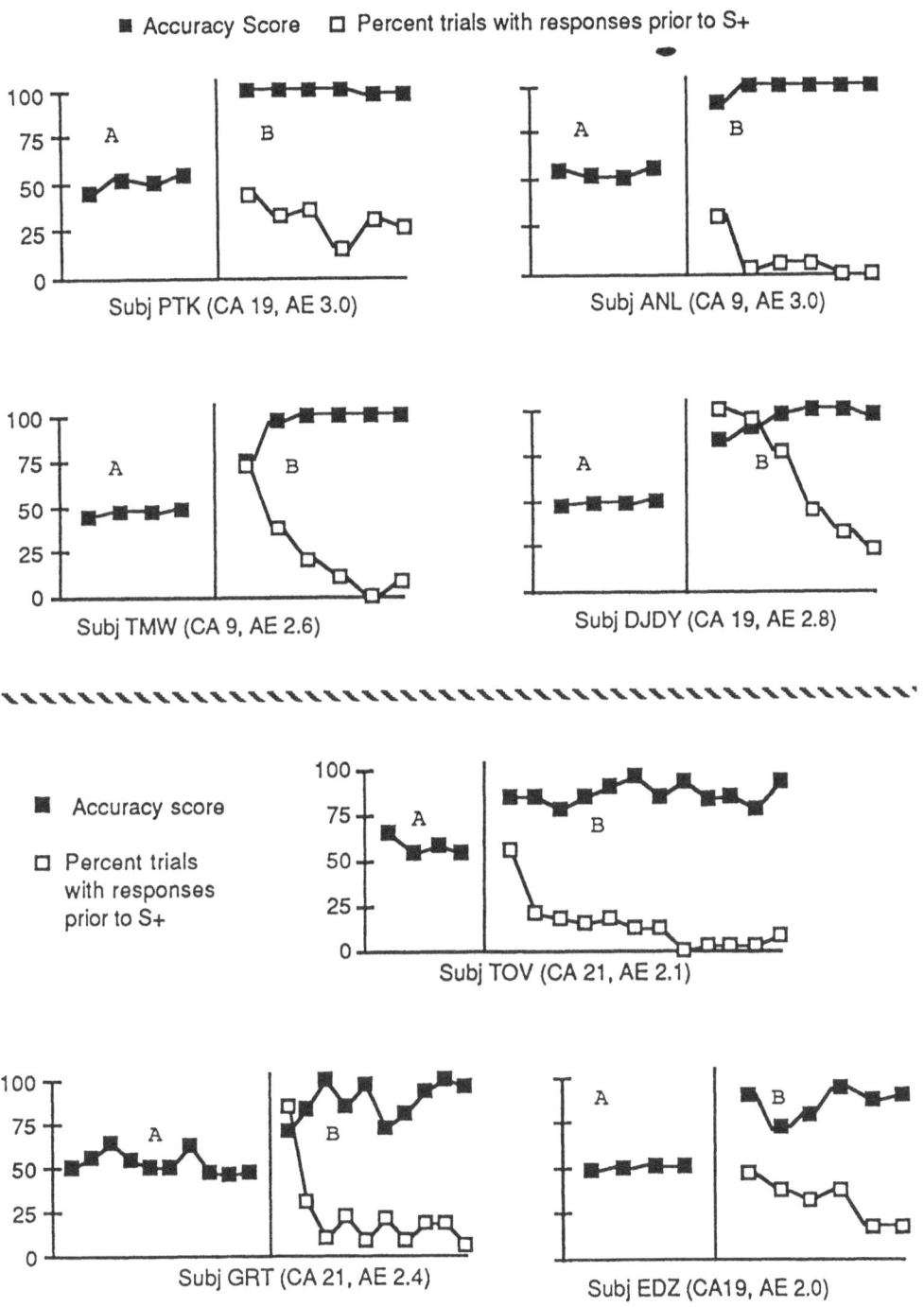

Figure 7. Delayed S+ data from seven individuals with severe mental retardation. Level of functioning is suggested by the Peabody Picture Vocabulary Test - Revised age-equivalent scores.

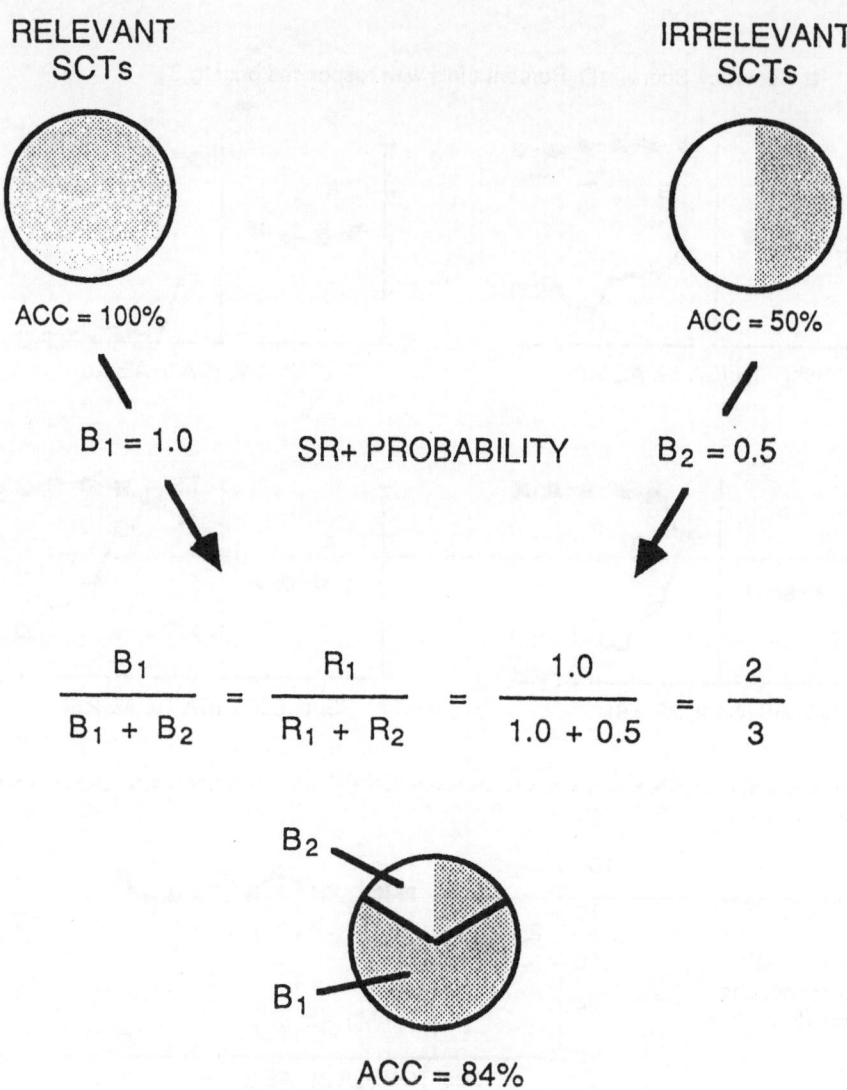

3.
Mechanics of the Animate

Peter R. Killeen

JOURNAL OF THE EXPERIMENTAL ANALYSIS OF BEHAVIOR 1992, 57, 429–463 NUMBER 3 (MAY)

MECHANICS OF THE ANIMATE

PETER R. KILLEEN

ARIZONA STATE UNIVERSITY

Behavior is treated as basic physics. Dimensions are identified and their transformations from physical specification to axes in behavioral space are suggested. Responses are treated as action patterns arrayed along a continuum of activation energy. Behavior is seen as movement along a trajectory through this behavior space. Incentives or reinforcers are attractors in behavior space, at the centers of basins of lowered potential. Trajectories impinging on such basins may be captured; repeated capture will warp the trajectory toward a geodesic, a process called conditioning. Conditioning is enhanced by contiguity, the proximity between the measured behavior and the incentive at the end of the trajectory, and by contingency, the depth of the trajectory below the average level of the potential energy landscape. Motivation is seen as the potential of an organism for motion under the forces impinging on it. Degree of motivation is characterized by the depth of the potential field, with low motivation corresponding to a flat field and a flat gradient of activation energy. Drives are the forces of incentives propagated through behavior space. Different laws for the attenuation of drive with behavioral distance are discussed, as is the dynamics of action. The basic postulate of behavior mechanics is incentive-tracking in behavior space, the energy for which is provided by decreases in potential. The relation of temporal gradients to response differentiation and temporal discrimination is analyzed. Various two-body problems are sketched to illustrate the application of these ideas to association, choice, scalar timing, self-control, and freedom.

Key words: dimensions, forces, drives, trajectories, conditioning, contingency, contiguity, association, choice, timing, self-control, system of behavior

---

Few have attempted to exhaust the power of a simple, physicalistic description of behavior; that is the goal of this paper. The treatment may seem abstruse in that it couches behavior in new and different terms. But the terms issue from a basic physical metaphor and are used in a simple way. As metaphors become more precise, they come to be called models. Few of the metaphors offered here are yet to that stage, but they may be brought to it by the efforts of our community. The benefit of ensconcing them in a system like the present one is the greater generality of application it will foster—the ability to utilize a model or approach developed for one dimension or force for other dimensions or forces.

Unlike physics, which started with compact rigid bodies subject to uniform forces, behavior analysis deals with soft bodies of articulated parts subject to forces that are seldom uniform. These difficulties are compounded by a historical emphasis on response rate as our fundamental datum; born of semiperiodic replications of movements, rate is more complicated

than uniform motion through space. Intermittent reinforcement schedules both modulate the strength and persistence of behavior and introduce new processes such as superstitious and adjunctive responding. Pursuit of these and other anomalies has often dictated our research programs, and, in the absence of a framework to guide inquiry, has dissipated our efforts. It is as though Galileo, in using inclined planes to study the behavior of falling bodies, found that at one inclination they would slide, at another roll, and at yet another bounce. In the face of such results it would be time for him to reconsider his procedures in light of his goals, not to shift his research interest to bouncing.

This article sketches the outlines of a mechanics of behavior, in the hope that it will encourage the reconsideration of our procedures from the vantage of a unified physicalistic perspective. It is only an outline; much is speculative, much will need to be added, much will need to be changed. But it is an approach that has the potential to unify various phenomena of behavior, reduce the profusion of data to common principles, and direct us to critical new problems whose solutions will clarify and stabilize the framework, until it has become a hospitable and ample abode for the theory that Skinner once envisioned.

This research was supported in part by NIMH Grant R01 MH43233 and in part by NSF Grant BNS-9021562. Address correspondence to Peter Killeen, Arizona State University, Tempe, Arizona 85287-1104; Bitnet ICPRK@ASUACAD.

DIMENSIONS

Space

Skinner spoke of behavior as "the movement of an organism or of its parts in a frame of reference provided by the organism itself or by various external objects or fields of force" (Skinner, 1938, p. 6). Modern technology makes it possible to picture this process (see Figure 1). There is much to be learned from such techniques. But Skinner also noted the difference between such "narration" and a scientific account; although the former may provide a near-exhaustive description, it does not become a scientific account until it specifies "the variables of which behavior is a function" (p. 8)—the forces and their influence on behavior.

The frame of reference for Figure 1 is provided by the experimental chamber and has an origin at the response key. The otherwise bland environment encourages behavior that is not oriented towards the key or hopper to be widely disbursed. For the study of key pecking, this may be useful. For other activities, such as preening or interaction with conspecifics or prey, other frames may be better. Just as mechanics may be simplified by considering gravitational forces as issuing from a point at the center of the object, psychology may be simplified by finding a center of gravity for actions. And as in physics, the best definitions of origins and distances will be those that respect the structure of the subject and make interpretation of its dynamics the simplest.

Organism-centered responses (such as grooming, sneezing, and scratching) are often difficult to condition. Categorizing them as reflexes does not explain this, because other reflexes (such as the startle reflex) are easily conditioned. Part of the problem stems from inappropriate definition of the response (Iversen, Ragnarsdottir, & Randrup, 1984). But it is generally the case that instrumental conditioning proceeds most rapidly when the organism's effectors are part of an allocentric frame of reference established by approachable signs of reinforcement. Such conditioning leads an organism to attend to and approach those signs of reinforcement. In turn, such approach diminishes the distance between the signs and the organism, and thus further enhances conditioning. Research has grown steadily on the psychology of origins and distances, beginning with the early work in the Tolman tradition on maze learning, and has been increasingly integrated with geometric models of perceptual space (Cheng, 1986; Crossman & Nichols, 1981; Gallistel, 1990; Killeen, 1974; Killeen & Riggsford, 1989; Wagner, 1985; Wilkie, 1989), and with neural models of the underlying brain structures (see, e.g., Pellionisz, 1989, and Schmajuk, 1990, for recent reviews).

Time

Time seems a more tractable dimension of behavior—a straight continuum with a clear origin and direction. This appearance is due to our incorporation of Newton's time into our phenomenology. For Newton, time, like mass and force, was a hypothetical construct to be understood in the way that made his system of mechanics the simplest and most powerful. For him, "absolute, true, and mathematical time, . . . flows equably without relation to anything external" (Newton, 1687/1934, p. 6). "It may be, that there is no such thing as an equable motion" with which to measure the flow of time precisely, but "the flowing of absolute time is not liable to any change" (p. 8). Our measurements of "common, sensible time" are approximations to this Platonic ideal, and must often be corrected in estimating it. This approach was more parsimonious for Newton than accepting "common" systems of time, because it permitted one system of mechanics, along with miscellaneous calibrations for the various instruments and contexts in which time was "sensed." Of course the rate of flowing of "absolute" time is not absolute but is relative to the acceleration of the inertial frame of reference, but this was a story for a subsequent century.

Following Newton's logic of science, rather than its implementation for inanimate bodies in Euclidian space, we also choose a definition of time (and the other dimensions as well!) that simplifies our system of behavior (see, e.g., Killeen, 1991a). We know, for example, that for rats, noon today is more similar to noon yesterday than it is to 9 a.m. today, as evidenced by their circadian generalization of behavior such as shock avoidance (Gallistel, 1990, provides a contemporary review of the organization of behavior around spatial and temporal dimensions). How can such rhythmicity be represented? Perhaps by drawing time not

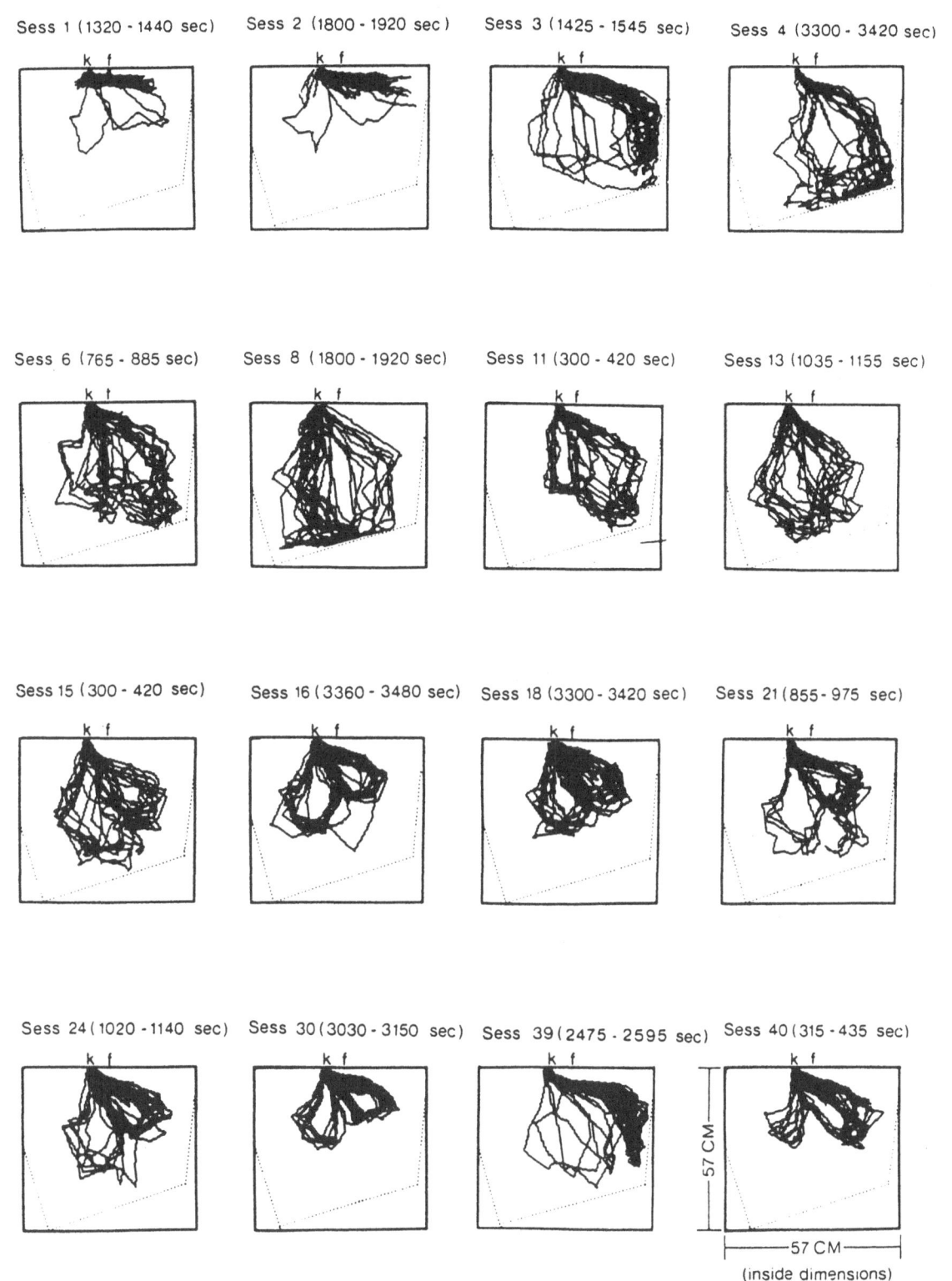

Sess 1 (1320 - 1440 sec) Sess 2 (1800 - 1920 sec) Sess 3 (1425 - 1545 sec) Sess 4 (3300 - 3420 sec)

Sess 6 (765 - 885 sec) Sess 8 (1800 - 1920 sec) Sess 11 (300 - 420 sec) Sess 13 (1035 - 1155 sec)

Sess 15 (300 - 420 sec) Sess 16 (3360 - 3480 sec) Sess 18 (3300 - 3420 sec) Sess 21 (855 - 975 sec)

Sess 24 (1020 - 1140 sec) Sess 30 (3030 - 3150 sec) Sess 39 (2475 - 2595 sec) Sess 40 (315 - 435 sec)

57 CM

57 CM
(inside dimensions)

Fig. 1. Spatial trajectories through a chamber of 1 pigeon at various stages of training on a VI 5-min schedule. The figure is from Pear (1985), and is reprinted with permission of the Society for the Experimental Analysis of Behavior.

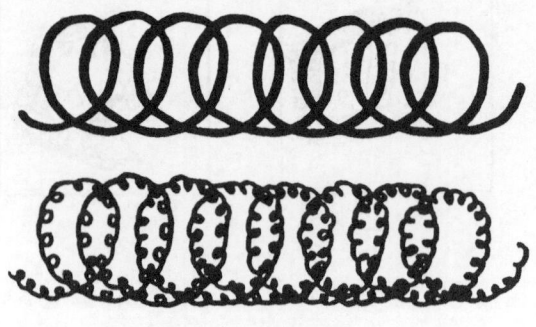

Subjective Time ——→

Fig. 2. Top: Time's helix. Bottom: A helix with curls, representing the nesting of imposed (e.g., schedule) periodicities upon the circadian rhythm.

as an arrow, but as a helix (Figure 2). A tight helix, like a coiled spring, indicates maximum circadian generalization, as points separated by one (~24 hr) cycle are closer to each other than points separated by only a few hours; a completely stretched spring indicates little or no such generalization. Newton's time runs through the spring steel in either configuration, flowing equably without relation to anything external; behavior's time respects the topography, as we in turn should respect it in order to understand behavior.

Do imposed periodicities such as those found in reinforcement schedules further bend time's helix, overlaying epicycles on the day's fundamental period? How should we portray quadridian cycles (Winfree, 1980, draws some interesting portraits)? Is there a unit of time, measured in scores or hundreds of milliseconds, about which all slower processes organize themselves as a harmonic? Does scalar timing, Weber's law applied to durations, suggest that time is "self-similar" with smaller intervals being condensed versions of larger ones? Is time's dimension fractal, imbuing all transits with an inherent path dependence? These are but a few of the questions and speculations that will arise in attempting to identify metrics for this primary axis of behavior space.

Stimulus

The objects manipulated by physicists are more than points on a line; they are coherent elements such as levers and planes, pulleys and pendula. It is the *action* of a lever within the system of mechanics—its mechanical advantage as a function of the distance from the ends

to the fulcrum, and how that transforms motion and force—that singles it out as a unit, not its physical form. In turn, forces are defined in terms of their actions on these simple elements. Similarly, it is the action of a stimulus within a system of behavior that singles it out as a unit. This was what Skinner (1935) meant when he spoke of the "generic nature of stimuli and responses."

Shepard (1987a) explored the logic of perception in organisms constrained by their evolution in the context of physical forces. Universalities of gravity, season, and tide have been "hard-wired" into the logic of the organism, whereas other less reliable regularities have been left to the more-or-less general-purpose learning abilities to model and thus predict. Entities with certain physical attributes (spatial contrast, motion, size, spectral composition) are candidates as potential stimuli because organisms have evolved sensitivity to those dimensions; they are good bets to be relevant to survival (Staddon, 1983). Stimulus generalization lets us infer how "close" various stimuli are to one another in psychological space, and from that we may infer the structure (the dimensions and rules for measuring distance along them) of the psychological space (see Figure 3). Shepard (1987b) has shown that a universal process of stimulus generalization—the ubiquitous exponential decay gradient between psychological distance and generalization—may be derived as a robust result of minimal inferences an organism must make when confronted with two stimuli and forced to judge whether they go together.

Special stimuli acquire special status as avatars of biologically important events; there is a rich literature on the evolution of sensitivity to those particular configurations of energy, called *sign stimuli* (see, e.g., Marler, Dooling, & Zoloth, 1980). Stimuli may also acquire additional distinctiveness as cues through their association with the attractors we call *unconditioned stimuli*. Von Uexküll (1921) spoke of the world as perceived by an organism, one with its own unique sensors and sensitivities, as its *Umwelt,* and a less knowable representation of its drives, motor preparedness, and stimulus input as its *Innenwelt*. These are useful terms, ones that will be appropriated and generalized here to refer to our reconstruction of the stimulus/response/time/incentive space of an organism.

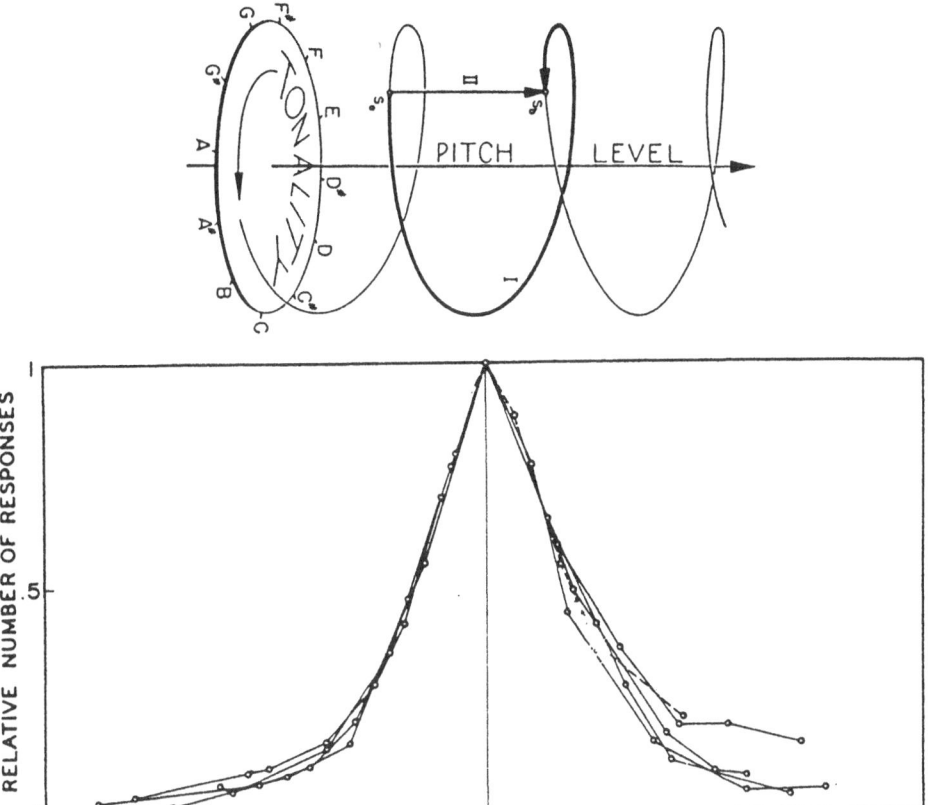

Fig. 3. Two stimulus dimensions. Top: Sinusoidal tones represented as points on a helix, with each cycle representing a new octave. Bottom: Hue generalization gradients around various training stimuli superimposed upon a wavelength dimension that has been adjusted to maximize the similarity of the gradients to one another. The adjusted points on the dimension (not shown) form the psychological dimension of hue. Note that the universal gradient is an exponential decay function around the training stimuli. Both figures are from Shepard (1965); the data in the bottom figure are from Guttman and Kalish (1956). Reprinted with permission of the Stanford University Press.

Tools such as levers have a logic to them: One must apply force to the beam, not the fulcrum, for the contraption to function as a lever. In like manner, stimuli must respect the logic of the nervous system, be that hard-wired or learned. Just as understanding the proper manipulation of tools such as levers teaches us about the nature of forces, learning how to manipulate stimuli effectively teaches us about the nature of behavior, and thus about the character of an organism's *Umwelt*.

Response

The attempt to specify the appropriate units of behavior has had a long history. We can clearly do better than recording all points on an animal's surface over time. Ethologists hoped that the *fixed-action pattern* would provide such a unit, and for many types of instinctive behavior it does, especially if we recognize the residual plasticity of even these units

by dropping the modifier "fixed," as is currently done, or by renaming them *motor programs* or *modal action patterns* (Barlow, 1977). In more sophisticated organisms, however, fewer instances of behavior are clearly identifiable as action patterns. The behavior of mammals often seems fluidly suited to need, with movements organized by their ends. "Behavior is only part of the total activity of an organism, [it] is what an organism is *doing*" (Skinner, 1938, p. 6). His theory of the "generic nature of the stimulus and response" recognized that neither could be considered independently of the other: Their essence depended on the correlation of stimuli and responses with each other *and* with unconditioned stimuli (UCS; these are treated here as functionally equivalent to rewards and reinforcers, although the latter often derive their force through a process of conditioning). A UCS not only releases unconditioned reflexive motor patterns but it also selects prior candidate *stimuli* that occasion the UCS and imbues them with some of the qualities of the UCS, whereupon they are called *conditioned stimuli* (CS), *discriminative stimuli,* or *sign stimuli* (Gould & Marler, 1984). It selects prior candidate *actions* that occasion the UCS and shapes (fine tunes) their topography, whereupon they are called *responses.* Just as the physicist's tools are convenient conceptual units for the application of the fundamental laws of motion, the ethologist's *action pattern,* the Pavlovian's *reflex,* and the Skinnerian's *operant* may be seen as candidate tools through which we may come to understand the fundamental laws of motion of animals.

Recent work on "constraints" on conditioning has shown that response topographies are often less ductile than once imagined. Quite apart from whatever shaping effects are exerted by contiguity with the UCS, that stimulus induces other actions variously called *unconditioned responses* (UCR), *consummatory responses,* and *adjunctive behavior.* "We begin to conceive of behavior, which we have always thought of as highly modifiable, as consisting of a lot of fixed packages, software programs as it were. These preformed packages can be shifted around from one application, or object, to another" (Bolles, 1983, p. 43). "Such a strategy, which involves building up complex motor behavior out of a 'library' of innate elements, has obvious advantages for certain tasks" (Gould & Marler, 1984, p. 66). The emerging picture is one of coherent modules—action patterns—with some limited degree of modifiability, including the important ability to be sequenced.

These packages are organized hierarchically (Dawkins, 1976). "Circuits at higher levels govern the operation of lower circuits by . . . raising the potential [for operation in some circuits] and lowering it in others—a higher unit establishes the overall pattern to be exhibited in the combined operation of the lower units, while leaving it to the lower units to determine the details of the implementation of this pattern" (Gallistel, 1981, p. 609). One picture of a hierarchy of action patterns associated with feeding in the rat is provided by Timberlake and Lucas (1990). Figure 4 shows a slightly rearranged version of one limb of their hierarchy. For our purposes, "higher" does not refer to the nested set of increasingly general conceptual categories in the left of the figure, but rather to the ordering along the spectrum of actions at the right of the figure, but rather to the ordering along the spectrum of actions at the right of the figure. We shall see that those lowest in this column may correspond to the actions that are easiest to motivate, whereas those highest in the column may correspond to actions that require more energy to motivate. Other vertical orderings of the action patterns, with insertions and deletions of various actions, accompany different incentives/drives. The wholesale ability to reorder and thus reprioritize our goals is characteristic of emotional control.

"Integrated behavior is a nested set of more or less coherent *processes* rather than a set of indivisible, independent, and separate *things.*" (Fentress, 1981, p. 624). Behavior evolves much like species, each requiring mechanisms of variation and of selection acting on units, but the units may be specified at various levels of generality (i.e., vertical slices through the hierarchy of Figure 4 at different abcissae), whereas selection will act concurrently at the various levels (e.g., by selecting predation/general search/scanning and the details of each). We are only beginning to appreciate how the hammer and anvil of ontogeny and phylogeny between them forge the units of behavior, at what points the selective forces impinge, and how they are transmitted to other levels of the hierarchy.

→

Fig. 4. A hierarchy of action patterns in the rat, adapted from Timberlake and Lucas (1990). The conceptual categories to the left provide an intuitive organization of the actions. When an organism is in another mode than predation (say, nesting), a different set of actions with a different ordering will prevail. The actions are arranged on the response continuum (right column) according to their activation energy. In the presence of stimuli that release an action, the rat will be attracted to an engage in that action. It will be differentially attracted to actions lower on the response continuum, and it will require energy to keep it at a higher level. Letting the rat approach the lower levels, or approach the stimuli that release them, will convert the potential activation energy into the kinetic energy of motion and conditioning.

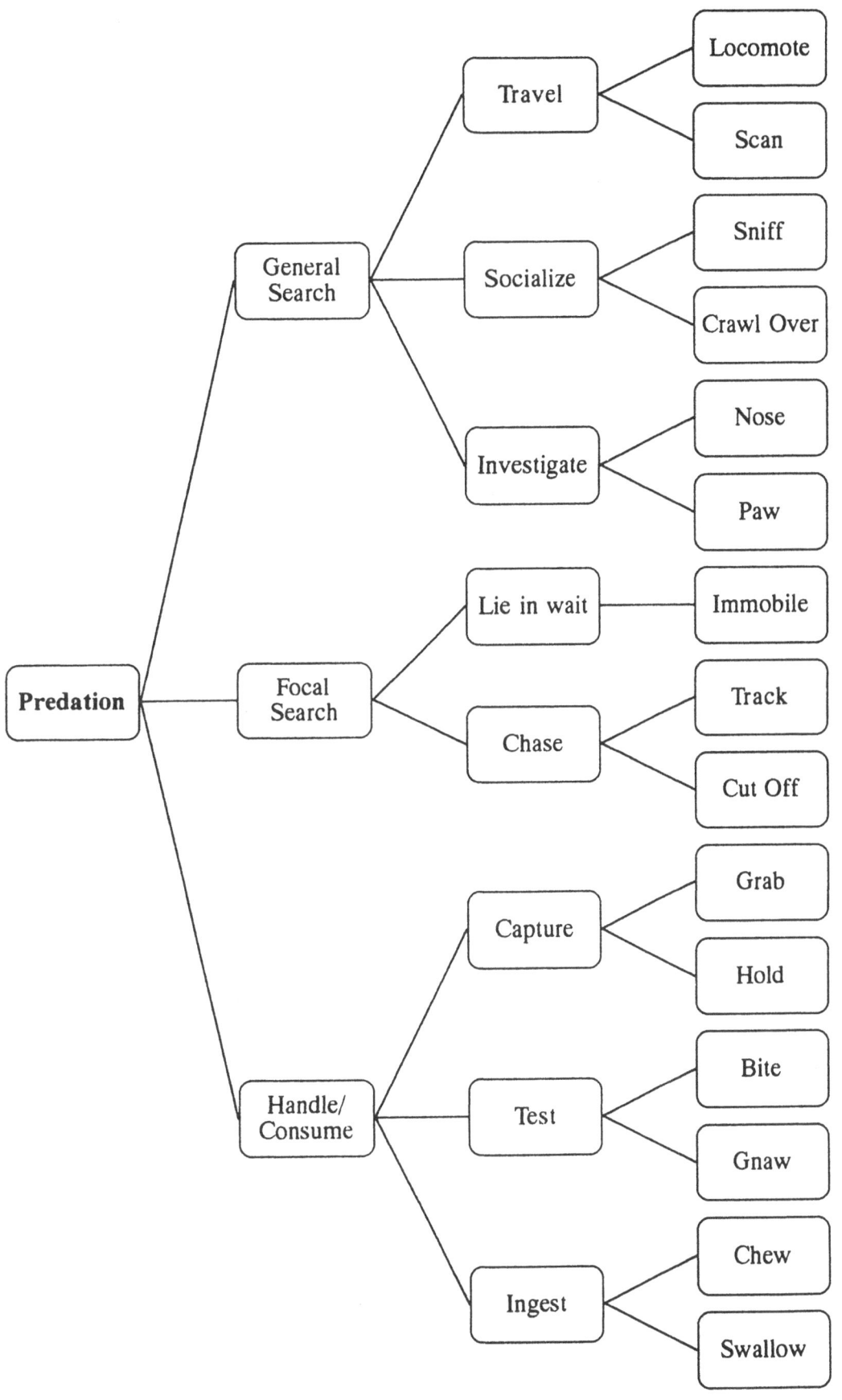

Sample Trajectories

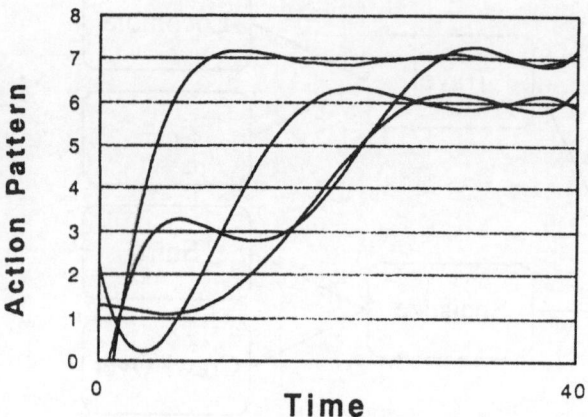

Fig. 5. Hypothetical trajectories through two dimensions of behavior space. The ordinates correspond to different action patterns such as those listed in the rightmost column of Figure 4. Each trajectory represents the path followed on a different trial.

The Dimensions Frame a Behavior Space

The dimensions invoked to encompass behavior will depend both on the controlling variables and on the level of representation desired (i.e., which column of descriptors we select from the hierarchy in Figure 4). They will often include things such as time, proximity to signs of reinforcement, orientation of sensors, level of deprivation, the levels of relevant hormones, and so on. A path, or *trajectory,* through this behavior space represents the movement of the animal through time, space, stimulus, and behavior, and also through the physiological changes it undergoes along that trajectory. Empirical construction of such a space must not only solve the problem of ordering of stimuli, responses, and reinforcers along their axes; it must also attend to their reordering under different motivational states, and to the appropriate offsetting of the axes so that special proximities (e.g., stimulus–response compatibilities, constraints on conditioning, mood-dependent sensitivities, and so on) are captured by the model. This can be done. Whether it can be done in a sufficiently parsimonious space to justify this larger view of behavior is an empirical question (see Appendix 1 for further ruminations).

As a cartoon example of such a space, let us center the temporal origin on the time a reinforcer is delivered, and identify a subset of that space as a "consequential region," a part of the behavioral space that the animal must enter to achieve another reinforcer. With time since the previous reward as the x axis and a response continuum such as that represented by the rightmost column of activities in Figure 4 as the y axis, we can specify consequential regions corresponding to the basic schedules of reinforcement. In the case of "time-place foraging" such as that engendered by a fixed-interval schedule where we reinforce the first response after a fixed time since the previous reinforcement, we may have the situation pictured in Figure 5. The picture is restricted to these two dimensions for convenience of representation, although we should remember that the trajectory will concurrently carry the animal through other dimensions not shown in this slice.

To limit our conceptualization of reinforcement to responses is to study only one of the many dimensions that reinforcement may affect. Many different trajectories will carry the animal into the consequential region, and it is those trajectories that are the candidates for reinforcement. However, not all trajectories in the sheaf of candidate paths may be equally amenable to the "strengthening" effects of reinforcement. We shall later suggest that there may be a logic to conditioning that moves the learned paths toward an optimal trajectory through behavior space—one that conforms to a principle of least action.

FORCES

Newton's plan for the *Principia* was straightforward: "The whole burden of philosophy seems to consist in this—from the phenomena of motions to investigate the forces of nature, and then from these forces to demonstrate the other phenomena" (Newton, 1687/1934, p. xvii). Thus the key hypothetical construct for Newton was *force,* just as for Skinner, for a little while, it was *drive.* The marvel is that Newton perfected his system of the world, explicating the motions of apples and comets alike, without understanding the intrinsic nature of the forces beyond their interaction with matter (it was this boldness that caused some contemporaries to disparage his work). He made no axioms concerning the nature of the forces, but he did frame hypotheses about them:

I am induced by many reasons to suspect that [the phenomena of nature] may all depend upon certain forces by which the particles of bodies, by some causes hitherto unknown, are mutually impelled toward one another, and cohere in regular figures, or are repelled and recede from one another. These forces [are] unknown, . . . but I hope the principles here laid down will afford some light to this. . . . (p. xviii; see also pp. 634, 671)

It is by analysis of their actions that we shall come to know the forces, not by a search for their essence.

Such is the case for behavior. Skinner noted that " 'drive' is a hypothetical state interpolated between operation and behavior and is not actually required in a descriptive system" (1938, p. 368); he quickly abandoned the construct, along with the hope of achieving more than a descriptive system—of achieving a science utilizing hypothetical constructs to achieve a parsimonious descriptive system whose elegance and economy in turn justifies reification of its constructs. Let us pick up where Skinner left off, with what he characterized as the then "traditional conception" of drive: "At one extreme, 'drive' is regarded as simply the basic energy available for the responses of an organism; at another it is identified with 'purpose' or some internal representation of a goal" (1938, p. 341). Cofer and Appley (1967) and Bolles (1975) reviewed the research on the energetic, instigational, and "inciting" properties of incentives. Craig (1918) emphasized the directive nature of instincts toward goals or away from antigoals: "Each instinct involves an element of appetite, or aversion, or both" (p. 91); both appetites and aversions were "states of agitation" that continued until a stimulus was received or removed. Thorndike operationally defined his key variable, "satisfiers," as a state of affairs "which the animal does nothing to avoid, often doing such things as attain and preserve it" (Thorndike, 1911, p. 245). Schneirla (1959) held that "*approach* and *withdrawal* are the *only* empirical, objective terms applicable to *all* motivated behavior in *all* animals" (p. 1). Hull noted that "The facts of adience and abience are so obvious in animal behavior that they cannot be overlooked" (Hull, 1943, p. 349), and wrote several influential theoretical accounts of them (Hull, 1952). Panksepp (1989) holds that

all of the diverse positively motivated behaviors exhibited by animals (e.g., thermoregulation,

feeding, drinking, salt-appetite, hoarding, predation, sexuality, maternal behavior, shelter-seeking) seem to be effected, to a substantial extent, by a common emotive brain circuit. The command impulse for all these goal-directed behaviors appears to arise from a shared foraging—expectancy command system which generates the primal tendency for an animal to move from where it is to where it must be to acquire materials needed for survival. (pp. 12–13)

These common themes of energization and motion toward a goal are developed in the following pages, where drives are treated as forces with both magnitude and direction. Drives are the fundamental forces, and incentives are the origins of those forces. Inciting an organism by introducing an incentive produces a potential for action, and releasing the organism to move through behavioral space to the incentive converts that potential to kinetic energy. Motivation is nothing other than motion, or the potential for motion, in this space. This is parallel with the physicist's treatment of gravity as a force with a massive body as its origin, of electrostatic forces with charged bodies as their origins. Incentives force behavior toward a consequential place. It was the attractions of organisms to incentives that motivated the statements of Thorndike, Schneirla, and Panksepp, who placed such a spatial force at the center of their conceptual systems. The spatial force of incentives is directly manifest in sign-tracking and goal-tracking. Incentives also force behavior toward a consequential time. We see this in the temporal control of behavior, in traditional research on schedules of reinforcement, and in the emphases on contingencies, the temporal relations between a response and its consequence. Incentives also force behavior toward a consequential (consummatory) response topography. We see this in action patterns and shaped responses, in adjunctive behavior and "misbehavior." Of course, none of these parts of the *Umwelt* exist independently of the others (or of the organism!). Most accurately, *incentives force organisms toward consequential regions in their stimulus-time-action space.* Incentives are attractors in behavior space. It is the force of incentives that mediates both performance (movement along a trajectory toward an incentive) and learning (displacement of the trajectory into a more efficient one).

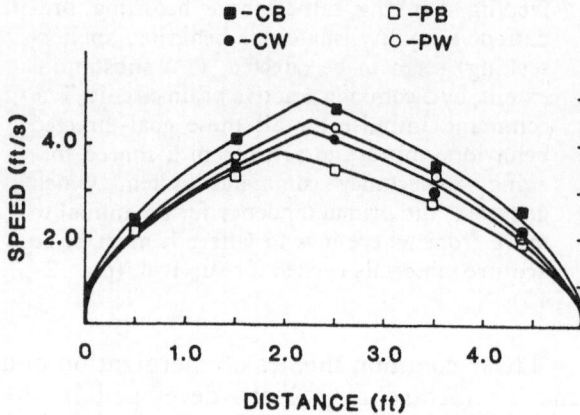

Fig. 6. The speed of rats through a runway. The curves are derived by assuming that the rats are uniformly accelerated toward the food cup until a midway brake point, whereafter they uniformly decelerate to come to rest over the cup. The partial-reinforcement between-group condition (PB) showed an earlier brake point and lower acceleration than the other conditions (continuous-between, continuous-within, and partial-within); this is also the only condition that reliably shows partial reinforcement extinction effects (i.e., prolonged responding in extinction with respect to the other groups). The figure is from Killeen and Amsel (1987), and is reprinted with permission of the American Psychological Association.

Forces Through Space

How do we learn about the forces that drive behavior? In physics, the procedures include directly measuring the force as a function of the distance from its source, as Coulomb measured the electrostatic force with a torsion balance; balancing one force against another, as Wheatstone measured the electromotive force with his "bridge" arrangement; measuring the acceleration caused by the force, and, invoking basic equations of motion, calculating backward to the forces. Galileo measured the accelerations, and those data were the "phenomena of motions" that Newton used "to investigate the forces of nature."

Analogues exist in psychology but have never been systematically pursued, as befits such fundamental research. Like Coulomb, Brown (1948) measured forces exerted by rats in approach/avoidance conflict using a strain gauge (unfortunately, only at two points; replication at multiple distances would provide invaluable data on the shape of the spatial force gradients); N. Miller (1971) provided a programmatic review of such research. Like Wheatstone, Warden (1931) measured forces such as hunger, thirst, and maternal drives by bal-

ancing them against opposing drives such as fear: He placed rats in boxes where they had to cross electrified grids to approach the incentive, and compared incentives at various levels of deprivation in terms of the number of grid crossings per session.

Incentives accelerate organisms. Speeds measured at different points of runways give different and inconsistent results upon manipulation of independent variables such as amount or probability of reward. But speed is a derived measure: Reinforcers *accelerate* animals along their spatial trajectory, they do not "speed" them. Amsel and I measured the speeds of rats in a runway and inferred from them the forces exerted by the food at the end of the alley (Killeen & Amsel, 1987). We hoped the data would appear more orderly if we chose acceleration as the dependent variable. To achieve this, we treated rats as falling bodies, under constant positive acceleration from the food cup until a brake point at which they began decelerating to come to rest over it. The treatment clarified and simplified the data, reducing them from overlapping curves of speed at various points through the runway to two numbers: accelerative force (measured as ft/s²), and brake point (see Figure 6).

None of these studies, however, were systematic enough to generate laws of forces as convincing as those of physics, nor did they evaluate other candidate versions of the forces. While waiting for more thorough experimental data, we may achieve some insight to the possible laws of behavioral forces by a *gedanken* experiment, starting with a simplistic example and moving to more interesting ones. Centuries ago the philosopher Buridan speculated that an ass placed perfectly between two piles of hay and equally attracted to each might never be able to move (see Figure 7). But we suspect from experience that a hungry pigeon placed an equal distance between two piles of grain would not long hesitate. What makes our real pigeon more decisive than Buridan's hypothetical ass? We can argue the pigeon's misperception of the piles, vagaries of its attention, or our inability to satisfy the conditions and place it perfectly central. But more productive arguments are possible.

Fields of force. Assume the ass is drawn to each incentive by forces that act like stretched springs (Figure 7 and Figure 8, Row *a*).

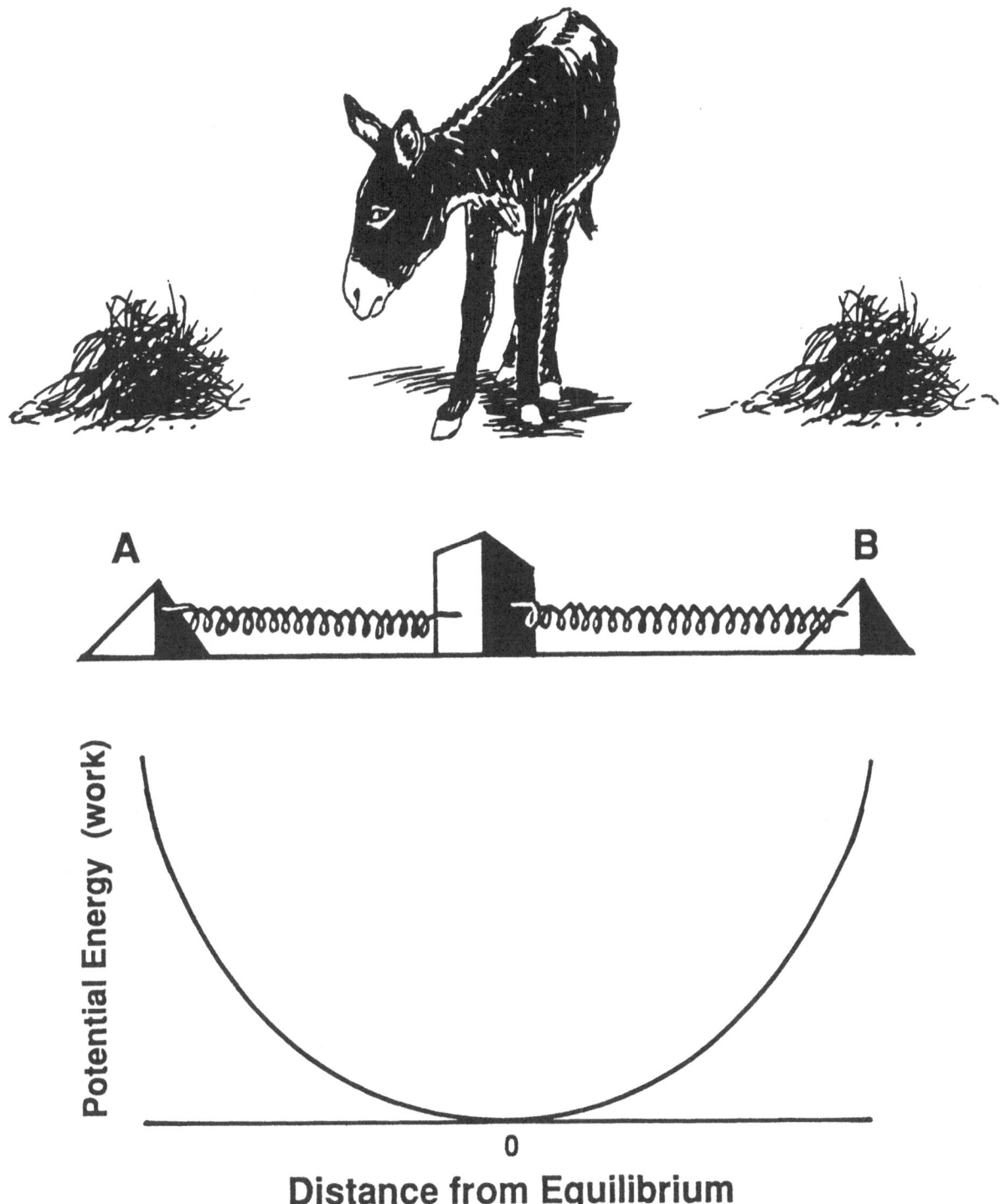

Fig. 7. Top: Buridan's ass. Middle: Hypothetical forces acting on the ass. Bottom: The potential function.

Hooke's law tells us that the force exerted by a spring is proportional to its extension. As the ass moves away from pile *A* toward pile *B,* the more it is drawn back to *A*. It is stuck, and no perturbations in its position will get it unstuck. If one pile is bigger than the other, the ass will come to rest a bit closer to it, but will still get stuck. Depending on frictional forces, perturbations will return it to equilibrium with overshoot, to oscillate between the

113

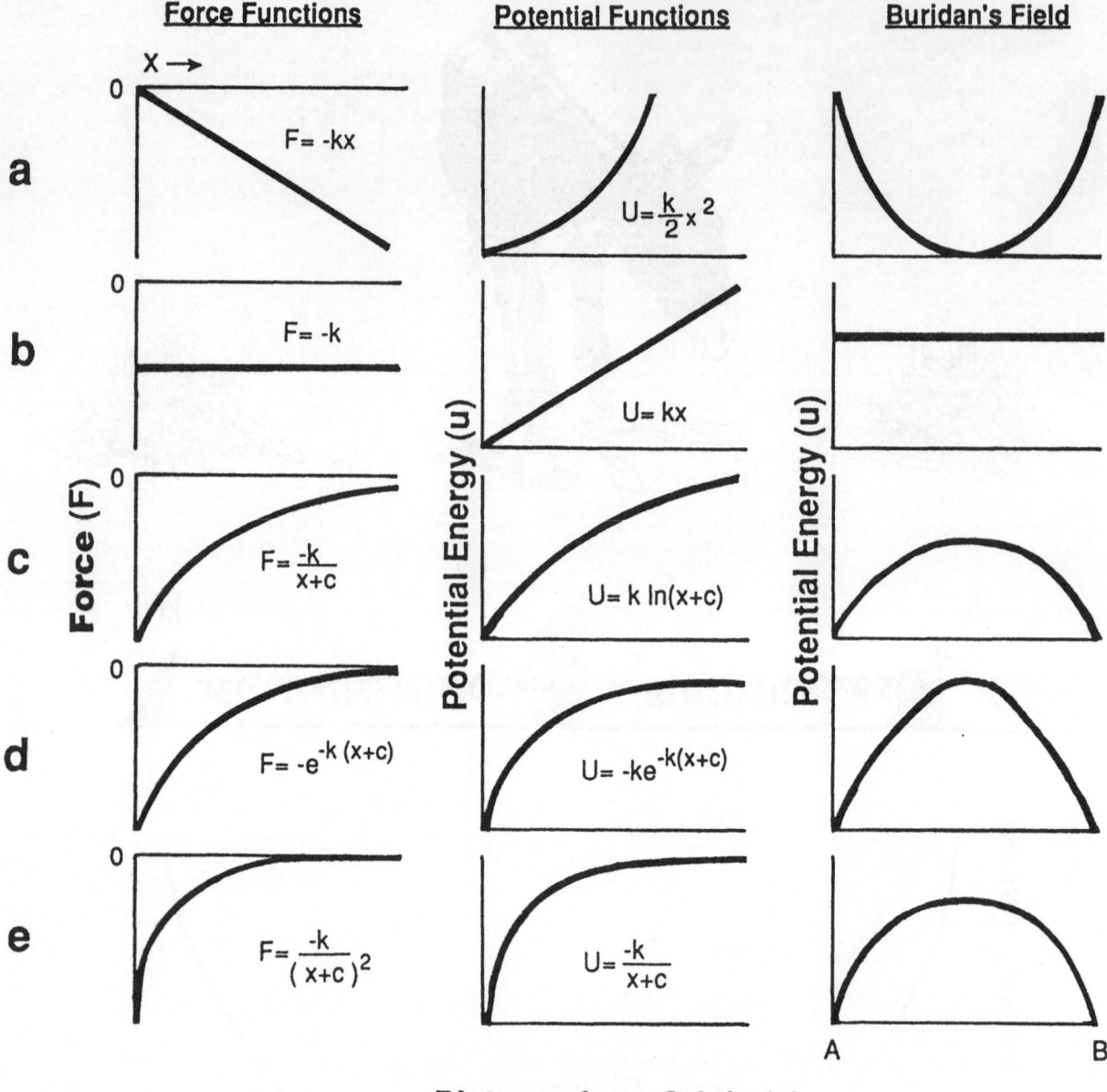

Distance from Origin (x)

Fig. 8. Columns: Force fields, potential functions, and field between two equal sources. The x axis of each figure is some psychological dimension (e.g., distance, time, stimulus, response). Forces are negative to indicate that they are attractors (move the object toward them). Potential energy is the work that may be accomplished by moving a body from its position (x) to the origin, and is calculated as the negative integral of the force along that trajectory. In the first two columns the attractor is at the origin; in the third column attractors are located at the origin and at B. Rows: Candidate force functions. Functions a and b are ruled out by elementary considerations; c and d have been suggested for force propagated along the temporal dimension (see Appendix 2), and d for the stimulus dimension; e has been proposed for the decrease in emotional involvement with geographical distances (Lundberg, Bratfisch, & Ekman, 1971).

two incentives, never attaining either. Although providing a model for Buridan's hypothetical ass, this type of force field corresponds to no known drives, with the possible exception of human mate selection.

The behavior of the pathetic object in Figure 7 may be further clarified by drawing its *po-*

tential function. A potential function is the negative integral of the force function, and tells us how much work is required to move an object through a distance in that force field. In the present case it is a parabola, with its lowest point midway between the incentives. This point of rest is called an *attractor*, because when the object is anywhere else, it is attracted to this stable equilibrium. It requires external energy to move the organism out of this *potential well.* Conversely, work can be accomplished by letting the organism respond to the restoring forces naturally, moving from off-center to its point of rest. We shall later suggest that it is the movement toward an attractor that accomplishes the work of conditioning.

Potential functions may be added to the behavior space of an organism by using an additional dimension. Although the potential functions drawn in Figures 7 and 8 are drawn as curves above the spatial dimension, they should be thought of as a multidimensional sheet, draped above the stimulus/response/time space. Movement from the high peaks to the low valleys converts potential energy to kinetic, and in the process accomplishes work.

Animals constructed with force fields that kept them trapped like Tantalus midway between incentives had little opportunity to create progeny, and are ill-represented among the animals we know of in the world today. Consider then forces similar to gravity at the surface of the earth (Figure 8, Row *b*). The distance to the center of gravity of the earth is so great relative to the objects of interest that the acceleration of gravity can be considered constant. Does a constant force field liberate Buridan's ass? Yes, but not to look like our real pigeon. If the incentives are of equal magnitude, the force exerted by one will exactly balance that exerted by the other, independent of the proximity of the organism to one or the other. Thus, the animal is free to "drift" into one incentive or the other, as it might if some external force helped it along; but its potential function is flat: No work can be accomplished by moving it closer to one pile than the other (and thus it cannot learn to approach one pile rather than the other). Even when it is much closer to one incentive than the other, there will be no differential attraction to it. Such flat force fields correspond to states of low arousal and flat emotion, and yield uncommitted organisms who are not "captured" by goals, but shift their direction whenever a slightly stronger attractor appears anywhere on the horizon.

Productive forces on behavior must have potential functions that encourage organisms to get close enough to an incentive to consummate reinforcement: The forces must increase, and the potential functions get deeper, the closer

one gets to an incentive. Fields in which force is a decreasing function of the distance away from the incentive (e.g., linear decreasing, inverse, inverse-square, etc.) capture the semblance of motivated organisms. Such potential fields are at a relative maximum midway between two equal incentives, so that any vicissitude that carries the organism off that unstable equilibrium an iota closer to one than the other will decide its fate: It will be accelerated with increasing speed and surety to the nearest incentive (see Figure 8, Rows *c, d,* and *e*). For some drives or conditions the potential function might be steep, for others it might be almost flat, but its magnitude must increase with proximity to the incentive according to function rules of the type shown in the first column of this figure.

Townsend and Busemeyer (1989) have modeled the intuitions of Miller (key papers are reprinted in N. Miller, 1971) and Lewin (e.g., 1933) concerning approach–avoidance gradients. Their work provides an excellent example of development of an explicit dynamic system from verbally stated intuitions. But we can go no further in specifying the nature of the force fields without systematic data. Experiments (e.g., analysis of trajectories and accelerations induced by one- and two-incentive arrangements) could readily generate the "phenomena of motions" that are necessary to instantiate these general approaches. But let us turn to other dimensions of our behavior space, for until now we have operated only along spatial dimensions. Just as the speed and direction of light vary with the medium through which it shines, and as magnetic fields induce electric potential when they move across conductors, the force of incentives may be propagated through dimensions other than space. When this happens, we may expect to encounter not only variation in the speed, range, and direction of forces, but also the induction of novel phenomena.

Forces Through Time

Forward. The classic "delay of reinforcement gradient" concerns the force of incentives as it diminishes over temporal distances. Numerous versions have been proposed, all of them treating force as some type of inverse function of distance. Mazur and others (e.g., Mazur, 1984; Rachlin, Raineri, & Cross, 1991) have proposed that the force is a simple inverse function of temporal distance from the reinforcer (see Figure 8, Row *c*); the strength of delayed reinforcers in capturing behavior changes as the reciprocal of the delay plus an additive constant. This yields a potential function (the negative integral of the force function) that is proportional to the logarithm of the delay.

Organisms' behavior may be sensitive to the potential function—they may have a sense of how much psychological effort it would take to move from one gradient to another (say, from choosing one delayed reward to choosing another). This analysis is developed in Appendix 2, where implications for stimulus discrimination and response differentiation are derived.

Other gradients have been proposed. I have suggested another type of inverse function, the negative exponential gradient, in treating the force an incentive exerts on a response, with the slope of the gradient depending on the arousal of the organism (Killeen, 1984). Its potential function is the exponential integral (see Figure 8, Row *d*).

Different types of behavior are differentially attracted by incentives (i.e., are differentially "reinforceable"): It is often easier to say "no" than to desist, to "misbehave" than to behave. If sensitivity to attraction is the mechanism of differentiation—a central thesis of this paper—we might also expect that different types of behavior will be more or less sensitive to our attempts to shape them along the time dimension. This is the case. Delays of reinforcement much too great to affect one behavior/incentive doublet may easily affect others (Killeen, 1985; Lejeune, 1990; Platt, 1984), with the phenomenon of conditioned taste aversion only the most salient of many examples.

Of course, when we speak of "attraction to future events," it is shorthand for all the influences, innate and learned, that permit animals to predict the appearance of mates, food, or predators, and thereby cause those projected events to control behavior. Reification of the attraction to future events may be a more effective tactic for a functional analysis of behavior than attempts to reconstruct the contextual stimuli that give rise to it. Indeed, the temporal dimension itself may eventually come to be seen as Newton's shorthand "mathematical time," glossing the story of a congeries

of stimuli and actions whose sequential association constitutes the essence of our sense of time (Killeen, 1991a; Revusky, 1977).

Backward. The termination of a reinforcement episode provides a marker on the temporal dimension, one that attracts postincentive behavior such as area-restricted search (Krebs, 1973). (We are so used to thinking of behavior organizing itself around forthcoming events that it may seem odd to think of a past event as continuing to attract orientation and behavior. But such reminiscence is commonplace: Nostalgia signals a forceful, and marketable, incentive.)

There is a competition between the attraction of past incentives and that of (predictably) forthcoming ones. As the interval between incentives is increased, these forces stretch out, or normalize, the distributions of behavior that fill the interval. For longer intervals, where the overlapping force fields are relatively weak, new types of behavior may intrude (Staddon, 1977, calls these "facultative" behavior). The nature of this expansion gives us fundamentally important information about the nature of the gradients. The expansion is apparently linear: Killeen (1975) overlays the distributions of adjunctive behavior and finds similarity of shape, if not congruence; Gibbon (1986) refers to the overlapping normalized distributions of terminal behavior such as key pecking and lever pressing as "superposition" and makes it an important part of his scalar expectancy theory of temporal control; Staddon and associates (Staddon & Higa, 1991; Staddon, Wynne, & Higa, 1991) note that animals' postreinforcement pause on interval schedules is a fixed proportion of the expected time to reinforcement and make this "linear waiting-time" part of a general model of temporal control. When attraction to the past is weakened by reducing its salience, attraction to forthcoming incentives propagates earlier in time (Staddon, 1974). When attraction to forthcoming incentives is weakened by reliably degrading them, attraction to the currently available incentive is enhanced (the "following schedule effect"; B. Williams, 1983). Working out the forms of the gradients that must relate various interim, facultative, and terminal types of behavior to incentives past and forthcoming, given the available data and the models noted above, remains a straightforward but unaccomplished exercise in theory construction.

Recent data (Perone & Courtney, 1992) reinforce this vision of schedule control as competition of past and future incentives for control of an organism that finds itself moving inexorably along the time line away from the former and toward the latter.

We see that a mechanical analysis of behavior has the potential to unite results from temporal discrimination experiments with those from delay of reinforcement experiments, and to relate both to temporal differentiation and schedule effects. In the process of realizing that potential, the nature of the forces through behavioral space will be clarified. Such unification holds the promise that aspects of one phenomenon will lead us, through characterization of the forces, to demonstrate the other phenomena. Much useful work will be accomplished in moving toward that goal.

Forces Through Stimuli

Classical conditioning establishes a neutral stimulus as an attractor. The rich outpouring of research on sign-tracking or autoshaping amply validates the persistence of approaches to signals for food and other incentives, when such approaches are ineffective (or even counterproductive!) in moving the animal toward the primary attractor along the temporal dimension (see, e.g., Locurto, Terrace, & Gibbon, 1981; Peden, Browne, & Hearst, 1977; Tomie, Brooks, & Zito, 1989). Sequences of stimuli that form a type of clock also attract behavior, and do so increasingly with their increasing proximity to the primary incentive (Palya & Bevins, 1990).

The prime example of forces transmitted through stimuli is found in stimulus generalization. If one stimulus is similar enough to a second, organisms will respond to the former as though it were the latter, even though the two are quite discriminable. Hearst (1965) systematically studied approach–avoidance gradients along the stimulus dimension, Rescorla and Furrow (1977) demonstrated the facilitation of conditioning as a function of the similarity of stimuli, and Steinhauer (1982) studied the facilitation of autoshaping as a function of stimulus similarity. Shepard (1987b) demonstrated that the exponential decay function is the universal law of stimulus generalization, once arbitrary differences in the stimulus spacing have been adjusted. This

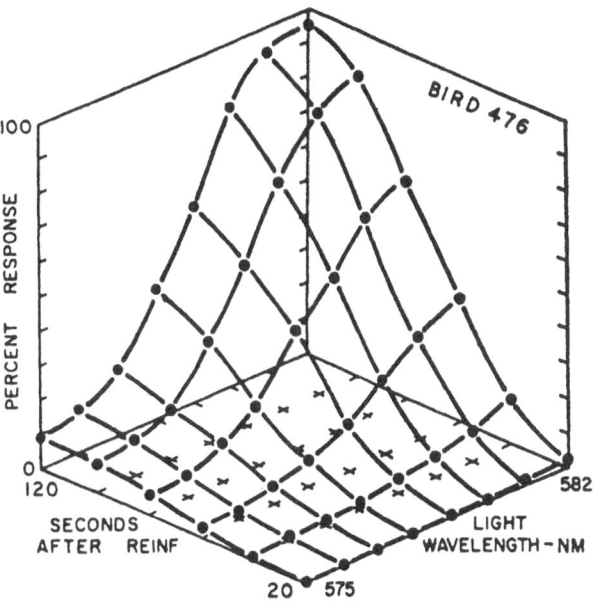

Fig. 9. The percentage of response made by a pigeon as a function of variation of stimuli along two dimensions: time and wavelength of the light illuminating the response key. Such conjoint measurement is a valuable technique for mutual calibration of the axes of behavior space. The figure is from Blough (1972), and is reprinted with permission of the Society for the Experimental Analysis of Behavior.

is strong evidence for Row *d* (Figure 8) as the rule by which force dissipates as it is transmitted through stimuli.

Fantino and associates have demonstrated that the psychological distance to reward must be measured in terms of both temporal and stimulus distances. Duncan and Fantino (1972) gave pigeons choices between two equal delays to reinforcement, one of which was marked with one stimulus, the other with a sequence of two or more stimuli. The animals always preferred the delay signaled by the less segmented set of stimuli. Indeed, space itself may be thought of as an array of stimuli; there is little surprise then that spatial contiguity facilitates Pavlovian second-order conditioning (Rescorla & Cunningham, 1979).

The notion that forces can be propagated through stimuli may be difficult to comprehend. It is clear that physical forces such as magnetism are affected by the materials through which they pass; electromotive forces are readily channeled by electric cords and modulated by semiconductor chips. Psychological forces also may be channeled and modulated by stimuli that are associated with incentives: Contiguity, similarity, and intensity have long been known to play potent roles in conditioning. To speak of stimuli as media for the propagation of the force of incentives is merely to suggest a coherent language in which similarities and differences are more easily discerned. Conjoint manipulation of different dimensions will provide an essential technique for calibrating the axes and measuring the forces flowing through them (see Figure 9). This perspective also raises empirical questions that have yet to be addressed, and provides a context for the answers obtained: If sequential segmentation of stimuli lengthens the psychological distance when time is held constant, would spatial segmentation of stimuli lengthen the psychological distance when time and physical distance are held constant? Would painting a path from one key to the source of food increase preference for that path over another, whose stimulus proximity was weakened by interposing a checkerboard grid? Would presentation of a constant tone when a rat moved from lever to food cup make that a preferred path over one accompanied by random melodies?

Forces Through Responses

A previous section suggested that responses may be ordered by their activation energies, giving a nonarbitrary continuum for behavior. What does it mean to say that the force of incentives may be propagated along this dimension? It means that behavior closest to the incentive—that is, proximate in the connections of the animal's nervous system—should feel the greatest force of the incentive, and those most remote should be least affected by it. There have already been demonstrations of such differences in susceptibility to reinforcement, most notably by Garcia (see, e.g., Garcia, McGowan, & Green, 1972), Seligman (1970), and other contributers to the literature on constraints on conditioning (e.g., Klein & Mowrer, 1990; Shettleworth, 1975, 1981). Reinforcement contingencies that do not respect preexisting action patterns will be ineffective or diversely effective, because they will force the trajectory through meanders that do not respect the natural proximity of actions in the organism's behavior space (Iversen et al., 1984; B. Moore, 1973). Killeen, Hanson, and Osborne (1978) suggested another measure of affinity between an incentive and behavior, the slope of the function relating the asymptotic probability of a response to the rate of incitement. The order of behavior along this continuum of susceptibility depends on the nature of the incentive; different forces attract different types of behavior differentially. We may order responses on our continuum, then, from those most engaged by a particular incentive to those least engaged.

Boltzmann curves for unconstrained behavior. A novel approach to specification of the dimensions of action is provided by Hanson (1991). From the assumption that it requires a certain amount of energy to activate independent units of behavior and from principles of statistical thermodynamics, he showed that the most likely distribution of behavior is one in which the frequency of each unit is a negative exponential function of its activation energy. Because we do not know the activation energies a priori, Hanson plotted the logarithm of the relative frequency of each behavior against its rank order. This is ad hoc and guarantees a monotonic function. But it does not guarantee a straight line in these coordinates, and a straight line fits the data much better than alternative candidates such as power functions (see Figure 10).

What does this mean? We may infer several things from Hanson's (1991) results. The first is that the assumptions of the model were satisfied. In particular, in cases in which the units of the behavior are clearly not independent, we expect systematic deviations from the negative exponential, and those were found. Next, they give some credence to treating units of behavior as requiring different levels of energy to activate them. Successful use of the rank order as the metric of action suggests that there is a quantal nature to activation, with each "higher" action pattern requiring exactly one more unit of energy to activate it. Even though variants of the action patterns may be shaped by reinforcement, their center of gravity must remain one unit above the next nearest. Finally, the exponential relation gives a test for the thoroughness of our observations and appropriateness of our categories. If at some point the line jags downward but maintains its linearity, we may have missed a category; if it shows a plateau, with two types of behavior having about the same frequency, it indicates the possibility of an inappropriately broad categorization.

Hanson (1980) found similar functions for

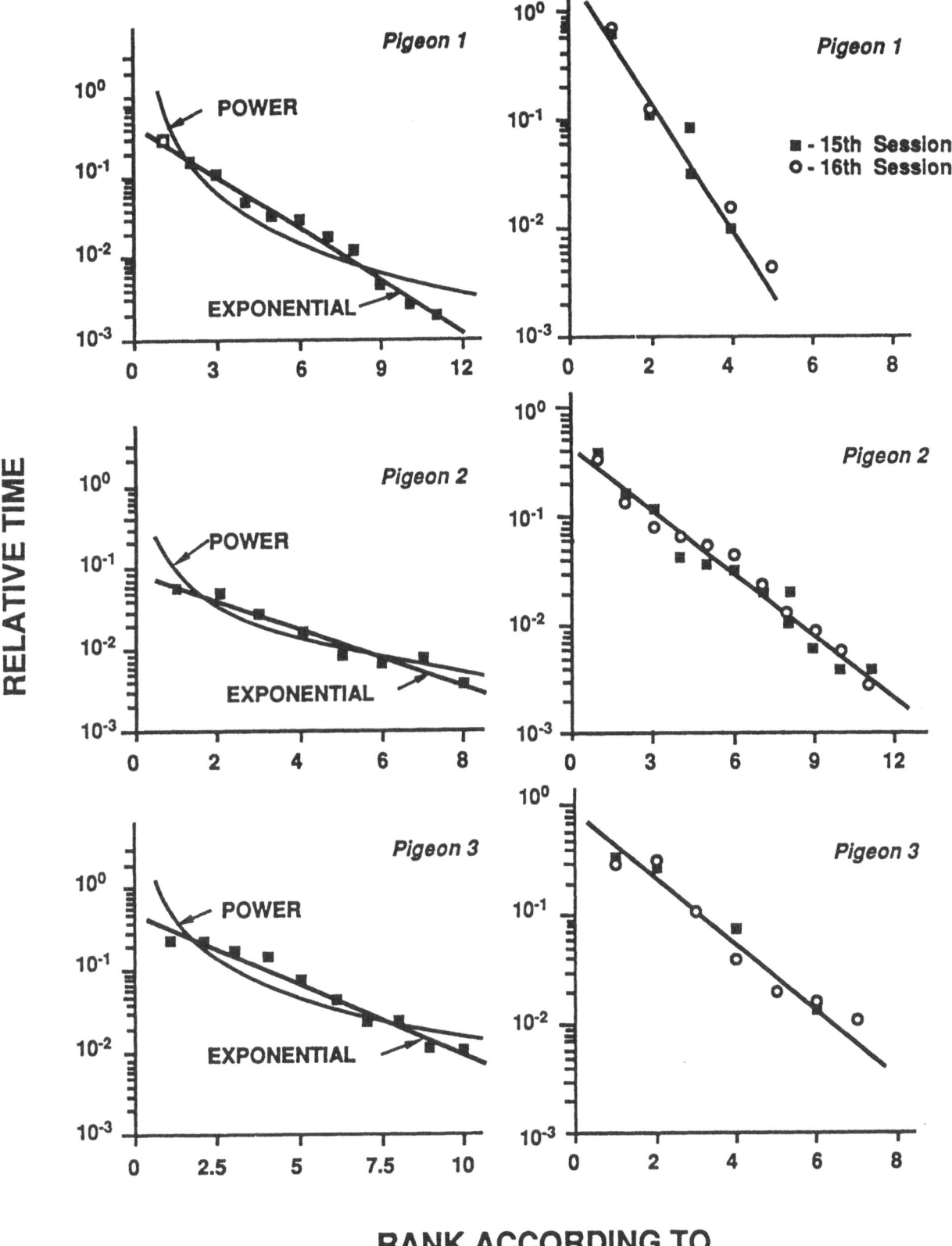

Fig. 10. Left column: Boltzmann curves (exponential decay functions, straight lines in these semilogarithmic coordinates) for 3 pigeons during habituation to the chamber. Symbols represent relative frequencies of various activities; curved lines represent best fitting power functions. Right column: Boltzmann curves for the same pigeons in a later condition of periodic feeding. Notice how the distributions for aroused animals steepen, indicating a decrease in diversity. The figure is from Hanson (1991), and is reprinted with permission of Lawrence Erlbaum and Associates.

the behavior of flies, crabs, and children. He also found that the activation energy of behavior changed when periodic incentives were introduced into the environment. We shall return to the implications of this reordering in a subsequent section. In summary, we see that we may treat the categories of behavior as belonging on an interval scale, as required for our system. However, the positions of different types of behavior on that dimension may change in the context of different incentives, as those attract different actions differentially. When no strong incentives exist in the behavior space, there will be no strong gradients, and behavior will be more subject to small local attractors, evincing more variability over time in stimuli attended to and actions taken (e.g., McSweeney, 1974).

The negative exponential relation over activities developed by Hanson (1980, 1991) is known in physics as Boltzmann's law. Let us tentatively identify the slope of the curves in Figure 10 as a measure of motivation. The reciprocal of the slope is the temperature of the system; in statistical mechanics this parameter is simply the degrees Kelvin, and in neural models such as the Boltzmann machine it is called the computational temperature. The variable in the exponent of Boltzmann's law is the energy required to activate the action pattern. At higher temperatures the slope of the exponential function gets flatter, indicating that more patterns with higher activation energies will be found in the mixture, because there is more thermal energy available to activate them. This situation corresponds to a relatively flat potential surface, one of many small attractors but no big ones. At states of lowered motivation, many goals may be contemplated because the organism is not strongly captured by any of them. Creative exploration of one's repertoire requires a calm organism with no imperious attractors channeling its behavior. As motivation increases, motion becomes possible while diversity becomes limited; at highest motivation, only the most salient action is possible and behavior is impetuously channeled toward it. This trade-off between motivational energy and productive focusing of it is the basis of the Yerkes–Dodson principle. Maslow's hierarchy of needs is another ranking of the magnitude of the force of different incentives, with the satisfaction of the most powerful a precondition for the organism

to navigate through its behavior space toward more "transcendental" ones.

If lower activation-energy types of behavior are more stereotyped and the higher ones more diverse, considerations of entropy become relevant (and with that, consideration of Gibbs' "free energy" version of Boltzmann's law). Soon the hypotheses become so speculative they may strike skeptics more as fiction than as science. That is the fate of all novel formulations. Similar metaphors have proven useful in designing computer simulations of concept formation by "simulated annealing" (Killeen, 1989). In the present case, data can easily be generated to test and either reject or adapt and appropriate such theoretical structures.

Instinctive drift. The y axis of Figure 4 represents the activation energy of each of the types of behavior, with units at the bottom of the spectrum having the lowest activation energy, and thus being the most likely to occur, given the opportunity. Attempting to keep the organism at higher levels given the opportunity for actions at lower levels involves a precarious balancing act that requires opposing energy, just as it would require continual adjustments to keep a stick balanced on its end or a pigeon balanced midway between two piles of grain. Thorndike noticed this when he reinforced licking by cats to gain exit from a box, and found "a noticeable tendency . . . to diminish the act until it becomes a mere vestige . . . the licking degenerates into a mere quick turn of the head with one or two motions up or down with tongue extended" (Thorndike, 1911, p. 48). Breland and Breland (1961) also found that some carefully shaped performances inevitably degenerated into "misbehavior" characteristic of the lower activation levels; they called this proclivity "instinctive drift" (Boakes, Poli, Lockwood, & Goodall, 1978; Timberlake, Wahl, & King, 1982). We may interpret instinctive drift as a response to the force of an incentive that is propagated along the dimension of response topography.

Another instance of instinctive drift is the hoarding of tokens of reward when they could more expeditiously be traded for the real thing. Possession of the token may carry the organism closer to the incentive in behavior space (where distances are measured along the stimulus and action dimensions, as well as along the temporal dimension) than it can get by any other route, including the passing of time. Only when hoarding sufficiently delays reward that it debases the CS properties of the token will it be dropped (or when the deliveries of reward are more predictable, so that greater proximity is possible along the temporal dimension, or when actions or stimuli that carry the organism even closer to the incentive are made available).

As an incentive comes closer in space or time, the gradient of attraction becomes steeper, and types of behavior that have ever higher activation energies may be elicited. As one moves closer to a wild rat, its first response is freezing, then fleeing, then threatening, then attack (Blanchard, Flannelly, & Blanchard, 1986; Fanselow & Lester, 1988). Positive incentives also have a hierarchy of species-specific actions associated with them, including those known as adjunctive behavior; their ordering throughout a fixed interval may give us additional clues to their activation energies.

The Provenance of Forces

A careful reader will by now have asked how force can emanate from an incentive, because it is the animal that gives value to objects, and whose satiation, illness, or change of emotional state will devalue those objects. A novel and undiscovered incentive cannot attract until an animal apprehends it; it is the consummatory behavior that is released by the incentive that is behaviorally potent, not the incentive itself. The mode of speaking in this paper makes exposition easier; however, we may equally well argue that incentives are attracted to the animal: For every action (of the animal) there will be an equal but opposite reaction. A food cup will be attracted to a rat through the action of the rat on the substrate: The rat's running would accelerate the cup, and the alley containing it, through space to him, but for the alley's mass. That it is the rat that changes its position in absolute space the most is an accident of relative masses and is of no other theoretical or psychological significance.

Limits of Attraction

As an object, such as a cosmic ray, moves from outside the surface of the earth to inside it, there is a change in the direction of the forces. As the distance to the center of gravity approaches zero, the forces do not approach infinity, as naive application of the inverse square law would predict, but rather approach zero. Inside a hollow shell, the forces of electrostatic attraction from surface charge cancel, so forces are uniformly zero. Close to an atom the attractive electromagnetic forces are outweighed by the repulsive nuclear forces. There are similar limits to attraction along the dimensions of behavior:

Time. In Mazur's (1984) model, the fixed unit in the denominator ($c \cong 1$ s) also provides a limit on the force of attraction, a temporal shell around the incentive. Such boundaries on attractive forces should, if our unification is to be productive, predict boundaries on discriminations. This seems to be the case: Below 1 s, Weber's law ceases to hold for temporal discriminations (Allan, 1979; Fetterman & Killeen, 1992; Kristofferson, 1976), where accuracy approaches a uniform limen.

Stimuli. Discrimination that is sharp up to the boundaries of a unit and flat within the unit is the hallmark of categorical perception (Harnad, 1987; Wasserman, Kiedinger, & Bhatt, 1988).

Responses. The operant is held to be a set of actions such that reinforcement of any of its exemplars strengthens all members of that set equally (Schick, 1971).

Treatment in terms of the origins and limits of potential fields may throw new light on these boundary conditions.

Extended Events

The force of an incentive has been treated as though concentrated at the instant and at the locus of its delivery (even while recognizing the spatio-temporal shell around it). What if the incentive is extended in time? Compare the effects of 2 s of eating to 10 s of eating. The latter exerts a greater force over behavior, although not five times as great a force. The force that each instant of the incentive exerts over behavior can be calculated and then summed to predict the aggregate effect. If the inverse temporal gradient is the correct form, the first instant contributes a unit mass at a unit distance (plus the distance to the shell), the second a unit mass at two units of distance, the nth a unit mass at n units of distance. The aggregate force is the integral of $1/(d + c)$ with respect to d, where d is the duration of the reward and c the radius of the shell. This integral is $\ln(d + c)$. Thus, the most obvious extension of the model predicts that the reinforcing strength of an incentive (i.e., the force of attraction it exerts; the depth of its potential well) is proportional to the logarithm of its duration, a conclusion that is in qualitative agreement with the facts. A similar extension of the exponential gradient model predicts a cumulative exponential relation between the duration of an incentive and its attractive force: $F = 1 - e^{-\lambda(d+c)}$. This model also agrees with the data, perhaps somewhat more closely than the logarithmic function (Killeen, 1985).

Animals responding on a schedule in which food is presented periodically show a Gaussian distribution of their response rates centered near the expected time of reinforcement. The traditional way of treating this is as an exact temporal location of which the animal is only approximately aware. Another way is to treat the incentive as diffused along the temporal dimension. Approach to this region of space accelerates until the organism enters the probability "cloud" of the incentive, whereupon the attraction begins to smoothly decrease until the organism passes its center of gravity, and the process reverses itself. What must be the nature of the diffusion, in conjunction with a standard gradient of the force over the temporal dimension, to give the observed response properties? Do similar analyses hold for extended stimuli and extended response topographies? Yet more problems that are set by this calculus of behavior.

Organisms themselves are extended entities: Incentives not only translate them through their behavioral space, they also exert a rotational torque on them. This is familiar to anyone who has trained a pigeon to turn circles: The hard part is the first 180°, where one is working against the torque; the last half is an automatic and enthusiastic slide down the gradient created by the work of the first half. As an extended, polarized entity, some ends of an organism are attracted to some incentives, others to other incentives. Skinner speaks of "self" control, in which one part of us controls another by putting our hands in our pockets to stop fidgeting, biting our tongue to thwart speech, and so on. For elementary purposes, an organism can be treated as a point particle; its systematic treatment as an extended body remains for the future.

CONDITIONING

Just as it is reinforcing to get closer to an incentive in space and in time, it is reinforcing to get closer to it in action—to move toward the bottom of Figure 4. Activities near the bottom of the axis will increase the probability of those above that lead to them. In Premack's (1965) terms, higher probability behavior reinforces lower probability behavior. But reflexive actions such as chase cannot just happen; they require stimulus releasers such as the sight of prey. Fortunately, instrumental behavior such as search often uncovers the necessary releasers; the nervous system of species that have survived sees to this. Under intense motivation, "vacuum" activities—reflexive behavior absent the typical releasing stimuli—sometimes occur, but these are rare; nervous systems that regularly permitted such gratuitous hedonism did not endure. The only reliable paths into the potential wells of reflexive actions are through the channels of releasing stimuli. Behavior that is successful in uncovering releasers is reinforced by the revealed proximity to behavior with lower activation energies, which in turn is closer along the response dimension to the nominal incentive. Thus our potential wells, when sliced across the behavioral dimension, are a series of terraces, with variants of behavior on one tier that lead over the lip to the next being the versions that are strengthened by the ensuing reduction in potential.

Reflexes. Because a stimulus may release or elicit a response does not guarantee that doublet a place on the activation hierarchy. "Certain simple reflexes are extremely difficult to condition. The abdominal, patellar, plantar and pupillary reflexes fall in this category" (Kimble, 1961, p. 51). A knee-jerk response is the paragon of unconditionable behavior. Kimble speculates that the reason these reflexes are not conditionable is that they are not centrally involved in the processes of motivation and reward. We may speculate that the process of evolution has disfavored organisms whose legs were easily conditioned to jerk. The important distinction becomes not one between operant and respondent, but one between reflexes whose force is heritable (i.e., is an incentive or disincentive) and those whose force is restricted to the eliciting stimuli.

Deprivation. In Figure 10 action patterns were scaled in terms of their activation energies in a relatively homogenous environment. This will tell us only about the ordering of a restricted part of the full range of activities available to the organism (say, those near the top of a diagram such as Figure 4). Because different actions run their course at different rates (i.e., their wells are more or less extended along the temporal dimension) and require different releasing stimuli to occur, observation of unconstrained behavior will take us just so far. Premack (1965) also recognized this when he qualified his principle by requiring that probabilities be measured in the context of realistic schedules of availability. Actions that have an appreciable probability of occurrence do so either because they are reinforcing in their own right—they are unconditioned attractors—or because they have been conditioned as part of a trajectory leading to an unconditioned attractor. In the former case, restricting access to an action holds the animal out of its well and thus generates a potential to bypass that restriction by moving behavior along existing trajectories (performance) or by forming new trajectories (learning), as in the frustration-induced variation in existing trajectories. But restricted wells change their depth over time—hunger deepens relatively uniformly, with some cyclicity, up to the point of extreme privation; sexual desire increases steeply at first, and then levels off and becomes shallower over time; different drugs each have their own time course. This changing topography will reorder activation energies, making one day's play the next day's work, and opening windows for commitments to future incentives that close irrevocably as the passage of time moves us under the thrall of other attractors.

Generalized sign-tracking. This picture of conditioning thus generalizes the concept of sign-tracking to all coordinates of behavioral space. A stimulus, time, or action that has been associated with a reduction in potential energy itself becomes an attractor, in just the same way that a signpost on an uncertain path is an attractor (Moore & Stickney, 1982; Rescorla, 1987). Just as animals move toward places associated with an incentive at some point in time, they move toward response topographies associated with that incentive-time. The process of this conferred attraction is what we have historically called conditioning. Seen in terms of a behavior space, we give it new meaning. The presence of an incentive generates a force field called drive; movement toward the incentive converts the potential energy into kinetic energy and thus may

accomplish work. This work is the attentional and associative process known as conditioning/reinforcement. As long as there is a more direct route through behavior space to the incentive, there is potential for additional conditioning. And now the concept of *direct* must be understood in reference to all dimensions of behavior space: The most direct temporal route may not be the most direct in space-time, and the introduction of sign stimuli or the opportunity for low-activation behavior may again bend the path through other dimensions (Bowe, Green, & Miller, 1987). Formulating theories of optimal performance that attend to only the temporal dimension is like doing geometry using only a straight edge.

Potential reduction, drive induction. Note that this not a theory of drive reduction. It does not identify the satisfaction of basic needs (such as hunger) as the cause of learning, nor drive as a state corresponding to a physiological need, nor to its reduction as the mechanism of learning. Conditions such as hunger are not drives, but rather preconditions for certain stimuli to be incentives, and thus function more like emotions. Drive is a force through behavior space issuing from incentives. It causes animals to approach the incentive (performance) and may change the location of the trajectory (learning). In approaching the incentive, the force (drive) actually increases; the potential is reduced, and it is this that maintains performance and may bring about learning if the current path gets the animal to the incentive more directly (in behavior space) than alternate paths. Whether or not the physiological need is then reduced may or may not affect learning; it will affect subsequent performance insofar as it affects motivation. If the incentive is never consummated, new trajectories may form to bypass conditioned incentives that are thus put in extinction. If the incentive is devalued—that is, the potential field containing it is flattened, either through satiation or conditioned aversion—responses in the chain will languish.

Geodesics. Animals are driven to incentives along trajectories in behavior space. The shortest path between two points in space is called a *geodesic*. Whereas conditioning is made possible by the energy liberated when a trajectory settles into a geodesic, conditioning in turn makes the geodesic a stable trajectory for the organism (Killeen, 1991b). Any shortening of the trajectory to an incentive has the potential to engender conditioning. But, because there are no shorter trajectories than a geodesic, the possibility of further conditioning decreases as the path more closely approximates the geodesic (when, in the terms of associative conditioning, the associative strength of a reinforcer has been fully allocated).

Just as it is not drive reduction that causes conditioning, neither is it surprise. One is naturally surprised when one slips over an edge down a gradient, but it is not the surprise that causes conditioning, as much as that is the earmark of an abrupt potential change that is itself the agent of conditioning.

Path lengths. A path that leads to an incentive in one unit of time is shorter (steeper) than a path that leads to the incentive in two units of time. Because the origin of the path is thus closer to the incentive, behavior is differentially attracted to this path. Thus, the shorter the delay and the greater the rate of reinforcement, the greater the ability to condition. But distance in behavior space is measured not just in time, but also in stimulus-time-action. The more stimulus support, the greater the ability to condition. The more appropriate the physical arrangements, the greater the ability to condition. The more appropriate the response units required, the greater the ability to condition. The more appropriate the response units required, the greater the ability to condition. The attraction of an incentive is greatest when propagated along the ("psychologically") shortest path to the organism; it is the variation of paths and the selection toward this geodesic that are the essence of learning.

Variability and the slope of the well. We have established that behavioral forces must increase with proximity to the incentive. The greatest opportunity for the work of conditioning is therefore closest to the incentive. Behavior proximate in the activation spectrum, stimuli proximate in similarity, and locations proximate in space, in conjunction with proximate times, are the most readily conditioned. At more distant areas of the behavior space, the forces are weak, and therefore the differences in the potential of trajectories that lead to the incentive are small. At these distances, there is inadequate energy to stabilize trajectories and therefore high variability in location and topography, as Hanson (1991) showed by measuring the entropy of behavior at various temporal distances from an incentive, and as

Nevin and associates showed in analysis of chained schedules (Nevin, Mandell, & Yarensky, 1981). Just as a river may meander over a plain to eventually cascade through a gorge, we see meanders in behavioral trajectories that are remote in place, time, or topography from an incentive, and an energetic and canalized execution close to the goal.

Contiguity and contingency. Drive, and thus potential reduction, increases most steeply near the incentive; this is why contiguity—proximity to the incentive along all dimensions—is so important: Small delays of reinforcement will undermine conditioning unless a shortcut is established through other dimensions, such as through a conditioned reinforcer. Contingency—the probabilistic relationship between a stimulus or response and incentive—is as important as contiguity in conditioning. The present treatment does not set these factors in opposition or try to reduce one to the other, but makes clear the distinctive roles such factors must play in the control of behavior.

Let us call emission of the response or attention to the stimulus that is being conditioned the *target trajectory*. The effects of varying contiguity may be ascertained once the target trajectory and the potential function along it are identified for that situation. Contingency may be manipulated by varying the probability of reinforcement outside the target trajectory (in the background; e.g., Dickinson & Charnock, 1985), that is, by creating multiple attractors in the landscape. Presenting incentives in the background lowers the potential field, and thus leads the trajectories of attention and behavior away from the target trajectory. Insofar as the background shares features with the target trajectory, the ambient potential (the lip around the potential well of the target) will be lowered, leaving less energy available for the work of conditioning the target trajectory.

Contingency may also be manipulated by reinforcing the target trajectory probabilistically. Just how this weakens the motivation along the trajectory has not yet been determined. If in general conditioning works through the back-propagation of the force of the incentive along the trajectory, as we have argued here, extinction may work by adding a random component to that vector, permitting obsolete extinguished paths to be left for more fruitful ones. Probabilistic extinction (partial reinforcement) will then condition a sheaf of paths through behavior space that terminate in the consequential region. Partially reinforced performance may be resistant to extinction because each of these parallel paths must be extinguished before attraction to that region ceases.

In both cases, these ways of manipulating contingency reduce the potential drop between the ambient landscape and the incentive at the end of the target trajectory, although the mechanisms differ (Galbicka & Platt, 1984, among others, come to analogous conclusions). Manipulating contiguity through delay of reinforcement reduces the potential drop by removing the target points on the trajectory back from the steepest part of the potential well of the incentive; this is different than reducing potential drop by debasing contingency, and we should be able to find distinctive side effects of each, beyond the reduced ability to condition a stimulus or response that all such manipulations share (Reed & Reilly, 1990). Approaches such as the comparator hypothesis of conditioned associations (e.g., R. Miller & Matzel, 1988) and attendant data will help to clarify further these interrelations, and may lead to an account that integrates the process of association in both Pavlovian and operant procedures (B. Williams, 1989).

Emotions. The emotions are an important class of conditioned responses. But anger, fear, and love are not just positions along the response dimension, nor are they attractors themselves; they change the topography of the behavior space. An organism in the predation mode pictured in Figure 4 may shift to a fear/escape mode upon the sight or sniff of its own predator. Other places, hiding holes, and redoubts immediately become more attractive. This may be the proper interpretation of the hypothetical categories that extend to the left of the action patterns in that figure: They identify the action hierarchy that is established by a change of emotional states.

Aversive control. Our treatment has largely been in terms of positive incentives; a complementary theory is necessary for aversive incentives. We may wish to treat them as repellers, but a good case has been made that they work through the establishment of an *Umwelt* in which signs of safety and relaxation become attractors (Denny, 1991; McAllister & McAllister, 1991). When there is no geo-desic established to avoid shock but a high drive to do so, performance will wander through a variety of trajectories, giving the appearance of agitation. As behavior settles into a geodesic, we expect responses to become canalized, even if that is in a nonoptimal canal such as the defensive postures of learned helplessness. Generating fear in an environment in which species-specific defense responses are precluded elevates the potential for action without providing a gradient along which it may be reduced in an adaptive trajectory of action.

DYNAMICS

Motion. Suppose we were to code the above considerations into a computer program and give organisms an initially random trajectory through behavior space. No motion or learning would occur. It is common to prime performance with free reinforcers, or to magazine train the subject; this engenders the forces that will get the animal moving. It has been said that animals are motivated not so much because we deprive them but because we feed them—*L'appetite vient en mangeant.* In earlier articles I called this drive *arousal* (Killeen, 1975; Killeen et al., 1978). Our model also needs machinery for variation, perturbations of the trajectories at various points that will stimulate the search for more direct routes to the goal. It is in an aroused organism that energy is available to stimulate high activation-energy behavior, and in which error may begin to play its creative role—lapses of attention, slips of action, adjunctive behavior, vacuum activity, miscues, and bad timing may all lead to better trajectories. The distraught activity we find in a potential field when movement toward the incentive is blocked—frustration—increases the likelihood of such productive errors.

Getting stuck. Once the potential field is established, the trajectory will naturally move toward a geodesic, for our mechanics provides the necessary machinery of selection (sign-tracking in behavior space driven by potential reduction). But the organism may get stuck before it achieves the geodesic, for several reasons. A force is like a pull; it is not inexorable, and its effectiveness depends on the presence of obstacles and of other forces—wrinkles in behavior space—as well as the organism's momentum and history of conditioning (Stokes & Balsam, 1991). The closer a trajectory is to

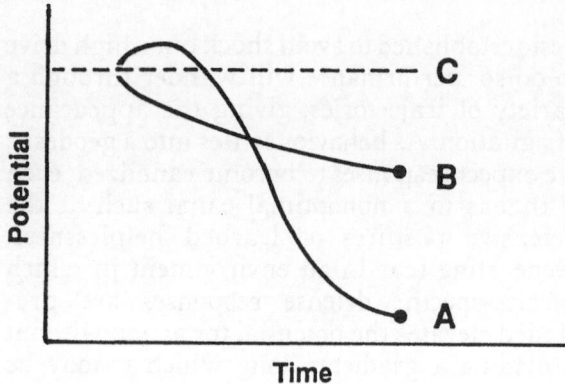

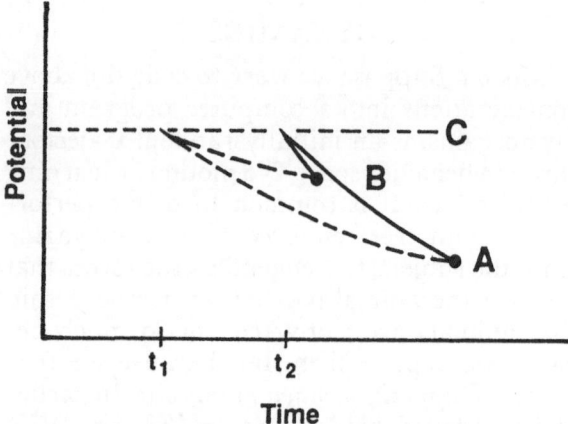

Fig. 11. Trajectories in a behavior space. *C* is the potential level; *B* is a weak attractor; *A* is a stronger attractor, as shown by its lower potential. Although the *x* axis is drawn as time, it may be any of the dimensions of behavior. Top: Choice of the path leading to *A* is made difficult by the energy required to carry the animal over the initial rise in that trajectory. The initial rise might be due to the need for involvement in a nonpreferred behavior, approach to a disliked stimulus and so forth. Motivational operations that shrink the potential axis (e.g., satiation of the organism) reduce the relative size of the potential hurdles and make diverse actions and trajectories possible. Bottom: At time t_2 the steepest gradient leads to the weaker attractor, making impulsive choice of it the geodesic. At time t_1 the steepest gradient leads to the stronger attractor. Committing an organism to a choice at t_1 will lead it to the deeper potential well, and is one technique of self-control.

the geodesic, the less likely that such variations will improve it further. It is this decreasing marginal utility of variation that makes conditioning proceed more slowly as it approaches asymptote, an insight captured by stimulus sampling theory and by the linear learning model at the core of the Rescorla–Wagner (1972) model. As the deviation from optimal approaches zero, so does the potential advantage of that move toward the geodesic, and so too does the probability that a random varia-

tion will be an improvement. As signposts to adequate routes attract the organism's attention, those increasingly familiar routes become the local minima. There will inevitably remain various stylistic differences between different performances when the potential advantage of further shifts in the trajectory becomes unrealizable because an adequate trajectory has become canalized. One of the unheralded advantages of being a "slow learner" may be the ability to avoid premature fixation on a suboptimal route through behavior space; evolutionary pressures against such fixation may have caused instrumental learning to generally run a slower course than found in imprinting and acquisition of phobias and taste aversions.

Another reason for suboptimal performance are traps in the behavior space, paths that have a lower potential than their nearest neighbors at some point, yet are thereafter much less direct than a geodesic (see Figure 11, top). Trajectories that start off well may wind up as dead ends, and the overall best trajectory may require seemingly prohibitive expenditure of effort, a kind of hill-climbing, at its start. Shock-maintained behavior (e.g., Galbicka & Platt, 1984) is but one of too many examples of the irrational press of incentives; irrational because a more direct overall route to one's goals can often be comprehended even as one is taking the immediately most attractive trajectory in a different direction. When the difference in outcomes is profound, we call such seemingly inexorable side-tracking "tragedy."

Shaping. It is possible to manipulate the system in ways that are more effective than random perturbation. Movement toward a geodesic may be effected by skilled shaping, whether by models that nature has evolved as signposts to incentives or by experimental psychologists. Shaping must respect the order of activation energies of the organism in the context of the relevant incentives, and draw the animal's attention toward the desired routes. Raising the criteria for performance too quickly may permit the organism to wander out of the shallow trajectory that has already been shaped. Savvy experimenters have developed a feel for this, and successful experimental programs have incorporated their tacit knowledge, both in software (e.g., percentile reinforcement schedules; see Galbicka, 1988) and hardware ("The analysis of learning has been divided into two parts, the principles of learning stated in textbooks, and the species-typical qualities of learning addressed in the design of the apparatus and procedures": Timberlake & Lucas, 1990, p. 240; "It is not so much that the rat learns to adapt to the apparatus we put it in, as it is that our apparatus has gradually evolved to suit itself to the rat's motor capabilities": Bolles, 1983, p. 43).

The form of the geodesic. Calculating where a curve is at its maximum or minimum is a problem for the differential calculus. Calculating what curve minimizes certain properties (e.g., what shape of ramp will get a rolling ball to the bottom in the minimum time) is a problem for the calculus of variations. Hanson (1977) used this technique to predict the changes in activity during a CS, based on the assumptions that each temporal epoch could be conditioned to the next, and the changes in activity from one epoch to the next followed a trajectory that minimized surprise. At this point, unfortunately, the uncertainty of the assumptions we must make concerning the metrics of the axes (and the difficulty of the calculations!) makes the mathematical specification of geodesics a nonelementary problem for the mechanistic analysis of behavior. Even so, Hanson showed that such specification is feasible in certain well-defined situations, and his solution may be directly applicable to other dimensions once we have ordered them. Most importantly, the concept of a geodesic lays the groundwork for the most ambitious principle to which such a system may aspire:

The principle of least action. One of the grand generalizations of mechanics is the principle of least action, from which many of its more specific laws may be derived. This variational principle states that systems evolve so as to minimize a weighted sum of kinetic and potential energies. Is a principle of least action possible for our system of behavior? There are reasons to think that changes in what Nevin (Nevin, 1992; Nevin, Mandell, & Atak, 1983) calls behavioral momentum will provide a measure of kinetic energy. Successful formulation of such a principle would provide powerful constraints on trajectories and a firm conceptual foundation for a science of behavior.

Two-Body Problems

If it is difficult to study gravity using three bodies, it is impossible to study it with only one. Similarly, our best understanding of the dynamics of behavior—how forces effect movements through behavior space—comes from analysis of interactions of pairs of units and their effects on trajectories. These analyses do not assume that the forces are confined to two dimensions, but only that our analytical interests (and, insofar as possible, our experimental constraints) are restricted to these planes.

Two times. Differential effects are most easily measured as the tendency to take one path, initiated by a response in one place, relative to an equivalent response in a nearby place. These paths may subsequently lead through different stimuli, places, or behavior, or to incentives of different strengths, with each path of different average temporal length. There is a long history of research on intertemporal choice, both with animals and humans. A popular paradigm for studying the control by identical incentives at equivalent places and various delays is the concurrent variable-interval schedule of reinforcement. The gradients between trajectories are blurred by varying the delays so that the relative advantage may vacillate over time and the behavior will follow suit, exposing the animal to each of the alternatives and encouraging a graded preference for one or the other. This preparation has provided a fertile climate for the development of models of choice—more "phenomena" with which we may "investigate the forces of nature."

The relative law of effect predicts the asymptotic rate of a measured response to be proportional to the rate of reinforcement for it, divided by all other rates of reinforcement in the animal's environment. The rate of reinforcement provides an index of the slope of the gradient along the temporal dimension. It is only an index because it averages physical measurements (time) rather than the force-transforms of them. However, Killeen et al. (1978, Appendix) showed that if the transform is an exponential decay, then for randomly delivered reinforcers, the average magnitude of the force will be proportional to the average rate of reinforcement. Fantino's model of choice (and insofar as the formal properties of the models are the same—see Killeen & Fantino, 1990—mine also) invokes the average temporal distance to reward, which determines the average potential in the experimental context.

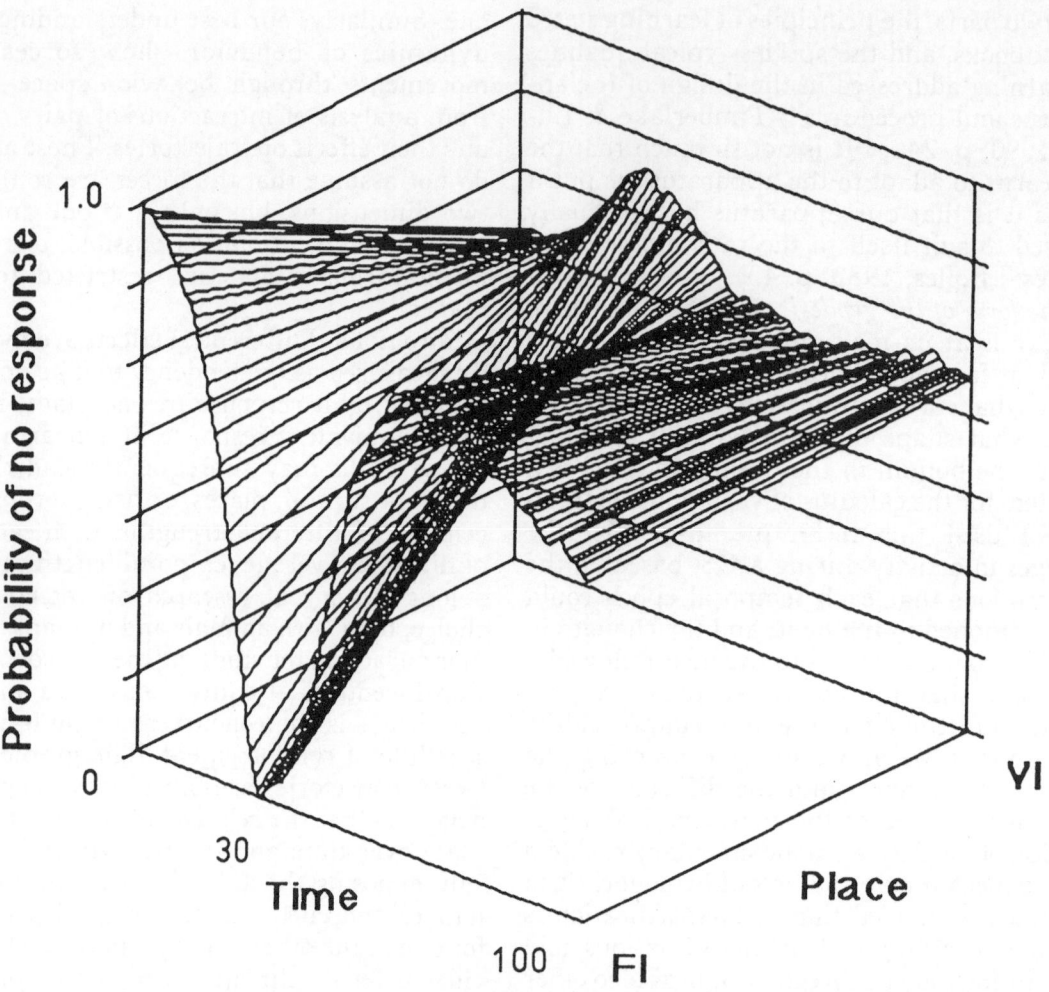

Fig. 12. Average data for 3 pigeons who received food with a probability of .50 after 30 s for a response to the FI key, and with a probability of .005 after every second for a response to the VI key. Nonreinforced trials ended with a blackout after 100 s. The vertical axis shows the probability of *not* making a response in any unit of time. A marble loosed on this surface and constrained only by the march of time would roll first to the VI side, then into the potential well of the FI, and then would be carried out by time back to the VI key. The data are from an unpublished study by the author.

Probability of choosing one alternative or the other is proportional to the depth of each alternative below this average.

Alternate sources of reward will pull the organism in other directions, reducing the potential for reinforcement of the measured response. For pigeons at least, "the alternative schedule chosen at any time is the one which offers the highest momentary reinforcement probability" (Hinson & Staddon, 1983, p. 25). The animals slide from one side to the other as each response briefly empties the potential well that the passage of time had recharged. Myerson (e.g., Myerson & Hale, 1988) has developed a kinetic model to govern such situations, one whose key assumptions are that

changes in the rate of switching from one trajectory to the other is a simple differential equation involving the rates of reinforcement along each trajectory and the rates of switching between trajectories (also see Bailey & Mazur, 1990). The sum of the rates of switching equals a constant, A, which he calls *attraction*, or directed arousal, and which he finds to be an increasing function of the overall rate of reinforcement in the context. We need merely understand his *attraction* as our *force* to see that a key model of movement between trajectories has been adumbrated in this work. It is closely related to the melioration approach, which has animals changing from one action to another whenever the likelihood of reinforcement is improved thereby (see Figure 12). Why should this potential reduction appear as the denominator in the relative law of effect, or as a power function of ratios of reinforcement rates in the generalized matching law, or as a difference in Fantino's model? Integration of these and other models may be facilitated by expressing them in the medium of behavioral mechanics.

When the different magnitudes of two incentives are played off against different delays, we have a paradigm for studying self-control (Green & Snyderman, 1980; Logue, 1988; Logue, Rodriguez, Peña-Correal, & Mauro, 1987; Rachlin & Green, 1972). Inserting a delay after a choice will flatten the gradient to the small incentive more than the large one, and bring behavior under the control of the deferred reward (see Figure 11). When a potent incentive comes after a less potent one, the effects on the trajectory to the earlier incentive will depend on whether the two exert a coordinated pull on the measured behavior, or whether the deeper basin around the larger deferred incentive leads behavior away from the control by the more immediate one. Flaherty and Rowan and Lucas and associates (Flaherty & Rowan, 1986; Lucas, Timberlake, Gawley, & Drew, 1990) review some of the literature on anticipatory contrast.

Two responses. Incentives exert their attraction on a range of component types of behavior, some of which may be performed in parallel, while others wait for the completion of predecessors or the uncovering of sign stimuli for their execution. Unfortunately, the development and blending of complex response topographies have until recently (Stokes & Balsam, 1991) been of greater interest to the trainer than the scientist. A rich field of motor control waits to be explored.

Two stimuli. Rescorla and Wagner's model of associative strengthening (Mackintosh, 1983; Pearce & Hall, 1980; Rescorla & Wagner, 1972) is a model of the potential of stimuli to lead to an incentive. The insights issuing from research on associative conditioning may be recast in terms of our behavioral mechanics. For example, once a path is established through a stimulus to an incentive, the stimulus forms part of the trajectory—and becomes a conditioned attractor. Once so established, it is difficult to move the trajectory through a new stimulus, especially if that offers no greater

reduction in potential. This phenomenon is termed *blocking*. A US presented just before another US provides a steep gradient to it along the temporal dimension (i.e., predicts it) and blocks paths through more remote parts of the stimulus/response/time space. At the heart of this theoretical approach is a linear learning model of the motion of the trajectory through the behavioral space. Associative strength in this model is the potential of the incentive, and the maneuver of summing the individual associative strengths to predict current overall associative strength is nothing more than asking how close do the two stimuli together bring the organism to the incentive along the stimulus dimension. Devaluing the reinforcer reduces its potential. Frey and Sears' (1978) addition of a dynamic attention rule formalizes the path dependence of trajectories through behavior space.

Although we expect blocking where one stimulus is well conditioned as part of the trajectory, if that stimulus is not yet well conditioned, or if it is faded out by reducing its salience, it is possible to transfer its control to a second stimulus. This can happen because all the other parts of the trajectory (e.g., the timing, the actions, the subsequent movement to consummate the primary incentive, etc.) have been moved toward their minimum, and the performance has achieved some momentum. One of the arts of training is the determination of when more is gained by conditioning the trajectory as a whole and when more is lost by overconditioning the stimulus that is eventually to be removed. If the original stimulus is never faded out, we expect no transfer of control to the new stimulus unless its character is innately associated with the incentive (e.g., as a red dot might be for the pecking response of a young gull).

Whereas behavioral mechanics will provide a broader context for the associative conditioning of stimuli, that research will itself serve as a model for conditioning along the other dimensions. For example, the force of incentives may be blocked not only by other stimuli but by other responses (B. Williams, 1975) and by other incentives (Catania, Sagvolden, & Keller, 1988); the decreasing marginal utility of extended incentives may be viewed as the overshadowing of attraction to the later epochs of the incentives by the earlier ones (Killeen & Smith, 1984); temporal control in general may be viewed as the overshadowing of distal epochs by proximal ones, or by the stimuli and responses contained therein.

Two incentives. What is the behavior of an organism in a field of force generated by two incentives? Much depends on whether the signs of the incentives are the same or different— that is, attractors or repellers. If both are attractors, then the resultant motion depends on the locations of incentives in behavior space. If they are different types of incentives, they will emanate different forces and attract through different actions. If they are at the same location in the spatio-temporal coordinates, their effects may add, but not completely, because the behavior they attract will not completely overlap, with each other or with the action (or set of actions) that we have chosen to measure (Ganesan & Pearce, 1988; Weiss, Schindler, & Eason, 1988). Furthermore, there may be competition among the action patterns that will make it difficult to predict the net effects. Thus water incentives may encourage topographies of behavior having aspects that are measured by our operandum but that are inconsistent with the aspects encouraged by food incentives. Water incentives may make the same spatial location an attractor at times when food reinforcement makes it a repeller (i.e., encourages search elsewhere).

Stimulus–response compatibility is a critical factor in establishing well-differentiated performance (Bowe, Miller, & Green, 1987). To a certain extent, these interactions may be incorporated within a framework of orthogonal axes for stimuli and responses, but adequately rich data may force the axes to nonorthogonality.

When different incentives or different locations of an incentive are correlated with different stimulus–response doublets, animals are better able to maintain good discriminative performance over long response–reinforcer delays. This is known as the *differential outcome effect,* and our knowledge of it is briefly reviewed and incremented by D. Williams, Butler, and Overmier (1990). These results suggest that the distinctiveness of two trajectories may be a function of their divergence in behavior space, and thus depend not only on the locations of the stimuli and responses but also on the locations of the incentives that constitute the terminus of the trajectories.

Two routes. If a path through one place is always steeper than that through another, then in the long run, and subject to an animal's ability to discriminate differences in gradients at various distances from the incentive, the animal will always choose the steepest descent. It is easy to manipulate steepness by varying the magnitude or delay of incentives. It can also be manipulated along the other dimensions (e.g., rats may find a path through a dark area or along a more tortuous route more attractive than a direct route through a brightly lit area; Timberlake, 1983).

When the route to the incentive is direct or well learned, increased motivation will get the animal to it more quickly. If the route is complicated or a new trajectory must be formed, high motivation may prematurely channelize behavior, locking organisms into nonoptimal trajectories, preclude them from reversing their path to a steeper descent, and possibly thwarting completion of the trajectory. Review the top of Figure 11; a relaxed organism might have a sufficiently flat potential surface to permit random perturbations to carry it over the hump to the deeper potential well, whereas the aroused organism may be forced headlong down the immediately more gratifying but ultimately inferior trajectory. A view of the incentive along one route may make that more attractive, even though it involves a greater distance along the temporal route than alternative paths (Chapuis, Thinus-Blanc, & Poucet, 1983), and even if it locks the animal into the local minimum of the *Umweg* problem.

Nonconservative Forces

It was earlier suggested that "noise" might be a productive part of the learning process, because it might liberate suboptimal trajectories to move closer to a geodesic and extinguished trajectories to wander out of the nonproductive part of behavior space. But such random effects are nonconservative, in that they dissipate energy in ways that cannot be recovered by simple retracing of the trajectories. This may not be as central a concern in psychology as it is in physics. But it leads to a phenomenon that has been inadequately studied by our field and provides one hypothetical perspective on it.

Consider organisms confronted with two paths whose temporal distances to reward are equal, and both of which contain one segmentation. On one of the paths the second segment provides two equal subpaths leading in parallel to the goal. What does this do to the attraction felt in the first segments of each? A simple dynamic model predicts no difference: The two subpaths will reduce the potential by the same amount, and the pigeons should be indifferent between the two trajectories. But consider what happens if we add dissipative forces to the model—if we assume that there is resistance, friction, reactance, along the paths. Two resistors in series will reduce the potential less than one of them by itself; we have seen that when a terminal link contains two stimuli in series, it is less attractive than a link containing only one stimulus. Two resistors in parallel will reduce the electrical potential more than one of them by itself. In like manner two stimuli in parallel may reduce the behavioral potential more than one by itself, and therefore be more attractive. Of course, unlike an electric current, an animal can choose only one of those stimuli and travel along one of those final paths at one time. In the case of behavior, the mechanism is probably the nonlinear concave relation between the frequency with which a stimulus is paired with an incentive and its conferred attractiveness. Distributing conditioning over two stimuli should then make the pair of them more attractive than a single, continuously reinforced stimulus. This is predicted by the Rescorla–Wagner (1972) model, and thus provides another point of contact between traditional theories and the current physicalistic reconstruction. In particular, the nonlinearity of the learning curve governing the movement of trajectories toward the geodesic introduces dissipative forces, and with them interesting if more difficult problems, such as path dependence.

A critical experiment by Catania (1975; see Kehoe, 1986, for comparable effects with CSs) shows that organisms do indeed prefer an alternative that gives them multiple subsequent paths to an incentive. This crucial study suggests that even after trajectories have been well learned, motion along them may be seen as encountering some resistance; providing multiple routes through alternate stimulus contexts, response topographies, or other dimensions will decrease the potential more than a single route can, and make that set of trajectories more attractive. It is for these reasons that Catania spoke of his work as showing that

organisms had a preference for keeping their options open, a preference for "freedom." Our mechanics of behavior may eventually provide a way to calculate just how much each new path will reduce the reactance to a set of trajectories, to calculate just what freedom means in terms of options for future action.

CONCLUSION

Incentives attract or repel organisms and thus change their motion. We have implicitly adopted Newton's second law of motion as our primary law of behavior: Change of motion is proportional to the resultant force, and in the direction in which that force is impressed. The resulting movement toward an incentive along spatial coordinates has been called sign-tracking. This article extends the concept of sign-tracking to a more general coordinate system that includes stimuli, actions, and time. The trajectory of motion through this extended coordinate system will shift with learning, as random variation steps the path closer to the optimal trajectory between starting point and incentive. Conditioning ends when variation can no longer shorten the path. This dynamic metaphor provides a pegboard for many of our empirical observations; as they become incorporated, they will exert constraints on other implications of the system, making predictions increasingly possible.

Some readers may object that such a treatment works for physics where there are real dimensions and forces, but for psychology it is a metaphor that will only frustrate our ultimate understanding of behavior in its own terms. But we understand nothing in its own terms. All understanding is a putting of things into other terms that we are comfortable with. We are content to say that reinforcement strengthens responses, yet we easily forget that is no less a metaphor than many of those that were presented in this paper. The forces that bend an organism toward an incentive are no less real than those that bend light toward the sun.

"A psychological system is an attempt to arrange and coördinate, in a logical and understandable fashion, the facts of the science into a meaningful and satisfying whole; to point to the weaknesses and gaps in our knowledge; and to show the way to future achievement" (Keller, 1937, p. 106). Behavioral mechanics is a protean system, constructed on a physicalistic framework that has the advantages of internal coherence, relative familiarity, and visualizability. It is little other than a working out of Skinner's definition of behavior as "the movement of an organism or of its parts in a frame or reference provided by the organism itself or by various external objects or fields of force." The exploration and calibration of the dimensions of that frame of reference, and the analysis of the fields of force, provide the next challenges in transforming this provisional approach into a psychological system of behavior.

REFERENCES

Allan, L. G. (1979). The perception of time. *Perception & Psychophysics, 26,* 340–354.

Bailey, J. T., & Mazur, J. E. (1990). Choice behavior in transition: Development of preference for the higher probability of reinforcement. *Journal of the Experimental Analysis of Behavior, 53,* 409–422.

Barlow, G. W. (1977). Modal action patterns. In T. A. Sebeok (Ed.), *How animals communicate* (pp. 98–134). Bloomington: Indiana University Press.

Blanchard, R. J., Flannelly, K. J., & Blanchard, D. C. (1986). Defensive behavior of laboratory and wild *Rattus norvegicus. Journal of Comparative Psychology, 100,* 101–107.

Blough, D. S. (1972). Recognition by the pigeon of stimuli varying in two dimensions. *Journal of the Experimental Analysis of Behavior, 18,* 345–367.

Boakes, R. A., Poli, M., Lockwood, M. J., & Goodall, G. (1978). A study of misbehavior: Token reinforcement in the rat. *Journal of the Experimental Analysis of Behavior, 29,* 115–134.

Bolles, R. C. (1975). *Theory of motivation* (2nd ed.). New York: Harper & Row.

Bolles, R. C. (1983). The explanation of behavior. *The Psychological Record, 33,* 31–48.

Bowe, C. A., Green, L., & Miller, J. D. (1987). Differential acquisition of discriminated autoshaping as a function of stimulus qualities and locations. *Animal Learning & Behavior, 15,* 285–292.

Bowe, C. A., Miller, J. D., & Green, L. (1987). Qualities and locations of stimuli and responses affecting discrimination learning of chinchillas (*Chinchilla laniger*) and pigeons (*Columba livia*). *Journal of Comparative Psychology, 101,* 132–138.

Breland, K., & Breland, M. (1961). The misbehavior of organisms. *American Psychologist, 16,* 681–684.

Brown, J. S. (1948). Gradients of approach and avoidance responses and their relation to level of motivation. *Journal of Comparative and Physiological Psychology, 41,* 450–465.

Catania, A. C. (1975). Freedom and knowledge: An experimental analysis of preference in pigeons. *Journal of the Experimental Analysis of Behavior, 24,* 89–106.

Catania, A. C., Sagvolden, T., & Keller, K. J. (1988). Reinforcement schedules: Retroactive and proactive effects of reinforcers inserted into fixed-interval performances. *Journal of the Experimental Analysis of Behavior, 49,* 49–73.

Chapuis, N., Thinus-Blanc, C., & Poucet, B. (1983). Dissociation of mechanisms involved in dogs' oriented displacements. *Quarterly Journal of Experimental Psychology, 35B,* 213–220.

Cheng, K. (1986). A purely geometric module in the rat's spatial representation. *Cognition, 23,* 149–178.

Cofer, C. N., & Appley, M. H. (1967). *Motivation: Theory and research.* New York: Wiley.

Craig, W. (1918). Appetites and aversions as constituents of instincts. *Biology Bulletin, 34,* 91–107.

Crossman, E. K., & Nichols, M. B. (1981). Response location as a function of reinforcement location and frequency. *Behaviour Analysis Letters, 1,* 207–215.

Dawkins, R. (1976). Hierarchical organisation: A candidate principle for ethology. In P. P. G. Bateson & R. A. Hinde (Eds.), *Growing points in ethology* (pp. 7–54). New York: Cambridge University Press.

Denny, M. R. (1991). Relaxation/relief: The effects of removing, postponing, or terminating aversive stimuli. In M. R. Denny (Ed.), *Fear, avoidance and phobias* (pp. 199–229). Hillsdale, NJ: Erlbaum.

Dickinson, A., & Charnock, D. J. (1985). Contingency effects with maintained instrumental reinforcement. *Quarterly Journal of Experimental Psychology, 37B,* 397–416.

Duncan, B., & Fantino, E. (1972). The psychological distance to reward. *Journal of the Experimental Analysis of Behavior, 18,* 23–34.

Fanselow, M. S., & Lester, L. S. (1988). A functional behavioristic approach to aversively motivated behavior: Predatory imminence as a determinant of the topography of defensive behavior. In R. C. Bolles & M. D. Beecher (Eds.), *Evolution and learning* (pp. 185–212). Hillsdale, NJ: Erlbaum.

Fentress, J. C. (1981). Network foci in integrated action: Units or something else? *Behavioral and Brain Sciences, 4,* 623–624.

Fetterman, J. G., & Killeen, P. R. (1992). Time discrimination in *Columba livia* and *Homo sapiens. Journal of Experimental Psychology: Animal Behavior Processes, 18,* 80–94.

Flaherty, C. F., & Rowan, G. A. (1986). Successive, simultaneous, and anticipatory contrast in the consumption of saccharin solutions. *Journal of Experimental Psychology: Animal Behavior Processes, 12,* 381–393.

Frey, P. W., & Sears, R. J. (1978). Model of conditioning incorporating the Rescorla-Wagner associative axiom, a dynamic attention process, and a catastrophe rule. *Psychological Review, 85,* 321–340.

Galbicka, G. (1988). Differentiating *The Behavior of Organisms. Journal of the Experimental Analysis of Behavior, 50,* 343–354.

Galbicka, G., & Platt, J. R. (1984). Interresponse-time punishment: A basis for shock-maintained behavior. *Journal of the Experimental Analysis of Behavior, 41,* 291–308.

Gallistel, C. R. (1981). Précis of Gallistel's *The Organization of Action*: A new synthesis. *Behavioral and Brain Sciences, 4,* 609–650.

Gallistel, C. R. (1990). *The organization of learning.* Cambridge, MA: MIT Press.

Ganesan, R., & Pearce, J. M. (1988). Interactions between conditioned stimuli for food and water in the rat. *Quarterly Journal of Experimental Psychology, 408,* 229–241.

Garcia, J., McGowan, B. K., & Green, K. F. (1972). Biological constraints on conditioning. In H. Black &

W. F. Prokasy (Eds.), *Classical conditioning II: Current research and theory* (pp. 3–27). New York: Appleton-Century-Crofts.

Gibbon, J. (1986). The structure of subjective time: How time flies. In G. H. Bower (Ed.), *The psychology of learning and motivation* (pp. 105–135). New York: Academic Press.

Gould, J. L., & Marler, P. (1984). Ethology and the natural history of learning. In P. Marler & H. S. Terrace (Eds.), *The biology of learning* (pp. 47–74). New York: Springer-Verlag.

Green, L., & Snyderman, M. (1980). Choice between rewards differing in amount and delay: Toward a choice model of self-control. *Journal of the Experimental Analysis of Behavior, 34,* 135–147.

Guttman, N., & Kalish, H. I. (1956). Discriminability and stimulus generalization. *Journal of Experimental Psychology, 51,* 79–88.

Hanson, S. J. (1977). *The Rescorla-Wagner model and the temporal control of behavior.* Unpublished master's thesis, Arizona State University, Tempe.

Hanson, S. J. (1980). *Studies of diversity in the pigeon.* Unpublished doctoral dissertation, Arizona State University, Tempe.

Hanson, S. J. (1991). Behavioral diversity, search and stochastic connectionist systems. In M. L. Commons, S. Grossberg, & J. E. R. Staddon (Eds.), *Quantitative analysis of behavior: Neural network models of conditioning and action* (pp. 295–344). Hillsdale, NJ: Erlbaum.

Harnad, S. (1987). *Categorical perception: The groundwork of cognition.* New York: Cambridge University Press.

Hearst, E. (1965). Approach, avoidance, and stimulus generalization. In D. I. Mostofsky (Ed.), *Stimulus generalization* (pp. 331–355). Stanford, CA: Stanford University Press.

Hinson, J. M., & Staddon, J. E. R. (1983). Hill-climbing by pigeons. *Journal of the Experimental Analysis of Behavior, 39,* 25–47.

Hull, C. L. (1943). *Principles of behavior.* New York: Appleton-Century-Crofts.

Hull, C. L. (1952). *A behavior system.* New Haven, CT: Yale University Press.

Iversen, I. H., Ragnarsdottir, A. G., & Randrup, K. I. (1984). Operant conditioning of autogrooming in vervet monkeys (*Cercopithecus aethiops*). *Journal of the Experimental Analysis of Behavior, 42,* 171–189.

Kehoe, E. J. (1986). Summation and configuration in conditioning the rabbit's nictitating membrane response to compound stimuli. *Journal of Experimental Psychology: Animal Behavior Processes, 12,* 186–195.

Keller, F. S. (1937). *The definition of psychology.* New York: Appleton-Century-Crofts.

Killeen, P. (1974). Psychophysical distance functions for hooded rats. *The Psychological Record, 24,* 229–235.

Killeen, P. R. (1975). On the temporal control of behavior. *Psychological Review, 82,* 89–115.

Killeen, P. R. (1982). Incentive theory II: Models for choice. *Journal of the Experimental Analysis of Behavior, 38,* 217–232.

Killeen, P. R. (1984). Incentive theory III: Adaptive clocks. In J. Gibbon & L. Allan (Eds.), *Timing and time perception* (pp. 515–527). New York: New York Academy of Sciences.

Killeen, P. R. (1985). Incentive theory IV: Magnitude of reward. *Journal of the Experimental Analysis of Behavior, 43,* 407–417.

Killeen, P. R. (1989). Behavior as a trajectory through a field of attractors. In J. R. Brink & C. R. Haden (Eds.), *The computer & the brain: Perspectives on human and artificial intelligence* (pp. 53–82). Amsterdam: Elsevier.

Killeen, P. R. (1991a). Behavior's time. In G. H. Bower (Ed.), *The psychology of learning and motivation* (Vol. 27, pp. 295–334). New York: Academic Press.

Killeen, P. R. (1991b). Behavioral geodesics. In D. S. Levine & J. S. Levin (Eds.), *Motivation, emotion, and goal direction in neural networks* (pp. 91–114). Hillsdale, NJ: Erlbaum.

Killeen, P. R., & Amsel, A. (1987). The kinematics of locomotion toward a goal. *Journal of Experimental Psychology: Animal Behavior Processes, 13,* 92–101.

Killeen, P. R., & Fantino, E. (1990). Unification of models for choice between delayed reinforcers. *Journal of the Experimental Analysis of Behavior, 53,* 189–200.

Killeen, P. R., & Fetterman, J. G. (1988). A behavioral theory of timing. *Psychological Review, 95,* 274–295.

Killeen, P. R., Hanson, S. J., & Osborne, S. R. (1978). Arousal: Its genesis and manifestation as response rate. *Psychological Review, 85,* 571–581.

Killeen, P. R., & Riggsford, M. (1989). Foraging by rats: Intuitions, models, and data. *Behavioral Processes, 19,* 95–105.

Killeen, P. R., & Smith, J. P. (1984). Perception of contingency in conditioning: Scalar timing, response bias, and the erasure of memory by reinforcement. *Journal of Experimental Psychology: Animal Behavior Processes, 10,* 333–345.

Kimble, G. A. (1961). *Hilgard and Marquis' conditioning and learning* (2nd ed.). New York: Appleton-Century-Crofts.

Klein, S. B., & Mowrer, R. R. (1990). *Contemporary learning theories: Instrumental conditioning theory and the impact of constraints on learning.* Hillsdale, NJ: Erlbaum.

Krebs, J. R. (1973). Behavioral aspects of predation. In P. P. G. Bateson & P. H. Klopfer (Eds.), *Perspectives in ethology* (pp. 73–111). New York: Plenum.

Kristofferson, A. B. (1976). Low-variance stimulus response latencies: Deterministic internal delays? *Perception & Psychophysics, 20,* 89–100.

Lejeune, H. (1990). Timing: Differences in continuity or generality beyond differences? In D. E. Blackman & H. Lejeune (Eds.), *Behavior analysis in theory and practice* (pp. 53–90). Hillsdale, NJ: Erlbaum.

Lewin, K. (1933). Environmental forces. In C. Murchison (Ed.), *A handbook of child psychology* (2nd ed., pp. 590–625). Worcester, MA: Clark University Press.

Locurto, C. M., Terrace, H. S., & Gibbon, J. (1981). *Autoshaping and conditioning theory.* New York: Academic Press.

Logue, A. W. (1988). Research on self-control: An integrating framework. *Behavioral and Brain Sciences, 11,* 665–709.

Logue, A. W., Rodriguez, M. L., Peña-Correal, T. E., & Mauro, B. C. (1987). Quantification of individual differences in self-control. In M. L. Commons, J. E. Mazur, J. A. Nevin, & H. Rachlin (Eds.), *Quantitative analyses of behavior: The effect of delay and of intervening events on reinforcement value* (pp. 245–265). Hillsdale, NJ: Erlbaum.

Lucas, G. A., Timberlake, W., Gawley, D. J., & Drew, J. (1990). Anticipation of future food: Suppression and facilitation of saccharine intake depending on the delay and type of future food. *Journal of Experimental Psychology: Animal Behavior Processes,* **16,** 169–177.

Lundberg, U., Bratfisch, O., & Ekman, G. (1971). *Emotional involvement and subjective distance: A summing up of the experiments and a final test of the inverse square root law.* Stockholm: The Psychological Laboratories, The University of Stockholm.

Mackintosh, N. J. (1983). *Conditioning and associative learning.* New York: Oxford University Press.

Marler, P. R., Dooling, R. J., & Zoloth, S. (1980). Comparative perspectives on ethology and behavioral development. In M. H. Bornstein (Ed.), *Comparative methods in psychology* (pp. 189–230). Hillsdale, NJ: Erlbaum.

Mazur, J. E. (1984). Tests of an equivalence rule for fixed and variable delays. *Journal of Experimental Psychology: Animal Behavior Processes,* **10,** 426–436.

McAllister, D. E., & McAllister, W. R. (1990). Fear theory and aversively motivated behavior: Some controversial issues. In M. R. Denny (Ed.), *Fear, avoidance, and phobias* (pp. 135–163). Hillsdale, NJ: Erlbaum.

McSweeney, F. K. (1974). Variability of responding in a concurrent schedule as a function of body weight. *Journal of the Experimental Analysis of Behavior,* **21,** 357–359.

Miller, N. E. (1971). *Neal E. Miller: Selected papers on conflict, displacement, learned drives and theory.* New York: Aldine/Atherton.

Miller, R. R., & Matzel, L. D. (1988). The comparator hypothesis: A response rule for the expression of associations. In G. H. Bower (Ed.), *The psychology of learning and motivation* (pp. 51–92). New York: Academic Press.

Moore, B. W. (1973). The role of directed Pavlovian reactions in simple instrumental learning in the pigeon. In R. A. Hinde & J. Stevenson-Hinde (Eds.), *Constraints on learning: Limitations and predispositions* (pp. 159–188). New York: Academic Press.

Moore, J. W., & Stickney, K. J. (1982). Goal tracking in attentional-associative networks: Spatial learning and the hippocampus. *Physiological Psychology,* **10,** 202–208.

Myerson, J., & Hale, S. (1988). Choice in transition: A comparison of melioration and the kinetic model. *Journal of the Experimental Analysis of Behavior,* **49,** 291–302.

Nevin, J. A. (1992). An integrative model for the study of behavioral momentum. *Journal of the Experimental Analysis of Behavior,* **57,** 301–316.

Nevin, J. A., Mandell, C., & Atak, J. R. (1983). The analysis of behavioral momentum. *Journal of the Experimental Analysis of Behavior,* **39,** 49–59.

Nevin, J. A., Mandell, C., & Yarensky, P. (1981). Response rate and resistance to change in chained schedules. *Journal of Experimental Psychology: Animal Behavior Processes,* **7,** 278–294.

Newton, I. (1934). *Philosophiæ naturalis principia mathematica* (3rd ed., F. Cajori, rev.). Berkeley, CA: University of California Press. (Original work published 1687)

Palya, W. L., & Bevins, R. A. (1990). Serial conditioning as a function of stimulus, response, and temporal dependencies. *Journal of the Experimental Analysis of Behavior,* **53,** 65–85.

Panksepp, J. (1989) The neurobiology of emotions: Of animal brains and human feelings. In H. L. Wagner & A. Manstead (Eds.), *Handbook of social psychophysiology* (pp. 5–26). New York: Wiley.

Pear, J. J. (1985). Spatiotemporal patterns of behavior produced by variable-interval schedules of reinforcement. *Journal of the Experimental Analysis of Behavior,* **44,** 217–231.

Pearce, J. M., & Hall, G. (1980). A model of Pavlovian learning: Variations in the effectiveness of conditioned but not of unconditioned stimuli. *Psychological Review,* **87,** 532–552.

Peden, B. F., Browne, M. P., & Hearst, E. (1977). Persistent approaches to a signal for food despite food omission for approaching. *Journal of Experimental Psychology: Animal Behavior Processes,* **3,** 377–399.

Pellionisz, A. J. (1989, June). Neural geometry. In *Proceedings of the International Joint Conference on Neural Networks.* Washington, DC: IEEE.

Perone, M., & Courtney, K. (1992). Fixed-ratio pausing: Joint effects of past reinforcer magnitude and stimuli correlated with upcoming magnitude. *Journal of the Experimental Analysis of Behavior,* **57,** 33–46.

Platt, J. R. (1984). Motivational and response factors in temporal discrimination. In J. Gibbon & L. Allan (Eds.), *Timing and time perception* (pp. 200–210). New York: New York Academy of Sciences.

Premack, D. (1965). Reinforcement theory. In D. Levine (Ed.), *Nebraska symposium on motivation* (pp. 123–180). Lincoln: University of Nebraska Press.

Rachlin, H., & Green, L. (1972). Commitment, choice and self-control. *Journal of the Experimental Analysis of Behavior,* **17,** 15–22.

Rachlin, H., Raineri, A., & Cross, D. (1991). Subjective probability and delay. *Journal of the Experimental Analysis of Behavior,* **55,** 233–244.

Reed, P., & Reilly, S. (1990). Context extinction following conditioning with delayed reward enhances subsequent instrumental responding. *Journal of Experimental Psychology: Animal Behavior Processes,* **16,** 48–55.

Rescorla, R. A. (1987). A Pavlovian analysis of goal-directed behavior. *American Psychologist,* **42,** 119–129.

Rescorla, R. A., & Cunningham, C. L. (1979). Spatial contiguity facilitates Pavlovian second-order conditioning. *Journal of Experimental Psychology: Animal Behavior Processes,* **5,** 152–161.

Rescorla, R. A., & Furrow, D. R. (1977). Stimulus similarity as a determinant of Pavlovian conditioning. *Journal of Experimental Psychology: Animal Behavior Processes,* **3,** 203–215.

Rescorla, R. A., & Wagner, A. R. (1972). A theory of Pavlovian conditioning: Variations in the effectiveness of reinforcement and nonreinforcement. In A. H. Black & W. F. Prokasy (Eds.), *Classical conditioning II: Current research and theory* (pp. 64–99). New York: Appleton-Century-Crofts.

Revusky, S. (1977). Learning as a general process with an emphasis on data from feeding experiments. In N. W. Milgram, L. Krames, & T. M. Alloway (Eds.), *Food aversion learning* (pp. 1–51). New York: Plenum Press.

Schick, K. (1971). Operants. *Journal of the Experimental Analysis of Behavior,* **15,** 413–423.

Schmajuk, N. A. (1990). Role of the hippocampus in temporal and spatial navigation: An adaptive neural network. *Behavioral and Brain Research,* **39,** 205–229.

Schnierla, T. C. (1959). An evolutionary and developmental theory of biphasic processes underlying approach and withdrawal. In M. R. Jones (Ed.), *Nebraska symposium on motivation* (pp. 1–42). Lincoln: University of Nebraska Press.

Seligman, M. E. P. (1970). On the generality of the laws of learning. *Psychological Review, 77,* 406–418.

Shepard, R. N. (1965). Approximation to uniform gradients of generalization by monotone transformations of scale. In D. I. Mostofsky (Ed.), *Stimulus generalization* (pp. 94–110). Stanford, CA: Stanford University Press.

Shepard, R. N. (1987a). Evolution of a mesh between principles of the mind and regularities of the world. In J. Dupré (Ed.), *The latest on the best: Essays on evolution and optimality* (pp. 251–275). Cambridge, MA: MIT Press/Bradford Books.

Shepard, R. N. (1987b). Toward a universal law of generalization for psychological science. *Science, 237,* 1317–1323.

Shettleworth, S. J. (1975). Reinforcement and the organization of behavior in golden hamsters: Hunger, environment, and food reinforcement. *Journal of Experimental Psychology: Animal Behavior Processes, 1,* 56–87.

Shettleworth, S. J. (1981). Reinforcement and the organization of behavior in golden hamsters: Differential overshadowing of a CS by different responses. *Quarterly Journal of Experimental Psychology, 33B,* 241–255.

Shull, R. L., & Spear, D. J. (1987). Detention time after reinforcement: Effects due to delay of reinforcement? In M. L. Commons, J. E. Mazur, J. A. Nevin, & H. Rachlin (Eds.), *Quantitative analyses of behavior: The effect of delay and of intervening events on reinforcement value* (pp. 187–204). Hillsdale, NJ: Erlbaum.

Skinner, B. F. (1935). The generic nature of the concepts of stimulus and response. *Journal of General Psychology, 12,* 40–65.

Skinner, B. F. (1938). *The behavior of organisms.* New York: Appleton-Century-Crofts.

Staddon, J. E. R. (1974). Temporal control, attention and memory. *Psychological Review, 81,* 375–391.

Staddon, J. E. R. (1977). Schedule-induced behavior. In W. K. Honig & J. E. R. Staddon (Eds.), *Handbook of operant behavior* (pp. 125–152). Englewood Cliffs, NJ: Prentice-Hall.

Staddon, J. E. R. (1983). *Adaptive behavior and learning.* New York: Cambridge University Press.

Staddon, J. E. R., & Higa, J. J. (1991). Temporal learning. In G. Bower (Ed.), *The psychology of learning and motivation* (Vol. 27, pp. 265–294). New York: Academic Press.

Staddon, J. E. R., Wynne, C. D. L., & Higa, J. J. (1991). The role of timing in reinforcement schedule performance. *Learning and Motivation, 22,* 200–225.

Steinhauer, G. D. (1982). Acquisition and maintenance of autoshaped key pecking as a function of food stimulus and key stimulus similarity. *Journal of the Experimental Analysis of Behavior, 38,* 281–289.

Stokes, P. D., & Balsam, P. D. (1991). Effects of reinforcing preselected approximations on the topography of the rat's bar press. *Journal of the Experimental Analysis of Behavior, 55,* 213–231.

Thorndike, E. L. (1911). *Animal intelligence.* New York: Macmillan.

Timberlake, W. (1983). Appetitive structure and straight alley running. In R. L. Mellgren (Ed.), *Animal cognition and behavior* (pp. 165–222). Amsterdam: North Holland Press.

Timberlake, W., & Lucas, G. A. (1990). Behavior systems and learning: From misbehavior to general principles. In S. B. Klein & R. R. Mowrer (Eds.), *Contemporary learning theories: Instrumental conditioning theory and the impact of constraints on learning* (pp. 237–275). Hillsdale, NJ: Erlbaum.

Timberlake, W., Wahl, G., & King, D. (1982). Stimulus and response contingencies in the misbehavior of rats. *Journal of Experimental Psychology: Animal Behavior Processes, 8,* 62–85.

Tomie, A., Brooks, W., & Zito, B. (1989). Sign tracking: The search for reward. In S. B. Klein & R. R. Mowrer (Eds.), *Contemporary learning theories: Pavlovian conditioning and the status of traditional learning theory* (pp. 191–223). Hillsdale, NJ: Erlbaum.

Townsend, J. T., & Busemeyer, J. R. (1989). Approach-avoidance: Return to dynamic decision behavior. In C. Izawa (Ed.), *Current issues in cognitive processes* (pp. 107–133). Hillsdale, NJ: Erlbaum.

von Uexküll, J. (1921). *Umwelt und Innenwelt der Tiere.* Berlin: Springer-Verlag.

Wagner, M. (1985). The metric of visual space. *Perception & Psychophysics, 38,* 483–495.

Warden, C. J. (1931). *Animal motivation: Experimental studies on the albino rat.* New York: Columbia University Press.

Wasserman, E. A., Kiedinger, R. E., & Bhatt, R. S. (1988). Conceptual behavior in pigeons: Categories, subcategories, and pseudocategories. *Journal of Experimental Psychology: Animal Behavior Processes, 14,* 235–246.

Weiss, S. J., Schindler, C. W., & Eason, R. (1988). The integration of habits maintained by food and water reinforcement through stimulus compounding. *Journal of the Experimental Analysis of Behavior, 50,* 237–247.

Wilkie, D. M. (1989). Evidence that pigeons represent Euclidean properties of space. *Journal of Experimental Psychology: Animal Behavior Processes, 15,* 114–123.

Williams, B. A. (1975). The blocking of reinforcement control. *Journal of the Experimental Analysis of Behavior, 24,* 215–226.

Williams, B. A. (1983). Another look at contrast in multiple schedules. *Journal of the Experimental Analysis of Behavior, 39,* 345–384.

Williams, B. A. (1989). Signal duration and suppression of operant responding by free reinforcement. *Learning and Motivation, 20,* 335–357.

Williams, D. A., Butler, M. M., & Overmier, J. B. (1990). Expectancies of reinforcer location and quality as cues for a conditional discrimination in pigeons. *Journal of Experimental Psychology: Animal Behavior Processes, 16,* 3–13.

Winfree, A. T. (1980). *The geometry of biological time.* New York: Springer-Verlag.

Zeiler, M. D. (1985). Pure timing in temporal differentiation. *Journal of the Experimental Analysis of Behavior, 43,* 183–193.

Received June 4, 1991
Final acceptance December 3, 1991

APPENDIX 1

Dimensions, Accelerations, and Forces

There are numerous places where precise analysis will involve complications. One may reasonably ask, for instance, how distance can be measured in a space where the dimensions have different units: How should two units of stimulus difference, three units of time, and four units of response difference be combined? Of course we combine "incommensurate" dimensions intuitively whenever we make decisions such as taking the stairs rather than waiting for an elevator. The problem is how to do this formally. The answer proposed here is to play Newton's game and impose metrics that will result in the simplest system of behavioral mechanics. The result will be psychological measures of distance along each of the dimensions that may have only approximate congruence with the "obvious" physical ones. Just as in physics, the general laws subsume a lot of particular calibrations and constants. Shepard (1987b) found a general law for propagation of force along the stimulus dimension, but it required the construction of often idiosyncratic maps between physical measures and psychological distances. Because space and time are sensed as sequences of stim-uli, the rules that hold for stimulus discrimination and concatenation will probably also predict distance along properly adjusted dimensions of physical space and time. In these cases, however, the maps are likely to be simple, smooth, and approximately monotonic functions of physical measures. The metrification of action patterns will be a more challenging enterprise.

Another place where there is room for more precision is in my treatment of the *accelerations* toward incentives as *forces*. Force equals the acceleration times the mass of the object accelerated, and is thus measured through the interaction of the source and the subject. This paper de facto treats the "mass" of the subject as unity, thus rendering the constructs equivalent. Behavioral mass thus constitutes an unutilized degree of freedom in the development of behavioral mechanics. It may be that the best way to treat operations such as satiation is as an increase in the behavioral mass, so that a given force is less able to accelerate the organism. It may be that Nevin's (e.g., Nevin et al., 1983) understanding of mass will be appropriate. These are just some of the issues that invite further theoretical and empirical work.

APPENDIX 2

Speculations on the Relation Between Potential Gradients, Discrimination, and Differentiation

We may measure the difference limen (DL) as the change in distance (temporal distance, t, here; distance along any of the dimensions in general) that is required to achieve a just noticeable difference (JND) in the potential field. Rearrange the potential function to isolate distance on the left, and take the derivative with respect to potential. We find thereby the distance that an organism must move in time (DL) to achieve a JND of change in the potential function.

For the simple inverse function (with an additive constant of 1.0 corresponding to the shell), the force is

$$F = \frac{-k}{t+1}, \qquad (1)$$

and the potential function, the negative integral of the distance over which it operates, is

$$U = k \ln(t+1) + C. \qquad (2)$$

If an animal is sensitive to the potential U along the temporal dimension as postulated, Equation 2 shows that the sensitivity will grow as a logarithmic function of temporal distance (Fechner's law).

Rearrange Equation 2 to isolate time and take the derivative with respect to U. This gives us the amount by which time must be incremented to change the potential by one unit:

$$\frac{dt}{dU} = \frac{1}{k}(t+1), \qquad (3)$$

which is simply the negative reciprocal of Equation 1. The quantity on the left is the DL; it is proportional to the temporal distance to the incentive. This is Weber's law. Rewriting dt/dU as Δt, and $1/k$ as w, the Weber fraction, we may cast it in the traditional form:

$$\frac{\Delta t}{t+1} = w, \qquad (4)$$

which, for temporal discrimination is known as *scalar timing*, and is generally found to hold true for retrospective temporal discriminations. Thus, if sensitivity to the potential difference between a past and current temporal location is the mechanism of discrimination, then the hypothesis of inversely decreasing forces predicts scalar timing.

What of prospective choice of various delays of reinforcement? The inverse hypothesis is effective in predicting single-trial choice responses. But, in general, there are two gradients that need to be accounted for in such paradigms: that of the delayed primary incentive and that of the immediate conditioned reinforcers consequent upon the choice. Stimuli correlated with the onset of the delay are attractors whose potential is set by the average immediacy of reinforcement they predict (Shull & Spear, 1987) and thus provide a steep gradient at the instant of choice when they become operative. Because immediacy is the inverse of delay, this gives the appearance of simple inverse gradients. In general, choice is most likely attracted both by conditioned reinforcers whose strength is an inverse function of the delay it signals and by an (exponential?) primary gradient.

The exponentially decaying force function is associated with an exponential-integral potential function. Unlike the logarithmic function, the exponential-integral function approaches an asymptote, suggesting that at sufficiently large distances, reinforcement will become absolutely ineffective in shaping behavior (consider the impossibility of maintaining expeditious behavior at the start of a fixed delay, even when such responses initiate the delay and thus affect rate of reinforcement). The derivative of the inverse potential function increases exponentially with the distance to reward, indicating exponentially increasing difficulty at long delays. And, in fact, the DL for temporal *differentiation* of responses does increase very steeply, perhaps exponentially, with temporal distance (Zeiler, 1985).

Different mechanisms may therefore underlie the estimation of elapsed time intervals (*discrimination*, where Weber's Law holds, mediated in some cases by sensitivity to elicited behavior; Killeen & Fetterman, 1988); the production of time intervals by the subject (*differentiation*, mediated by sensitivity to the differential attraction of force fields, where Weber's law does not hold), and the choice of delayed reinforcers (controlled by both an exponential gradient between the choice response and the delayed primary incentive and by the conditioned reinforcers whose strength is an inverse function of the rate of reinforcement they signal; Killeen, 1982, 1991a).

4.
The Response Dimension

Peter R. Killeen and Lewis A. Bizo

The Response Dimension

Peter R. Killeen and Lewis A. Bizo
Arizona State University

In the *Mechanics of the Animate,* Killeen (1992) argued for the spread of the effects of reinforcement through behavioral space along dimensions such as stimuli, time, and responses. He made the case for an exponential gradient of diffusion of effects along stimulus dimensions, and possibly along temporal and response dimensions as well. The domain of these functions cannot be arbitrary measures of the events, but rather must be psychologically scaled dimensions, in the fashion of Shepard (1987). For a behavioral theory, the key dimension is the response; but the requisite scaling of the behavioral repertoire has not begun. It is possible to proceed with the development of the mechanics for responses without dallying with psychometric scaling if we confine our attention to a single class of responses. Even within this limited domain, there is much to be said.

The discourse begins with the three basic processes that are central to the field of learning and performance: motivation, association, and response constraints. A recent theory of reinforcement (Killeen, 1994; 1995)—epitomized these processes in three principles:

1. Activation

The first principle states that an incentive activates a seconds of responding. If R incentives are delivered per hour, it follows that:

$$S = aR/\delta \qquad (1)$$

responses are activated per hour, where the symbol S stands for response strength. The parameter a is a constant called the specific activation-- "specific" because it represents the total number of response-seconds that a single incentive will generate under those motivational conditions, and δ is the duration of a response. Figure 1 shows the asymptotic rate of activity responses made by pigeons that were given brief access to mixed grain R times per minute (Killeen, 1975). These asymptotes are the response rates that were projected to occur in the absence of competing responses. The linearity of these data support the first principle.

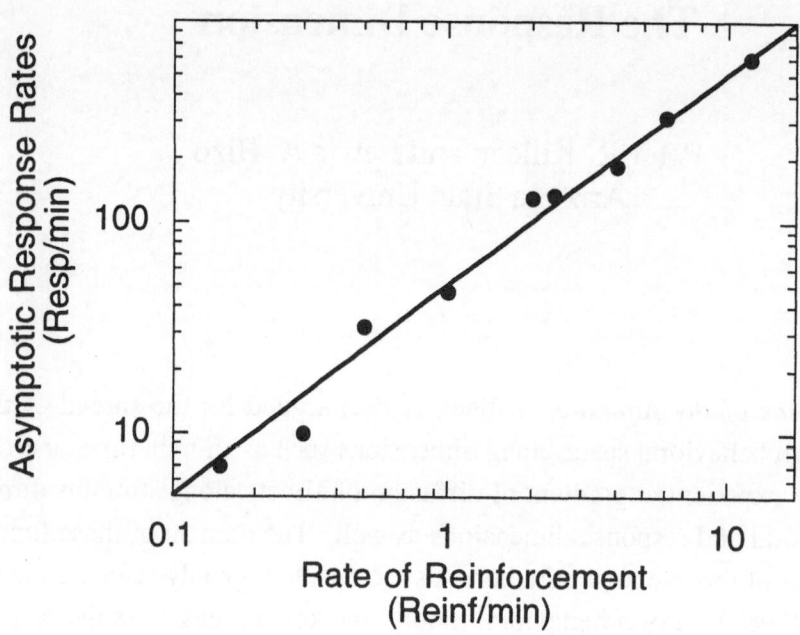

Figure 1. The asymptotic rate of general activity averaged over four pigeons who were fed periodically at different rates. The data are from Killeen (1975).

<u>Arousal versus Association</u>. The repeated delivery of incentives engenders a heightened state of arousal; depending on the degree of coupling to one or another response, it may also reinforce associations. The distinctive action of the two processes —arousal and association— must be recognized in our theories of behavior. Incentives that are not coupled to a particular response arouse an animal, but are unlikely to reinforce the instrumental response of interest to the experimenter--that is, are unlikely to reinforce the *target response*. Instead we may see the elicitation of substantial "adjunctive", "superstitious", or "frustrative" responses; in fact, it is the emission of such desultory behavior that we interpret as the hallmarks of arousal.

Figure 2 (from Killeen, 1996) shows an intriguing set of data produced by Hoffman and associates (1961). After 54 sessions of conditioning, rats still showed a substantial increase in responding throughout the course of a session. The authors showed definitively that the increase had nothing to do with learning: when subjects deprived of the opportunity to respond during the first 40 trials of the session were finally given access to the lever, they responded at exactly the same rate as the control animals. Killeen, Hanson and Osborne (1978) provide a basic model for this warm-up process; the asymptote of their model is given by Equation 1.

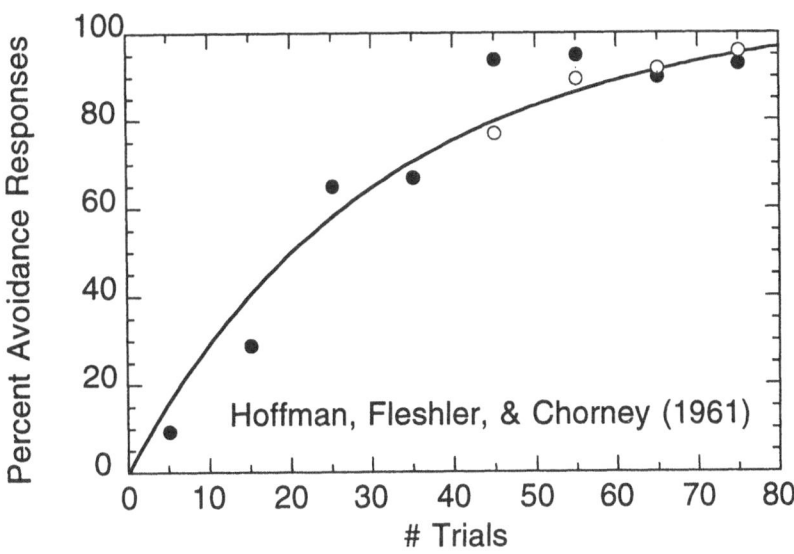

Figure 2. The percentage of trials with a response in an aversive conditioning experiment conducted by Hoffman and associates (1961). The subjects contributing the data denoted by unfilled circles had no opportunity to respond during the first 40 trials, yet responded at rates comparable to the control subjects (filled circles), indicating that the initial rise of the curve was due to warmup, not associative conditioning. The curve through the data comes from the basic theory of behavioral mechanics (Killeen, 1996).

2. Association

According to these principles, reinforcement occurs when an incentive occupies the same time-frame as a response. The more responses in memory at the time the incentive is delivered, the more that responses of that class will be reinforced. If memory is filled with target responses, the incentive will have a large impact on that class; if it is filled with non-target responses, it will strengthen those to the detriment of the target class. To the extent that the design of the chamber or of the contingencies of reinforcement encourage non-target responses, the motivation due to the incentive will be wasted on them. To be maximally effective in conditioning a particular target response, incentives should be delivered when the subject's memory is filled with that response. Experimental contingencies linking responses to the incentives determine the degree to which this is achieved. In particular, the efficacy of reinforcement depends on the degree of correlation between our definition of the act to be reinforced and the animal's memory of its behavior at the instant the incentive is delivered.

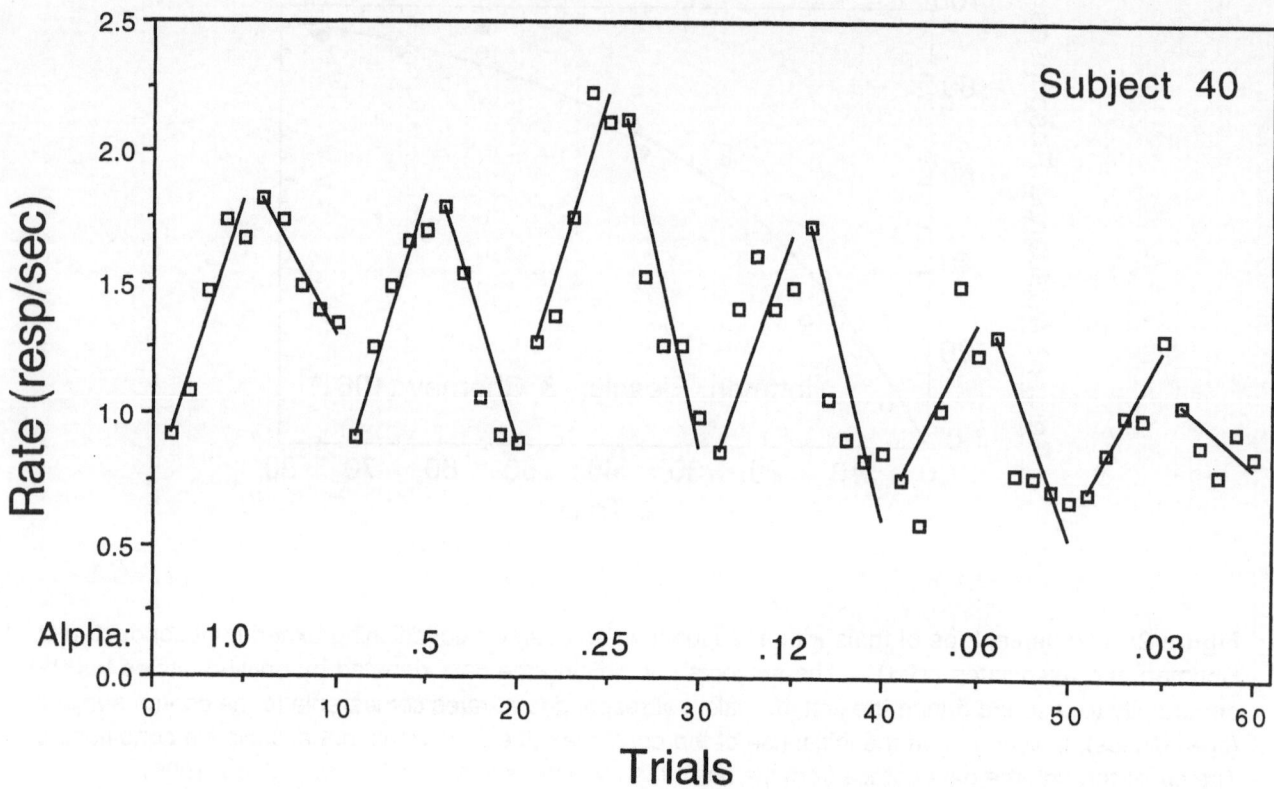

Figure 3. Learning curves for a pigeon whose response rate was alternately driven high and then low by experimental contingencies. The ordinate is response rate in pecks per second. The abcissae are the nominal session numbers (after two blocks of five sessions, the conditions were replicated and are plotted as averages over the nominal session numbers). In the first block of five sessions response rate was driven high by experimental contingencies, with a value for α equal to 1.0. For the second block of five sessions rate was driven low by the same experimental contingencies. (For the third and fourth, blocks, not explicitly shown, these conditions were replicated and are combined with the earlier data to yield the figure as plotted). The value for α was then decreased to 0.5 and the procedure continued. Each symbol corresponds to the response rate over the course of an experimental session, and derives from the average of two sessions spaced two weeks apart. The straight lines are regressions anchored in the last session of the previous condition. Note that the slopes of these lines are greatest around α's of 0.25.

Figure 3 demonstrates this claim with data from a pigeon for whom the criterion for reinforcement was that its rate of key-pecking exceed (or, the next week, be less than) the average response rate on the previous trial. Rate was measured as an exponentially-weighted moving average of the recent inter-response times. "Recent" means falling within the exponential window, whose average width is the inverse of the rate constant (see Figure 4). For instance, at its steepest, the experimenter's criterion requires that the most recent inter-response time exceed (or, the next week, be less than) the previously reinforced time. Whatever happened before the last response time is irrelevant. Conversely, for a very flat function, each response time only affects the experimenter's average by a small proportion (3% being the smallest value of alpha studied here). Note that at intermediate widths of the window ($1/\alpha = 4$ events) response rates increase and decrease most rapidly. It is inferred that this is the memory window also employed by the subject. This inference is supported by simulations of stat-rats with known windows.

Having identified the memory window of individual subjects, can we predict the slopes of the learning curves? Figure 4 shows what happens when an animal with a window of 3.3 events is being shaped by an experimenter who takes a longer view of its performance. (Think of the miscommunication between a runaway dog who has just found its way home, and a master focused on its earlier departure). The heavy weight the experimenter puts on the distant past will be wasted on the animal; conversely the experimenter gives insufficient weight to the most recent past. The correlation of the two perspectives is given by the intersection of the areas under the curves. In this case it is the area under the shallower (experimenter's) curve up to the point t^* at which they cross, and the area under the subject's curve thereafter. This correlation is easy to calculate by solving for t^* and then integrating. Call the rate constant of the subject β and that of the experimenter α. When the experimenter's definition extends farther into the past than the animal's memory, then the correlation (zeta, ζ, the *coupling coefficient*) is:

$$\zeta = 1 + \left(\frac{\alpha}{\beta}\right)^{\frac{\beta}{\beta-\alpha}} - \left(\frac{\alpha}{\beta}\right)^{\frac{\alpha}{\beta-\alpha}} \qquad\qquad \beta > \alpha > 0 . \qquad\qquad (3)$$

If $\beta \gg \alpha$, the last term goes to 1 and $\zeta \sim \alpha/\beta$.

When the subject's window on the past is deeper than the experimenter's, the rate constants are exchanged to give:

$$\zeta = 1 - \left(\frac{\alpha}{\beta}\right)^{\frac{\beta}{\beta-\alpha}} + \left(\frac{\alpha}{\beta}\right)^{\frac{\alpha}{\beta-\alpha}} \qquad\qquad \alpha > \beta > 0 . \qquad\qquad (4)$$

If $\alpha \gg \beta$ the middle term goes to 1 and $\zeta \sim \beta/\alpha$.

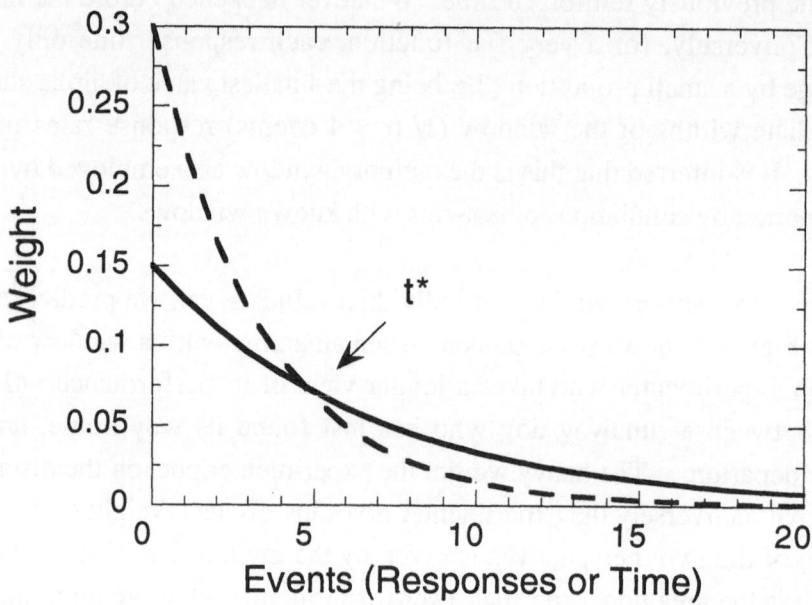

Figure 4. The weighting of the past, which extends to the right on the x-axis, by the animal (continuous curve) and the experimenter (dashed curve). The coupling between the subject's and the experimenter's perspectives is given by the area under the lowest curve segments. It is assumed that the subject's weighting is fixed, whereas the experimenter's is arbitrary, and thus might be adjusted to match the subject's.

When the two processes are perfectly matched, the coupling is given by the area under either curve, corresponding to perfect coupling:

$$\zeta = 1 \qquad\qquad \alpha = \beta . \qquad (5)$$

If we assume that the speed of learning is proportional to the coupling coefficient, zeta, then we should be able to predict the slopes of the learning curves for all values of alpha given the animal's value of β and the constant of proportionality. Figure 5 shows the average of the ascending and descending slopes of the learning curves for this and the other three pigeons in the experiment. There was no significant difference between the recovered values of β for the ascending and descending series. The tuning curves through the data were derived from Equations 3-5. The left limbs of the curves are drawn by Equation 3 and the right limbs by Equation 4. The abscissae of the maxima are the subjects' rates of memory decay, β.

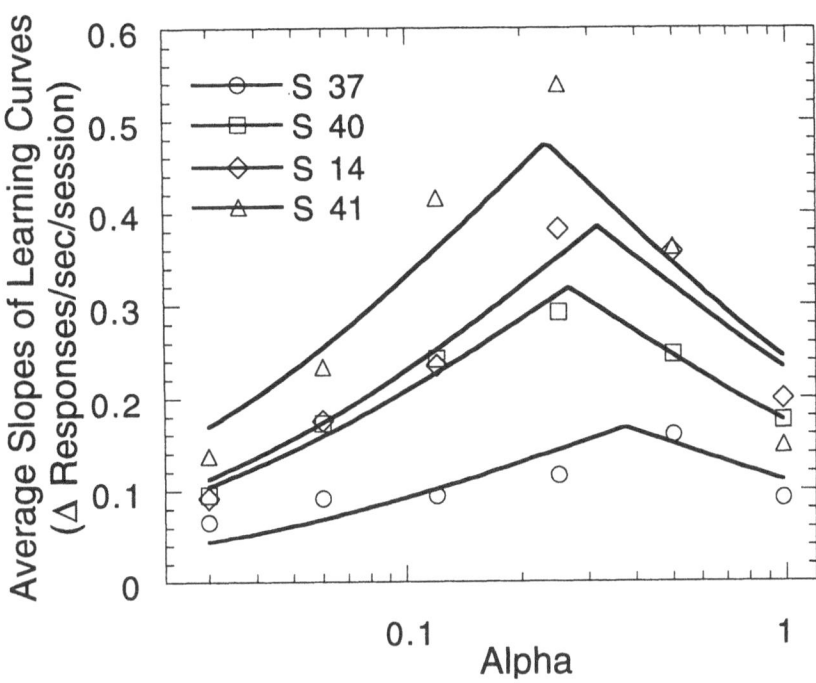

Figure 5. The slopes of the learning curves shown for four pigeons. The absolute values of the ascending and descending slopes for each value of alpha are averaged and displayed for each subject. The tuning curves through the data are from Equations 3-5, with scale free to vary with subject. The abscissae of the peaks are the recovered values of β. Recovered values are (from top to bottom): 0.23, 0.32, 0.27, and 0.37; the curve through the aggregate data takes a value of $\beta = 0.28$.

Different criteria for delivering reinforcement —the familiar "schedules of reinforcement" of Ferster and Skinner (1957)— generate different coupling between responses and reinforcement; for some schedules, such as ratio schedules, the coupling is close to 1, whereas for other arrangements it is much lower. Whenever reinforcement is given, whether by behaviorists or by cognitivists, there is *de facto* a schedule in effect, and a corresponding coupling coefficient may be calculated. Surprisingly, reinforcing every response does not guarantee the greatest coupling. Since memory can encompass more than one response, reinforcing a single lever-press only uses a fraction of the effectiveness of the reinforcer; the rest is squandered, reinforcing the prior consummatory response (which needed little help!), and whatever post-prandial behaviors were elicited by it. Greater coupling is provided by "partial", or probabilistic reinforcement.

Killeen (1994) provides rules for calculating the correlation between reinforcement contingency and memory for various experimental arrangements and schedules of reinforcement.

These rules permit a priori prediction of response rates in novel situations, once ceilings on response rates are taken into account.

3. Constraints on Responding

Response strength S is not the same as rate of responding, because there may be inadequate time in a session to emit all the responses that are activated by the incentive. The execution of other responses, including responses of the same type, may block the emission of a response. If it requires delta (δ) seconds to emit a response, then the maximum response rate is obviously $1/\delta$. Less obvious is how rates change as they approach their asymptote. Killeen (1994) employed a blocked-counter model of the approach, one which generated a hyperbolic increase in response rates to their asymptotes. The following more transparent treatment generates the same function: Let us take S to measure the instantaneous rate of initiation of a response, with each response requiring δ seconds for its complete emission. Then the average time between responses will be $1/S + \delta$, and the measured response rate will be the reciprocal of that inter-response time:

$$B = \frac{S}{1 + \delta S}, \qquad\qquad \delta > 0 . \qquad (6)$$

If we insert Equation 1 into Equation 6, we get the key model governing response rates:

$$B = \frac{\zeta aR}{\delta(1 + aR)}, \qquad\qquad \delta > 0 . \qquad (7)$$

Not all of the behavior elicited by incentives is measured as the target response. ζ (zeta, $0 < \zeta < 1$) is the coupling coefficient designating the proportion of the total behavior incited by the incentives that is actually entrained as measured target behavior by the contingencies of reinforcement. The parameter a is the incentive value of a reinforcer—determined by the quality of the reinforcer and the drive level of the organism, R is the rate of reinforcement, and δ is the minimum time required for a response. The value of the coupling coefficient depends on how the characteristics of the response interact with the schedule of reinforcement. When applied to response rates under conventional schedules of reinforcement, Equation 7 provides a good fit to the data, as shown by Figure 6.

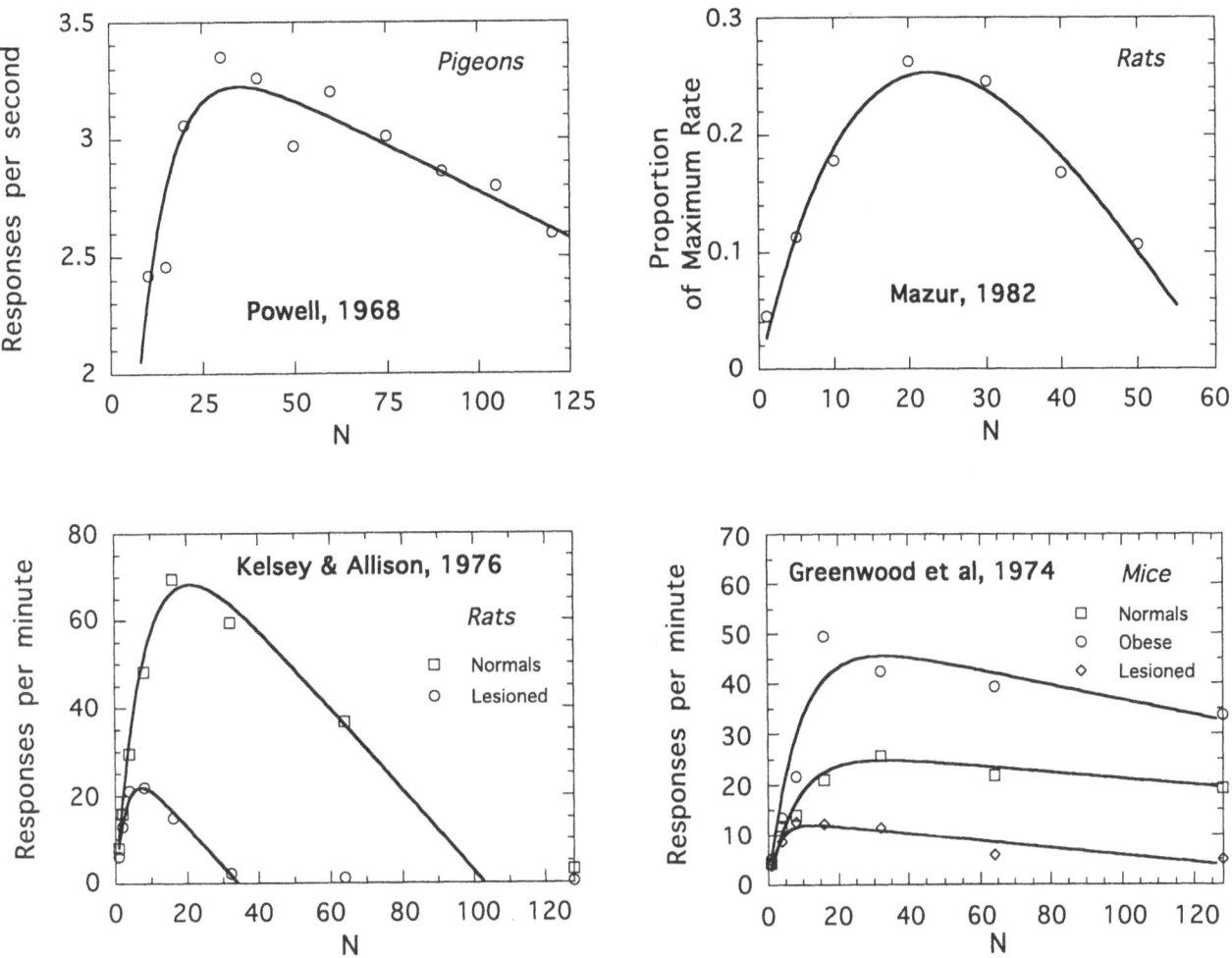

Figure 6. The mean response rate on ratio schedules as a function of ratio size. The first panel show the average "running rate" (i.e., rate excluding the post-reinforcement pause) from three pigeons. The curve through the points comes from Equation 9; see Table 1 for parameters. The one to its right shows the average data from 3 rats. The lower left panel shows the data from 7 control (sham-lesioned) rats and 5 rats made obese by surgical lesions of the ventromedial hypothalamus. The last panel shows data from three groups of mice, each comprised of 4 subjects: Normal black mice, normal mice made obese by aurothioglucose lesions, and genetically obese mice. This figure is from Killeen (1994), and is reprinted with permission of Behavioral and Brain Sciences.

In the case where the response requirement on ratio schedules is used as the independent variable, as in the above graphs, Equation 7 reduces to the elegant Equation 8:

$$B = \frac{\zeta}{\delta} \cdot \frac{N}{a}, \qquad\qquad N < a/\delta; \quad \delta, \alpha > 0, \qquad (8)$$

with the coupling coefficient zeta being a cumulative exponential function of N.

Coupling coefficients for various schedules of reinforcements, along with illustrations of their fit to archival data, are reported in Killeen (1994). This ability to parse complex schedule performance into three orthogonal parameters — Motivational (carried by a), topographical (carried by δ), and memorial (carried by β which determines the coupling coefficient for any particular schedule of reinforcement) — permits us to determine the role of these fundamental factors in the determination of any particular performance.

In the Context of Other Responses

Figures 3 – 5 provide relatively direct —if not yet conclusive— evidence as to the form of the force function through a field of homogenous responses; like the function for the propagation of force through stimuli, it is exponential. Rather than test the power of that function against other candidate functions, we proceeded to use the exponential in a general theory of reinforcement. The success of the theory provides a kind of additional synthetic, as opposed to analytic, evidence for the exponential diffusion of the effects of reinforcement.

Behavior is not unidimensional: it constitutes a trajectory through a field of attractors. We may construct a behavior-space in which to couch our analysis, as was done by Killeen (1994). In this space the ordinate is the cumulative number of target responses within a trial, with the other dimensions representing the other modal action patterns available to the organism. The attractor for behavior under an $FR\ N$ schedule is a horizontal line from an ordinate of N target responses. Why should the horizontal line "attract" behavior? Assume a uniform probability of making the target response $p_i = \rho$, and consider the response that occurs after the 15th such response on an FR 16 schedule. If it is also a target response, it is strengthened by the amount $w_1 = \beta$, so that its expected coupling is $\zeta_{Target} = \rho\beta$; if it is some other response followed by a target response, which occurs with probability $(1-\rho)\rho$, it is strengthened by the amount $w_2 = \beta(1-\beta)$; if it is two other responses followed by a target response, which occurs with the probability $(1-\rho)^2\rho$, it is strengthened by the amount $w_3 = \beta(1-\beta)^2$. The sum of this infinite series gives an expected:

$$\zeta_{Other} = \frac{\rho\beta(1-\rho)(1-\beta)}{1 - (1-\rho)(1-\beta)} . \qquad\qquad \beta, \rho > 0 \qquad (9)$$

Note that if ρ is close to 1, the association of reward with Other responses goes to zero, and the target response receives the preponderance of coupling. If ρ is close to 0, a target response is unlikely to occur; but if it does, is will receive a strength of β. This will feed back into the calculations by increasing the value of ρ, carry behavior over the consequential line, and thereby increase the likelihood of a target response on the next trial. We may calculate the advantage to the target response as $\zeta_{Target}/\zeta_{Other}$. The derivative of this ratio with respect to ρ is $1/[(1-\rho)^2(1-\beta)]$: Whenever ρ gets larger the advantage gets larger, and the gains are a strongly increasing function of ρ.

The target response may be unlikely and hard to get going, but once it "catches," it is self-sustaining. A trainer may expedite the process by shaping the organism to make the first response, but as long as ρ is non-zero, the process will eventually catch by itself. This attraction of behavior toward the target response propagates each step back in the sequence, but with diminishing effectiveness. Too large a requirement of low probability responses affords many junctures at which the trajectory is strained to the breaking point. The attraction exerted by an incentive on behavior is not that of a magnet to a metal. Rather, it is the increased likelihood, because of the conditioning process, of the trajectory rotating into a more vertical position after each reinforcement.

The metric of behavior-space. Responses often are exclusive: At any one time an animal may be either grooming or lever-pressing, but not both. This means that it is generally impossible to travel through behavior space on a diagonal. The trajectory moves in quantal steps up or to the right. Because of this, its metric is not Euclidean, but rather "city-block". Thus, the distance from the origin to the consequential line is the same whatever sequence of target and other responses are emitted, and all are equally attracted.

For "integral" responses that are continuous blends, however, the metric is Euclidean, and the shortest trajectory from the origin forms a right angle with the consequential line. This is important under ratio schedules, where the continuous components of a response are shaped toward their most efficient form, causing the value of δ to approach its true minimum.

Depiction of reinforcement as a linear attractor is a computational approximation. Extended incentives strengthen target responses even while they are displacing them from memory. Thus, we should speak of planes of attractors in behavior-space which attract first instrumental responses, and then consummatory responses. It is not surprising that instrumental responses often share some properties of the consummatory responses toward which they are directed, and that coupling is greatest in experimental arrangements that engineer a high correlation between instrumental and consummatory response elements. What the Brelands (1961) called "instinctive drift" in the form of the response is recognized here to be the competing attraction toward consummatory versus instrumental response forms that is maximum at the instant before reinforcement.

Noise. Noise is often a creative element in complex systems (e.g., Killeen, 1992b). Consider the force of attraction exerted on a target response by an FR 32 schedule. It may be drawn as an exponential decay function with a maximum at 32 and trailing off to either side of that maximum—a version of the Laplace distribution. If the decay through responses occurs over an exponential window of only 3 or 4 responses, behavior remote from reinforcement cannot easily become engaged by it. But let us suppose that, through inattention, not every response enters memory. If there is a 1/3 chance of missing any particular response, this adds a random walk to the abcissae. The smooth function shown in Figure 7 is the average of the gradients on this jiggled axis from 30 trials; the curve is a Gaussian density.

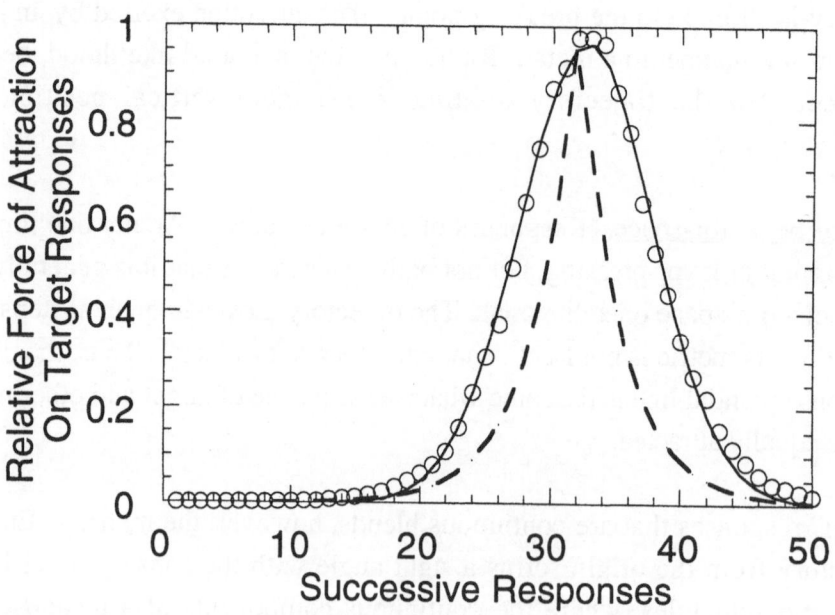

Figure 7. Exponential decay of the force of attraction to the 32nd response (dashed lines), and the diffusion that occurs when there is a constant probability of 1/3 that any particular response will not have entered memory (circles, fit with a Gaussian density).

Because psychological scales may be variable from one trial to the next, as is the case with this Markov model, the underlying forces of attraction may only be inferred, rather than directly measured. Gaussian densities such as the one shown in Figure 7 which are often fit to distributions of responses may mark even at this early stage the ultimately inevitable transition from a mechanics to a statistical mechanics of the animate.

Envoi

Many of the above equations are equilibrium solutions for behavior that has settled into a steady-state. In physics the study of systems at equilibrium is called statics; analogously, the above equations are part of a statics of behavior. Much of the recent research in "learning" theory concerns such asymptotic behavior. It derives from a tradition of descriptive behaviorism; whenever a cumulative record is displayed or a regression is fit through a scatter of data, the goal is description. This is a first step toward a more general science: "Galileo was concerned not with the causes of motion but instead with its description. The branch of mechanics he reared is known as kinematics; it is a mathematically descriptive account of motion without concern for its causes" (Frautschi, Olenick, Apostol & Goodstein, 1986, p. 114). It follows in the Pythagorean tradition which "approached phenomena in terms of order and was satisfied to discover an exact mathematical description" (Westfall, 1971, p. 1). There are many examples of such a tradition in psychology today, including descriptive statistics, the laws of psychophysics, and the original matching law (Herrnstein, 1961).

The study of forces that cause objects to move is called dynamics -- dynamics constitutes "a theory of the causes of motion" (Frautschi et al., 1986, p. 114). Behavior is the motion of organisms, and the study of changes in behavior as a function of motivation, learning and other causal factors constitutes a dynamics of behavior. A framework that embraces all of the above special cases is called a mechanics. This term no longer refers to hypothetical internal mechanical linkages; such machinery is the vestige of the Cartesian tradition in which Newton labored when he began to establish the modern science of mechanics. That mechanical tradition sought to provide causal explanations of phenomena—although such causes were often narrowly construed as material causes involving the motions of particles or aggregations of matter underlying the phenomena. It was one of Newton's chief disappointments that he was never able to provide a mechanical substrate for forces such as gravity, and he finally repudiated knowledge of such hypothetical causes in his famous "hypotheses non fingo", offering instead a precise mathematical description of the effects of those forces. His dynamical theory reconciled "the tradition of mathematical description, represented by Galileo, with the tradition of mechanical philosophy, represented by Descartes" (Westfall, 1971, p. 159). As noted in the last section of this paper, we have begun to extend the statics to a full dynamics. At the same time, we recognized that variability in causal factors will lead us into the realm of statistical mechanics. But these are stories for another year.

References

Breland, K., & Breland, M. (1961). The misbehavior of organisms. <u>American Psychologist</u>, <u>16</u>, 681-684.

Ferster, C. B., & Skinner, B. F. (1957). <u>Schedules of Reinforcement</u>. New York: Appleton-Century-Crofts.

Frautschi, S. C., Olenick, R. P., Apostol, T. M., & Goodstein, D. L. (1986). <u>The mechanical universe: Mechanics and heat, advanced edition</u>. New York: Cambridge University Press.

Herrnstein, R. J. (1961). Relative and absolute strength of response as a function of frequency of reinforcement. <u>Journal of the Experimental Analysis of Behavior</u>, <u>4</u>, 267-272.

Hoffman, H. S., Fleshler, M., & Chorny, H. (1961). Discriminated bar-press avoidance. <u>Journal of the Experimental Analysis of Behavior</u>, <u>4</u>, 309-316.

Killeen, P. R. (1975). On the temporal control of behavior. <u>Psychological Review</u>, <u>82</u>, 89-115.

Killeen, P. R. (1992). Mechanics of the animate. <u>Journal of the Experimental Analysis of Behavior</u>, <u>57</u>, 429-463.

Killeen, P. R. (1992b). Psychophysics: Plus ça change . . . . Commentary on Lockhead's *Psychophysical scaling*. <u>Behavioral and Brain Sciences</u>, <u>15</u>, 569.

Killeen, P. R. (1994). Mathematical principles of reinforcement. <u>Behavioral and Brain Sciences</u>, <u>17</u>, 105-172.

Killeen, P. R. (1995). Economics, ecologics and mechanics: The dynamics of responding under conditions of varying motivation. <u>Journal of the Experimental Analysis of Behavior</u>, in press.

Killeen, P. R. (1996). The first principle of reinforcement. In J. E. R. Staddon & C. Wynne (Eds.), <u>Models of Action</u>, in press.

Killeen, P. R., Hanson, S. J., & Osborne, S. R. (1978). Arousal: Its genesis and manifestation as response rate. <u>85</u>, 571-581.

Shepard, R. N. (1987). Toward a universal law of generalization for psychological science. <u>Science</u>, <u>237</u>, 1317-1323.

Westfall, R. S. (1971). <u>The construction of modern science: Mechanisms and mechanics</u>. New York: Cambridge University Press.

Authors' Note: Preparation of this article was supported by NSF Grants ISBN-9408022 and BNS 9021562.

5.
Nonlinear Phenomena in Learning Processes

Michael Stadler

Nonlinear Phenomena in Learning Processes

Michael Stadler, Günter Vetter, John D. Haynes & Peter Kruse
Institute of Psychology and Cognition Research
and Center for Cognitive Sciences,
University of Bremen, D-28334 Bremen/Germany

Learning as self-organization

On the 1st Appalachian Conference we had the opportunity to present a synergetic view of order - chaos - order transitions in multistable perceptual processes (Stadler & Kruse 1993, Kruse & Stadler 1995). This time we shall deal with learning processes in which order - chaos - higher order phase transitions have to be expected.

The existence of nonlinear phenomena in the behavior of a system is a necessary condition and a strong hint at self-organization processes. All the fascinating examples of self-organization discovered and elaborated in the fields of physics, chemistry, and biology during the last three decades started with irritating and unexpected observations of spontaneous and sudden changes in the behavior of natural systems. In spite of the basic theoretical assumption of continuity in nature ("natura saltum non facit") the complex dynamics of self-organizing systems is able to jump from one to another stable state of order. These non-equilibrium phase transitions are caused by autocatalytic amplification of elementary fluctuations which enable the emergence of new macroscopic behavioral states of order in self-organizing systems. From a microscopic viewpoint of the nervous processes the systems are unpredictable. In phases of instability they are open to minimal influences causing maximal behavioral effects.

The brain is a system which seems to be able to produce a nearly endless variety of ordered macroscopic (behavioral and phenomenal) states and it is probably the most complex system in nature. There is much neurophysiological evidence that the brain has to be understood as a self-organizing system and this is an obvious and by no means new idea. Already in 1925 Wolfgang Köhler, the important protagonist of Gestalt theory, wrote: "The somatic processes underlying static visual fields are stationary equilibrium distributions developed from the inner dynamics of the optical system itself". Though the holistic approach of Gestalt theory was constricted by the concepts of linear thermodynamics of those days, Gestalt theory can be understood as a precursor of the modern theories of self-organizing systems, especially synergetics. In contrast to this early conceptualization of a theory of self-organization in brain, behavior and cognition, nonlinear phenomena are only rarely reported in the psychological literature. Which are the possible reasons for the discrepancy between the theoretical evidence of self-organization processes in behavior and cognition and the lack of empirical reports of nonlinear phenomena in psychological research?

One basic reason may be that up to now conceptualizations in psychology are directed more or less strictly towards homeostatic modelling. This preference reflects the fact, that the least questionable goal of a living system is to survive. Life of a biological system is strictly bound to homeostasis. In homeostatic modelling the preservation of a clearly defined stable state is the basic mode of operation of the system. Unpredictable spontaneous changes can only be regarded as a breakdown of normal functioning. Therefore also in cognitive and behavioral organization nonlinearities and phase transitions seemed to be marginal phenomena and were not in the focus of interest. Yet, a purposeful reanalysis of already existing empirical results of psychological research might bring up more nonlinear phenomena than found at first sight.

A second closely related reason can be seen in the narrow band in which phase transitions occur. In the behavioral space of a biological system phase transitions are necessarily limited to well-defined border conditions. Regimes of linear system behavior are the rule and nonlinearities have to be the exception to guarantee a stable basis of living and action. For cognitive functions like perception or thinking the problem is even further intensified. To guarantee a stable basis of action a phase transition from one stable state to another - if existent at all - has to be very quick and no or not much conscious capacity should be wasted on recognizing the process of cognitive order formation. Therefore, by principle, it will not be easy to detect and measure phase transitions in psychological experiments. Psychological measurements are dependent on highly indirect methods.

Like the second, also the third reason refers to a theoretical and a methodological aspect. In experimental psychology a tendency exists to try to reduce the complexity of behavioral and cognitive phenomena by analyzing elementary processes under very restricted conditions. This is due to the requirements of the experimental methods used on the one hand and on the other hand it results from the assumption that the complexity may be rebuilt by connecting the elementary findings. In some cases the research is even limited by the technical equipment available. To observe the full dynamics of a complex living system it is necessary to look for experimental methods which allow a systematic analysis of behavior without reducing and constraining the system too much. Self-organization characteristics will only appear when the system behavior is unconstrained. The system must be able to follow its own inner dynamics more or less freely.

The reevaluation of the significance of nonlinearities in empirical psychology is the starting point and the consequence of a self-organization theory of learning. In the theoretical framework of **synergetics** (Haken 1977) the first step of analysis is to demonstrate the existence of phase transitions in a complex system. If such phase transitions can be shown, a number of distinct theoretical expectations have to be satisfied to catagorize the phenomenon as a consequence of a self-organization process. For the spontaneous reorganization a certain **control parameter** has to be defined which releases the sudden transition from one to another stable state of order when continuously enhanced. Approaching the point of change by gradually increasing the control parameter the system behavior should show a tendency to persist in the previous stable state (**hysteresis**). Before the phase transition an autocatalytical destabilization of the system appears. This destabilization is manifested by so called **critical fluctuations** and by a **critical slowing down** of the process in order to conserve the existing stable state of order. In synergetics the autonomous reorganizations are explained by the appearance of different modes of behavior competing until one of them predominates the other by slaving the behavior of the elementary components of the system (see Fig. 1). This predominating mode is called an **order parameter**. The self-organization process is characterized by a certain circular causality between the microscopic and the macroscopic level of system behavior. The macroscopic stable states (i.e. the order parameters) organize the microscopic interactions of elementary components of the system from which they have emerged.

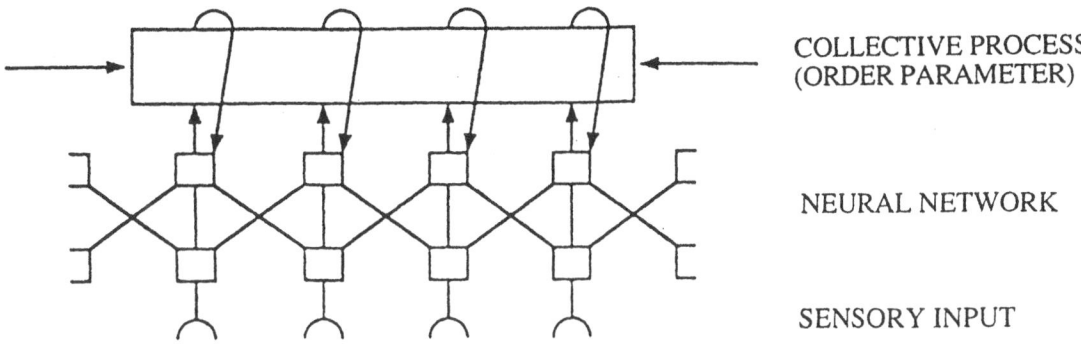

Figure 1. The relation between microscopic and macroscopic processes in the brain.

Many investigations have shown these characteristics of phase transitions in the field of perception (Kruse & Stadler 1995). In this field the instabilities and the phase transitions are directly represented and can be measured by psychophysical procedures. Unfortunately the phenomenal representation of phase transitions lacks in learning processes. So in this lecture we shall try to reanalyse some ancient experimental results of research on learning from a synergetic perspective.

Theoretically certain types of learning can be understood as processes of self-organized order formation. The constant learning effort of the learner while difficulty of the learning material is increased can be taken as a control parameter and the learning achievements are the results of emerging cognitive structures. If this learning is a process of self-organization it has to be expected that the achievement curve of the learner does not linearly increase over time. The learning curves should show phases of linear increasing, phases of stagnation, and phases of significant sudden improvement of performance. During the linear increase a cognitive structure is optimized and transferred into performance. Maximum optimization at a given time of the learning process shows up as a stagnation. After some time this stagnation is followed by a destabilization of the existing structures, which is followed by a phase of sudden improvement in performance correlated to the emergence of a higher state of order. Three empirical hypotheses can be derived if such a process takes place. (1) When measured over sufficiently long time, with nearly constant learning effort and linear increase of task difficulty, learning processes will show characteristic nonlinearities in performance (**phase transitions**). (2) The stability of some parameters of the performance will break down at the end of each phase of stagnation (**critical fluctuations**). In contradiction to naive expectation one hypothesis should be that at the end of the highly optimized performance the rate of errors increases. (3) Again in contradiction to naive expectation the learner will also show a significant increase in sensitivity to disturbances at the end of this phase of optimized performance (**critical slowing down**).

Phase transition in the learning curve

Sometimes in learning of complex tasks like piano-playing, reading or even typewriting people report of stages in which no advance is made despite their high effort. This is often said to be followed by sudden improvement in which a new level of control is reached. In learning curves these periods of arrested progress are indicated by **plateaus**.

The phenomenon was first studied a hundred years ago by Bryan and Harter (1897, 1899) whose subjects had to learn to send as well as to receive telegrafic language (Morse code). The achievement over a long period showed a normal learning curve in sending - but in receiving a significant plateau was found (see Fig. 2). In a following experiment Bryan and Harter varied their stimulus material for receiving systematically in its degree of semantic information: (a) they sent series of letters that formed words and sentences, (b) series of letters that formed words but no sentences and (c) series of letters that didn't form words (Fig. 3). In single-letter-reception no plateaus occurred. Single words produced slight and whole sentences distinct plateaus. This was interpreted as an asynchronous improvement of different hierarchical habits. Freezing of attention to lower order habits (recognition of letters) until they had become automatic impeded the improvement of higher order habits (recognition of words and sentences). Bryan and Harter´s results are controversial. Several attempts to replicate them failed (Taylor 1943, Cook 1957) which leads Keller (1958) to the conclusion that the plateau is no more than a "phantom". Pfisterer's (1988) recent studies in Morse code do reveil plateaus but they are interpreted by him as a natural consequence of the testing procedure because arrested progress is observed at levels where subsections are completed.

There are some methodological problems in the measurement of plateaus: they are difficult to trace in experimental settings because of several factors. The definition of what could be titled a "plateau" has not been clarified. Learning curves are influenced by statistical effects such as dispersion. Normally a plateau is defined as a phase in which all results lie below a previously reached level which ignores the fact that this better score could have been a mere incident. It is also difficult to predict which kind of tasks will lead to periods of arrested progress. These only seem to occur in complex skills where learning stretches over long periods of time (several months) which makes it complicated to provide constant conditions. The distinction between plateaus due to situational and motivational influences on one side and plateaus due to characteristics of the learning process itself on the other is often impossible. This is tried to be solved by taking subjects' introspections after every trial which leads to other methodological problems. Motivation and situation should affect the learning progress unsystematically leading to clean curves when several scores are superimposed. But a systematic tiring-effect (e.g. a gradual loss of concentration) still remains a possible interference. In marathon running e.g. a phase of "hitting the wall" is often experienced when physical exhaustion is reached. This period of depression is followed by improved performance and the feeling of a "high". In a study of Summers (1983) this kind of plateau was reported by 40% of subjects, to most of which it occured at a similar distance between 19 and 24 miles.

The last mentioned problem does not apply to the motor tasks reported of here, because they do not exceed an average every day level of physiological strain. Batson (1916) occasionally found long period plateaus in a task of manipulating an apparatus carrying balls that had to be shot into a pocket. Chapman and Hills (1919) found short plateaus in typewriting. Trow and Sears (1927) studied one subject´s progress in card dealing where they discovered one plateau which they interpreted as due to conflicting methods of practice, as a period of trial-and-error with selection and rejection of methods.

Smith (1930) investigated several tasks that included motor coordination: ball-throwing, ball-guiding and shorthand. In all of these he occasionally found plateaus which were obviously attributable to factors inherent in the learning process rather than to motivational or situational factors. He explains

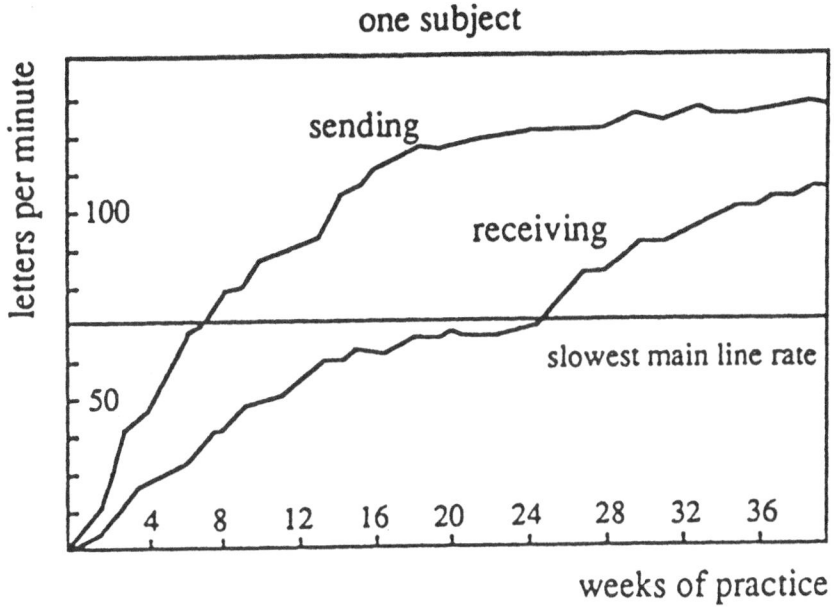

Fig. 2. A plateau in the learning curve of receiving telegrafic language.

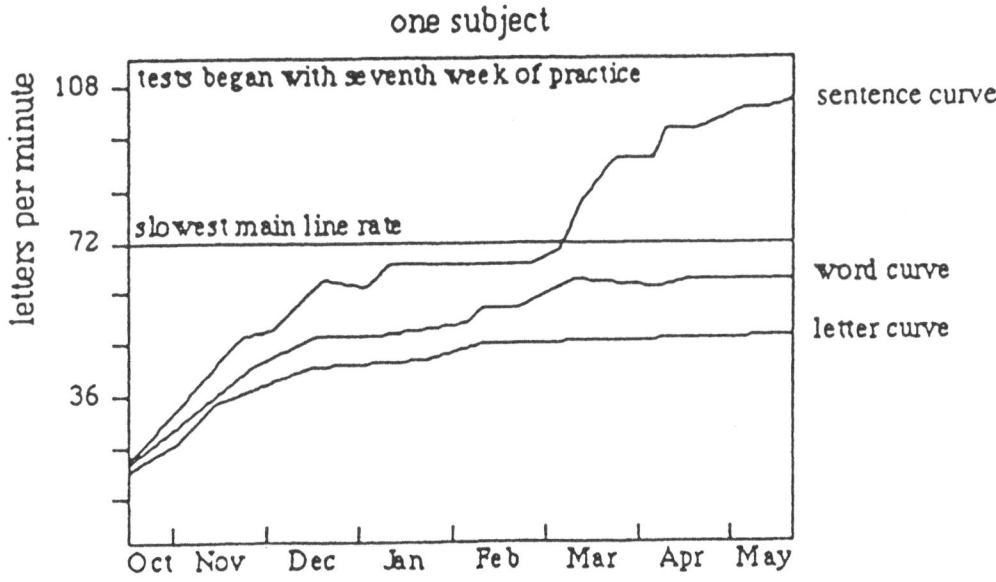

Fig. 3. Learning to receive telegrafic language: letters that form no words (lower curve); letters that form words, but no sentences (middle curve) and letters that form words and sentences (upper curve); see text.

161

them as difficulties in the coordination of different components of a task. The way attention was directed either to the parts or to the integrated movement had a crucial effect on improvement. He found that sometimes performance was best when the subjects were not paying conscious attention to the movement at all.

Kao (1937) examined two tasks: one that involved the simultaneous control of the three variables timing, angle and force of a pendulum, and one that afforded a certain successive organization of movement for manipulating a ball-throwing-device. She investigated how far the prior learning of the individual components influenced the acquisition of the complex task. Most of the plateaus she believes to have found can be interpreted as merely statistical irregularities. Yet, there are short but distinct plateaus in the complex pendulum-task - only for subjects with prior training of the individual components (Fig. 4). The periods also show large daily fluctuations. This effect can be clearly attributed to some kind of reorganization of the simple skills. Without that training the subjects produce normal learning curves. That means, if the simple skills have to be integrated on a higher level, a self-organization process must take place that results in a plateau followed by a destabilization just before the integration. This fits exactly Kao´s interpretation: "another group of short periods of little progress with large daily fluctuations is due to difficulty in **changing pattern**. After certain processes had come to be regarded as patterns in themselves, welding them into a single complex pattern proved too difficult at first." Kao's plateaus can be regarded as free of motivational and situational effects. They occur at constant stages in the learning process rather than being unsystematically spread. Physiological exhaustion as a cause is unlikely because the pendulum is controlled by only small movements of the finger. So we can conclude that phase transitions seem to exist in complex learning tasks at least as long as there is a hierarchy of components to be integrated on higher levels. It is interesting to read the interpretations of the plateaus given by the authors cited: Bryan and Harter (1899) suggested hierarchical acquisition of habits which can be seen as equal to

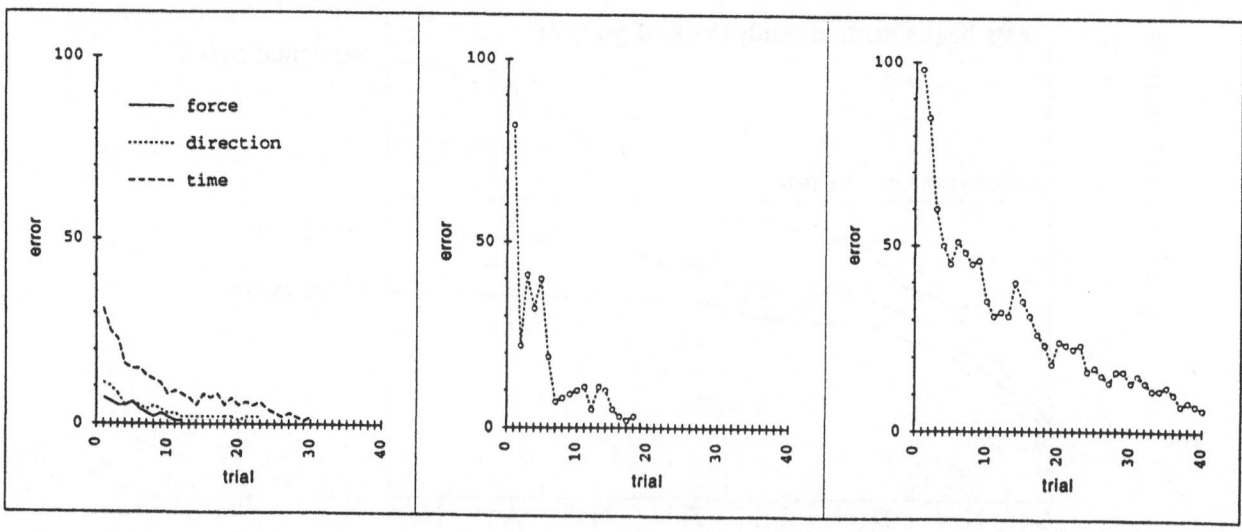

Figure 4. Learning the individual components of a complex task (left diagram); the integration of the individual components exhibits plateaus (middle diagram); learning of the complex task from the beginning one shows a normal learing curve (right diagram).

162

nonlinear emergence of cognitive skill. Trow and Sears (1927) proposed conflicting methods of practice and Batson (1916), Smith (1930) and Kao (1937) suggested problems with coordination. This can be interpreted as the coexistence of single incompatible motor programs which require reorganization in form of temporal and spatial coordination in order to be able to emerge into a single complex unit. This proved most clearly in the experiments of Kao (1937): simple skills that had been learned as units in themselves produced a period of arrested progress when execution was required simultaneously in a complex task. The strong fluctuations in performance during this order formation are typical indicators of a nonlinear phase transition.

The way attention is directed either to the individual parts or to the complex skill strongly influences the process of learning. Both strategies can be effective at different stages. It seems that after sufficient training of the parts some kind of divided attention is necessary to integrate them into a higher level unit. This is when the phase transition occurs. Smith (1930, p. 23) reported that his subjects stated to have been "in a sort of trance", "thinking of other things" or "experiencing a dizzy kind of feeling" and hadn't beeing paying conscious attention to the movement at all. This reminds of the "aha"-experience in problem solving where often solutions occur to a subject suddenly and passively at a period when after intensive effort no more conscious attention is payed to the problem itself.

Behavioral fluctuations in the learning process

Abrupt discontinuities in the learning curve really were not that unusual for learning psychologists in former times. As early as 1932 Tolman noted: "The fact of sudden drops in the learning curve is of cause familiar..." (Tolman 1967, p. 216), referring especially to Thorndike (1911), Yerkes (1916) and Koffka (1928). As already described, this nonlinear behavioral covariation of the learning process can be interpreted as a phase transition from one ordered state to another ordered state.

Fluctuations are, as we have seen, a necessary prerequisite of synergetic phase transitions. In the instable phase (i.e. the learning phase) the system exhibits fluctuating behavior, which increases up to critical fluctuations, and thereafter to a new stable state. Are there any behavioral correlates of such fluctuations in learning situations? Numerous experimenters indeed have reported a more or less typical and frequent pattern of behavior which occurs at the point of choice in the discrimination box, in a maze or during the choice process in visual discrimination studies employing jumping stands. This pattern has been variously described as "looking to the right or left before choice", "running back and forth", "head movements" etc. To this general pattern of behavior Münzinger and Fletcher (1936) have given the name **"vicarious trial-and-error"**, abbreviated "**VTE**". Although first used to label choice point behavior of rats prior to spatial or nonspatial discriminative responses, the term VTE has subsequently been extended to vacillatory behavior in other learning situations and today refers to oscillating behavior of various types of subjects at points of choice in a wide range of learning situations.

And what about critical fluctuations? If the nonlinear drops in the learning curve are due to a phase transition, then shortly before and during such a drop VTE-behavior should be expected to increase to a maximum (and then fall again or even disappear). Again it was Tolman who observed such a relationship and emphasized its importance for learning theory: "I have seen at certain stages in their learning very patent instances of such hesitation at the point between two alleys. The rat stops and wiggles its nose from side to side and then finally chooses" (1926, p. 367). Later he suggested that there might be a relation between such choosing behavior and learning and this in a very specific way: "The fact of sudden drops in the learning curve is of course familiar... but what we are seeking, now, is not this mere fact of sudden drops but rather a correlation between the appearance of such drops and the appearance of just preceding 'running back and forth'..." (1967, p. 216).

In another paper (1938), Tolman suggested to call these new types of activity **catalyzing behavior**, regretting that "the rat psychologists have to date rather pigheadedly... ignored such catalyzing behavior" (1938, p. 27). The term catalyzing is so near to the concept of critical fluctuations that one might speculate, that - had this theory been developed some 50 years earlier - Tolman would have adopted a synergetic interpretation. Anyway, he expressed his belief, "that in the future technological advances in recording will bring to the fore many other instances of such catalyzing behavior for study" (1938, p. 27). And it was still another great psychologist, who almost foresaw the theoretical importance of this kind of behavior. Describing the VTEing of dancing mice at the choicepoint of his discrimination experiments, Yerkes commented as early as 1907 that "could we but discover what the psychical states and the physiological conditions of the animals were during this period of choosing, comparative psychology and physiology would advance by a bound" (1907, pp. 130/131).

In the following there indeed was some research on VTE-behavior and learning, and we shall remind the reader of the data, which are in accordance with the hypothesis derived from self-organization theory formulated above.

Münzinger (1938) scored VTEs in relation to mastery of a habit. Adopting two consecutive series of 10 errorless trials as a criterion of learning he got the following results. Conceding, that any criterion of learning is an arbitrary measure, he notes, that nevertheless there seems to be a specific psychological event, which corresponds to the chosen criterion: "The frequencies of VTE in the two series immediately preceding the criterion and during the two series of the criterion are higher than those further removed. The conclusion suggests itself that the frequency of VTE attains its peak during the phase of learning in which the mastery of the habit is established" (pp. 78/79). Theoretically not unimportant is a further observation of Münzinger. At the beginning of his research on VTE and learning he assumed that the main function of VTE-behavior was to enable the animal to compare the cues to be discriminated, and that without discriminable cues present in the choice alleys there would be only sporadic occurrance of VTE. He tested this by training rats to respond to diffuse stimuli in a tone experiment. The source of the sound was suspended 1 m above the point of choice and the animals learned to turn in one alley, when the tone was sounded, and in the other alley, when it was silent. The result was, that all animals exhibited VTE and "- what was still more surprising, - *there was the same relationship between frequency of VTE and learning efficiency as in the case of a visual discrimination*" (1938, p. 82). Obviously, he notes, VTEing has another function than the mere comparison of cues in the different alleys. However, "the effect [of it] is likewise a facilitation of learning" (1938, p. 82). This observation seems to strengthen the assumption, that VTE in fact is a behavioral correlate of critical system fluctuations shortly before reaching a new stable state.

Further evidence for this kind of relationship is exhibited in learning curves with an elevated discrimination setup obtained by Honzik and presented by Tolman (1938). The animals had to discriminate between a black and a white face each on a door. There was a partition projecting out between the doors. The rats were required to jump a gap just in front of a door. If chosen incorrectly, these animals had to jump back again to the starting platform and then make a second jump to the correct door. The results of the error curves and the VTE-curves are depicted in Fig. 5. Each point represents an average of 10 trials. Clearly, VTE-behavior is at its maximum immediately preceding errorlessness and declines shortly thereafter.

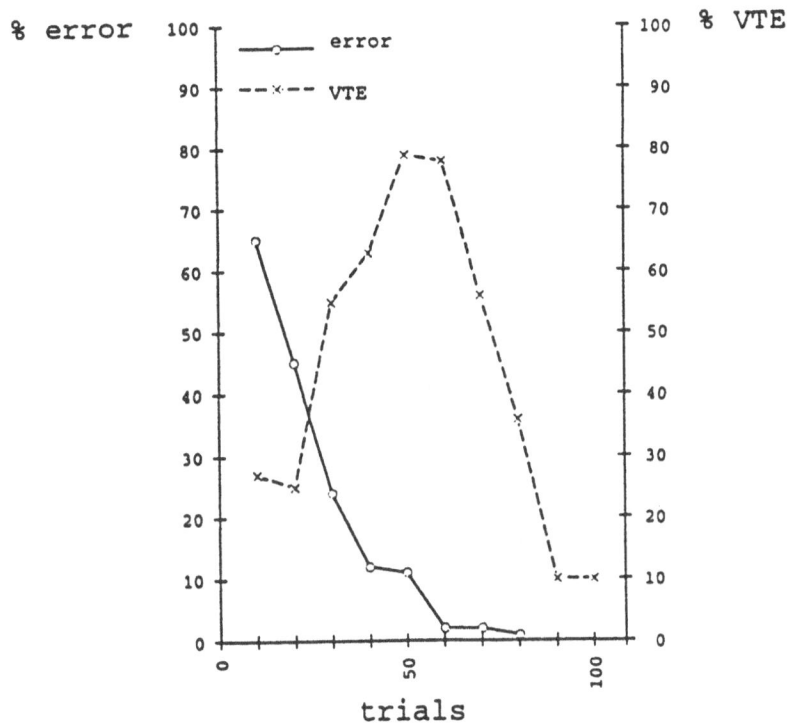

Figure 5. The relation of VTE and learning success: see text.

In still another setting (maze learning), Peterson (1917) observed that "just before entrance [of the blind alley] is eliminated completely, there frequently occurs a peculiar and rapid vibration of the rats' head between the directions of the true part and that of the tempting blind alley" (1917, p. 52). Similarly, for animals learning a horizontal versus vertical bar discrimination in a modified Grice-apparatus, Lane (1952) reported the appearance of "much VTE-behavior from the 5th day of training on until just before the animal learned the discrimination" (1952, p. 48).

Very interestingly, there seems to be a difference between easy and difficult discriminations. Thus it has been observed that the described relationship (i.e. maximum VTE just before mastery, then a drop) holds for easy, not however for difficult discriminations. When more difficult discriminations are required, VTE frequency tends to remain at a certain level for at least a while even after reaching the criterion (Goss & Wischner 1956). As already mentioned, any learning criterion is an arbitrary measure. From the point of view of synergetics one would argue, that the learning criterion is reached when the appropriate order state, the new stability, is **fully established**, i.e. when no more behavioral fluctuations are being exhibited.

We can switch our hypothesis the other way round and predict, in what phase of the learning process VTE-behavior (i.e. fluctuations) should not be expected. In discrimination and other choice situations animals occasionally start the learning trials with what is called positions preferences. In a white/black discrimination for instance, when the stimuli are randomly presented either on the right or on the left side during learning, these animals start with a habit, namely always responding to one specific side. If it is true that VTE indicates system fluctuations in the destabilized transition phase, then one should expect absence of VTE in clearly stable phases, like those position preference habits. One would describe this situation as a phase transition from one ordered state to another, i.e. VTE should be absent at the beginning of learning. Actually this is the case.

Goss & Wischner (1956) summarized Klüver's observations (1933) that monkeys had shown no VTE-behavior in weight discrimination training when responding in terms of a right position preference. And inspection of curves presented by Tolman (1939) indicates, that in the early stages of discrimination learning, if position habits were operating, there was very little VTEing. As the animals began to respond with fewer than chance errors, VTE frequency increased. So, it may be summarised, that VTE-behavior of animals occurs in the destabilized phase transitions of the learning process and not in the stable states, as it would be expected by the synergetic phase transition theory.

Critical slowing down

We come to the third hypothesis concerning characteristics of synergetic phase transitions in learning: if self-organization theory holds for the reported learning situations, then one might expect to observe still another phenomenon. The theory assumes, that the time needed for a system being in an instable phase to find a new stable state, increases, the closer the point of sudden transition is approached. Consequently, if VTE is a behavioral correlate of system fluctuations, one would expect a positive correlation of VTE and response latencies. Indeed, Crannell (1942) reported an increase in VTEing and **hesitation time** as rats approached criterion in multiple-path-mazes. And this is not a unique result. In their review on VTE, Goss and Wischner (1956) pointed out, that a number of investigators found that VTE frequency and hesitation time were positively correlated in discrimination, trial and error, and delayed response situations (Tolman 1939, Jackson 1943, McCord 1939). Unfortunately there are no studies measuring the fluctuations and the critical slowing down of animals in learning processes independently. So, in most cases the slowing down of the learning process just before the phase transition might be a simple consequence of the increase of VTE-behavior which takes certain additional time for the animals to be exhibited.

Consequences of nonlinear phase transitions for education

If some forms of learning are processes of self-organization rather than self-gratification as it has been made probable by the data presented some conclusions can be drawn that may even stimulate further research. Thus, the first conclusion may be that the dominance of conditioning and compulsory education by means of reward and punishment is reduced and more attention is paid to individual differences. **Let them learn to find their own self-organized rhythm**. Systematic observations of babies' habits have shown, that they find their rhythm without being driven by their parents. We made such observations already a long time ago, before chaos and synergetic theory became adopted by psychologists. Families were asked to feed their babies at any time they cried for food and never, when their mothers thought they should be hungry. Such a diagram of the feeding times of a baby between the 5th and the 11th week of life is depicted in Fig. 6 (first published by Metzger 1975). The diagram shows that the feeding times following only the needs of the baby seem to have a chaotic distribution over the daily hours in the first weeks. Beginning from about the 8th week of life the feeding times begin to organize regularly and seem to have found their stable distribution after the 10th week when a daily order of five meals is found. Later, T. Elbert from Münster found a very similar development of the sleeping times of babies, which at first followed a chaotic attractor and then showed a self-organized phase transition, to a periodic attractor. So it looks as if the cyclic needs in human development find their self-organized order in a fashion that their rhythm is synchronized and adapted to the circadian rhythm.

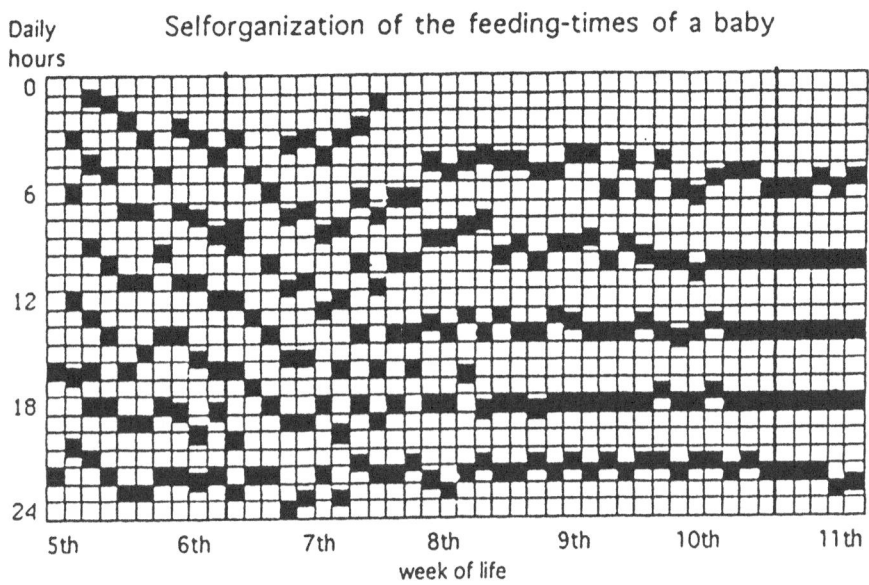

Figure 6. Record of the feeding-times of a baby between the 5th and 11th week of life: see text.

The second conclusion concerning children's education might be: **let them make mistakes**, don't punish them for errors. As we have seen in this contribution an increased number of mistakes appear in the instable phase of fluctuations just before a learning process becomes stabilized at a higher order state. Thus mistakes might be a necessary feature of the accomplishment of the learning process. As an example one may take children learning to read in the first two classes at school. In such a learning process they follow certain plateaus very similar to those found by Brian and Harter. First children learn to combine letters to words and then, when they recognize the word at a glance they learn to combine words to sentences. Reading by words means for the children not to combine the single letters but instead to organize them in terms of meaning. In the transition between the two strategies mistakes may occur. A child may for instance read "tone" instead of "tune" being mistaken in one letter only but changing the meaning of the word entirely. Such mistakes may occur more frequently just before the plateau of letter-reading is left passing a period of fluctuations to the higher level of word-reading. Empirical research on the Gestalt organization of human errors has shown that typical mistakes occur when behavior is suddenly governed by higher ordered states (Wehner & Stadler 1994). The view on learning as a self-organizing process not only explains the peculiarities of behavior in the destabilized phase transition but also underlines the creative power of these behavioral "pathologies".

Reference

Batson, W. H. (1916): Acquisition of Skill. Psychological Monographs, 21, No. 3.

Bryan, W. L. & Harter, N. (1897): Studies in the physiology and psychology of the telegraphic language. Psychological Review, 4, 27-53.

Bryan, W. L. & Harter, N. (1899): Studies on the telegraphic language: the acquisition of a hierarchy of habits. Psychological Review, 6, 345-375.

Chapman, J. C. & Hills, M. E. (1919): The learning curve in typewriting. Journal of Applied Psychology, 3, 252-268.

Cook, D. A. (1957): Message Type as a Parameter of Learning to Receive International Morse Code. Paper read at Eastern Psychological Association Meetings, New York, April 1957.

Crannell, C. W. (1942): The choice point behavior of rats in a multiple path elimination problem. Journal of Psychology, 13, 201-222.

Elbert, T. (1993): Personal communication.

Goss, A. E. & Wischner, G. J. (1956): Vicarious trial and error and related behavior. Psychological Bulletin, 53, 35-54.

Haken, H. (1977): Synergetics. Berlin: Springer.

Jackson, L. L. (1943): VTE on an elevated T-maze. Journal of Comparative Psychology, 36, 99-108.

Kao, D.-L. (1937): Plateaus and the Curve of Learning in Motor Skills. Psychological Monographs 79, No. 219/3.

Keller, F. S. (1958): The phantom plateau. Journal of Experimental Analysis of Behavior, 1, 1-13.

Klüver, H. (1933): Behavior Mechanisms in Monkeys. Chicago: University of Chicago Press.

Köhler, W. (1925): Gestaltprobleme und Anfänge einer Gestalttheorie. Jahresberichte für die gesamte Physiologie und experimentelle Pharmakologie, 3, 512-539.

Koffka, K. (1928): The Growth of the Mind. 2d ed. rev. New York: Harcourt, Brace and Company.

Kruse, P. & Stadler, M. (1995)(eds.): Ambiguity in Mind and Nature. Berlin: Springer.

Lane, P. A. (1952): A Comparison of the Correction versus the Non-Correction Methods of Discrimination Learning. Unpublished master´s thesis, University of Massachusetts.

McCord, F. (1939): The delayed reaction and memory in rats. II. An analysis of the behavioral dimension. Journal of Comparative Psychology, 27, 175-210.

Metzger, W. (1975): Psychologie in der Erziehung. Bochum: Kamps-Verlag.

Münzinger, K. F. & Fletcher, F. M. (1936): Motivation in learning: VI. Escape from electric shock compared with hunger-food tension in the visual discrimination habit. Journal of Comparative Psychology, 22, 72-91.

Münzinger, K. F. (1938): Vicarious trial and error at a point of choice: I. A general survey of its relation to learning efficiency. Journal of Genetic Psychology, 53, 75-86.

Peterson, J. (1917): The effect of length of blind alleys on maze learning: an experiment on 24 white rats. Behavioral Monographs, 1917, 3, No. 4.

Pfisterer, P. (1988): Kognitive Prozesse beim Decodieren von Morsezeichen. Doctoral dissertation, Zürich.

Pomm, H. P. (1973): Eine mathematische Interpretation von Lernplateaus. Psychologische Beiträge, 15, 387-395.

Smith, D. (1930): Periods of arrested progress in the acquisition of skill. British Journal of Psychology, 21, 1-28.

Stadler, M. & Kruse, P. (1993): Neurodynamics and synergetics. In K. H. Pribram (ed.), Rethinking

Neural Networks: Quantum Fields and Biological Data. Hillsdale, N.J.: Lawrence Erlbaum Associates.

Summers, J. J. & Machin, V. J. & Sargent, G.I. (1983): Psychosocial factors related to Marathon running. Journal of Sport Psychology, 5, 314-331.

Taylor, D. W. (1943): Learning telegraphic code. Psychological Bulletin, 40, 461-487.

Thorndike, E. L. (1911): Animal Intelligence. New York: The Macmillan Company.

Thorndike, E. L. (1913): Educational Psychology. Vol. 2. The Psychology of Learning. New York: Teachers College.

Tolman, E. C. (1926): A behavioristic theory of ideas. Psychological Review, 33, 252-269.

Tolman, E. C. (1938): The determiners of behavior at a choice point. Psychological Review, 45, 1-41.

Tolman, E. C. (1939): Prediction of vicarious trial and error by means of the schematic sowbug. Psychological Review, 46, 318-336.

Tolman, E. C. (1967): Purposive Behavior in Animals and Men. New York: Appleton-Century-Crofts.

Trow, W. C. & Sears, R. (1927): A learning plateau due to conflicting methods of practice. Journal of Education Psychology, 18, 43-47.

Wehner, T. & Stadler, M. (1994): The cognitive organisation of human errors: A Gestalt theory perspective. Applied Psychology: An International Review, 43, 565-584.

Yerkes, R. M. (1907): The Dancing Mouse. New York: Macmillan.

Yerkes, R. M. (1916): The Mental Life of Monkeys and Apes. A Study of Ideational Behavior. Behavioral Monographs, 3, No. 12.

6.

The Attractor of the Intentional Learning System

Lillian Greeley

LEARNING AS SELF ORGANIZATION NOT SELF GRATIFICATION
Radford University's Center for Brain Research
and Informational Sciences

September 22-25, 1995

The Attractor of the Intentional Learning System

Lillian Greeley, Ed.D.
Center for the History and Philosophy of Science
Boston University

Learning operates by a system of attention in the cognitive generative learning process, a process one uses to probe and explore in order to learn about any unknown. Intentionality, the personally discriminative activity of learning, directs a person's exploration of an unknown, which is why learning can be regarded as a universal function of the person's Intentional Learning Process. It is now possible to track and graph this learning process, which opens up new research vistas, including the area of Intentionality.

The significance of the role of attention in Intentionality was originally found by analysis of philosophical dialectics, where the phenomenon of Spacing has been identified as the driver of the attention system. Spacing is a break or space away from the analytical work of the exploration. It serves as the physical place where complex dynamics are negotiated and effected by providing an open feedback loop for the dynamics of the emotional and motivational controls of attention. Spacings are comprised of seven elements: Emotion [E], Being Present [BP], [Emotion & Being Present] [E&BP], Content [CT], Rest [R], Confusion [CN] and Closure [CE], which are used in particular combinations within each Spacing.

Three separate samples, separated by a span of 2,500 years, were used to probe the philosophical dialectical learning system: the Early Socratic Dialectics [ESD], the Mat Lipman Dialectics [MLD] [contemporary philosophical novels for children (K-12)], and the entire Euthydemus Dialogue. Analysis of the philosophical dialectics was done with a simple component coding system: Question [Q], Answer [A], Spacing [S] and New Understanding [N]. The coding system was tested and found viable with interrater reliability tests and the dialectics were then coded for sequence of order operations. Rouelle's and Taken's Trick Protocol was used with the order of sequence list to reconstruct the attractor of the system. This protocol generates the attractor of the system, its graphic functional form, in three dimensional graphs. The attractor of the Euthydemus, found by using this protocol, is shown in Figure 1.

(See fig. 1)

[Other dynamical analyses were done, and by the reconstruction of the state-space of the attractor analysis, the state-space of the attractor was found to be low dimensional and fractal. The original analysis to demonstrate the entropy flow of the system was also consistent with a dynamical system, and another analysis to better show the flow of the system is in process. The order of component events appears to proceed in a nonrandom intermittency pattern, with intermittent changes between deterministic chaotic episodes and apparently random episodes. This pattern is similar to patterns found in the genetic sequencing coding system of DNA molecules, where a "pervasive patchiness" of DNA "junk" sequences between meaningful sequences has been found. "Junk" sequences are regarded as noise that serves to amplify and strengthen the system itself.]

The attractor depicted in the graph is remarkable for several reasons:

1] A clear multi-scaled attractor for the Intentional Learning Process is defined with three attractors, one global and the other two smaller replicas that are "self-similar." It is a symmetrical structure with two small replicas of itself within it, reverse-tucked into opposite ends of the larger attractor, as shown in Figure 1. A fully formed self-similar scaled replica resides in the area that I have named "Galaxy 7" [or "G7"], and a partially formed self-similar smaller scaled replica, lacking a "Galaxy 6" [G6], resides in Galaxy 6. Because the "stable state" of the system within the attractor [where it resides most of the time] is manifested in G7, it appears that the larger attractor, as well as the partially formed, or possibly deteriorated, replica, is a replicate of the "powerhouse" of the smaller attractor in G7, although an historical determination is not yet possible.

Scaling, or self-similarity, represents energy changes in the system. At this time, we are not clear whether the scales shown in the attention system of learning represent bifurcations, i.e., parameter changes in the system that alter competing boundary basins in the attractor, or whether they represent wings on a global attractor that is visited when the system is forced out of the basal space of a collapsed hypertorus by a change in the input, not parameter of the system. This understanding, from Ichiro Tsuda's model of "chaotic itinerancy," a visit to historical "attracted ruins" of the system's attractor, is the one that Walter J. Freeman uses to understand the self-similar scaling learning attractor that he has found in his olfactory studies with rabbits. By understanding the replicates as chaotic itinerancy migrations, wings onto which a trajectory can migrate without a fixed order, a self-similar space transition rather than a bifurcation, would allow an attractor great flexibility.

Enough is not yet known about the Spacing system to determine whether the self-similar replicates are due to bifurcations among different attractors within the system, or are wings within a global attractor accessed by local chaotic itinerancies. It is known that imperceptible discriminations of thought and language representing trajectory excursion routes of an attractor can be tracked, and that dynamical analyses are more sensitive to the discrimination of qualitative differences than are our average perceptive abilities. This understanding is suggested by an unexpected finding, illustrated by tracking the dynamics of a dialectical passage, as described below.

2] The attractor for the Intentional Learning System demonstrates a sensitivity to discriminitive intentional behavior. Figure 2 represents the points, or values, of the Euthydemus attractor. There are eight clusters of points, or values, called "Galaxies [G]," representing rest points that define junction points of broken trajectories. They are arbitrarily numbered from G1 to G8.

(*See fig. 2*)

On analysis, each of the 8 Galaxies is comprised of a pair of Components, either Q and A, or S and N. At this time it is not yet known why the Galaxies are paired this way. We distinguish a "local" orbit as a small excursion originating in and returning to the "stable state" of Galaxy 7 [Q and A], where activity most often occurs. The route it most often takes is from G7 to G1 to G8 to G4 to G7, or, simply, G71847. This is not a whole orbital excursion of the entire system but it is the local orbit that the system most frequently takes. Of these four Galaxies, three are Q and A Galaxies and one is an S and N Galaxy. The following example illustrates how the Component of Q, Questioning, occurs in two different Galaxies, G1 and G8:

Table 1. Example of How Question [Q] Occurs in Two Galaxies.

ESD Component and No.		Galaxy	Text
A [Answer]	[508]	7	[Socrates: And the good and the bad experience pain and pleasure to a like degree, though perhaps the bad even more so.]
A [Answer]	[509]	7	Callicles: /Yes.\
Q [Question]	[510]	1	Socrates: /Then the evil man becomes just as bad and good as the good man, or even more good. Is not this the result, alone with what we said before, if anyone identifies the pleasant and the good?\
Q [Question]	[511]	8	/Must not this be so, Callicles?\
S [Spacing]	[512]	4	Callicles: /I have been listening to you for a long time, Socrates, and agreeing with you, as I reflected that, if one concedes something to you even in play, you gladly seize hold of it like a child. Just as if you really think that I or anyone else does not hold some pleasures to be better and others worse!
			Socrates: Ho, ho, Callicles! What a rascal you are, treating me like a child and deceiving me by saying the same things are now thus, now different. And yet I did not think at the beginning that you would willingly deceive me, since you are my friend. But now I have been misled, and apparently, as the old proverb goes, I must make the best of the
A [Answer]	[513]	7	circumstances and take just what you give me.\ /What you now say, it seems, is that some pleasures are good, and some bad.\

I did not code for qualitative differences within Components, except for the Component Spacing, when this passage was coded, nor, despite avid interest in the attempt, could I figure out how to do it, for coding qualitative differences in areas such as Q and A presents formidable problems. Therefore, no differential treatment was accorded to any qualitative differences of Components, except, of course, in the separate Spacing Element analysis. Yet a qualitative difference in the Qs has been sensitively discriminated by the dynamical analysis, demonstrated by the appearance of Q in both Galaxies 1 and 8, graphically and practicably reflecting sensitivity to its historical and futuristic influences, and captured by Ruelle and Taken's ingenious technique to ferret out the function of the system. The technique has also discriminated differences in the other Components, A, S, and N. Although the Qs in this example belong to different Galaxies, G1 and G8, reflecting a difference in the quality of the questions, this consideration was not in any way a part of the analyses. Apparently, this potentially powerful dynamical analytic technique is sensitive to thought and language discrimination, for some qualitative difference in the system is being reflected in the linguistic analysis before a change in the system is actually perceived. It is, perhaps, at this level that the system begins to reflect its newly created and evolving constraints that have been developed by the reciprocal interaction of environmental events and the learner's cognitive generative processes, driven and guided by his attention system, which, it is hypothesized, is driven and guided by his Intentionality.

Little, if anything, is known about how Intentionality guides a biological system, but I think that physically felt neural constraints, physically experienced as pain, the feeling of a necessity for closure, and, as well, a neural experience of beauty, are integral parts of the system. While a clear perceptual discrimination of language may still be lacking, its emotional content is generally well discriminated in person, perhaps another reason that individual Spacing Elements, evolved to, or from, greater importance in attention, and consequently, in the cognitive generative learning process, including the greater usage of the Element, Being Present [BP] that has been observed. Is it possible that the cognitive generative learning process generates a correlation of meaning to information by use of the emotional controls of the attention system?

While a qualitative difference in the Questions may be evident in the above illustration, the capacity of Ruelle and Taken's Trick protocol to sensitively detect apparently imperceptible differences within a complex system makes it a potentially powerful analytical tool, the value of which cannot be overestimated. Probes to analyze the dynamics of these self-similar replicate attractors must yet be developed, for they pose intriguing physical and metaphysical questions which are of important practical and heuristic value.

3] The attractor for the Intentional Learning System is robust, as demonstrated not only by its continuity over a 2,500 year span of samples, but even in a system with as few as nine points, as shown in a dialectic in Matthew Lipman's contemporary philosophical novels for children, the attractor demonstrates the intentional learning attractor pattern, as shown in Figure 3.

(See fig. 3)

4] The attractor for the Intentional Learning System appears to be universal. In a literacy study of pre-schoolers' interactions with parents during reading aloud sessions, David Yaden, at UCLA, used a coding system of six components, with a correlative Spacing component. When the data were analyzed, the same Intentional Learning System attractor was found, shown in Figure 4, supporting the hypothesis that it is a universal system.

(See fig. 4)

The attractor in Figure 5, without a discernible pattern, is hypothesized to be that of a child or parent who has not yet mastered this specific learning process.

(See fig. 5)

In conclusion, by analysis of the attractor in the attention system of the cognitive generative learning process, we now know not only that Intentional Learning Systems exist, but that they operate to affect cognitive processing at well below the conscious level. Moreover, because its attractor appears to be multi-scaled, intentional, robust and universal, we are compelled to conclude that the Intentional Learning System has these qualities as well. This study supports Karl Pribram and Diane McGuiness' finding that the emotional and motivational controls on attention are prime players in cognitive processing.

[1] Lillian Greeley, Philosophical Spacing [PS]: Key to the Nonlinear Complex Dynamics of the Attentional System of the Cognitive Learning Process in the Philosophical Dialectic Method, Doctoral Dissertation, Harvard University, 1990.

[2] Greeley; also: Greeley, Lillian, "Complexity in the Attention System of the Cognitive Generative Learning Process," in Chaos and Society, ed. A. Albert, IOS Press and Presses de l'Universite du Quebec, Amsterdam, 1995.

[3] Greeley, Philosophical Spacing, also: Baranger, Michel and Lillian Greeley, to be published.

[4] Samuel Karlin and Volker Brendel, "Patchiness and Correlations in DNA Sequences," Science, v. 259, 29 Jan., 1993, pp. 677- 679.

[5] Ichiro Tsuda, "Chaotic Itinerancy as a Dynamical Basis of Hermeneutics in Brain and Mind," World Futures, v. 32, 1991, pp. 167-184; also: Walter J. Freeman, "Controlled Chaos in the Basal Forebrain, Bifurcation during Learning; Itinerancy during Perception," 2nd International Conference on Fuzzy Logic and Neural Networks, Iizuka, July 20, 1992, p. 4.

[6] Walter J. Freeman, "The Physiology of Perception," Scientific American, February 1991; also: Christine A. Skarda and Walter J. Freeman, "How Brains Make Chaos in Order to Make Sense of the World," Brain and Behavioral Sciences ,10:2 (1987), 10;

[7] I am grateful to Prof. Walter J. Freeman for guiding me to this understanding.

[8] Gorgias, in The Collected Dialogues of Plato, ed. by Edith Hamilton and Huntington Cairns (Princeton University Press, Princeton, NJ, 1961); [Philosophical Dialectic Sample 5, Page 282, Line 9-Page 282, Line 26] in: Greeley, "Philosophical Spacing."

[9] Lillian Greeley, "The Bumper Effect Dynamic in the Creative Process: The Philosophical, Psychological and Neuropsychological Link," The Journal of Creative Behavior," v.20, n.4, Fourth Quarter, 1986, pp. 261-275; also: Greeley, "Philosophical Spacing," pp. 241, 253.

[10] Lipman, Matthew, Elfie, Book Three (Upper Montclair, New Jersey, The Institute for the Advancement for the Advancement of Philosophy for Children, 1988), [Philosophical Dialectic Sample MLD 37, Page162,Line1-Page164,Line10] in: Greeley, "Philosophical Spacing."

[11] Diane McGuiness and Karl Pribram, "The Neuropsychology of Attention: Emotional and Motivational Controls," in The Brain and Psychology, ed. by M.C. Wittrock (New York, Academic Press, 1986), pp. 95-139; also: Greeley, pp. 184-222.

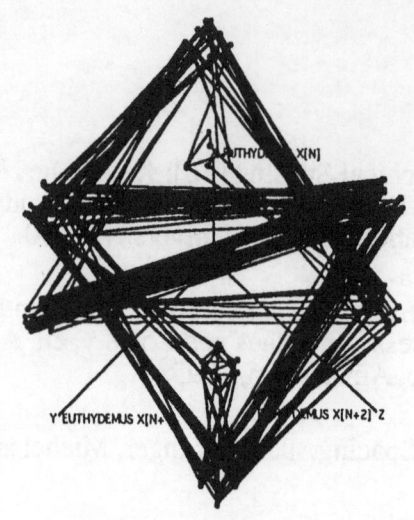

FIGURE 1. ATTRACTOR OF THE EUTHYDEMUS,
755 EVENTS/POINTS; TRAJECTORIES, VIEW B
[LILLIAN GREELEY, 1990].

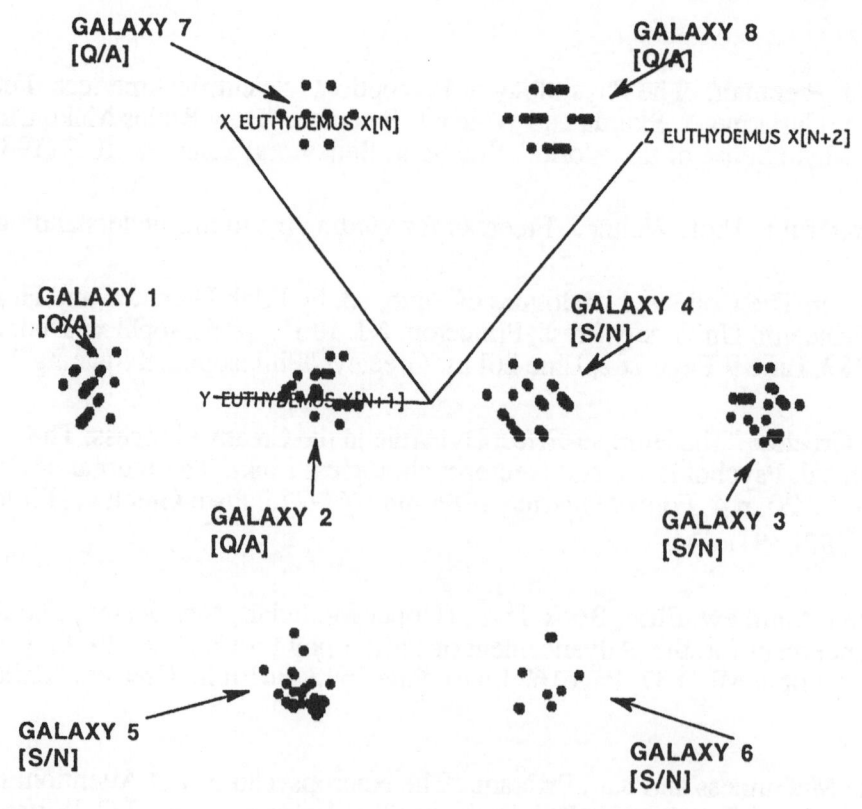

FIGURE 2. GALAXIES OF THE EUTHYDEMUS ATTRACTOR,
755 EVENTS/POINTS, VIEW C
[LILLIAN GREELEY, 1994].

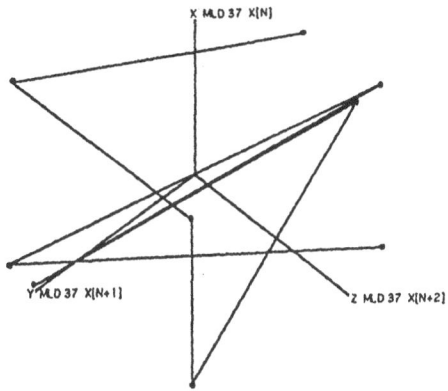

FIGURE 3. ATTRACTOR OF MATTHEW LIPMAN'S PHILOSOPHICAL
DIALECTIC IN HIS PHILOSOPHICAL NOVEL FOR CHILDREN,
ELFIE [SAMPLE MLD 37, 9 EVENTS/POINTS, TRAJECTORIES,
VIEW B; LILLIAN GREELEY, 1993].

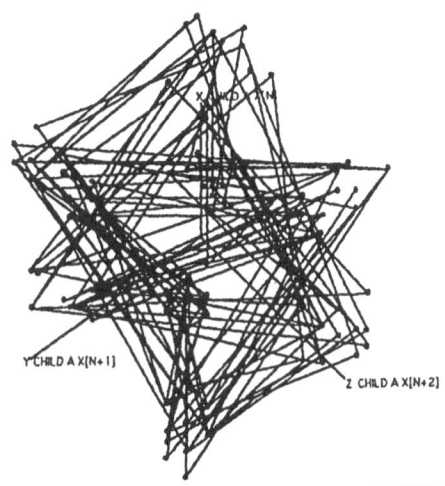

FIGURE 4. ATTRACTOR OF LITERACY LEARNING PROCESS WITH
PRESCHOOLERS AND PARENTS, CHILD A, FEMALE, 4.1 YEARS,
227 POINTS/EVENTS [DAVID YADEN, CODING; LILLIAN
GREELEY, ANALYSIS; 1994].

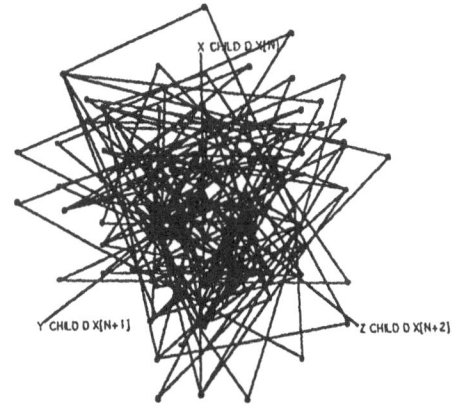

FIGURE 5. ATTRACTOR OF PRESCHOOLER IN LITERACY LEARNING WITH
PARENTS STUDY SHOWING SUSPECTED NONMASTERY OF
THE LEARNING PROCESS, CHILD D, MALE, 4.8 YEARS, 233
EVENTS/POINTS [DAVID YADEN, CODING; LILLIAN GREELEY,
ANALYSIS, 1994].

The Network Level: Self-Organization

7.

The Three Languages of the Brain: Quantum, Reorganizational, And Associative

Subhash C. Kak

The Three Languages of the Brain: Quantum, Reorganizational, and Associative

Subhash C. Kak
Department of Electrical & Computer Engineering
Louisiana State University
Baton Rouge, LA 70803-5901
Email: kak@gate.ee.lsu.edu

September 22, 1995

1 Introduction

Progress in science is reflected in a corresponding development of language. The vistas opened up by the microscope, the telescope, tomography and other sensing devices have resulted in the naming of new entities and processes. Quantum theory has led to the supersession of the classical atomic picture and one speaks in terms of tangled processes and non-binary logic. Quantum theory has also led to deep questions related to the definition of the observer and the observed. This has been one path to the examination of the mystery of mind. The other paths are rooted in ancient philosophical traditions and the psychological theories of the past century.

The language for the description of the mind in scientific discourse has not kept pace with the developments in the physical sciences. The mainstream discussion has moved from the earlier dualistic models of common belief to one based on the emergence of mind from the complexity of the parallel computer-like brain processes. The two old paradigms of determinism and autonomy, expressed sometimes in terms of separation and interconnectedness, show up in various guises. Which of the two of these is in favor depends on the field of research and the prevailing fashions. Although quantum theory has provided the foundation for physical sciences for seventy years, it is only recently that holistic, quantum-like operations in the brain have been considered. This fresh look has been prompted by the setbacks suffered by the various artificial intelligence (AI) projects and also by new analysis and experimental findings. It is being recognized that stimulus-response constructs such as "drive" are often inadequate in providing explanations; and one invokes the category "effort" to explain autonomous behavior. Karl Pribram's classic *Languages of the Brain* (1971) describes many paradoxes in the standard linguistic and logical categories used in describing brain behavior. Since that book was written many new approaches have been tried and found wanting in resolving these paradoxes.

The languages used to describe the workings of the brain have been modeled after the dominant scientific paradigm of the age. The rise of mechanistic science saw the conceptualization of the mind as a machine. In our present computer age, the brain is often viewed as a computing machine; Neural networks are the engines of this machine. The Appalachian Conferences on Behavioral Neurodynamics have been in the forefront in questioning long-standing assumptions at the basis of brain science. These meetings and several books (e.g. Pribram, 1991, 1993, 1994; Penrose, 1994; King and Pribram, 1995) have shown that the standard paradigm based on the computer-like signal processing model of the brain is flawed in many ways.

Although the neural network approach has had considerable success in modeling many counterintuitive illusions, there exists other processes in human and nonhuman cognition that appear to fall outside the scope of such models. Scholars have expressed the opinion that brain processing cannot be described by Turing machines. We do not wish to go into the details of these arguments; rather we will examine the question in the broadest terms.

Briefly, the classical neural network model does not provide a resolution to the question of binding of patterns: How do the neuron firings in the brain come to have specific meanings or lead to specific images? The proposal that 40-Hz waveforms are characteristic of consciousness and may somehow bind the activity in different parts of the brain is too vague to be taken seriously. Furthermore, machines have been unable to match many computing capabilities of nonhumans. Is that because computers lack the self-organizational feature of biological systems?

In unified theories of physics one speaks of a single force that, upon symmetry breaking, manifests itself into three or four distinct forces. Analogously, we argue that the quantum language of the brain manifests itself in terms of other languages. In this paper I consider the computational aspects of the problems of perception and adaptation in light of dual and associative processes. First, I summarize the limitations of computational models by considering questions raised by new researches in animal intelligence. The central insight obtained from the study of animal intelligence is that it is predicated on continual self-organization, as seen, for example, in superorganisms. That biological processing has a quantum basis has been argued by several authors. Quantum models provide a natural explanation for the unity of awareness in addition to explaining other puzzling features of brain behavior. In one class of such models, quantum behavior is postulated within neurons. But this does not resolve the question of the continuing self-organization of a biological system. My own proposal (Kak, 1992, 1995b) looks at organization and information as new quantum mechanical variables and I call this holistic view as quantum neural computing. This topic is reviewed and its implications are described.

For any quantum phenomenon there should be classical approximate representations. Since self-organization is the basic feature of biological processing, one needs to consider an explicit signaling scheme for this. This additional signal provides a dual to the usual neural transmissions in parallel with the many-component vectors of a quantum description. This bottom-up dual signaling regime for the brain may be taken to complement the top-down quantum view.

This paper also considers the *associative learning* problem, that deals with the most basic linguistic category of how associations are implemented. When a pattern is presented to a

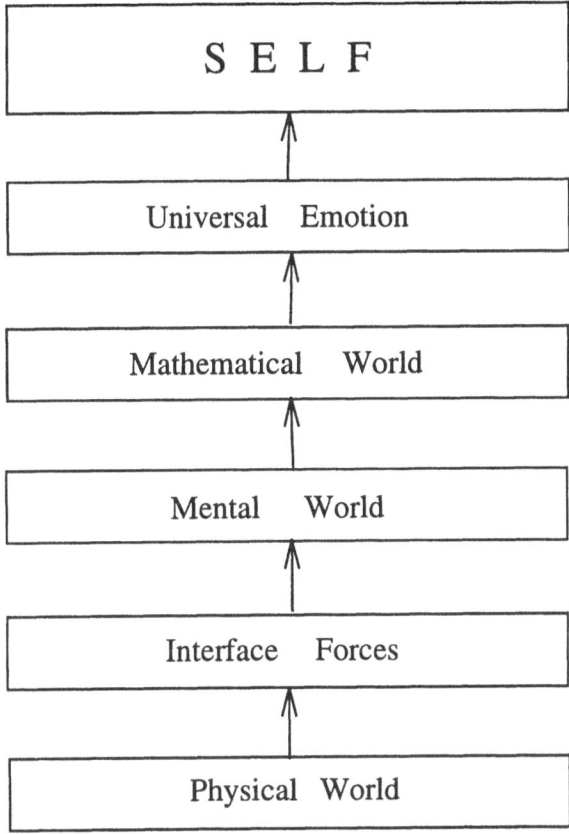

Figure 1: The Vedic model for the mind

system for the first time, the synaptic weights that can resonate, or generate, this in the neural circuitry don't exist. So notions of supervised learning cannot be realistic biologically. A scheme is presented that can find synaptic weights instantaneously.

In summary, our paper speaks of the three languages of brain: quantum, reorganizational, and associative. Our learning scheme provides a basis for the interiorization of the last two languages. How these languages fit into the overarching quantum framework remains to be investigated.

2   Old and New Models of Mind

The puzzle of cognition was undoubtedly a part of the discourse of all ancient civilizations. Often the mystery was expressed by the notion of god or spirit: the gods were the cognitive centers of the mind. In at least one culture, namely the Vedic civilization of India, an astonishingly sophisticated cognitive theory emerged.

The Vedic model of mind, that goes back to at least 2000 B.C., provides a hierarchical structure with a twist that allows it to transcend the categories of separation and wholeness. Figure 1 presents this structure. Notice that the lowest level is the physical world or body, with higher levels of interface forces, the mind, scientific intuition, emotion, with the universal self sitting atop. The lower levels are machine-like whereas the self is the sole entity of

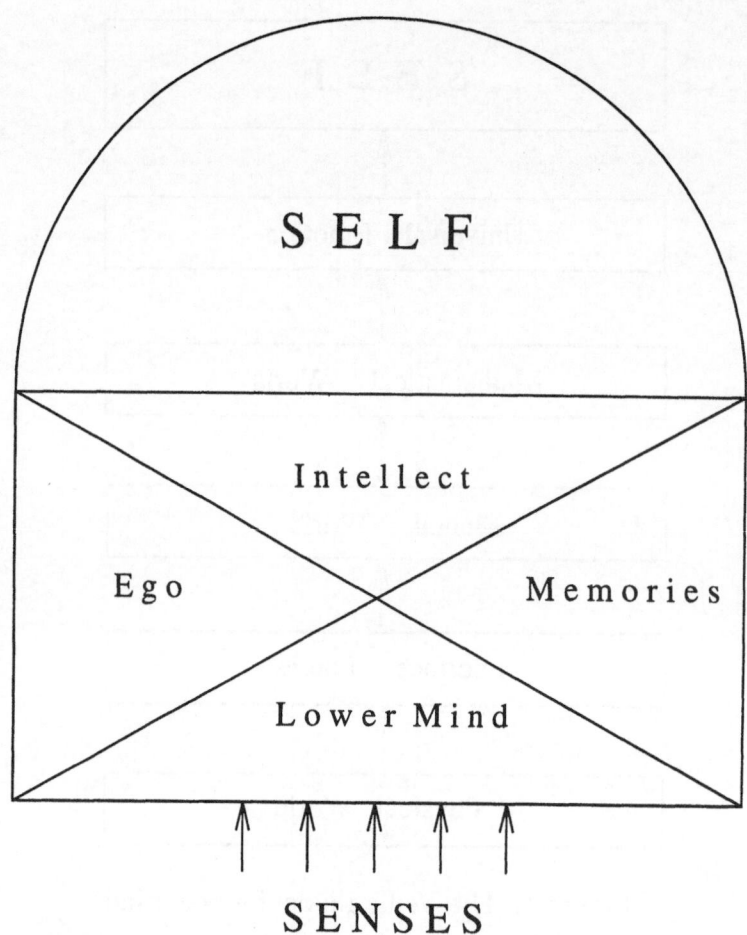

Figure 2: Mind's structure

consciousness. The individual's idea of consciousness arises out of associations with events, illuminated by the consciousness principle.

Figure 2 provides greater details in terms of the relationship of memories, I-ness or ego that provides unity, intellect for judgment and the universal self. The I-ness is taken to arise from associations; and intellect represents decision making whose basis is also taken to be associative. In this view the lower mind is machine like, and it is the self that provides the binding to the associations of the lower mind.

The most striking part of this model is the nature of the universal self. Considered to transcend time, space and matter, the self engenders these categories on the physical world. For this reason the Vedic model is often taken to be an idealist model but that is not an entirely correct interpretation. According to the Vedic view any description of reality is inevitably paradox ridden. This is so since consciousness, through which we observe the universe, is also a part of it. The original sources speak of consciousness transcending categories of space, time and matter. There are two interpretative traditions that have a long history in India:

- A theory where consciousness is the ground-stuff from which time, space and matter emerge.

- A theory where consciousness is a field separate from time, space and matter. The duality in this conception is more apparent than real having taken form after the first separation of the two categories.

According to these traditions mind itself must be seen as a complex structure. Whereas mind is emergent and based on the capabilities of neural hardware, it cannot exist without the universal self. One implication of these ideas is that machines, which are based on classical logic, can never be conscious (Kak, 1993a).

It is not well known that this model had an important influence on the development of quantum mechanics. In 1925, before his creation of wave mechanics, Erwin Schrödinger wrote:

> This life of yours which you are living is not merely a piece of this entire existence, but in a certain sense the "whole"; only this whole is not so constituted that it can be surveyed in one single glance. This, as we know, is what the Brahmins express in that sacred, mystic formula which is yet really so simple and so clear: *tat tvam asi,* this is you. Or, again, in such words as "I am in the east and the west, I am above and below, *I am this entire world.*" (Schrödinger, 1961 (1925); Moore, 1989, page 170-3)

Schrödinger's influential *What is Life?* also used Vedic ideas. According to his biographer Walter Moore, there is a clear continuity between Schrödinger's understanding of Vedanta and his research:

> The unity and continuity of Vedanta are reflected in the unity and continuity of wave mechanics. In 1925, the world view of physics was a model of a great machine composed of separable interacting material particles. During the next few years, Schrödinger and Heisenberg and their followers created a universe based on superimposed inseparable waves of probability amplitudes. This new view would be entirely consistent with the Vedantic concept of All in One. (Moore, 1989, page 173)

For a summary of the Vedic theory of mind and its later developments see Kak (1993a); and for chronological issues see Kak (1994a), Klostermaier (1994), and Feuerstein et al (1995). During its evolution, a strand of the Vedic tradition took consciousness to be the sole reality. This anticipates several contemporary speculative ideas of physics.

The Vedic theory of mind is part of a recursive approach to knowledge (e.g. Kak, 1994a). The Vedas speak of three worlds, namely the physical, the mental, and that of knowledge. Consciousness if the fourth, transcending world. There is also reference to four kinds of language: gross sound, mental imagery, gestalts, and a fourth that transcends the first three and is associated with the self (Kak 1993a).

Plato's ideas are less comprehensive than the Vedic model, but they go beyond the common-sensical dichotomy of body and mind. He enlarged this dichotomy by speaking of a third world of ideas or forms. In his parable of the cave, Plato speaks of ideas that have independent existence of which our senses only see the traces or shadows on the wall.

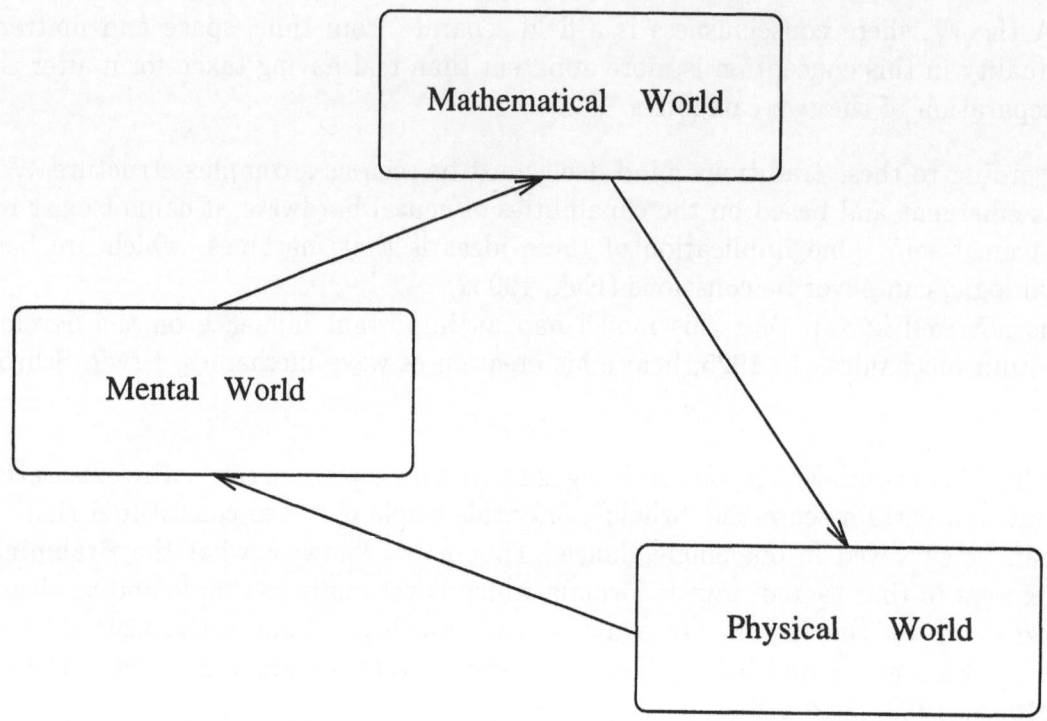

Figure 3: The three worlds: physical, mental, mathematical

Plato's model represents the lower four levels of the Vedic model of Figure 1, with the second level of interface forces subsumed into the mental world. It has exercised great influence on Western philosophy and its variants have been examined by many writers. In contemporary debate (Popper and Eccles, 1977; Goswami, 1993; Penrose, 1994) the world of ideas has been interpreted as objective knowledge. Consider Figure 3. This shows three worlds: the physical, the mental, and the world of mathematical or scientific ideas. The puzzle is: How is a part of the physical world which generates the mental picture which, in turn, creates the scientific theory is able to describe nature so well. This has been termed by one scientist as the unreasonable effectiveness of mathematics. In truth, objective knowledge consists of many paradoxes (Kak, 1986). Accumulation of knowledge often amounts to making ad hoc choices in the underlying formal framework to conform to experience. The most fundamental, and a very ancient, antinomy is that between determinism and free will. Formal knowledge can at best be compared to a patchwork.

2.1 New models

Considering that the physical world is described at its most basic level by quantum mechanics, how can classical computational basis underlie the description of the structure (mind) that in turn is able to comprehend the universe? How can machines, based on classical logic, mimic biological computing? One may argue that ultimately the foundations on which the circuitry of classical computers is based is at its deepest level described by quantum mechanics. Nevertheless, actual computations are governed by a binary logic which is very different from the tangled computations of quantum mechanics. And since the applicability of quantum mechanics is not constrained, in principle, by size or scale, so classical computers do appear to be limited.

Why cannot a classical computer reorganize itself in response to inputs? If it did, it will soon reach an organizational state associated with some energy minimum and will then stop responding to the environment. Once this state has been reached the computer will now merely transform data according to its program. In other words, a classical computer does not have the capability to be selective about its inputs. This is precisely what biological systems can do with ease.

Most proposals on considering brain function to have a quantum basis have done so by default. In short the argument is: There appears to be no resolution to the problem of the binding of patterns and there are non-local aspects to cognition; quantum behavior has non-local characteristics; so brain behavior might have a quantum basis. But these models do not explain the self-organizing part of brain structure. Other quantum models consider the organization of the brain itself to be a quantum variable.

Since a model of mind (i.e. Vedic) was influential in the early development of quantum theory, it is very fitting that quantum mechanical ideas should, in turn, shape the unfolding of new ideas in brain science.

How one might go about devising a system that is more capable than a classical system? If one takes the parallel with the early development of quantum mechanics, it is necessary to speak of vectors—rather than scalars— carrying information. This was recognized early by Karl Pribram who has long been a proponent for moving beyond the classical computer paradigm for the brain. In Pribram (1971) a dual signal model that stresses the significance of reorganization was proposed:

> The unit of analysis for brain function has classically been the neuron. The present proposal for a two-process mechanism recognizes an additional unit: the neuron junction, whose activity can become part of an organization (the slow potential microstructure) temporarily unrelated to the receptive field of any single neuron. Neural junctions are thus much more than just way stations in the transmission of nerve impulses; they compose, at any moment, a neural state that is operated upon by arriving nerve impulses. In turn, nerve impulses generated by neurons are influenced by this state. (page 25)

Newer analysis has led to the understanding that one needs to consider reorganization as a primary process in the brain— this allows the brain to define the context. Does a field govern the process of reorganization? The signal flows now represent the processing or recognition done within the reorganized hardware. Such a change in perspective can have significant implications. Dual signaling schemes eventually need an explanation in terms of a binding field; they do not solve the basic binding problem in themselves but they do make it easier to understand the process of adaptation.

3 Machine and Biological Intelligence

For all computational models, the question of the emergence of intelligence is a basic one. Solving a specified problem, that often requires searching or generalization, is taken to be a sign of AI, which is assumed to have an all or none quality.

From an evolutionary point of view it may be assumed that intelligence has gradation. If such gradation exists, does it manifest itself on top of a minimum that is common to life? If this minimum intelligence cannot be replicated by machines then it would follow that machine intelligence, based on classical logic, can never match biological intelligence.

In the biological realm, we find that all animals are not equally intelligent at all tasks; here intelligence refes to performance of various tasks, and this performance may depend crucially on the animal's normal behavior. It may be argued that all animals are *sufficiently* intelligent because they survive in their ecological environment. Nevertheless, even in cognitive tasks of the kind normally associated with human intelligence animals may perform adequately. Thus rats might find their way through a maze, or dolphins may be given logical problems to solve, or the problems might involve some kind of generalization (Griffin, 1992). These performances could, in principle, be used to define a gradation.

If we take the question of AI programs, it may be argued that the objectives of each define a specific problem solving ability, and in this sense AI programs constitute elements in a spectrum. But machine intelligence has not been predicated on some basic, benchmark tests.

3.1 Can machines think?

If we define thinking in terms of language or picture understanding then, by current evidence, machines cannot think. As we will see in the next subsection, machines cannot even perform abstract generalization of the kind that is natural for birds and other animals. But the proponents of strong-AI believe that, notwithstanding their current limitations, machines will eventually be able to simulate the mental behavior of humans. They suggest that the Turing (1950) test should suffice to establish machine intelligence.

We first show that Turing test is not suitable to determine progress in AI (Kak, 1995d). According to this test the following protocol is used to check if a computer can think: (1) The computer together with a human subject are to communicate, in an impersonal fashion, from a remote location with an interrogator; (2) The human subject answers truthfully while the computer is allowed to lie to try to convince the interrogator that it is the human subject. If in the course of a series of such tests the interrogator is unable to identify the real human subject in any consistent fashion then the computer is deemed to have passed the test of being able to think. It is assumed that the computer is so programmed that it is mimicking the abilities of humans. In other words, it is responding in a manner that does not give away the computer's superior performance at repetitive tasks and numerical calculations.

The asymmetry of the test, where the computer is programmed to lie whereas the human is expected to answer truthfully is a limitation of the test that has often been criticized. This limitation can be overcome easily if it is postulated that the human can take the assistance of a computer. In other words, one could speak of a contest between a computer and a human assisted by another computer. But this change does not mitigate the ambiguity regarding the kind of problems to be used in the test. The test is not objectively defined; the interrogator is a human.

It has generally been assumed that the tasks that set the human apart from the machine are those that relate to abstract conceptualization best represented by language understanding. The trouble with this popular interpretations of the Turing test, which was true to its intent as best as we can see, is that it focused attention exclusively on the cognitive abilities of humans. So researchers could always claim to be making progress with respect to the ultimate goal of the program, but there was no means to check if the research was on the right track. In other words, the absence of intermediate signposts made it impossible to determine whether the techniques and philosophy used would eventually allow the Turing test to be passed.

In 1950, when Turing's essay appeared in print, matching human reasoning could stand for the goal that machine intelligence should aspire to. The problem with such a goal was that it constituted the ultimate objective and Turing's test did not make an attempt to define gradations of intelligence. Had specific tasks, which would have constituted levels of intelligence or thinking below that of a human, been defined then one would have had a more realistic approach to assessing the progress of AI.

The prestige accorded to the Turing test may be ascribed to the dominant scientific paradigm in 1950 which, following old Cartesian ideas, took only humans to be capable of thought. That Cartesian ideas on thinking and intelligence were wrong has been amply established by the research on nonhuman intelligence of the past few decades (Griffin, 1992).

To appreciate the larger context of scientific discourse at that time, it may be noted that interpretations of quantum mechanics at this time also spoke in terms of observations alone; any talk of any underlying reality was considered outside the domain of science. So an examination of the nature of "thought", as mediating internal representations that lead to intelligent behavior, was not considered a suitable scientific subject. Difficulties with the reductionist agenda were not so clear, either in physical sciences or in the study of animal behaviour.

3.2 Animal intelligence

For considerable time it was believed that language was essential ground for thought; and this was taken as proof that only humans could think. But nobody will deny that deaf-mutes, who never learnt a language, do think. Language is best understood as a subset of a large repertoire of behavior. Research has now established that animals think and are capable of learning and problem solving.

Since nonhumans do not use abstract language, their thinking is based on discrimination at a variety of levels. If such conceptualization is seen as a result of evolution, it is not necessary that this would have developed in exactly the same manner for all species. Other animals learn concepts nonverbally, so it is hard for humans, as verbal animals, to determine their concepts. It is for this reason that the pigeon has become a favorite with intelligence tests; like humans, it has a highly developed visual system, and we are therefore likely to employ similar cognitive categories. It is to be noted that pigeons and other animals are made to respond in extremely unnatural conditions in Skinner boxes of various kinds. The abilities elicited in research must be taken to be merely suggestive of the intelligence of the animal, and not the limits of it.

In a classic experiment Herrnstein (1985) presented 80 photographic slides of natural scenes to pigeons who were accustomed to pecking at a switch for brief access to feed. The scenes were comparable but half contained trees and the rest did not. The tree photographs had full views of single and multiple trees as well as obscure and distant views of a variety of types. The slides were shown in no particular order and the pigeons were rewarded with food if they pecked at the switch in response to a tree slide; otherwise nothing was done. Even before all the slides had been shown the pigeons were able to discriminate between the tree and the non-tree slides. To confirm that this ability, impossible for any machine to match, was not somehow learnt through the long process of evolution and hardwired into the brain of the pigeons, another experiment was designed to check the discriminating ability of pigeons with respect to fish and non-fish scenes and once again the birds had no problem doing so. Over the years it has been shown that pigeons can also distinguish: (1) oak leaves from leaves of other trees, (ii) scenes with or without bodies of water, (iii) pictures showing a particular person from others with no people or different individuals.

Other examples of animal intelligence include mynah birds who can recognize trees or people in pictures, and signal their identification by vocal utterances—words—instead of pecking at buttons, and a parrot who can answer, vocally, questions about shapes and colors of objects, even those not seen before (for references see Griffin, 1992). The intelligence of higher animals, such as apes, elephants, and dolphins is even more remarkable.

Another recent summary of this research is that of Wasserman (1995):

> [Experiments] support the conclusion that conceptualization is not unique to human beings. Neither having a human brain nor being able to use language is therefore a precondition for cognition... Complete understanding of neural activity and function must encompass the marvelous abilities of brains other than our own. If it is the business of brains to think and to learn, it should be the business of behavioral neuroscience to provide a full account of that thinking and learning in all animals—human and nonhuman alike.

An extremely important insight from experiments of animal intelligence is that one can attempt to define different gradations of cognitive function. It is obvious that animals are not as intelligent as humans; likewise, certain animals appear to be more intelligent than others. For example, pigeons did poorly at picking a pattern against two other identical ones, as in picking an A against two B's. This is a very simple task for humans.

Wasserman (1993, 1995) devised an experiment to show that pigeons could be induced to amalgamate two basic categories into one broader category not defined by any obvious perceptual features. The birds were trained to sort slides into two arbitrary categories, such as category of cars and people and the category of chairs and flowers. In the second part of this experiment, the pigeons were trained to reassign one of the stimulus classes in each category to a new response key. Next, they were tested to see whether they would generalize the reassignment to the stimulus class withheld during reassignment training. It was found that the average score was 87 percent in the case of stimuli that had been reassigned and 72 percent in the case of stimuli that had not been reassigned. This performance, exceeding the level of chance, indicated that perceptually disparate stimuli had amalgamated into a new category. A similar experiment was performed on preschool children. The children's score was 99 percent for stimuli that had been reassigned and 80 percent for stimuli that had not been reassigned. In other words, the children's performance was roughly comparable to that of pigeons. Clearly, the performance of adult humans at this task will be superior to that of children or pigeons.

Another interesting experiment related to the abstract concept of sameness. Pigeons were trained to distinguish between arrays composed of a single, repeating icon and arrays composed of 16 different icons chosen out of a library of 32 icons (Wasserman, 1995; Wasserman et al, 1995). During training each bird encountered only 16 of the 32 icons; during testing it was presented with arrays made up of the remaining 16 icons. The average score for training stimuli was 83 percent and the average score for testing stimuli was 71 percent. These figures show that an abstract concept not related to the actual associations learnt during training had been internalized by the pigeon.

Animal intelligence experiments suggest that one can speak of different styles of solving AI problems. Are the cognitive capabilities of pigeons limited because their style has fundamental limitations? Can the relatively low scores on the sameness test for pigeons be explained on the basis of wide variability in performance for individual pigeons and the unnatural conditions in which the experiments are performed? Is the cognitive style of all animals similar and the differences in their cognitive capabilities arise from the differences in the sizes of their mental hardware? And since current machines do not, and cannot, use inner representations, is it right to conclude that their performance can never match that of animals? Most importantly, is the generalization achieved by pigeons and other nonhumans beyond the capability of machines?

Donald Griffin (1992) expresses the promise of animal intelligence research thus:

> Because mentality is one of the most important capabilities that distinguishes living animals from the rest of the known universe, seeking to understand animal minds is even more exciting and significant than elaborating our picture of inclusive fitness or discovering new molecular mechanisms. Cognitive ethology presents us with one of the supreme scientific challenges of our times, and it calls for our best efforts of critical and imaginative investigation. (Page 260)

3.3   Recursive characteristics

A useful perspective on animal behavior is its recursive nature, or part-whole hierarchy. Considering this from the bottom up, animal societies have been viewed as "superorganisms". For example, the ants in an ant colony may be compared to cells, their castes to tissues and organs, the queen and her drones to the generative system, and the exchange of liquid food amongst the colony members to the circulation of blood and lymph. Furthermore, corresponding to morphogenesis in organisms the ant colony has sociogenesis, which consists of the processes by which the individuals undergo changes in caste and behavior. Such recursion has been viewed all the way up to the earth itself seen as a living entity. Parenthetically, it may be asked whether the earth itself, as a living but unconscious organism, may not be viewed like the unconscious brain. Paralleling this recursion is the individual who can be viewed as a collection of several "agents" where these agents have sub-agents which are the sensory mechanisms and so on. But these agents are bound together and this binding defines consciousness.

A distinction may be made between simple consciousness and self-consciousness. In the latter, the individual is aware of his awareness (Eccles, 1979, 1989). It has been suggested that while all animals may be taken to be conscious, only humans might be self-conscious. It is also supposed that language provides a tool to deal with abstract concepts that opens up the world of mathematical and abstract ideas only to humans.

Edelman (1992) suggests that selection mechanism might be at work that has endowed brains, in their evolutionary ladder, with increasing complexity. But this work does not address the question of holistic computations at all. From an evolutionary perspective if the fundamental nature of biological computing is different from that of classical computers then models like that of Edelman cannot provide the answers we seek. We cannot also accept the line of reasoning according to which complexity, once it crosses a certain threshold, leads to consciousness and, furthermore, beyond another threshold leads to self-consciousness.

4   Holistic Processing and Quantum Models

Neural activity in the brain is bound together to represent information; but the nature of this binding is not known. The brain constantly reorganizes itself based on the information task. Now quantum mechanics has provided a new understanding of the physical world although its philosophical implications are quite contentious and murky. Quantum mechanics is a theory of "wholes" and in light of the fact that the eye responds to single photons (Baylor et al, 1979)—a quantum mechanical response—and that the mind perceives itself to be a unity, one would expect that its ideas would be applied to examine the nature of mind and of intelligence. But for several decades the prestige of the reductionist program of neurophysiology made it unfashionable to follow this path.

Meanwhile, the question of the nature of information, and its observation, has become important in physics. The binding problem of psychology, and the need to postulate a mediating agent in the processing of information in the brain, has also brought the "self" back into the picture in biology. Oscillations and chaos have been proposed as the mechanisms to explain this binding. But we think that the strongest case can be made for a quantum mechanical substratum that provides unity to experience. Such quantum mechanical models of consciousness have attracted considerable attention; this work is reviewed in Kak (1995 a,b), Penrose (1994).

Quantum computing in the style of Feynman (1986) is considering the use of lattices or organo-metallic polymers as the apparatus; but the idea here is to perform computations in a tangled manner that can provide speedup over classical computers. This research does not consider the question of modeling of mind. It has been argued (Feynman, 1982; Kak, 1995c) that it is not possible to simulate quantum mechanics on a traditional computer. If it is accepted that intelligence has a quantum mechanical basis, then it follows that Turing-machine models of intelligence are inadequate. This, in turn, leads to several questions: What hardware basis is required before intelligence can emerge out of a quantum structure? Does intelligence require something more than a quantum basis, the presence of the notion of self?

Microtubules, the skeletal basis of cells that consist of protein polymers, have been proposed by Hameroff and others (Jibu et al, 1994; Penrose, 1994) as supporting quantum mechanical processes. It has been suggested by Fröhlich (1975) that Bose-Einstein condensation might be responsible for quantum coherence in biological structures. Fröhlich's model requires large energy of metabolic drive and extreme dielectric properties of the materials. The large scale quantum coherence is predicted to appear in the frequency range of 10^{11} to 10^{12} Hz and there is some evidence that such oscillations actually take place (Grundler and Keilmann, 1983). Hameroff and his associates (Jibu et al, 1994) have suggested that water in the microtubules provides this quantum coherence. But it is not shown how quantum coherence can leap across the synaptic barrier. This work also does not deal with the issue of what structures are needed before consciousness can arise. Most significantly, this does not address the issue of the self-organizing ability of biological systems.

My own work (Kak, 1976, 1984, 1995a,b) has examined the basic place of information in a quantum mechanical framework and its connections to structure. This work shows that although many processes that constitute the mind can be understood through the framework of neural networks, there are others that require a holistic basis. Furthermore, if each computation is seen as an operation by a reorganized brain on the signals at hand, this has a parallel with a quantum system where each measurement is represented by an operator. I also suggest that the macrostructure of the brain must be seen as a quantum system.

Study of animal intelligence provides us with new perspectives that are useful in representing the performance of machines. For example, the fact that pigeons learn the concept of sameness shows that this could not be a result of associative response to certain learnt patterns. If evolution has led to the development of specialized cognitive circuits in the brain to perform such processing, then one might wish to endow AI machines with similar circuits. Other questions arise: Is there a set of abstract processors that would explain animal performance? If such a set can be defined, is it unique, or do different animal species represent collections of different kinds of abstract processing that makes each animal come to achieve a unique set of conceptualizations?

One striking success of the quantum models is that they provide a resolution to the determinism- free will problem. According to quantum theory, a system evolves causally until it is observed. The act of observation causes a break in the causal chain. This leads to the notion of a participatory universe (Wheeler, 1979-1981). Consciousness provides a break in the strict regime of causality.

It would be reasonable to assume that this freedom is associated with all life. But its impact on the ongoing processes will depend on the entropy associated with the break in the causal chain.

4.1   A universal field

If one did not wish for a reductionist explanation as is inherent in the cytoskeletal model, one might postulate a different origin for the quantum field. Just as the unified theories explain the emergence of electromagnetic and weak forces from a mechanism of symmetry breaking, one might postulate a unified field of consciousness-unified_force-gravity from where the individual fields emerge.

Eugene Wigner (1961) spoke of one striking analogy between light and consciousness:

"Mechanical objects influence light—otherwise we could not see them—but experiments to demonstrate the effect of light on the motion of mechanical bodies are difficult. It is unlikely that the effect would have been detected had theoretical considerations not suggested its existence, and its manifestation in the phenomenon of light pressure." He also acknowledged one fundamental difference between light and consciousness. Light can interact directly with virtually all material objects whereas consciousness is grounded in a physico-chemical structure. But such a difference disappears if it is supposed that the physico-chemical structure is just the instrumentation that permit observations.

In other words, the notion of a universal field requires acknowledging the emergence of the individual's I-ness at specialized areas of the brain. This I-ness is intimately related to memories, both short-term and long-term. The recall of these memories may be seen to result from operations by neural networks. Lesions to different brain centers effect the ability to recall or store memories. For example, lesions to the area V1 of the primary visual cortex lead to blindsight. These people can "see" but they are unaware that they have seen. Although such visual information is processed, and it can be recalled through a guessing game protocol, it is not passed to the conscious self.

5 Quantum Neural Structures

5.1 The principles

We begin with a brief resume of quantum mechanics. Quantum mechanical objects are described by complex numbers known as amplitudes. The probability of a process is the square of the magnitude of the amplitude. The most fundamental principle in the theory is that of superposition: To compute the total amplitude for a process add the amplitudes of its component processes.

Quantum mechanics presents a view of reality which is a radical departure from the previous mechanistic viewpoint. Quantum mechanical objects do not have an objective existence before measurement.

In the path integral formulation of quantum mechanics, one needs to compute the sum of an infinite number of paths at each step. To summarize:

1. The probability $P(a, b)$ of a process moving from stage a to stage b is the square of the absolute value of a complex number, the transition function $K(a, b)$:

$$P(a, b) = |K(a, b)|^2.$$

2. The transition function is given by the sum of a certain phase factor, which is a function of the action S, taken over all possible paths from a to b:

$$K(a, b) = \sum_{paths} k e^{i2\pi S/h},$$

where the constant k can be fixed by

$$K(a, c) = \sum_{paths} K(a, b) K(b, c),$$

and the intermediate sum is taken over paths that go through all possible intermediate points b.

This second principle says that a particle "sniffs" out all possible paths from a to b, no matter how complicated these paths might be. In quantum theory a future state, within the bounds of time and space uncertainty, can also influence the present.

Information has a central role in quantum theory. Although Schrödinger's equation is linear, the reduction of the wave packet, upon observation, is a nonlinear phenomenon. Observation is thus tantamount to making a choice. From another perspective, there is an asymmetry in the preparation of a state in a quantum system and that of its measurement, because the measurement can only be done in terms of its observables. In a classical system also there is intervention at the beginning and the end of the physical process that defines the computation. Time-reversible equations of physics cannot, in themselves, explain communication of information. Creation of information requires reduction in entropy. Owing to the fact that these models must carry the input along, reversible models can become extremely slow, so as to become unable to solve the problem in any reasonable period of time.

Since measurement in a quantum system is a time-asymmetric process, one can speak of information transfer in a quantum observation. Let v, with eigenvalue equation

$$v|\psi_n\rangle = v_n|\psi_n\rangle,$$

be the dynamical variable of the system S being measured. Let the measurement apparatus A be characterized, correspondingly, by the eigenvalue equation:

$$M|A_n\rangle = M_n|A_n\rangle.$$

Let the system and the apparatus be in the states $\psi\rangle$ and $A_0\rangle$, respectively at the beginning of the measurement, where $\psi\rangle = \sum_m a_m|\psi_m\rangle$. The state of the system and the apparatus, $S+A$, will be $\psi\rangle A_0\rangle$. The Schrödinger equation $i\hbar\frac{d}{dt}|\psi\rangle = H|\psi\rangle$ will now define the evolution of this state.

If one uses the reduction postulate, the state $|\psi\rangle|A_0\rangle$ collapses to $|\psi_m\rangle|A_m\rangle$ with the probability $|a_m|^2$. But if one uses the Schrödinger equation then the initial state evolves into the unique state:

$$\sum_m a_m|\psi_m\rangle|A_M\rangle$$

If one were to postulate another apparatus measuring the system plus the original apparatus the reduction problem is not solved. In fact one can visualize such a process in the so-called von Neumann chain. According to Wigner (1961), the reduction should be ascribed to the observer's mind or consciousness.

According to the orthodox view, namely the Copenhagen interpretation, the workings of the measuring instruments must be accounted for in purely classical terms. This means that a quantum mechanical representation of the measuring apparatus in not correct. In other words, a measurement is contingent on localization.

In such a view the slogan that "computation is physics" loses its generality. It is not surprising that certain types of computations are taken to indicate the participation of conscious agents. Thus if we were to receive the digits of π in transmissions from outer space, we will take that as indication of life in some remote planet. The same will be true of other arithmetic computations.

In brief, signals or computations that simulate the world using models much less complex than the real world indicate intelligence.

Quantum mechanics may be viewed as a theory dealing with basic symmetries. The wavefunction can only be symmetric or antisymmetric, defining bosons and fermions, respectively. In a two-dimensionally constrained world there are other possibilities that have been named anyons (Wilczek 1990) but that need not concern us.

5.2   Quantum knowledge

Quantum theory defines knowledge in a relative sense. It is meaningless to talk of an abjective reality. When we talk, for example, that electric fields exist in a field, this implies that such measurements can be made. Knowledge is a collection of the observations on the reductions of the wavefunction ψ, brought about by measurements using different kinds of instrumentations.

The indeterminacy of quantum theory does not reside in the microworld alone. For example, Schrödinger's cat paradox shows how a microscopic uncertainty transforms into a microscopic uncertainty. Brain processes are not described completely by the neuron firings; one must, additionally, consider their higher order bindings, such as thoughts and abstract concepts, because they, in turn, have an influence on the neuron firings. A wavefunction describing the brain would then include variables for the higher order processes, such as abstract concepts as well. But such a definition will leave a certain indeterminacy in our description.

5.3   The fallacy of knowing the parts from the whole

If we knew the parts completely, one can construct a wavefunction for the whole. But as is well known (Schrödinger 1980; originally in 1935): "Maximal knowledge of a total system does not necessarily include total knowledge of all its parts, not even when these are fully separated from each other and at the moment are not influencing each other at all." In other words, a system may be in a definite state but its parts are not precisely defined.

To recapitulate, we claim that without going into the question of how the state function associated with the brain is to be written down there is a fundamental indeterminacy associated with the description of its component parts. This is over and above the reasons of complexity that one cannot discover the details of the workings of a brain. Now, in a suitable configuration space, where the state function is described in the maximal sense, quantum uncertainty will apply. Since the results of the interactions between the environment and the brain are in terms of the self-organization of the latter, clearly the structure, chosen out of the innumerably many possibilities, represents one of the quantum variables.

5.4   Structure

We must first distinguish between the structures of nonliving and living systems. By the structure of a nonliving system we mean a stable organization of the system. The notion of the stability may be understood from the perspective of energy of the system. Each stable state is an energy minimum.

But the structure in a living system is not so easily fixed. We may sketch the following sequence of events: As the environment (the internal and the external) changes, the living system reorganizes itself. This choice, by the nervous system, of one out of a very large number of possibilities, represents the behavioral or cognitive response.

We might view this neural hardware as the classical instrumentation that represents the cognitive act. This might also be viewed as a cognitive agent. Further processing might be carried out by this instrumentation. We may consider the cognitive categories to have a reality in a suitable space.

A living organism must have entropy in its structure equal to the entropy of its environment. If it did not, it will not be able to adapt (respond) to the changing environment. *Principle*: The position of the organism in its ecological environment is determined by the entropy of its information processing system.

This defines a hierarchy. According to this view the universe for an organism shows a complexity and richness corresponding to the complexity of the nervous system. This idea should be contrasted from the anthropic principle where the nature of the universe is explained by the argument that if it was different there would not have been man to observe it. According to our view, the universe might come to reveal new patterns if we had the capacity to process such information. Computer assisted processing will then reveal new patterns.

It is characteristic of neurophysiology that activity in specific brain structures in given a primary explanatory role. But any determination of the brain structure is impossible. If the brain has 10^{11} neurons and 10^{14} synapses, then even ignoring the gradations in the synaptic behavior, the total number of structures that could, in principle, be chosen exceeds $2^{10^{14}}$, which is greater than current estimates of all elementary particles in the universe.

Assume a system that can exist in only two states. Such a system will find its place where the environment is characterized by just two states. So we can speak of an information theoretic approach to the universe.

Any structure may be represented by a graph as in Figure 4, which may, in turn, be represented by a number, or a binary sequence. Thus in a one dimension, the sequences

$$00111001, 10001101010, 11000001111$$

represent three binary-coded structures.

Assume that a neural structure has been represented by a sequence. Since this representation can be done in a variety of ways, the question of a unique representation becomes relevant.

Definition 1 *Let the shortest binary program that generates the sequence representing the structure be called p.*

The idea of the shortest program gives us a measure for the structure that is independent of the coding scheme used for the representation. The length of this program may be taken to be a measure of the information to be associated with the organization of the system. This

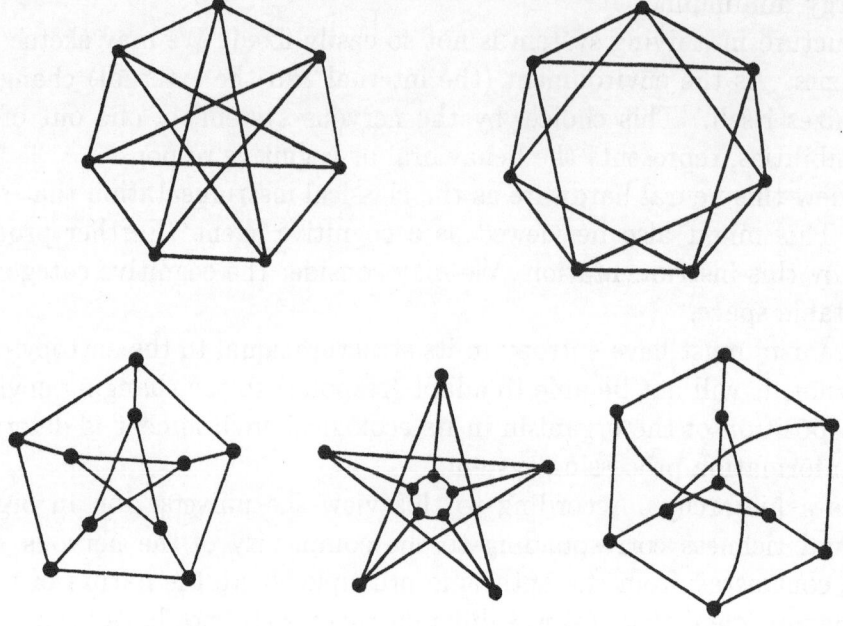

Figure 4: Structures represented as graphs

length will depend on the class of sequences that are being generated by the program. Or in other words, this reflects the properties of the class of structures being considered. Evidence from biology requires that the brain be viewed as an active system which reorganizes itself in response to external stimuli. This means that the structure p is a variable with respect to time.

Assuming, by generalized complementarity, that the structure itself is not defined prior to measurement, then for each state of an energy value E, we may, in analogy with the Heisenberg's uncertainty principle, say that

$$\delta E \delta t \geq k_1 \tag{1}$$

where k_1 is a constant based on the nature of the organizational principle of the neural system.

The external environment changes when the neural system is observed, due to the interference of the observer. This means that as the measurement is made, the structure of the system changes.

This also means that at such a fundamental level, a system cannot be associated with a single structure, but rather with a superposition of several structures. Might this be a reason behind pleomorphism, the multiplicity of forms of microbes?

The representation described above may also be employed for the external environment.

Definition 2 *Let the shortest binary program that generates the external environment be called x.*

If the external environment is a eigenstate of the system, then the system organization will not change; otherwise, it will.

We may now propose an uncertainty principle for neural system structure:

$$\delta x \delta p \geq k_2 \qquad (2)$$

This relation says that the environment and the structure cannot be simultaneously fixed. If one of the variables is precisely defined the other becomes uncontrollably large. Either of these two conditions implies the death of the system. In other words, such a system will operate only within a narrow range of values of the environment and structure.

We conjecture that $k_1 = k_2 = k$.

One may pose the following questions:

- Are all living systems characterized by the same value of k?

- Can one devise stable self-organizing systems that are characterized by a different value of k? Would artificial life have a value of k different from that of natural life?

- What is the minimum energy required to change the value of p by one unit?

- Does a Schrödinger type equation define the evolution of structure? It appears that in the original configuration space this indeed is true. But how might such an evolution be represented in terms of structure alone?

It is also clear that before a measurement is made, one cannot speak of a definite state of the machine, nor of a definite state of the environment.

Clearly, we can only talk in terms of generalities at this stage. In order to make further advance in our understanding it is essential to consider the notion of structure as a classical variable first. This we do by speaking of signals that might be exclusively dedicated to altering the organization of the system. These signals may be taken to be the dual to the neuron firings that constitute the better studied response of brains.

6   Reorganizing Signals

Pribram (1971) suggests a state composed of the "local junctional and dendritic (pre- and postsynaptic) potentials." Pribram also considered holographic models of memory which also require dual signaling of a certain kind. But the motivation here was more from the point of view of capacity to store information rather than self-organization.

A specific type of 40 Hz oscillation as a dual signal has been proposed (e.g. Niebur et al, 1993) to explain binding. But this model is too vague at this point to provide a satisfactory resolution to the problem of self-organization.

Living systems are characterized by continual adaptive organization at various levels. The reorganization is a response to the complex of signal flows within the larger system. For example, the societies of ants or bees may be viewed as single superorganisms. Hormones and other chemical exchanges among the members of the colony determine the ontogenies of the individuals within the colony. But more pronounced than this global exchange is the activity amongst the individuals in cliques or groups (Moritz and Southwick, 1992).

Paralleling trophallaxis is the exchange of neurotransmitters or electrical impulses within a neural network at one level, and the integration of sensory data, language, and ideas at other levels. An illustration of this is the adaptation of somatosensory cortex to differential inputs. The cortex enlarges its representation of particular fingers when they are stimulated, and it reduces its representation when the inputs are diminished, such as by limb deafferentation.

Adaptive organization may be a general feature of neural networks and of the neocortex in particular. Biological memory and learning within the cortex may be organized adaptively. While there are many ways of achieving this, we posit that nesting among neural networks within the cortex is a key principle in self-organization and adaptation. Nested distributed networks provide a means of orchestrating bottom-up and top-down regulation of complex neural processes operating within and between many levels of structure.

There may be at least two modes of signaling that are important within a nested arrangement of distributed networks. A fast system manifests itself as spatiotemporal patterns of activation among modules of neurons. These patterns flicker and encode correlations that are the signals of the networks within the cortex. They are analogous to the hormones and chemical exchanges of the ant or bee colonies in the example mentioned above. In the brain, the slow mode is mediated by such processes as protein phosphorylation and synaptic plasticity. They are the counterparts of individual ontogenies in the ant or bee colonies. The slow mode is intimately linked to learning and development (i.e., ontogeny), and experience with and adaptation to the environment affect both learning and memory.

By considering the question of adaptive organization in the cortex, our approach is in accordance with the ideas of Gibson (1976, 1979) who has long argued that biological processing must be seen as an active process. We make the case that nesting among cortical structures provides a framework in which active reorganization can be efficiently and easily carried out. The processes are manifest by at least two different kinds of signaling, with the consequence that the cortex is viewed as a dynamic system at many levels, including the level of brain regions. Consequently, functional anatomy, including the realization of the homunculus in the motor and sensory regions, is also dynamic. The homunculus is an evolving, and not a static representation, in this view.

From a mathematical perspective, nesting topologies contain broken symmetry. A monolithic network represents a symmetric structure, whereas a modular network has preferential structures. The development of new clusters or modules also represents an evolutionary response, and a dual mode signaling may provide a means to define context. It may also lead to unusual resiliences and vulnerabilities in the face of perturbations. We propose that these properties may have relevance to how nested networks are affected by the physiological process of aging and the pathological events characterizing some neurobiological disorders.

Reorganization explains the immeasurable variety of the response of brains. This reorganization may be seen as a characteristic which persists at all levels in a biological system. Such reorganization appears to be the basis of biological intelligence. It was a mistaken emphasis on the characterization of life in terms of reproducibility by John von Neumann that led the AI community astray for decades.

6.1 Adaptive organization

Active perception can be viewed as adapting to the environment. In the words of Bajcsy (1988) : "It should be axiomatic that perception is not passive, but active. Perceptual activity is exploratory, probing, searching; percepts do not simply fall onto sensors as rain falls onto ground. We do not just see, we look."

It is not known how appropriate associative modules come into play in response to a stimulus. This is an important open question in neural computing. The paradigm of "active" processing in the context of memory is usually treated in one of two ways. First, the processing may be pre-set. This is generally termed "supervised learning", and it is a powerful but limited form of active processing. A second type of processing does not involve an explicit teacher, and this mechanism is termed "unsupervised learning". It is sensitive to a number of constraints, including the structure and modulation of the network under consideration.

We posit that active memories are inherently connected to the structure of the network. In particular, we propose that active memories can be defined in terms of a selection of a substructure of the nested organization. We see that this is complementary to associations in the manner of Hebb, where a memory is seen as an attractor in the states of the neurons. The Hebbian view considers the stable arrangements of firing neurons as memories. In a nested structure, the firing neurons contribute to multi-level memories. Individual neurons obey local rules, but because there are anatomical and functional boundaries between clusters, higher level memories may emerge from the combinations of lower level memories. Higher memories are actively formed by correlations among lower level memories and clusters. From this perspective, selecting a structure through adaptation to the stimulus is not really a departure from Hebb's perspective.

7 Nested Networks

There are different ways that biological memory may be self-organizing, and in this section, we suggest that the nesting of distributed neural networks within the neocortex is a natural candidate for encoding and transducing memory. Nesting has interesting combinatorial and computational features (Sutton et al, 1988, 1990; Sutton, 1995), and its properties have not been fully examined. The seemingly simplistic organization of nested neural networks may have profound computational properties, in much the same way as recent deoxyribonucleic computers have been put to the task of solving some interesting fundamental problems. However, we do not claim that nesting is the only important feature for adaptive organization in neural systems.

A natural consideration is to examine the structural properties of the forebrain, including the hippocampus and neocortex, which are two key structures in the formation and storage of memory. The hippocampus is phylogenetically an ancient structure, which among other functions, stores explicit memory information. To first approximation, this information is then transferred to the neocortex for long term storage. Implicit memory cues can access neocortical information directly.

The neocortex is a great expanse of neural tissue that makes up the bulk of the human brain. As in all other species, the human neocortex is made up of neural building blocks. At a rudimentary level, these blocks consist of columns oriented perpendicular to the surface of the cortex. These columns may be seen as organized in the most basic form as minicolumns of about 30 μm in diameter. The minicolumns are, in turn, organized into larger columns of approximately 500 - 1000 μm in diameter. Mountcastle estimates that the human neocortex contains about 600 million minicolumns and about 600,000 larger columns. Columns are defined by ontogenetic and functional criteria, and there is evidence that columns in different brain regions coalesce into functional modules (Mountcastle, 1978). Different regions of the brain have different architectonic properties, and subtle differences in anatomy are associated with differences in function.

The large entities of the brain are "composed of replicated *local neural circuits*, modules which vary in cell number, intrinsic connections, and processing mode from one large entity to another but are basically similar within any given entity." In other words, the neocortex can be seen as several layers of nested networks (Mountcastle, 1978, Sutton and Trainor, 1990). Beginning with cortical minicolumns, progressive levels of cortical structure consist of columns, modules, regions and systems. It is assumed that these structures evolve and adapt through the lifespan. It is also assumed that the boundaries between the clusters are plastic: they change slowly due to synaptic modifications or, more rapidly, due to synchronous activity among adjacent clusters.

Results from the study of neural circuits controlling rhythmic behavior, such as feeding, locomotion, and respiration, show that the same network, through a process of "rewiring" can express different functional capabilities (Simmers et al, 1995).

In a study of the pattern generator of the pyloric rhythm in lobster, it has been found that the behavior is controlled by fourteen neurons in the stomatogastric ganglion. The predominant means of communication between the neurons is through inhibitory synapses. The reshaping of the output of the network arises from neuromodulation. More than fifteen different modulatory neurotransmitters have been identified. These allow the rewiring of the network. Turning on the pyloric suppressors restructures the otherwise three independent networks in the stomatogastric nervous system into a single network, converting the function from regional food processing to coordinated swallowing.

> Rather than seeing a system as a confederation of neatly packaged neural circuits, each devoted to a specific and separate task, we must now view a system in a much more distributed and fluid context, as an organ that can employ modulatory instructions to assemble subsets of neurons that generate particular behaviors. In other words, single neurons can be called on to satisfy a variety of different functions, which adds an unexpected dimension of flexibility and economy to the design of a central nervous system (Simmers et al 1995, page 268).

7.1 Adaptive organization and modularity

We consider the issue of the reorganization of the structure or activity, in response to a stimulus, in more detail. We will motivate the reader in terms of visual inputs but the discussion is valid for other types of inputs as well. We sketch the following process:

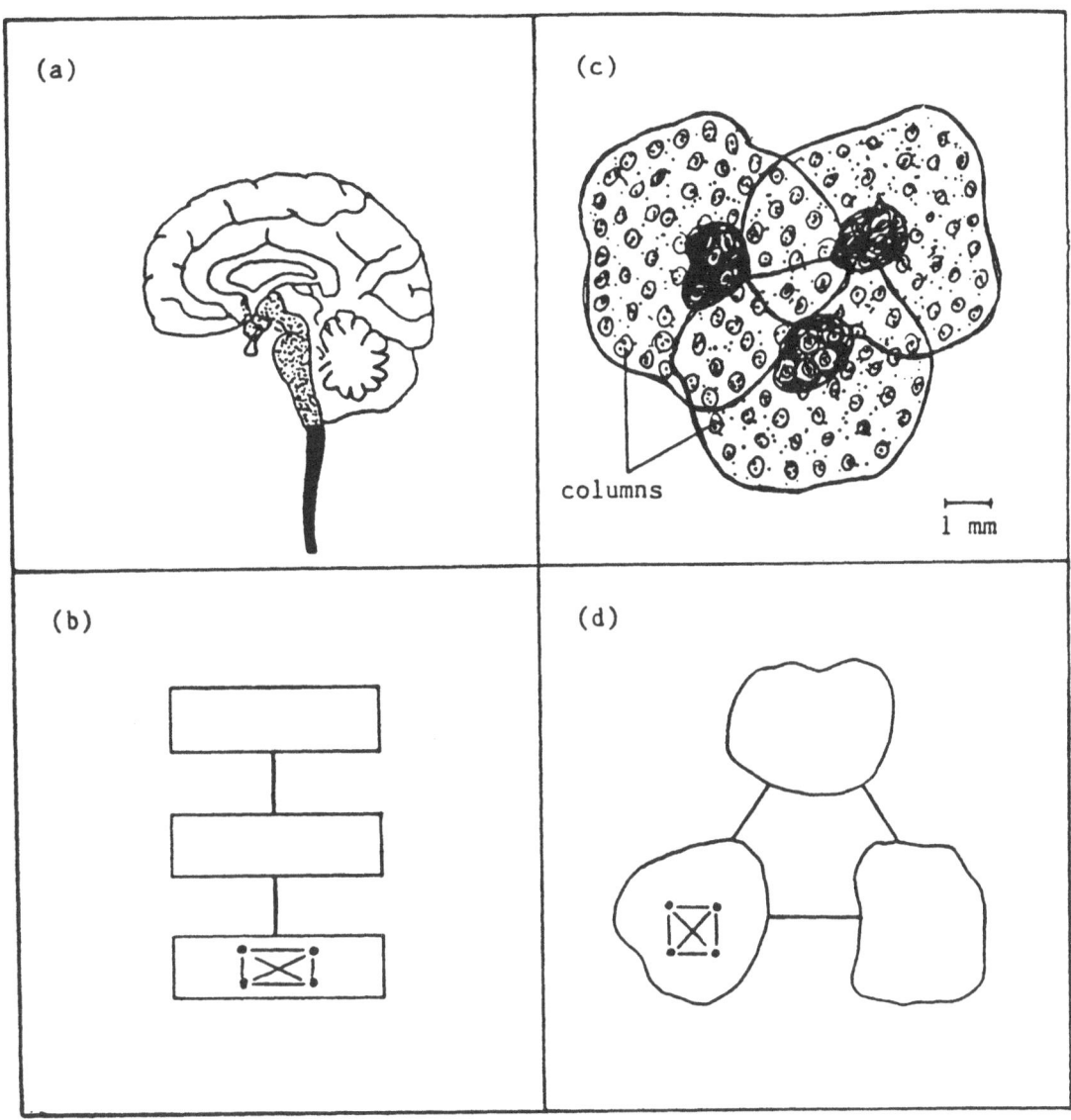

Figure 5: Neocortex as a hierarchical system (2a,b) with clusters (2c) and nesting (2d). From Sutton and Trainor (1990).

1. The overall architecture of the nested system is determined by the external or internal stimulus, this represents the determination of the connections at the highest hierarchical levels and progresses down in a recursive manner. The learning of the connections in each layer is according to a correlative procedure. The sensory organs adjust to the stable state reached in the nearest level.

2. The deeper layers find the equilibrium state corresponding to the input in terms of attractors. Owing to the ongoing reorganization working in both directions, up to the higher levels as well as toward the lower levels, the attractors may best be labeled as being dynamic.

Superorganisms also have nested structures in terms of individuals who interact more with certain members than others. In the case of ants, the castes provide further "modular" structure (Sudd and Franks, 1987). For the case of honeybees:

> [It is important to recognize] subsystems of communication, or cliques, in which the elements interact more frequently with each other than with other members of the communication system. In context, the dozen or so honeybee workers comprising the queen retinue certainly communicate more within their group (including the queen) than they do with the one or two hundred house bees receiving nectar loads from foragers returning from the field. The queen retinue forms one communication clique while the forager-receiver bees form another clique. (Moritz and Southwick, 1992, page 145)

The parallel for two distinct pathways of communication is to be seen in superorganisms as well:

> [The] superorganism is a self-organizing system incorporating two very distinct pathways of communication. One mode is via localized individual worker interactions with low connectedness, and the other one via volatile or semiochemical pheromones with high connectedness. If we examine the communication modes at the functional level, we see that the pheromones reach the entire superorganism, more or less: a global message with a global reaction (for example, the queen pheromones simultaneously and continuously signal to every worker in the colony that it is queenright). (Moritz and Southwick, 1992, page 151)

An index of average system connectedness (ASC) has been defined as

$$ASC\ index\ =\ Actual\ dyads/Possible\ dyads. \tag{3}$$

For a fully connected network the index is 1. According to one study the linkages in the communication network of honeybees accepting water or dilute nectar from returning foragers under heat stress conditions was 0.031. An example is shown in Figure 6.

Another fundamental communication within the superorganism is the one that defines its constitution. This is a much slower process which can be seen, for example, when a queen ant founds her colony. The queen governs the process of caste morphogenesis (Brian, 1983; Hölldobler and Wilson, 1994). Within the new colony, the queen, having just mated with

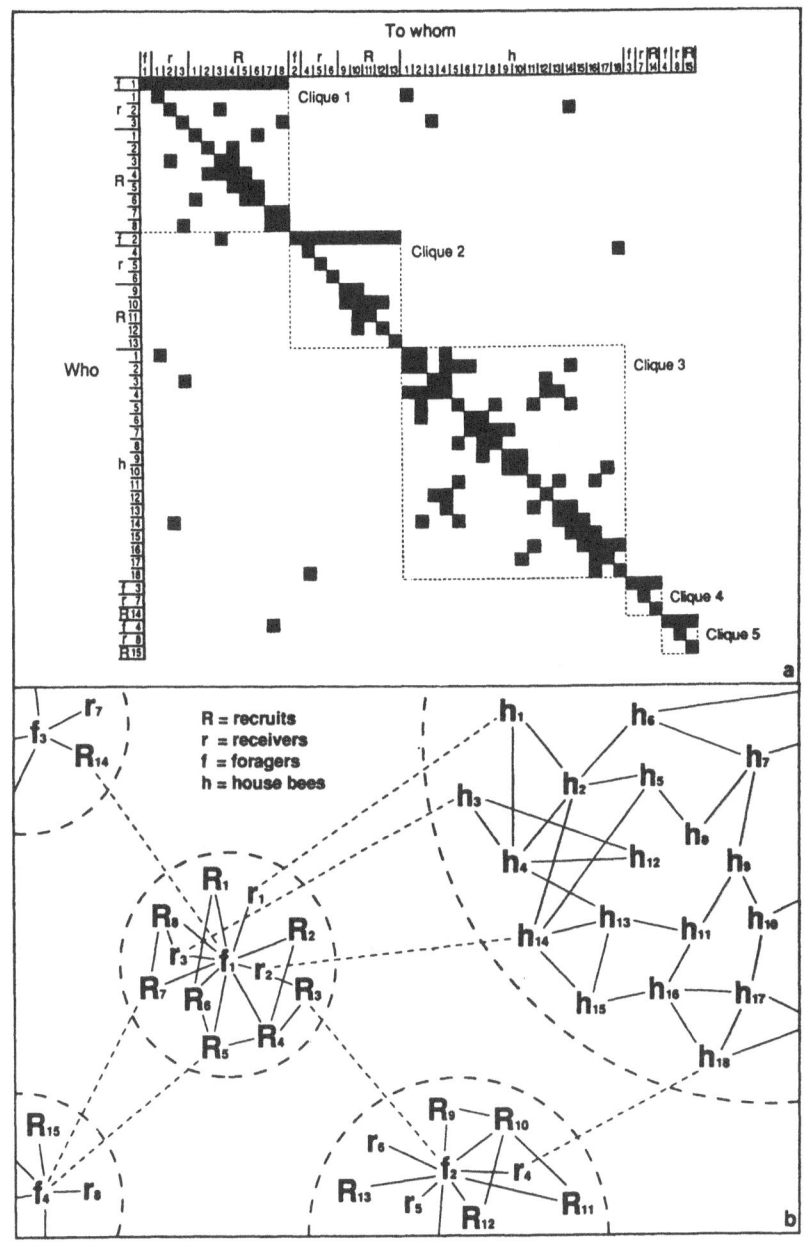

Figure 6: Histogram of communication linkages for honeybees. From Moritz and Southwick (1992).

her suitors and received more than 200 million sperm, shakes off her wings and digs a little nest in the ground, where she now is in a race with time to produce her worker offspring. She raises her first brood of workers by converting her body fat and muscles into energy. She must create a perfectly balanced work force that is the smallest possible in size, yet capable of successful foraging, so that the workers can bring food to her before she starves to death.

The queen produces the workers of the correct size for her initial survival and later, after the colony has started going, she produces a complement of workers of different sizes as well as soldier ants in order to have the right organization for the survival of the colony. When researchers have removed members of a specific caste from an ongoing colony, the queen compensates for this deficit by producing more members of that caste. The communication process behind this remarkable control is not known.

The communication mechanisms of the ant or the honeybee superorganisms may be supposed to have analogs in the brain.

8   Self-Organization by Association

Now we consider the question of the learning by association by the neural hardware. We assume here that such structures are realized as feedforward networks. Self-organization by association may also be viewed as learning.

As shown in Figure 7, the eyes wander in the process of perception; they jump, then come to rest momentarily which produces a dot on the record. This process skips areas with little details. These eye movements suggest that there in no fixing of any particular pattern in perception. It appears then that schemes such as backpropagation, where the synaptic weights are adjusted upon considerable training, are not realistic. We would expect that, to be biologically plausible, learning by association should be instantaneous.

A new approach to training such neural networks for binary data was proposed by me three years ago (Kak, 1993b, 1994b). Such a method might be relevant in the learning of organization. This is based on a new architecture that depends on the nature of the data. It was shown that this approach is much faster than backpropagation and provides good generalization. This approach, which is an example of prescriptive learning, trains the network by isolating the corner in the n-dimensional cube of the inputs represented by the input vector being learnt. Several algorithms to train the new feedforward network were presented. These algorithms were of three kinds. In the first of these (CC1) the weights were obtained upon the use of the perceptron algorithm. In the second (CC2), the weights were obtained by inspection from the data, but this did not provide generalization. In the third (CC3), the weights obtained by the second method were modified in a variety of ways that amounted to randomization and which now provided generalization. During such randomization some of the learnt patterns could be misclassified; further checking and adjustment of the weights was, therefore, necessitated. Various comparisons were reported in (Raina, 1994; Madineni, 1994). The comparisons showed that the new technique could be 200 times faster than the fastest version of the backpropagation algorithm with excellent generalization performance.

Here we show how generalization can be obtained for such binary networks just by inspection. We present a modification to the second method so that it does provide generalization.

Figure 7: The shifting focus of the eye. In the girl's picture the record is for one minute, whereas in the pine forest scene the two records are for 2 and 30 minutes. From Yarbus (1967)

This technique's generalization might not be as good as when further adjustments are made, but the loss in performance could, in certain situations, be more than compensated by the advantage accruing from the instantaneous training which makes it possible to have as large a network as one pleases.

8.1 Hidden neurons

It is well known (Mirchandani and Cao, 1989; Georgiou, 1991) that the number of hidden neurons, H, needed to separate M number of regions in a $d-$dimensional space is given by

$$M(H, d) = \sum_{k=0}^{d} \binom{H}{k} \tag{4}$$

where

$$\binom{H}{k} = 0, \ H < k.$$

Let the number of regions that the hidden neurons separate be equal to C, where $C \leq M$. Since the number of classes at the output is only equal to 2, these C regions coalesce into the 2 classes at the output.

Let the input space dimension be d and let each dimension be quantized so that the total number of of binary variables is n. Not each dimension may require the same precision. If the average number of bits used per input dimension is q then $n = q \times d$.

If the number of training samples is T, then $M \leq T$. For $d = 1$, $H = M - 1$, for $d = 2$, $H = (\sqrt{(8M - 7)} - 1)/2$, and for $d \geq H$, $H = log_2 M$.

When the data points are binary, as in our case, then these formulas require modification. The set of 2^n data points can now be separated by a total of n hidden neurons. But the outputs of these hidden neurons need to be combined using various logical operations to pass specific input patterns. This is not a desirable strategy to adopt if the learning is supposed to be decentralized with a cumulative response to all the training data.

Our network has H nearly equal to T (or M), therefore, our algorithms consider the data as effectively one-dimensional.

8.2 Prescriptive training

We assume that the reader is familiar with the background papers (Kak, 1993b, 1994b). We consider the mapping $Y = f(X)$, where X and Y are $n-$ and $m-$dimensional binary vectors. But for convenience of presentation, it will be assumed that the output is a scalar, or $m = 1$. Once we know how a certain ouput bit is obtained, other such bits can be obtained similarly. We consider binary neurons that ouput 1 if and only if the sum of the inputs exceeds zero. To provide for effective non-zero thresholds to the neurons of the hidden layer an extra input $x_{n+1} = 1$ is assumed. The weights in the output layer all equal 1.

A hidden neuron is required for an input vector in the training set if the output is 1. We might say that the hidden neuron "recognizes' the training vector. Consider such a vector for which the number of 1's is s; in other words, $\sum_{i=1}^{n} x_i = s$. The weights leading from the input neurons to the hidden neurons are:

$$w_j = \begin{cases} h & if \; x_j = 0, \; for \; j = 1, ..., n, \\ +1 & if \; x_j = 1, \; for \; j = 1, ..., n, \\ r - s + 1 & for \; j = n + 1. \end{cases} \tag{5}$$

The values of h and r are chosen in various ways. This is a generalization of the expression in (Kak, 1993b, 1994b) where w_j for $j = n+1$ is taken to be $(r - s + 1)$ rather than $(1 - s)$, and where $h = -1$.

This change allows the learning of the given training vector as well as others that are at a distance of r units from it (for $h = -1$); in other words, r is the *radius of the generalized region*. This may be seen by considering the all zero input vector. For this $w_{n+1} = r$. Since, all the other weights are -1 each, one can at most have $(r - 1)$ different $+1$s in the input vector for it to be recognized by this hidden neuron.

The choice of r will depend upon the nature of generalization sought. If no generalization is needed then $r = 0$. For exemplar patterns, the choice of r defines the degree of error correction.

If the neural network is being used for function mapping, where the input vectors are equally distributed into the 0 and the 1 classes, then $r = \lfloor \frac{n}{2} \rfloor$. This represents the upper bound on r for a symmetric problem. But the choice will also depend on the number of training samples.

The choice of h also influences the nature of generalization. Increasing h from the value of -1 correlates patterns within a certain radius of the learnt sequence. This may be seen most clearly by considering a 2−dimensional problem. The function of the hidden node can be expressed by the separating line:

$$w_1 x_1 + w_2 x_2 + (r - s + 1) = 0. \tag{6}$$

This means that

$$x_2 = \frac{-w_1}{w_2} x_1 + \frac{-(r - s + 1)}{w_2}. \tag{7}$$

Assume that the input pattern being classified is (0 1), then $x_2 = 1$. Also, $w_1 = h$, $w_2 = 1$, and $s = 1$. The equation of the dividing line represented by the hidden node now becomes:

$$x_2 = -h x_1 - r. \tag{8}$$

When $h = -1$ and $r = 0$, the slope of the line is positive and only the point $(0, 1)$ is separated. To include more points in the learning, $h < 0$, because the slope of the line becomes negative.

8.3 Relationship between h and r

Consider the all zero sequence (0 0 ... 0). After the appending of the 1 threshold input, we have the corresponding weights $(h \; h \; ... \; h \; r + 1)$. Sequences at the radius of p from it will yield the strength of $ph + r + 1$ at the input of the corresponding hidden neuron. For such signals to pass through

$$ph + r + 1 > 0. \tag{9}$$

In other words, generalization by a Hamming distance of p units is achieved if

$$h < \frac{-(r+1)}{p}. \tag{10}$$

When $h = -1$; $p = r$. When $h = positive$, all the input patterns where the $0s$ have been changed into $1s$ will also be passed through and put in the same class as the training sample.

8.4   Training samples

The total number of sequences 2^n equals the number of classification classes M times the average number of members in each class.

Let the radius of the class i be r_i. The number of elements in this class will be

$$\sum_{k=0}^{r_i} \binom{n}{k}. \tag{11}$$

If all the classes are of the same size and each class is represented by a single training sample:

$$T \times \sum_{k=0}^{r} \binom{n}{k} = 2^n. \tag{12}$$

The following table gives the size of the training set for the example of $n = 10$; $2^n = 1,024$.

Table 1: Generalization and training set size

r	T
0	1,024
1	32
2	12
3	9

Since the probability that each training sample belongs to a different class could be small, the above numbers represent very rough estimates.

Experiments were conducted on a real-world time-series. Half of the time series was used for finding the weights; the remaining half was used to test the model. Satisfactory results were obtained.

One way to improve generalization is by varying the radius of generalization with the training sample. This may be done easily if each training sample can be characterized by a measure of quality. Such a characterization may be possible for certain situations.

The generalized prescriptive rule presented in this article makes it possible to train a neural network instantaneously. There is no need for any computations in determining the weights. This allows the building up of neural networks of unlimited size.

It may be useful to use a two-step strategy when the learning method described here is used. In the first step use a separate network to determine the mode of the data. The instantaneous method of neural network training described here could be the method at work in biological systems.

Although this section presents the theory of instantaneous training of neural networks in the context of binary data, this can be adapted, in a variety of ways, to deal with continuous signals. For example, continuous data can be quantized into discrete levels. Whether a continuous signal generalization of this training exists is not known.

9 Conclusions

We have reviewed different kinds of evidence in favor of an underlying quantum basis to brain behavior: response to single photons by the vision system, the unity of the awareness process, and the fact that the process of self-organization is best seen as triggering a reduction of the wavefunction corresponding to the thought process. The self-organizational signals are a response to a combination of the inner and the external sensory signals. If biological processing is distinct from classical computing then it is easy to see why machines are unable to do certain kind of holistic processing that is easily done by animals. We suggest that biological organization may be supposed to be a quantum macrostructure that is constantly interacting with the environment and reorganizing. This reorganization may parallel perception representing a reduction into an eigenstructure.

In several writings, Neils Bohr stressed how the principle of complementarity must include life. Although complementarity as a philosophy is not much in vogue these days due to the ascendancy of the computer metaphor, it is the only consistent interpretive approach to quantum mechanics. Schrödinger's cat paradox shows how indeterminacy is associated with macroscopic systems if they are interacting with quantum systems. Life cannot exist without light; from this perspective alone we are compelled to consider quantum models.

An implication of this reasoning is the rejection of the materialist position that considers the identity of the neural and thought processes. A complete description of the individual must be in an suitable dimensional space which includes thoughts and concepts.

The structure of system may be described in terms of a binary sequence. One can then speak of complementary variables relating the structure and the environment. Such a reformulation of quantum mechanics may allow it to include living organisms.

Brain processes may also be seen in terms of two kinds of communications: one faster and the other slower. This is illustrated by the example of superorganisms where we have localized individual worker interaction with low connectedness and a faster communication using semiochemical pheromones. A specific signaling that regulates organization could provide important clues to the development of the quantum mechanics of living systems.

A quantum theoretical basis to life provides resolution to several thorny questions although it raises other fresh problems. The most pleasing feature is that it acknowledges the reality of "effort," and "intention," "free will," which have no place in materialist or causal schemes. Neither can we consider consciousness as an epiphenomenon. If consciousness has independent existence then it is a universal function and a brain is to be considered as simply the hardware that reduces this function.

We have considered the most basic behavior in our description of the three languages of the brain. At higher levels of description we must speak of other languages.

References

Bajcsy, R. (1988). Active perception. *Proceedings of the IEEE*, 78: 996-1005.

Baylor, D.A., Lamb, T.D., and Yau, K.-W. (1979). Responses of retinal rods to single photons. Journal of Physiology, 288, 613-634.

Brian, M.V. (1983). *Social Insects.* Chapman and Hall, London.

Eccles, J.C. (1979). *The Human Mystery.* Springer-Verlag, Berlin.

Eccles, J.C. (1989). *Evolution of the Brain.* Routledge, London.

Edelman, G.M. (1992). *Bright Air, Brilliant Fire: On the Matter of the Mind.* BasicBooks, New York.

Feuerstein, G., Kak, S.C., Frawley, D. (1995). *In Search of the Cradle of Civilization.* Quest Books, Wheaton.

Feynman, R.P. (1982). Simulating physics with computers. *International Journal of Theoretical Physics,* 21, 467-488.

Feynman, R.P. (1986). Quantum mechanical computers. *Foundations of Physics,* 16, 507-531.

Fröhlich, H. (1975). The extraordinary dielectric properties of biological materials and the action of enzymes. *Proc. Natl. Acad. Sci. USA,* 72, 4211-4215.

Georgiou, G. (1991). Comments on 'On Hidden Nodes.' *IEEE Trans. on Circuits and Systems,* 38, 1410.

Gibson, J.J. (1966). *The Senses Considered as Perceptual Systems.* Houghton-Mifflin, Boston.

Gibson, J.J. (1979). *The Ecological Approach to Visual Perception.* Houghton Mifflin, Boston.

Goswami, A. (1993). *The Self-Aware Universe.* G.P. Putnam's Sons, New York.

Griffin, D.R. (1992). *Animal Minds.* The University of Chicago Press, Chicago.

Grundler, W. and Keilmann, F. (1983). Sharp resonances in yeast growth proved nonthermal sensitivity to microwaves. *Phys. Rev. Letts.,* 51, 1214-1216.

Herrnstein, R.J. (1985). Riddles of natural categorization. *Phil. Trans. R. Soc. Lond.,* B 308, 129-144.

Herrnstein, R.J., W. Vaughan, Jr., D.B. Mumford, and S.M. Kosslyn. (1989). Teaching pigeons an abstract relational rule: insideness. *Perception and Psychophysics,* 46, 56-64.

Hölldobler, B. and Wilson, E.O. (1994). *Journey to the Ants.* Harvard University Press, Cambridge.

Jibu, M., Hagan, S., Hameroff, S., Pribram, K.H., Yasue, K. (1994). Quantum optical coherence in cytoskeletal microtubules: implications for brain function, *BioSystems*, 32, 195-209.

Kak, S.C. (1976). On quantum numbers and uncertainty. *Nuovo Cimento*, 33B, 530-534.

Kak, S.C. (1984). On information associated with an object. *Proc. Indian National Science Academy*, 50, 386-396.

Kak, S.C. (1986). *The Nature of Physical Reality.* Peter Lang, New York.

Kak, S.C. (1992). Can we build a quantum neural computer? *Technical Report ECE-92-13*, December 15, Louisiana State University.

Kak, S.C. (1993a). Reflections in clouded mirrors: selfhood in animals and machines. Symposium on *Aliens, Apes, and AI: Who is a person in the postmodern world*, Southern Humanities Council Annual Conference, Huntsville, AL, February 1993.

Kak, S.C. (1993b). On training feedforward neural networks. *Pramana -J. of Physics*, 40, 35-42.

Kak, S.C. (1994a). *The Astronomical Code of the Ṛgveda.* Aditya, New Delhi.

Kak, S.C. (1994b). New algorithms for training feedforward neural networks. *Pattern Recognition Letters*, 15, 295-298.

Kak, S.C. (1995a). On quantum neural computing. *Information Sciences*, 83, 143-160.

Kak, S.C. (1995b). Quantum neural computing. In *Advances in Imaging and Electron Physics*, Peter Hawkes (ed.). Academic Press, New York, pp. 259-313.

Kak, S.C. (1995c). Information, physics and computation. *Technical Report ECE-95-04*, April 19, Louisiana State University.

Kak, S.C. (1995d). Can we define levels of artificial intelligence. *Journal of Intelligent Systems*, in press.

Kandel, E.R., Hawkins, R.D. (1992). The biological basis of learning and individuality. *Scientific American,* 267(3): 79-86.

King, J.S. and Pribram, K.H. (eds.) (1995). *Scale in Conscious Experience: Is the Brain too Important to be Left to Specialists to Study?* Lawrence Erlbaum Associates, Hillsdale, New Jersey.

Klostermaier, K. (1994). *A Survey of Hinduism.* State University of New York Press, Albany.

Madineni, K.B. (1994). Two corner classification algorithms for training the Kak feedforward neural network. *Information Sciences,* 81, 229-234.

Meyrand, P., Simmers, J., Moulins, M. (1994). Dynamic construction of a neural network from multiple pattern generators in the lobster stomatogastric nervous system. *J. of Neuroscience,* 14: 630-644.

Mirchandani, G. and Cao, W. (1989). On hidden nodes for neural nets. *IEEE Trans. on Circuits and Systems,* 36, 661-664.

Moritz, R.F.A., Southwick, E.E. (1992). *Bees as Superorganisms.* Springer-Verlag, Berlin.

Mountcastle, V.B. (1978). An organizing principle for cerebral function: The unit module and the distributed system. In G.M. Edelman and V.B. Mountcastle, eds., *The Mindful Brain.* The MIT Press, Cambridge.

Niebur, E., Koch, C. and Rosin, C. (1993). An oscillation-based model for the neuronal basis of attention. *Vision Research,* 18, 2789-2802.

Penrose, R. (1994). *Shadows of the Mind.* Oxford University Press, Oxford.

Popper, K.R and Eccles, J.R. (1977). *The Self and its Brain.* Springer-Verlag, Berlin.

Pribram, K.H. (1971). *Languages of the Brain: Experimental paradoxes and principles in neuropsychology.* Brandon House, New York.

Pribram, K.H. (ed.) (1991). *Brain and Perception: Holonomy and Structure in Figural Processing.* Lawrence Erlbaum Associates, Hillsdale, New Jersey.

Pribram, K.H. (ed.) (1993). *Rethinking Neural Networks: Quantum Fields and Biological Data.* Lawrence Erlbaum Associates, Hillsdale, New Jersey.

Pribram, K.H. (ed.) (1994). *Origins: Brain & Self Organization.* Lawrence Erlbaum Associates, Hillsdale, New Jersey.

Raina, P. (1994). Comparison of learning and generalization capabilities of the Kak and the backpropagation algorithms. *Information Sciences,* 81, 261-274.

Schrödinger, E. (1961). *Meine Weltansicht.* Paul Zsolnay, Wien.

Schrödinger, E. (1980). The present situation in quantum mechanics. *Proc. of the American Philosophical Society,* 124, 323-338. Also in Wheeler and Zurek (1983).

Simmers, J., Meyrand, P., Moulins, M. (1995). Dynamic networks of neurons. *American Scientist,* 83: 262-268.

Sudd, J.H., Franks, N.R. (1987). *The Behavioural Ecology of Ants.* Blackie, Glasgow.

Sutton, J.P. (1995). Neuroscience and computing algorithms. *Information Sciences,* 84, 199-208.

Sutton, J.P., Beis, J.S., Trainor, L.E.H. (1988). Hierarchical model of memory and memory loss. *J. Phys. A: Math. Gen.*, 21: 4443-4454.

Sutton, J.P. and Trainor, L.E.H. (1990). Real and artificial neural hierarchies. In C.N. Manikopoulos (ed.), *Cybernetics and Systems*, Vol. 1, 229-236. NJIT Press, Newark.

Turing, A.M. (1950). Computing machinery and intelligence. *Mind*, 59, 433-460.

Turney, T.H. (1982). The association of visual concepts and imitative vocalizations in the mynah. *Bulletin Psychonomic Society*, 19, 56-62.

Wasserman, E.A. (1993). Comparative cognition: Beginning the second century of the study of animal intelligence. *Psychological Bulletin*, 113, 211-228.

Wasserman, E.A. (1995). The conceptual abilities of pigeons. *American Scientist*, 83, 246-255.

Wasserman, E.A., Hugart, J.A., Kirkpatrick-Steger, K. (1995). Pigeons show same-different conceptualization after training with complex visual stimuli. *J. of Experimental Psychology*, 21, 248-252.

Wheeler, J.A. (1979-81). Law without law. In Wheeler, J.A. and Zurek, W.H. (eds.) (1983). *Quantum Theory and Measurement*. Princeton University Press, Princeton.

Wheeler, J.A. and Zurek, W.H. (eds.) (1983). *Quantum Theory and Measurement*. Princeton University Press, Princeton.

Wigner, E.P. (1961). In *The Scientist Speculates*, I.J. Good (editor). Basic Books, New York.

Wilczek, F. (1990). *Fractional Statistics and Anyon Superconductivity*. World Scientific, Singapore.

Yarbus, A.L. (1967). *Eye Movements and Vision*. Plenum Press, New York.

8.
Automatic Formation of Wavelet- and Gabor-Type Filters in an Adaptive-Subspace SOM

Teuvo Kohonen

Automatic Formation of Wavelet- and Gabor-Type Filters in an Adaptive-Subspace SOM

Teuvo Kohonen

Helsinki University of Technology

Neural Networks Research Centre

Rakentajanaukio 2 C, FIN-02150 Espoo, Finland

Abstract. It is shown in this paper that a particular adaptive self-organizing network is able to create sets of wavelet- and Gabor-type filters when randomly displaced or moving input patterns are used as training data. No analytical functional form of these filters is thereby postulated. It is plausible that the same kind of adaptive system could create many other kinds of invariant-feature filters, if there exist corresponding transformations in the training data. Such a system, called ASSOM (Adaptive-Subspace SOM), can act as a learning feature-extraction stage for pattern recognizers, being able to adapt to an arbitrary sensory environment.

1 Introduction: The Problem

In target recognition as well as implementation of genuine associative memory, one of the most important objectives and at the same time the most difficult problem is identification of sensory patterns under various *transformations* of their representations. In order to understand biological neural networks more deeply, it would be very timely to create new theories for *self-organizing invariant-feature analyzers*, whereby also the artificial organisms could use such analyzer and be better adapted to noisy and changing environments.

This paper describes a novel, dissident approach to neural networks. Closeby neurons in such a network are made to co-operate as *groups* that correspond to *manifolds* (such as linear subspaces) in the signal space. It is possible to construct neural units of the above kind that produce invariant outputs for various elements of observations in spite of different transformations (such as translation, rotation, scaling, perspective transformation etc.) being applied to their representations, or their being in motion. Two examples of adaptive preprocessing discussed in this paper are the automatically formed filters that resemble *wavelet* and *Gabor functions*, respectively.

2 Wavelet and Gabor Transforms

Consider a scalar-valued time function $f(t)$. It can be transformed using kernels called *wavelets*. Like in Fourier transformation, there exist cosine and sine wavelets, denoted ψ_c and ψ_s, respectively; if they are centered around instant t_0, they are of the form

$$
\begin{aligned}
\psi_c(t - t_0; \omega, \sigma) &= e^{-\frac{\omega^2(t-t_0)^2}{2\sigma^2}} \cos\left(\omega(t - t_0)\right), \\
\psi_s(t - t_0; \omega, \sigma) &= e^{-\frac{\omega^2(t-t_0)^2}{2\sigma^2}} \sin\left(\omega(t - t_0)\right).
\end{aligned}
\tag{1}
$$

The *wavelet transforms* of $f(t)$ evaluated at t are defined as

$$\Psi_c(t; \omega, \sigma) = \int_\tau \psi_c(t - \tau; \omega, \sigma) f(\tau) d\tau ,$$

$$\Psi_s(t; \omega, \sigma) = \int_\tau \psi_s(t - \tau; \omega, \sigma) f(\tau) d\tau . \tag{2}$$

Let us define the square of the *wavelet amplitude transform A* as

$$A^2 = \Psi_c^2(t; \omega, \sigma) + \Psi_s^2(t; \omega, \sigma) . \tag{3}$$

It can be shown that the sinusoidal oscillation is eliminated from A^2, and this expression is almost invariant to time shifts around t, provided that these shifts are smaller than σ/ω.

A set of wavelet amplitude transforms A, taken for different ω and σ, can then serve as a *feature vector that has approximate invariance with respect to locally limited time shifts.*

In image analysis, the two-dimensional "wavelets" are called the *Gabor functions* [1,2], and the corresponding translationally invariant features are the *Gabor amplitude transforms.*

If we could make the wavelet and Gabor transforms *emerge* automatically in a learning process, we would obtain a set of analyzers for locally invariant features. A severe problem exists if no functional forms for the kernels can be preassumed; the feature analyzers must then be determined by the signals. We shall show that filters resembling (1) will indeed emerge.

3   Linearly Dependent Sample Vectors

In systems theory, sets of successive samples from the same waveform are frequently represented as *sample vectors*: Define

$$x = x(t) = [f(t - t_1), f(t - t_2), \ldots, f(t - t_n)]^{\mathrm{T}} , \tag{4}$$

whereby $x \in \Re^n$ is a column vector, and the $t_1 \ldots t_n$ are the sampling instants *relative to time t.*

Assume tentatively that $f(t)$ is a sinusoidal function. It is then a trivial fact that at least three successive sample vectors are *linearly dependent*, provided that their time difference is not a multiple of π/ω. Consider the two sample vectors (referring to times $t^{(1)}$ and $t^{(2)}$, respectively)

$$b_1 = [f(t^{(1)} - t_1), \ldots, f(t^{(1)} - t_n)]^{\mathrm{T}} ,$$
$$b_2 = [f(t^{(2)} - t_1), \ldots, f(t^{(2)} - t_n)]^{\mathrm{T}} , \tag{5}$$

and call $b_1, b_2 \in \Re^n$ the *basis vectors*. Now it holds for an arbitrary sample vector $x(t)$ from $f(t)$ that

$$x = \alpha_1 b_1 + \alpha_2 b_2 , \tag{6}$$

where α_1 and α_2 are scalar parameters. In other words, for a sinusoidal $f(t)$, an arbitrary sample vector is a *linear combination* of two basis vectors constructed by similar sampling.

Consider henceforth that $f(t)$ is nonsinusoidal, but reasonably well describable as a *superposition of a finite number of sinusoidal components*, in local intervals at least, such as the speech signal in which different superpositions usually exist in different intervals. Our aim is to construct a *competitive-learning process* wherein pairs of basis vectors (b_{i1}, b_{i2}) are formed adaptively at different neural units in response to the signal components present in $f(t)$. The process relates to a set of competitive adaptive filters, and each of the latter, by a winner-take-all function, selects a particular one of the signal components, to which it is then tuned even better. The learning law is such that only signal components that are linearly dependent in the above sense are able to tune the basis vectors. We can start the learning process with arbitrary basis vectors, whereby they assume a similar role as the adaptive weight vectors of neural networks in general do. These vectors are updated in an unsupervised learning process toward their "optimal" values. The mathematical formalism underlying this process is the *theory of linear subspaces* discussed next.

4   Manifolds as Neural Units

A *manifold* is a topological space that has certain specific properties. A *linear subspace* is a specific manifold: if the $b_1, \ldots, b_N \in \Re^n$ are N linearly independent (basis) vectors, then $\mathcal{L} = \mathcal{L}(b_1, \ldots, b_N)$ is the linear subspace, the general element of which is the linear combination $\alpha_1 b_1 + \cdots + \alpha_N b_N$, where the $\alpha_1, \ldots, \alpha_N$ are free scalar parameters.

If $\mathcal{L}$ is now regarded as a *class*, then the *distance* of an arbitrary vector $x \in \Re^n$ from $\mathcal{L}$ can be defined in terms of *orthogonal projections*. It is always possible to decompose x as

$$x = \hat{x} + \tilde{x} \, , \qquad \cdot \qquad (7)$$

where $\hat{x} \in \Re^n$ is a particular linear combination of the $b_1 \ldots b_N$, whereas the Euclidean norm of $\tilde{x} \in \Re^n$, denoted $||\tilde{x}||$ in the sequel, is minimum. Further $\hat{x}$ and $\tilde{x}$ are orthogonal, i.e., their dot product is zero. The orthogonal projections $\hat{x}$ and $\tilde{x}$ can conveniently be computed by the traditional Gram-Schmidt method [3].

Whereas $\hat{x}$ and $\tilde{x}$ are vectorial variables, the distance of x from $\mathcal{L}$ is the scalar entity $||\tilde{x}||$. When we then define different classes $\mathcal{L}_i$ using different basis vectors $b_{i1} \ldots b_{iN}$, input x is considered to belong to that class $\mathcal{L}_i$ from which its distance $||\tilde{x}_i||$ is minimum.

The above definition of classes is not optimal in the sense of the Bayesian decision theory, but it has the invariance property that is not considered with the Bayesian classifiers. This principle seems to work very well in many practical applications in which the data are high-dimensional and very noisy, like in neural-network applications in general. One has to notice that *the subspace classifier is also invariant with respect to the length of x, i.e., to the average intensity of the pattern.* We have applied the subspace classifier with success to speech [4] and textures [5].

We might illustrate the subspace classifier as in Fig. 1: an external input pattern x is broadcast in parallel to a number of neural units, each consisting of a linear input layer followed by a quadratic output "neuron" denoted by Q. The "weight" vectors of the neurons in the input layer correspond to the basis vectors of the respective linear subspace. If the basis vectors of each input unit are orthonormalized, each output "neuron" only has to form the sum of squares of its inputs.

5   The Learning Subspace Method (LSM)

This author, in 1978, invented the algorithm named the *Learning Subspace Method (LSM)*; it was the first *supervised competitive-learning method* thereby known. The idea was to improve class-separation of the basic subspace classifier: if the algorithm produces errors, the basis vectors of the closest subspace ("winner"), and those of the next-to-closest subspace ("runner-up"), were *rotated* in order to increase $||\tilde{x}_c||$, the distance of x from the "winner" subspace $\mathcal{L}_c$, and to decrease $||\tilde{x}_r||$, the distance of x from the "runner-up," provided that the latter would classify x correctly. The logic behind this supervised competitive learning scheme was very much the same as in the present LVQ2 (Learning Vector Quantization Type Two) algorithm [3]. The rotation of the basis vectors b_{ih} of class $\mathcal{L}_i$ was defined as the *operator product*

$$b_{ih}(t+1) = \left(I - \alpha(t) \frac{x(t)x^{\mathrm{T}}(t)}{||x(t)||^2} \right) b_{ih}(t) , \tag{8}$$

where I is the identity matrix, and $0 < \alpha(t) < 2$. Preferably $\alpha(t) \ll 1$. This rotation turns the basis vectors in such a direction that $||\tilde{x}_i||$ is increased; if we change the sign of $\alpha(t)$, the rotation is toward x, whereby $||\tilde{x}_i||$ is decreased. In this work, however, we only consider rotations toward x (unsupervised learning).

6   A Particular Generalization of the SOM: The ASSOM

The Self-Organizing Map (SOM) can be generalized in many ways [3]. One of them is to define the map units as subspaces discussed above. We shall show that such a SOM will learn to define subspaces in which the training samples mostly lie. With a modification of the winner and a few other additional details, the units of this kind of SOM, dubbed the *Adaptive-Subspace SOM (ASSOM)*, will become *invariant filters* for meaningful components of the input signal.

6.1   Adaptive Wavelet-Type Filters for Temporal Sequences

In this subsection we shall again pick up the patterns from the time domain of signals. According to Ch. 3, define the input x to the SOM as a set of samples taken from some continuous signal $f(t)$, where the $t_1 \ldots t_n$ are constant displacements in time, and call this particular sample vector the *sequence vector* in the sequel:

$$x = x(t) = [f(t-t_1), f(t-t_2), \ldots, f(t-t_n)]^{\mathrm{T}} \in \Re^n . \tag{9}$$

The central result in the ASSOM is that if pieces of almost sinusoidal components exist in the signal $f(t)$, each of the neural units will be tuned to one of them, independent of its phase (for more detailed justification, cf. [3]).

Assume that each node i (neural unit) in the SOM array contains k basis vectors $b_{ih} \in \Re^n$, $h = 1, 2, \ldots, k$. These vectors span and define the respective subspace $\mathcal{L}_i$. The "winner" node c for a single sequence vector x is defined in terms of the orthogonal components of x,

$$\hat{x}_c = \max_i \{\hat{x}_i\} \text{ or } \tilde{x}_c = \min_i \{\tilde{x}_i\} . \tag{10}$$

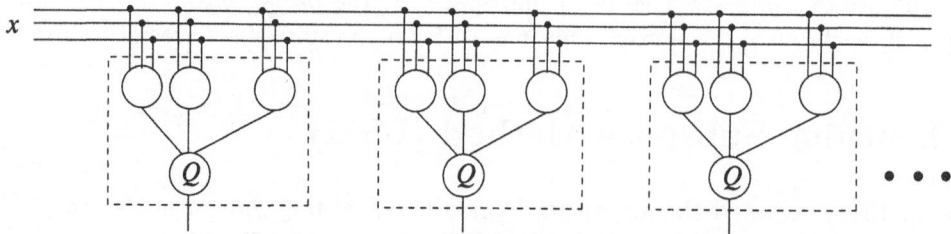

squared lengths of projections of x on the due subspaces

Figure 1: Architecture of the subspace classifier. The "neural units" are shown with dashed lines.

Another idea we need is the concept of *a "representative winner" over a set of subsequent sequence vectors $x(t)$ ("episode")*. For every $x(t)$ we shall, of course, define the usual "winner" node $c = c(t)$, but what we actually want a particular node and its topological neighborhood in the network to do is *to learn the general linear combination of adjacent sequence vectors $x(t)$.*

There may exist several possibilities for the definition of the "representative winner" relating to a set of sequences adjacent in time. Consider the sequences $x(t_1), x(t_2), \ldots, x(t_k)$ where each vector x is defined as a different set of samples taken from the same signal waveform. Let the projections on subspace $\mathcal{L}_i$ be $\hat{x}_i(t_1), \ldots, \hat{x}_i(t_k)$. For the definition of the "representative winner" c_r we may take

$$c_r = \arg\ \max_i \left\{ \sum_{p=1}^{k} ||\hat{x}_i(t_p)||^2 \right\} . \tag{11}$$

There now seems to exist a paradox that whereas the "representative winner" was already determined by one set of sequences, the samples in these sequences are no longer available for learning later on. This paradox, however, is more apparent than real. In many practical cases, e.g., when analyzing ergodic waveforms, the correlation of having the same "representative winner" during the closest subsequent episodes is high, at least when the waveforms are changing slowly. Thus one could use one period of time for determining the "winner" and the next one for learning, respectively. In technological constructs one might use delay lines or other memories to buffer the same training samples to be used for learning after determination of the "representative winner." *In the simulations reported below, learning was simply based on the same set of samples that was used for finding the "representative winner" and the corresponding neighborhood N_{c_r} in the array.*

Learning in the SOM is in general defined to occur only in the "topological neighborhood" N_c of cells in the neural network, around the winner cell c. In the present case, the basis vectors in the neighborhood set N_{c_r} around the "representative winner" are *rotated* in learning toward input x:

$$b_{ih}(t+1) = \prod_{t_d \in E} \left[I + \alpha(t - t_d)x(t - t_d)x^{\mathrm{T}}(t - t_d) \right] \cdot b_{ih}(t - t_d) , \tag{12}$$

where E is the set of sequence vectors used for this training step ("episode"), and the t_d are the relative displacements of the training vectors in time. The sample vector $x(t)$ should preferably be normalized; this can be done for each sample separately, or for some larger set of samples (i.e., dividing by their average norm). Below we shall always take $h \in \{1, 2\}$ corresponding to two-dimensional subspaces. In principle, to define the subspaces $\mathcal{L}_i$, the b_{ih} need not be orthogonalized or normalized; however, if we want the b_{i1} and b_{i2} represent the cosine and sine pair of a wavelet, then orthonormality of these components is desirable.

One of the peculiarities of the ASSOM compared with the usual SOM is the size of the learning step, which we see best if we imagine that the subspaces are one-dimensional, spanned by a single basis vector b_i. If the learning episode also consists of one unnormalized sample vector x only, the rotation operation can be expressed as

$$b_i(t+1) = \left(I + \alpha(t)\frac{x(t)x^{\mathrm{T}}(t)}{||x(t)||^2} \right) b_i(t) = b_i(t) + \alpha(t)\frac{b_i^{\mathrm{T}}(t)x(t)}{||x(t)||^2}x(t) , \tag{13}$$

and $b_i(t+1)$ should still be orthonormalized after (13). The essential difference here with respect to the basic "Dot-Product SOM" [3] is that the correction is now proportional to the cosine of the angle between $b_i(t)$ and $x(t)$. This factor, diminishing corrections at large errors (close to $\pi/2$), may give rise to extra instability. After some considerations described in [3] we decided to take for the rotation operator the expression

$$P = I + \alpha(t - t_d) \frac{x(t - t_d) x^{\mathrm{T}}(t - t_d)}{\|\hat{x}_i(t - t_d)\| \, \|x(t - t_d)\|} \, , \tag{14}$$

where P is the bracketed expression in (12), and $\hat{x}_i$ is the projection of x on $\mathcal{L}_i$.

In the experiment reported below we automatically generated wavelet-type filters for time-domain speech waveforms obtained from the TIMIT database. These signals have been sampled at 12.8 kHz. We formed the training vectors $x \in \Re^{64}$ from the samples by picking them up from randomly selected time windows of 64 points, equivalent to time intervals of 5 ms. We had a SOM with 24 units arranged in a one-dimensional array. The training process consisted of 30000 steps, whereby the radius of N_{c_r} changed linearly from twelve to one lattice spacings, and $\alpha(t)$ was of the form $A/(B + t)$, where $A = 18$ and $B = 6000$. At each step the episode consisted of eight windows displaced randomly (relating to a binomial distribution) over 64 adjacent sampling instants. In order to create good filter banks, a number of further details had to be introduced to the ASSOM process. First, in order to enchance signal energy at high frequencies, the signals in the TIMIT database were *high-pass filtered* by taking the differences of successive values. The data vectors x were then normalized. Second, the b_{ih} vectors were orthonormalized after each rotation operation. (This orthonormalization can also be made more seldom, say, at each tenth episode.) Third, if we would not *weight* the input samples in any way, the wavelet function might be formed anywhere within the sampling window; so, in order to stabilize the wavelets into the center of the window, the middle samples were provided with higher weights, and a Gaussian weighting function over the sampling "window" was used. The weighting was started using a narrow Gaussian weighting function having the full width at half maximum of eight sampling points (equivalent to 0.7 ms), and centered in the window. The narrow Gaussian stabilizes the high-frequency filters first, whereby also all filters are coarsely ordered. When the Gaussian is thereafter let to broaden up to 50 sampling points during the learning phase, the low-frequency filters acquire their final correct values. Fourth, after all previous precautions we still noticed a peculiar instability: some filters, especially those close to the minima of the speech spectrum, were prone to be tuned to *two* frequency bands. (In principle, if competitive learning were not present, the filters could freely be tuned to several frequencies.) We found out that the following simple *dissipative effect* was a sufficient remedy for producing smooth, well-distributed and asymptotically stable filters with a single peak: after each orthonormalization of the basis vectors, a small constant value (equivalent to $\alpha/50$) was subtracted from the *amplitude* of each component of the basis vectors (when they were orthonormalized at each step). This subtraction was saturable: the components did not change their signs.

Figure 2 shows the pairwise basis vectors formed in the self-organizing process at each "neural" unit. *Notice that the filters, especially at higher frequencies, do not have the identical form as stipulated by (1). Instead, the range of translational invariance is now much less dependent on frequency than in (1), being coarsely of the same order of magnitude over the whole frequency scale! The "Gaussian" envelopes depicted in Fig. 2c were determined totally in the learning process.*

6.2 Automatic Formation of Gabor-Type Filters

For two-dimensional images with intensity $I(r)$ at the image coordinate $r \in \Re^2$, the input vector to the SOM is

$$x = x(r) = [I(r - r_1), I(r - r_2), \ldots, I(r - r_n)]^{\mathrm{T}} \in \Re^n, \tag{15}$$

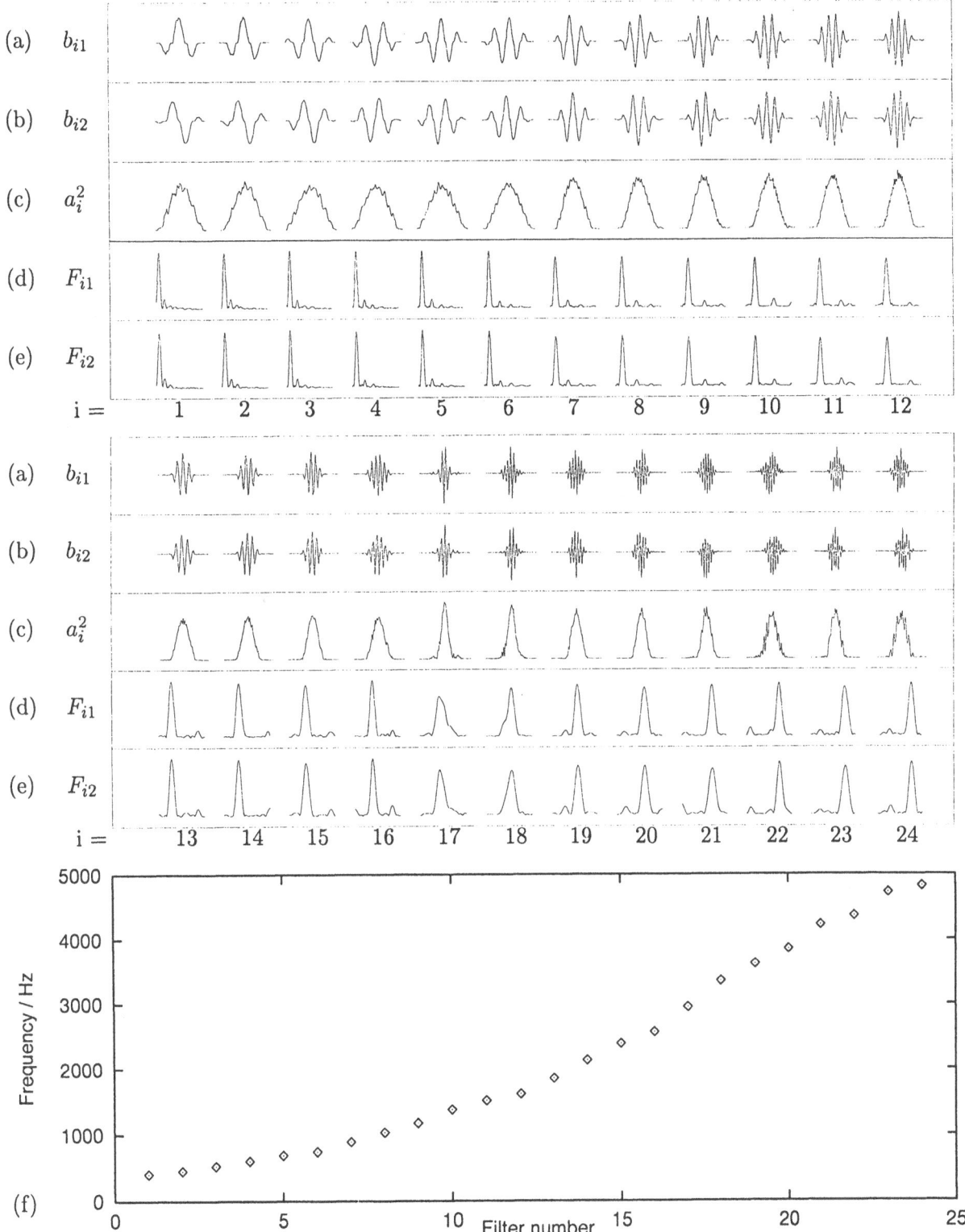

Figure 2: Notice that the diagrams (a) through (e) are in two parts. (a) Cosine-type wavelets (b_{i1}), (b) Sine-type wavelets (b_{i2}), (c) $a_i^2 = b_{i1}^2 + b_{i2}^2$, (d) Fourier transforms (F_{i1}) of the b_{i1}, (e) Fourier transforms (F_{i2}) of the b_{i2}, (f) Distribution of the average frequencies of the filters. The steeper part around 2500 Hz corresponds to a broad minimum of the speech spectrum!

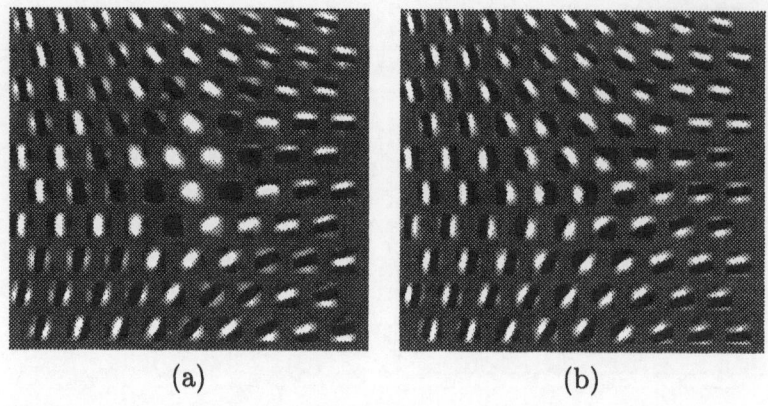

(a)   (b)

Figure 3: An ASSOM that has formed Gabor-type filters: (a) The b_{i1}, (b) The b_{i2}.

where now the $r_1, \ldots, r_n \in \Re^2$ form a two-dimensional *sampling lattice*. In the present experiment we used equidistant sampling with 225 sampling points over a square area of the closest pixels. At each learning cycle, this lattice was shifted randomly in a few (say, four) closeby positions, and the pattern vectors $x(t)$ thereby obtained constituted the two-dimensional "episodes" over which determination of the "representative winner", and updating of the corresponding N_{c_r}, were made. The learning cycle was repeated for other parts of the image, and for different images.

Discretized photographic images were used as training data. High frequencies were emphasized by subtracting local averages from the pixels. The sampling lattice was further weighted by a Gaussian function for better initial concentration of the filters. The Gaussian was flattened during learning. Fig. 3 illustrates the pairs of Gabor-type filters thereby produced.

7 Conclusion

There is nothing in the ASSOM principle that would restrict its use to translationally invariant pattern analyzers. Wavelet- and Gabor-type filters were formed above because the episodes consisted of translated patterns only. Different kinds of analyzers will be formed, if the input data exhibit, say, rotational or perspective transformations, eventually being in motion.

Sometimes it may be beneficial to introduce three or more basis vectors for each subspace, and the ASSOM array might be, say, three-dimensional. The ASSOM architecture may be used as a genuine, adaptive *feature-extraction layer* in a pattern recognition system. Its decision-making part then receives the set of outputs $||\tilde{x}_i||$ and can be trained to classify targets by a supervised learning process; the classifier could be, e.g., LVQ [3]. It is also possible to use decision-directed training of the feature-extraction layer (see the so-called FASSOM architecture discussed in [3]).

References:

[1] D. Gabor, J. IEE 93, 429 (1946).
[2] J. Daugman, IEEE Trans. Syst., Man, Cybern. 13, 882 (1983).
[3] T. Kohonen, *Self-Organizing Maps.* Springer 1995.
[4] T. Kohonen, G. Nemeth, K.-J. Bry, M. Jalanko, H. Riittinen, Proc. 1979 ICASSP, p. 97.
[5] E. Oja, T. Kohonen, Proc. ICNN'88, p. I-277.

9.

Democratic Reinforcement: Learning via Self-Organization

Dimitris Stassinopoulos and Per Bak

Democratic Reinforcement: Learning via Self-Organization

Dimitris Stassinopoulos* and Per Bak†

*Center for Complex Systems, Florida Atlantic University
Boca Raton, Florida 33431, USA
†Department of Physics, Brookhaven National Laboratory
Upton, New York 11973, USA

ABSTRACT

The problem of learning in the absence of external intelligence is discussed in the context of a simple model. The model consists of a set of randomly connected, or layered integrate-and fire neurons. Inputs to and outputs from the environment are connected randomly to subsets of neurons. The connections between firing neurons are strengthened or weakened according to whether the action is successful or not. The model departs from the traditional gradient-descent based approaches to learning by operating at a highly susceptible "critical" state, with low activity and sparse connections between firing neurons. Quantitative studies on the performance of our model in a simple association task show that by tuning our system close to this critical state we can obtain dramatic gains in performance.

I. Introduction

One of the most remarkable properties of biological neural networks is their ability to learn via *self-organization*. Simply put, this means that animals acquire experience and make sense of their environment without the aid of a "teacher" or some other form of external intelligence. To any non-expert that has ever seen a toddler acquiring a new skill with virtually no guidance or an animal adapting to a novel situation the statement would seem obvious. Yet for all its simplicity and common sense this idea has long remained on the fringe of the experimental and theoretical research concerning brain function, in all likelihood because of the severity of the contraints it imposes on brain modelling.

The term self-organization has been used in many different disciplines such as physics, chemistry, biology, and psychology, and often to convey different underlying mechanisms. Here, however, we will discuss self-organization solely in the context of learning[1].

The oldest and perhaps still most dominant approach to understanding the brain is what we call the "engineered brain" paradigm. According to this paradigm, brain function emerges because nature, in the role of the engineer, has created all the necessary mechanisms by establishing an intricate web that brings billions of pieces together. But how can evolution achieve such an engineering feat? We do not deny the role of evolution in many aspects of brain function – the very fact that our brain is different than a lobster's brain has to be attributed to evolution. Nevertheless, it cannot possibly account for the brain's ability to deal with unforseen situations, specific to an individual's experience, or for novel ones that evolution had never had the opportunity to confront.

In providing an alternative to this view, the field of artificial neural networks (ANN) has been instrumental (For reviews see Hertz *et al*[2] and Haykin[3]). The major contribution of ANN was that it demonstrated how non-trivial tasks can be achieved with networks composed of many simple computing elements. It also offered the first evidence that principles for brain function can be captured with models that have simple structure. Despite all the important insights ANN offered, however, they have not eliminated the need for an external intelligence. In the widely used *supervised learning* paradigm this takes the form of a "teacher" providing the system with a detailed scheme for the update of the synaptic weights based on knowledge of the goal to be achieved. Furthermore, most models for learning use gradient-based update rules, such as back-propagation, which are biologically implausible because they impose strong constraints on the architecture and they require computation that cannot be performed by the neural network itself. Thus, again the network is formed by design rather than by self-organization.

The issues of self-organization have been addressed in the context of reinforcement learning models[2,3]. These models are more realistic in the sense that there is no teacher explaining how to modify the synaptic weights, but only a "critic" telling the system whether its performance is successful or not. Most reinforcement learning models, however, still rely on back-propagation[4], or some other overseeing agency possessing prior knowledge of the problem, for the update of the synaptic weights. There is one exception, however. Barto[5], in one of the first variants of his *Associative Reward-Penalty*[6] (A_{R-P}) algorithm, discusses the idea of "self-interested" elements which do not have access to information other than a feedback signal from the environment broadcast simultaneously to all elements. We very much agree with his view that the difficulties in solving the problem of learning under the severe constraints imposed by self-organization are fundamental.

Recently, we have proposed *Democratic Reinforcement* (*DR*) as a new approach to the long-standing issues of learning via self-organization[7]. A similar approach was originally used to solve a non-trivial tracking problem by a continuous modification of its synaptic weights[8]. An evaluative feedback is sent democratically to all neurons simultaneously. The reinforcement rule operates in two modes: a "learning" mode when the evaluative feedback is positive, and an "exploration" mode when the evaluative feedback is negative. The reinforcement rule depends on the firing states of the presynaptic and postsynaptic neurons only. The novel feature of our model is that the threshold for firing is regulated in order to keep the output activity minimal. This sets the system up at or near a critical state, which turns out to be crucial for the performance of the network.

To the best of our knowledge, the A_{R-P} and the DR represent the only attempts to address the problem of learning via self-organization. However, the two algorithms are fundamentally different. While the A_{R-P} is a gradient-descent algorithm, DR solves problems by operating at or near a highly-susceptible "critical" state in which the system becomes very sensitive to modification of the synaptic weights. To make our point more concrete we shall discuss the two algorithms in the context of a specific association task.

II. The Model

The problem we are addressing can be summarized as follows: How can many "agents", be it neurons or some other kind of computing element, operating under local rules and receiving input only from a small fraction of other agents, cooperatively perform macroscopic tasks imposed on them by their environment? For the agents to perform a task, they are given the freedom to tune some local parameters, such as their synaptic weights, to appropriate values. The severity of the problem arises from the requirements: i) that there can be no external intelligence with prior knowledge of the problem to instruct the agents on how to tune their parameters, only some form of overall evaluative feedback that tells the agents when some parameters essential for "survival" fall out of bounds, and ii) that the system is robust and versatile, allowing for solutions that are not task or architecture-specific. From an agent's point of view any effective adjustment can be made only when the adjustment has a detectable impact on the overall, collective behavior of the system, allowing for a robust feedback. This, however, is a non-trivial requirement. Most neural networks are rather insensitive to small changes in their parameters. Then we are left with a situation where no agent can "learn" from its actions because there is no way for it to know whether it should get "credit" or "blame" in the final evaluation. This is known as the *credit-assignment problem*[9].

The above problem remains largely unresolved. Of course, real biological networks serve as examples par-excellence that solutions to this problem exist, but how do we mimic this in a simple model?

Our model imposes no constraints on the architecture whatsoever. In the most general case neurons are connected randomly to each other via unidirectional connections[7]. Each neural unit receives input and sends output to a small fraction of the total number of units. This ensures that the majority of units in our system are hidden, i.e., units that interact with the environment indirectly through other units. The system interacts with the outer world in three ways (Fig. 1); via i) its input units which receive input from the outside and thus provide the system with information about the state of the world; ii) its output units which allow the system to act on the environment; and iii) a binary yes/no feedback that is broadcast to all units and indicates whether the action to the environment was successful or not.

Although from a conceptual point of view the random architecture is the most appealing, we found the layered architecture of Fig.1 to work better[10]. Here each unit is connected to its three nearest neighbors in the next layer. The units can be either in a firing state, $n_i = 1$, or in a quiescent state, $n_i = 0$. In standard fashion, each unit integrates the input of its presynaptic units, $h_i = \sum_j J_{ij} n_j$. The unit fires if the input exceeds a threshold T. Input patterns, indicated in Fig. 1 by dark disks, are presented to the system by setting the corresponding units into a firing state. The system acts on the environment via its output sites, shaded disks in Fig. 1. The feedback, r, broadcast by the environment, takes two values; positive, r_0, if the action was evaluated as successful and negative, $-r_0$, if not. Both the synaptic modification and the regulation of the threshold depend on the evaluative feedback, r.

The update rule affects only connections between firing neurons, $J_{ij} \rightarrow J_{ij} + [rJ_{ij} + h_{ij}]n_i' n_j$, where n_i' denotes the state of the i'th neuron at the next time step and h_{ij} is a random noise between $-h_0$ and h_0. The outgoing weights are normalized, $J_{ij} \rightarrow J_{ij}/\sum_i J_{ij}$. The rule differs from standard gradient-descent based update rules in one crucial aspect: When $r > 0$ the system operates in a "learning" mode in which connections are being strengthened and the performance improves but when $r < 0$ the system operates in an "exploratory" mode in which strong connections are being weakened and weaker connections are being strengthened. Typically, during this phase the performance deteriorates. In contrast, standard reinforcement schemes, such as A_{R-P}, rely on an improvement of the performance both for positive and negative r and perform the exploration stochastically.

In addition to the synaptic update rules, r regulates the threshold, T. The objective is that the output activity is kept to a minimum. This is essential to ensure that the system attributes credit and blame to the minimum possible number of active units, in order to keep the network intact for other problems. In our first versions of the DR algorithm (Ref. 7, 8) the output activity was regulated to a small but arbitrarily chosen level of activity. Choosing a value for this parameter, however, assumes prior knowledge of the task to be completed which runs counter to the self-organization philosophy. More recently[11] we have introduced a threshold mechanism that depends solely on r, $T \rightarrow T + \delta(r)$, where δ assumes a positive value, δ_+, if $r > 0$, and a negative value, δ_-, if $r < 0$ ($|\delta_-| >> \delta_+$).

The typical criterion for success, $r > 0$, is that the selected output sites are active and for failure, $r < 0$, that at least one of the selected output sites is inactive. On first thought, that might sound like nonsense: the system can trivially obtain a positive feedback and thus get its reward by lowering its threshold, for instance, and keeping all of its output units active! However, the solution that the system opts for is the one where the selected output sites are active and all the rest are inactive. But even if we accept that such a rule makes computational sense what sense does it make in terms of biology? It is true that in some simple situations we may view this reward/penalty coming from the environment. This was the case in Ref. 7 where we used the analogy of a monkey that presses one or more buttons. In such a situation the environment indeed provides food as long as the selected buttons are among the ones pressed by the monkey. However, the monkey can not be considered successful merely because now and then it happens to press the right buttons. It is important to reduce the incorrect actions. In that respect it makes more sense to view $r > 0$ not as an external reward but rather as the default mode of operation; an innate tendency of the system to minimize its efforts, while still having success. In contrast, one might view the $r < 0$ signal as an external wake-up call announcing that something is wrong, for instance when some parameter that is crucial to survival exceeds a certain value. The preference for passivity is sharply interrupted when $r < 0$. There is no symmetry between $r < 0$ and $r > 0$.

How does the system solve problems? How does it successfully attribute credit and blame where it is due? Our studies indicate that this involves a build-up process in which the synaptic "landscape" reaches a near critical, highly susceptible state in which small changes in the synaptic weights can have a big effect on the collective activity. In such a state the system can establish efficiently causal relationships between changes in the synapses and the output. To achieve such a critical state the system: i) assigns credit and blame only to connections between active neurons, ii) keeps the activity low by means of the global regulation of the threshold and the local learning rules. By combination of these two mechanisms the system attributes credit and blame selectively by driving the system to the interface between success and failure.

III. Self-organization in a Simple Association Task

In an association task we ask the system to generate a certain input/output pattern. The insets in Figs. 2a, b offer examples of a simple association task. The system accomplishes the task by "carving" paths between the input sites and the output sites. In previous work[7,8,12] we have investigated the performance of DR in a variety of situations: multiple input/output patterns, recovery from "damage", tracking, conditioning, and so on. Here we will be concerned with the question of degradation of performance as the size of the association task grows.

We consider an $L_1 \times L_2$ layered network (Fig. 2a, inset). The number of layers in this networks is kept fixed, $L_1 = 16$, while the lateral dimension is varied, $L_2 = 16, 32, 64, \ldots$ The number, c, of input and output sites in the input/output pattern is varied accordingly, $c = L_2/16$. Here the input is confined to the top row and the output to the bottom row. To minimize crossover between paths we keep the input and output sites in pairs, well separated from each other. More precisely, each of the c columns of a given network contains a single input and a single desired output site.

One motivation for this analysis is to demonstrate in a convincing manner the difference between DR and A_{R-P}. We were not so interested in the absolute performances of the two algorithms, which tend to be sensitive to the tuning of the various parameters, but rather to the scaling of the performance with the size of the network. For our simulations with A_{R-P} the same layered architecture was chosen (Fig. 1) but the number of layers was set to $L_1 = 4$ (more layers would degrade the performance of A_{R-P} too much). The lateral dimension is varied, $L_2 = 4, 8, 12, \ldots$ The input/output patterns were chosen with similar considerations as in the DR case (see insets in Figs. 2a,b for a comparison). Figures 2a and 2b show examples of the performance, P, for DR and A_{R-P} as a function of time. P is defined as the temporal average of the activity at the selected output sites minus the activity at the rest of the output sites. Appropriate normalization assures that best performance ($P = 1$) corresponds to persistent firing at the selected output sites only, whereas worst performance ($P = -1$) corresponds to persistent firing everywhere except at the selected sites.

The DR is characterized by intervals of rapid improvement in performance, interrupted by sudden dips. This behaviour is a signature of the dual mode of operation of the algorithm: i) the learning or exploitation mode in which the system strengthens connections and "weeds out" irrelevant paths, and ii) the exploration mode in which the system tends to spread the activity in an attempt to explore new possibilities with subsequent decrease in the performance P. The A_{R-P} performance versus time, although also highly irregular, seems to have a very different structure. It is dominated by very fast fluctuations at the smallest time scale (not seen here due to averaging of P). At longer time scales it seems to be dominated by long periods during which the system seems trapped at a certain level of performance. Once the system escapes this barrier the transition to a new performance level appears to be very fast. We would like to point out that at the individual level, and with the limited information available to it, each neuron *always* opts for the change that it expects will increase the collective performance. In its decisions, however, it cannot take into account the positive or negative contributions of the other neurons. Therefore, it is only in a statistical sense that the system senses the gradient towards a better performance and can tune its synapses accordingly. The stochastic nature of the A_{R-P} can also be witnessed in Fig. 3, ($\triangle$). Here we depict the time to completion of the task, $\bar{t}_s$ (averaged over many runs obtained with different initialization) as a function of the number, c, of input/output pairs. In a first order approximation, it seems that $\bar{t}_s$ scales exponentially with c, $\bar{t}_s \sim e^{\alpha c}$, with $\alpha \simeq 1.6$.

DR (Fig. 3, $\bigcirc$) has a significantly better scaling behavior. When plotted in a log-log plot (inset of Fig. 3) $\bar{t}_s$ might follow a power law, $\bar{t}_s \sim c^\gamma$, which subsequently breaks down around $c = 8$. If this is true it would not be inconsistent with our suggestion that the algorithm operates near a "critical", highly susceptible regime. Although evidence of such a critical regime have previously been seen in the dynamics of our system[7], this scaling behavior offers the first quantitative evidence. Clearly, this initial data seems to be amenable to more than one interpretation, therefore it is imperative that the direct consequences of our critical-state hypothesis are tested further.

The convergence towards the critical state is accomplished by ensuring that the patterns of activity for different input signals do not overlap, on one hand, while, on the other hand, not being too sparse to connect inputs with desired outputs. In "sand" models of self-organized criticality,[13] overlap of events ("avalanches") is avoided by keeping the input rate low. Here, criticality is achieved by keeping the output low.

It turns out that one can improve the efficiency, and carry the system closer to the critical point by further ensuring that changes in the activity, due to threshold modulation do not overlap in time, while not happening too infrequently with respect to the synaptic modification rate. We do so by allowing a variable rate δ_+ for increasing the threshold T. More precisely the rise of the threshold is governed by $\delta_+(t, r)$, $\delta_+ \rightarrow a\delta_+$, where $a \gtrsim 1$. Notice that now δ_+ is time dependent in the sense that, while $r > 0$, it is constantly increased and r dependent in the sense that it is reset to a small value whenever r becomes negative, $\delta_+ \rightarrow a^{-L_1}\delta_+$. The rate of decrease of T, δ_- is kept constant as before. The modified algorithm leads to a significant improvement of the performance (Fig. 3, ($\square$)). Furthermore, the new curve, $\bar{t}_s(c)$, gives a stronger indication for the existence of a power law with exponent $\gamma \simeq 1.3$. The modification was chosen for its simplicity rather than its performance and, based on our experience, it appears to be straightforward to obtain further improvements.

IV. Conclusions

In this paper we have been concerned with the issues of self-organization which must play a central role in brain function. We propose a new mechanism through which efficient self-organized learning takes place. The central element of is a build-up process that allows the system to operate at a "critical" state, characterized by high sensitivity to small modifications of the synaptic weights and low output activity. The combination of those features allows the system to establish strong cause-effect relationships that allow the coexistence of many input/output patterns.

We suggest that the mechanisms that enables self-organizion in our model might also underlie real brain function, so DR can serve as an excellent testbed for further exploration of the consequences of such a hypothesis. It might be worthwhile exploring whether some signature of the critical state described in our model can also be observed in actual experiments.

References

1. A qualification is necessary here to avoid confusion of our use of the term "self-organization" as this used in *unsupervised learning*. In this paradigm the system modifies its synaptic ways with no feedback or any supervision whatsoever from the environment. Our view is that the consideration of environmental constraints is essential in biological learning and intelligent behavior in general, therefore, unsupervised learning has been entirely omitted from our discussion.

2. J. Hertz, A. Krogh, & R. G. Palmer, *Introduction to the Theory of Neural Computation* (Addison-Wesley, Redwood, 1991).

3. S. Haykin, *Neural Networks, a Comprehensive Foundation* (Macmillan, New York, 1994).

4. Based on the partial information provided by the critic a target pattern is determined and the output-weight errors computed. The rest of the weights can then be updated by back-propagating this error-signal through the network (see Hertz *et al* in Ref. 2).

5. A. G. Barto, Human Neurobiology **4**, 229 (1985).

6. A. G. Barto & P. Anandan, IEEE Transactions on Systems, Man, and Cybernetics, **15**, 360 (1985).

7. D. Stassinopoulos & P. Bak, Phys. Rev. E **51**, 5033 (1995).

8. P. Alstrøm & D. Stassinopoulos, Phys. Rev. E **51** 5027 (1995).

9. M. L. Minski, *Proceedings of the Institute of Radio Engineers* **48**:8-30.

10. This is primarily due to computer time limitations since the layered architecture gives substantially better performance[7] but also due to our interest in making comparisons with the A_{R-P} algorithm which is implemented in feedforaward architectures.

11. D. Stassinopoulos & P. Bak, for submission to J. Theoret. Biol.

12. D. Stassinopoulos, P. Bak, & P. Alstrøm, in *Proceedings of the Second Appalachian Conference on Behavioral Neurodynamics - Radford*, edited by K. Pribram, (Lawrence Erlbaum, New Jersey, 1994), 172-195.

13. P. Bak, C. Tang, & K. Wiesenfeld, Phys. Rev. Lett. **59** 381 (1987).

Figure captions

Figure 1. Block diagram of the model, here shown for the layered architecture. The input sites that receive signals from the environment are shown as dark discs. The output sites are shown as shaded disks. Periodic boundary conditions are assumed for the layers.

Figure 2. a) DR: Performance vs. time, for a 16×64 system and, for the input/output pattern shown in the inset. Light sites denote quiescent units and dark sites denote firing ones. The firing sites connect the input and output sites by effectively forming wires. The parameters of the algorithm have been set to: $r_0 = 0.1, \delta_+ = 0.01/16, \delta_- = -0.05/16$, and $h_0 = 0.01$. The performance is obtained by averaging over 50 time steps. b) A_{R-P}: Performance vs. time (measured in 'trials'), for a 4×16 system, and for the input/output pattern shown in the inset. The connections between input and output sites are more complicated. The central element of the algorithm is the update rule for the synaptic weights, $J_{ij} \rightarrow J_{ij} + \eta(r[n_i- < n_i >] + \lambda(1 - r)[-n_i- < n_i >])n_j$, where η is the 'learning' coefficient, λ is the 'penalty learning rate factor', $< n_i >=$ $\tanh(\beta \sum_j J_{ij}n_j)$ is the average firing state, and r is the evalutation feedback (for details see Hertz *et al* in Ref. 2). The parameters have been set to: $\eta = 0.5, \lambda = 0.001$, and $\beta = 0.5$. The performance is obtained by averaging over 100 trials. *Insets:* Typical successful ($P = 1$) activity patterns for DR and A_{R-P}.

Figure 3. Average time elapsed, $\bar{t}_s$, to completion of an association task vs. number of input/output pairs, c. ($\triangle$) A_{R-P}: systems of size $L_1 \times L_2$, $L_1 = 4$ and $L_2 = 4, 8, 12$, and 16 have been considered. For each case twenty runs were performed (with the exception of the 4×16 system for which we conducted five runs only, due to computing time limitations) for the same association task but with different initialization; ($\bigcirc$) DR: systems of size $L_1 \times L_2$, $L_1 = 16$ and $L_2 = 16, 32, 64, 128$, and 192 have been considered. For the same association task fifty runs, with differing initializations, were considered; ($\square$) DR with variable δ_+ ($a = 1.05$): systems of size $L_1 \times L_2$, $L_1 = 16$ and $L_2 = 16, 32, 64, 128$, and 256 have been considered. For the same association task, fifty runs with differing initializations, were considered. Vertical bars denote standard deviation. *Inset:* Same data in log-log plot. The solid lines represent least-squares fits. ($\triangle$): $\sim e^{1.6c}$; ($\square$): $\sim c^{1.3}$.

Figure 1

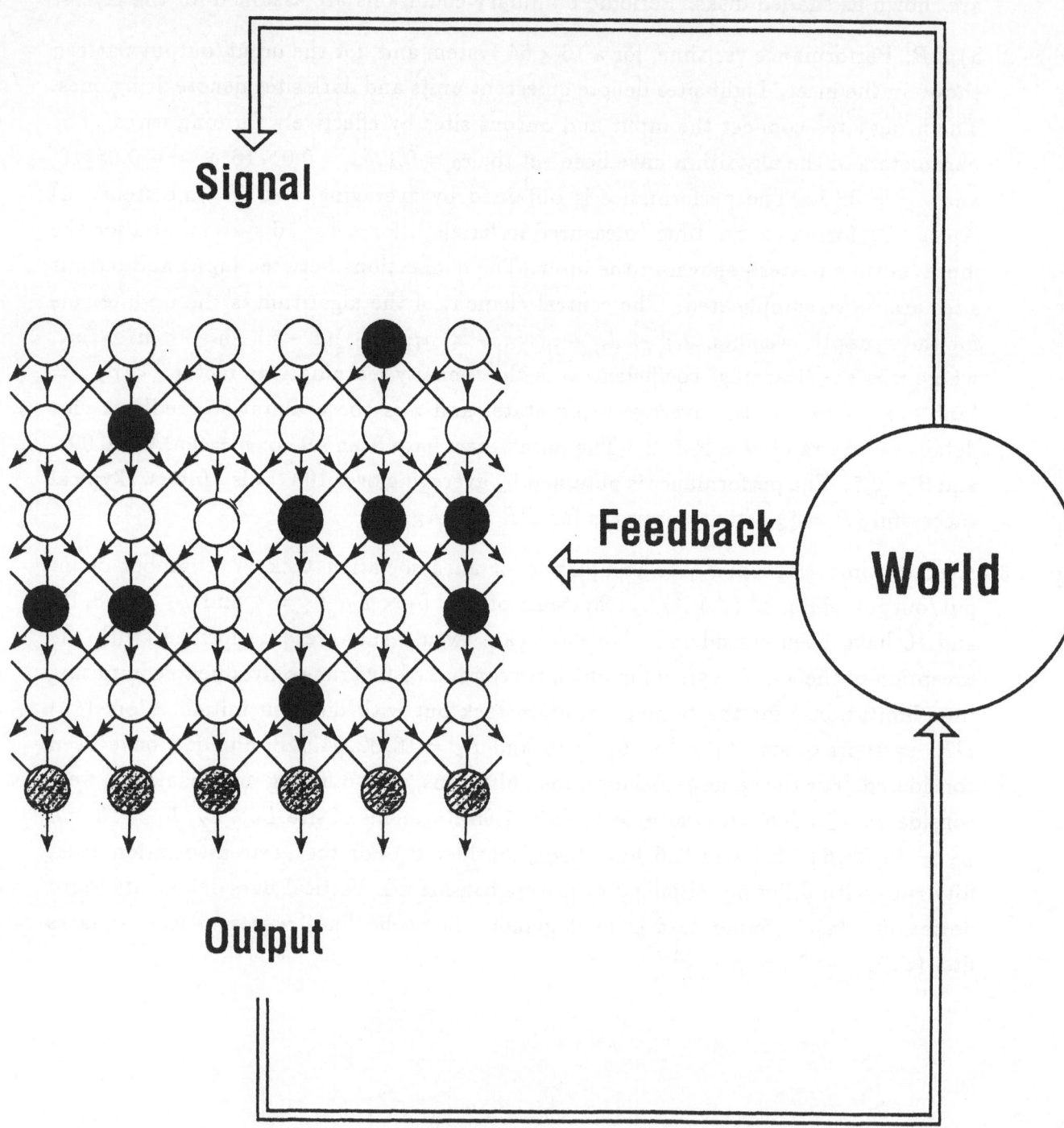

Signal

Feedback

World

Output

Figure 2

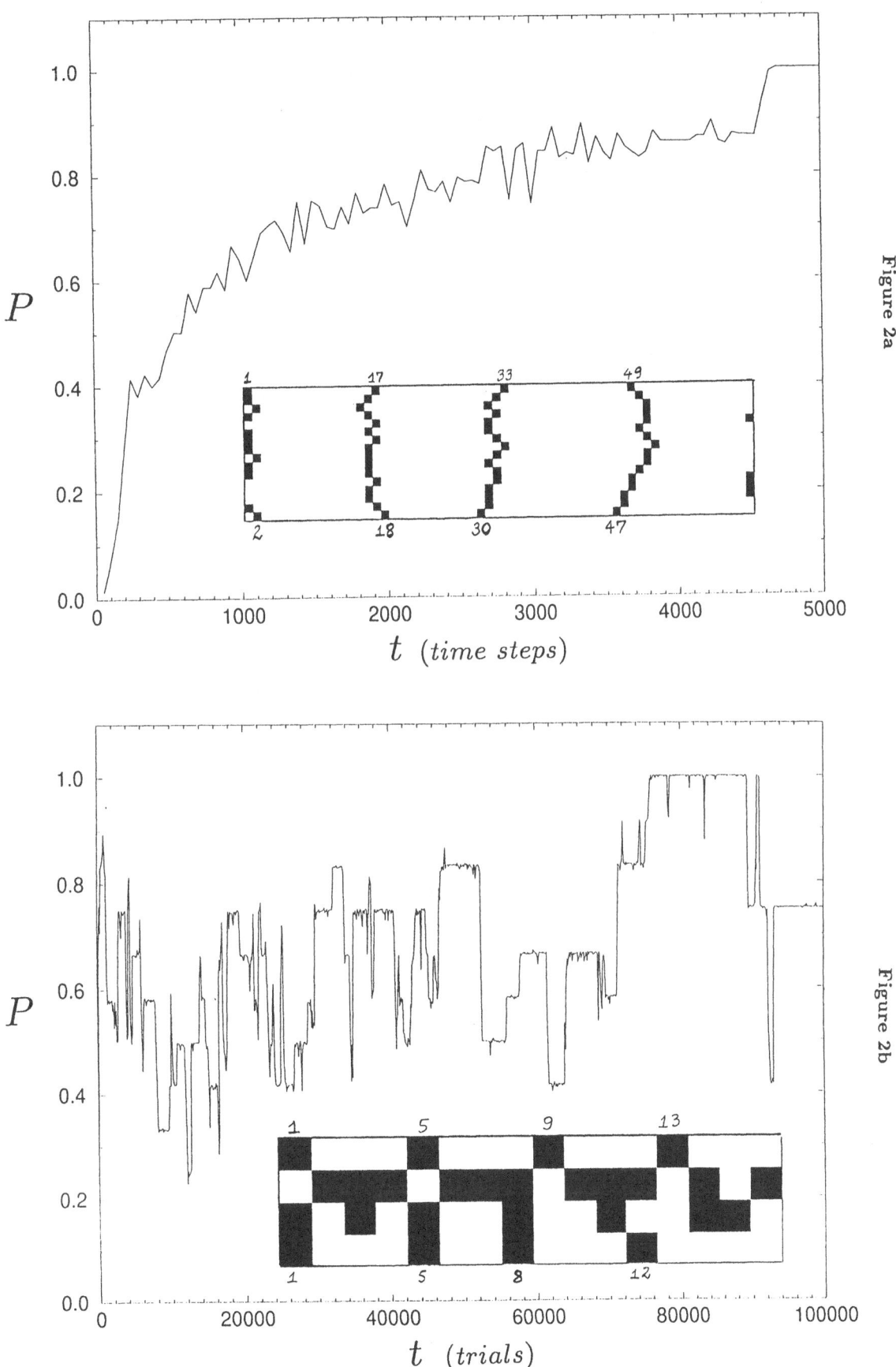

Figure 3

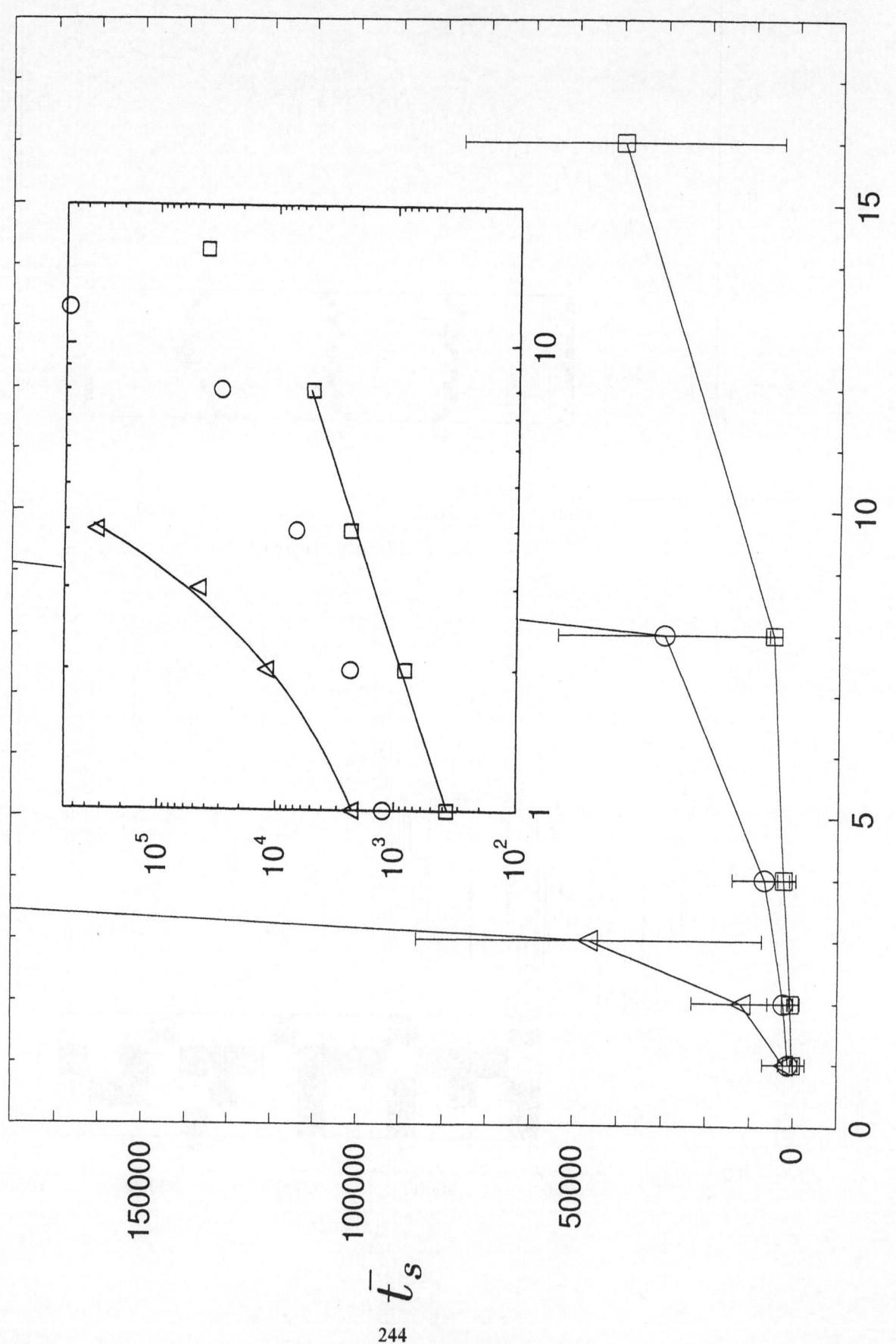

10.
Biological Plausibility of Synaptic Associative Memory Models

Daniel L. Alkon, Kim T. Blackwell, Garth S. Barbour, Susan A. Werness, and Thomas P. Vogl

.

Biological Plausibility of Synaptic Associative Memory Models

Daniel L. Alkon, Kim T. Blackwell, Garth S. Barbour,
Susan A. Werness, and Thomas P. Vogl

Laboratory of Adaptive Systems, NINDS and The Environmental Institute of Michigan

Abstract—*Observations in brains of neuronal networks that subserve associative learning in living organisms have been exceedingly sparse until the past decade. Recently, some fundamental biophysical and biochemical properties of biological neural networks that demonstrate associative learning have been revealed in the marine mollusc, Hermissenda crassicornis. In mammals, we have localized distributed changes, specific to associative memory, in dendritic regions within biological neural networks. Based on these findings, it has been possible to construct an artificial neural network, Dystal (dynamically stable associative learning) that utilizes non-Hebbian learning rules and displays a number of useful properties, including self-organization; monotonic convergence; large storage capacity without saturation; computational complexity of $O(N)$; the ability to learn, store, and recall associations among arbitrary, noisy patterns after four to eight training epochs; a weak dependence on global parameters; and the ability to intermix training and testing as new training information becomes available. The performance of the Dystal network is demonstrated on problems that include face recognition and hand-printed Kanji classification. The computational linearity of Dystal is demonstrated by its performance on a MasPar parallel hardware computer.*

Keywords—Associative learning, Biological neural networks, Dystal, Non-Hebbian learning rules, Self-organization, Saturation, Face recognition.

1. INTRODUCTION

Attempts to implement artificial neuronal networks on computers, which work millions of times faster than the human brain, do not effectively approach the brain's ability to sense, store, recognize, and recall patterns from specific experiential incidents. Parallel computing, by itself, produces none of the crucial elements of learning: association, abstraction, and generalization.

Most currently popular neural networks (Hopfield, 1982; Hopfield & Tank, 1986; Anderson, 1983, 1986; Grossberg, 1982, 1987; Rumelhart & McClelland, 1986; Rumelhart & Zipser, 1985; McClelland & Rumelhart, 1981; McClelland, 1985; Kohonen & Makisara, 1989; Fukushima, 1988; Sejnowski & Kienker, 1986; Sejnowski & Rosenberg, 1986; Kienker et al., 1986; Lippmann, 1987) use basic neuronal models that are variants of an idea originally proposed by

Hebb (1949) in which 1) "synaptic" inputs to an element are summed and the element fires if a threshold is exceeded, and 2) the weight or strength of a synaptic junction is increased only if both the presynaptic and postsynaptic elements fire. In essence, these neural network designs implicitly or explicitly use the output of a neuronal element in the adjustment of the weights on its input. In most implementations, a nonlinear optimization or relaxation algorithm is used in setting weights. These algorithms are all computationally intensive, because for each set of inputs to be learned, many iterations are required before convergence is achieved and the network has learned the input patterns. Depending on the details of the implementation, the computational complexity of these algorithms [reflected in the number of iterations needed to achieve convergence is $O(N^2)$ to $O(N^3)$], where N is the number of weights (connections) to be determined. Consequently, the total computational effort *per connection* increases as the number of connections is increased.

Observations of brain networks that subserve pattern recognition in living organisms have, until the last decade, been remarkably sparse. It is understandable, therefore, that efforts to reduce the computational intensity and to improve the performance of computer-based networks relied mainly on the ingenuity of math-

Acknowledgements: The partial support of this research effort by ONR contracts N00014-88-K-0659 and N00014-02-C-0018, and NIH/NINDS contracts N01NS02389 and N01NS32304 is gratefully acknowledged.

ematicians and physicists for ad hoc ideas derived from concepts of physics and engineering, although key biological and psychological insights have been exploited (e.g., Grossberg, 1982, 1987; Klopf, 1988). Recently, some fundamental properties of biological neuronal networks that demonstrate associative learning have been revealed. Acquisition and storage of associative memories have been shown to involve changes in the regulation of ionic flux through specific channels, molecular transformations that control ionic flux, and structural changes, all accompanying shifts in the effective weight of synaptic interactions among neurons that function as integrated systems or networks. These changes, underlying associative learning and memory, have been elucidated both in the marine snail *Hermissenda crassicornis* (Alkon, 1983, 1984, 1987) and more recently in the hippocampus of the rabbit (Alkon, 1984, 1989; Bank et al., 1986, 1988; LoTurco et al., 1988; Disterhoft et al., 1986; Alkon & Rasmussen, 1988; Olds et al., 1989).

Distributed, associative memory-specific changes within biological neuronal arrays have been localized to dendritic regions throughout neuronal arrays of such brain structures as the mammalian hippocampus or olfactory cortex (Olds et al., 1989, 1990, 1994). Electrophysiologic, imaging, and biochemical studies have suggested how functioning biologic networks transform signaling patterns to represent imaging patterns sensed from the environment. These functions involve non-Hebbian synaptic transformations, molecular regulation of ion channels, and long-term changes of neuronal architectures.

To suggest how computer-based network designs can be derived from these newly revealed biological learning networks, we will begin with a very brief overview of major information processing steps during pattern recognition by mammalian brains.

The brain pathway that mediates recognition of visual images, for example, begins with sensory transduction. Patterns of light intensity and wavelength are distributed across spatial fields at various distances from the viewer's eye. Light patterns collected by the retina elicit patterns of electrical signals. Immediately following transduction, the transduced patterns begin to be processed both to preserve spatial referencing and to extract a host of features within the patterns. From retinal cells to, successively, the cells of the lateral geniculate, primary, and secondary visual cortical areas and stations extending to the inferior temporal lobe, features are extracted and then apparently reconfigured to construct images. Feature extraction and image reconstruction are then followed by transmission in parallel to parietal cortex and to the hippocampus. Information received by the hippocampus and transmitted to the frontal cortex is thought to concern recognition of "what" an object is while information received by the parietal cortex is thought to concern where that object is. Within the cortex, we can infer that information patterns are collected to form classes, and classes are grouped to construct progressively more abstract representations.

Behavioral observations suggest that brain networks acquire and store relationships among increasingly complex patterns. These begin with relationships between discrete stimuli such as a conditioned (neutral) stimulus and unconditioned (reflexive) stimulus in a Pavlovian conditioned paradigm. The spatial relationships of features are learned as familiar faces or the appearance of a written word that names the face. Relationships between patterns, such as between the written or vocalized name and a facial image, are also stored, presumably by networks in other brain structures.

Thus, it seems possible that biological networks, at many stages of pattern information processing, are learning relationships, and these networks may share common architectural and mechanistic features. Extensive analyses of associative learning networks within invertebrates and mammalian brains have motivated our design of artificial networks according to currently available biological principles.

2. NEUROBIOLOGY

2.1. Neuroanatomical Considerations

For any associative mnemonic process, there must exist brain loci where signals from two or more temporally associated stimuli interact. In the mammalian brain, the overwhelming preponderance of synapses occur on vast dendritic arborization of neurons. A smaller but still significant number occur on somata. While axo-axonal and dendro-dendritic synapses do occur they are extremely infrequent. Examples of presynaptic terminals ending on other presynaptic terminals have been identified or inferred in the spinal cord, the olfactory bulb, and the retina, but represent exceptions rather than the rule. The anatomy of the vertebrate central nervous system, therefore, dictates that signals from associated stimuli must first interact on and/or within postsynaptic sites, most often the dendritic branches.

Although the above considerations imply that associative mnemonic processing must commence postsynaptically, there is no neurobiological reason why such processing must subsequently be confined to postsynaptic sites. The changes might reasonably begin on postsynaptic dendritic branches and/or spines and later, in turn, affect presynaptic boutons. Data from brain networks that change with associative learning have implicated persistent alteration of postsynaptic K^+ channels (Alkon, 1984), activation of protein kinase C (Olds et al., 1989; Alkon, 1989), and phosphorylation of low molecular weight G-protein substrates of protein kinase C (Nelson et al., 1989, 1990). Experiments employing models of short-term synaptic change that might occur during memory, such as long-term potentiation (LTP) and long-term depression (LTD) have implicated both postsynaptic and presynaptic changes (Changeux et al., 1987).

2.2. Electrophysiological Correlates

The vast majority of the work on short-term neuronal modifications (LTP, LTD) have suggested interpretations consistent with either the classical or expanded Hebb synapse models. Whether LTP is produced by single inputs activated at high frequency but in parallel, or by associated inputs, the targets and sites for change have always been conceived of as single synapses. The above notion is in sharp contrast to more recent observations of an explicitly understood biological associative network, the *Hermissenda* visual–vestibular system (Figure 1), that have suggested entirely new interpretations, consistent with more than one site for synaptic change even for a single discrete association. Thus, for example, in vitro studies of isolated *Hermissenda* nervous systems allow simulation of in vivo conditions for associative learning. Light steps presented to living animals can also be presented to the visual–vestibular network that, together with the central nervous system, has been isolated from the rest of the animal. Vestibular stimuli can be presented by rotation of the nervous system, vibration of the vestibular organ

or statocyst, injection of positive current into statocyst sensory neurons called hair cells, or by pulsed ejection of the hair cells' neurotransmitter, GABA, onto the terminal receiving branches of the postsynaptic visual sensory neuron called the B cell.

Such in vitro studies have recently revealed that neuronal signals from the statocyst interact with the signals of the B cell by releasing GABA onto terminal branches of the B cell. When hair cell and type B cell signals are not temporally correlated, that is, when the light step and vestibular stimulus occur separately, GABA released by hair cells causes inhibition of the B cells. When the hair cell signals and B photoreceptor signals are correlated, that is, when the onset of a light step occurs 0.5–1.0 s before the onset of a coterminating vestibular stimulus, hair cells initially cause inhibition followed by excitation of the B cells. If the correlated presentation of visual and vestibular signals is repeated just three times, the release of GABA by hair cells causes pure excitation of the B cells. Correlated stimulus presentations, therefore, transform classical GABA-mediated synaptic inhibition into persistent GABA-mediated synaptic excitation. With three paired presentations of a 4.0-s light step with a 3.0-s GABA application (or current-induced firing of the presynaptic hair cell), this synaptic transformation typically

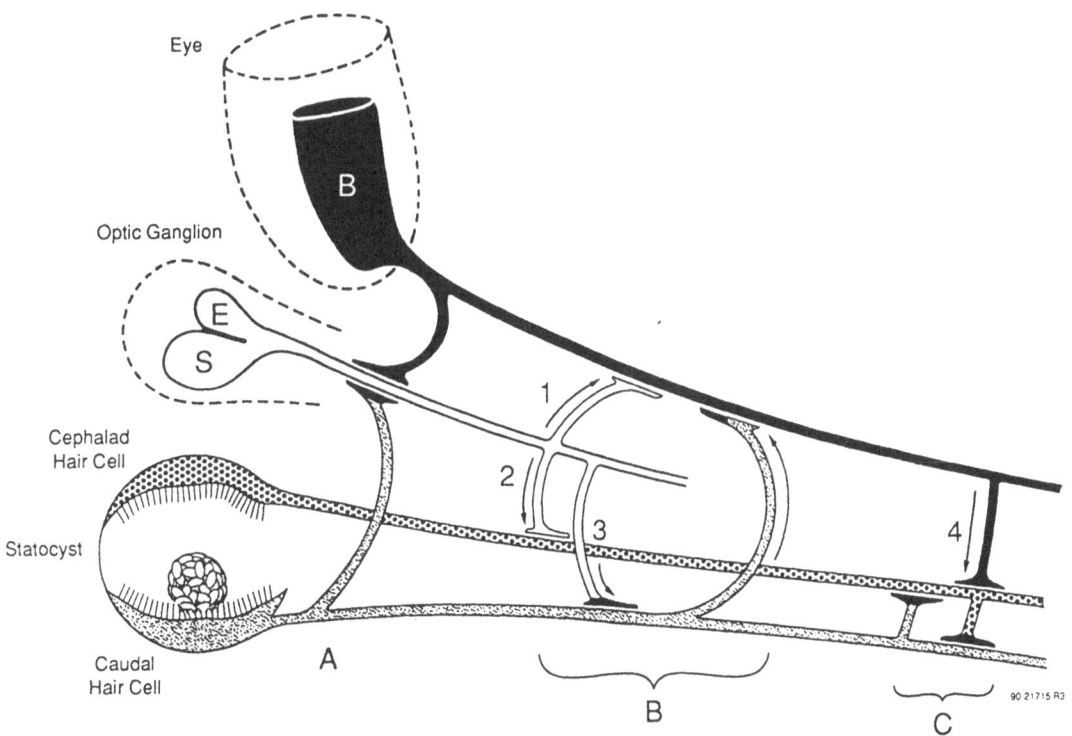

FIGURE 1. Intersensory integration by the *Hermissenda* nervous system. (A) Convergence of synaptic inhibition from B and caudal hair cells (part of the statocyst) on the E cell. (B) Positive synaptic feedback onto a B photoreceptor. 1, direct synaptic excitation, 2, indirect excitation: E-cell excites the cephalic hair cell that inhibits the caudal hair cell and thus disinhibits the B cell; 3, indirect excitation: E inhibits the caudal hair cell and thus disinhibits the B cell. (C) Intra- and intersensory inhibition. Cephalic and caudal hair cells are mutually inhibitory. The B cell inhibits mainly the cephalic hair cell. All filled endings indicate inhibitory synapses; open endings indicate excitatory synapses; half filled-half open endings indicate variable synapse. (Adapted from Tabata & Alkon, 1982.)

lasts more than 1 h. When the animal is trained in vivo, 90 such pairings occur during a 2-h training interval on three consecutive days; the transformation can be expected to have long persistence as has been confirmed by recent observations, discussed below.

Current and voltage-clamp experiments have helped reveal the mechanisms responsible for the GABA-mediated transformation. Briefly summarized, the hair cells release GABA that combines with postsynaptic GABA-A and GABA-B receptors on the B photoreceptor cell. Before pairing, activated GABA-A receptors open Cl^- channels and activated GABA-B receptors open K^+ channels, to cause hyperpolarization of the B cell. When the B cell is depolarized sufficiently to cause significant elevation of intracellular Ca^{2+} (i.e., at membrane potentials ≥ 20 mV above the resting level of -60 mV), GABA-B receptor activation also causes prolonged reduction of steady-state K^+ conductance. This reduction results when GABA-B receptor activation causes elevation of intracellular Ca^{2+} by releasing it from intracellular stores. Subsequent to several pairings of GABA-activated A and B receptors with depolarization induced elevation of intracellular Ca^{2+}, GABA no longer activates GABA-A receptors to open Cl^- channels or GABA-B receptors to open K^+ channels. GABA does, however, continue to release Ca^{2+} from intracellular stores. In fact, fura measurements of intracellular Ca^{2+} reveal that, after pairing, Ca^{2+} elevation induced by GABA remains prolonged. This prolonged GABA-mediated release of Ca^{2+} also appears to persist on days following paired training of the living animals but not controls (Ito et al., 1993). Transformation of synaptic inhibition to excitation, then, is due to a shift from a net increase of conductance to a net decrease. Before pairing, GABA increases conductance to Cl^- and K^+ to cause inhibition. After pairing, GABA decreases conductance to K^+ (via elevation of intracellularly released Ca^{2+}) to cause excitation. It is particularly interesting that to generate the transformed synaptic response, GABA works via a mechanism previously implicated in the production of memory-specific reduction of K^+ currents for the days and weeks during which the animal retains its learned conditioned response (Alkon, 1984, 1989).

Taken together, the nature of the Ca^{2+}-mediated K^+ currents during memory retention and the Ca^{2+}-mediated transformation of GABA synaptic response, suggest an entirely new role for the site of unconditioned stimulus (UCS) synaptic input. Heretofore, only one site, that of the conditioned stimulus (CS) synaptic input, was considered to change. Now, the UCS site also appears to change and this change could, in turn, facilitate the change of the CS site. If during and for some days after training, GABA causes depolarization and more prolonged elevation of intracellular Ca^{2+}, it could, indirectly, cause more reduction of K^+ currents at the CS site that is in physical proximity. A more prolonged Ca^{2+} elevation might spread by local diffusion or via a wave of Ca^{2+}-mediated release of Ca^{2+} from intracellular stores. More K^+ current reduction at the CS site could, in turn, cause more voltage-dependent influx of Ca^{2+}, elevation of intracellular Ca^{2+}, and spread of Ca^{2+} to the UCS site. GABA-induced synaptic transformation, together with learning-induced reduction of voltage-dependent K^+ currents, might functionally link the CS and UCS sites together via prolonged and spreading elevation of intracellular Ca^{2+} and, presumably, the Ca^{2+} triggered phosphorylation pathways previously identified.

Such CS-UCS linkage, of course, need not be limited to two synaptic sites, that is, of single CS and UCS inputs. As previously formulated (for pattern recognition), multiple CS sites together with the UCS site could interact in an analogous manner.

3. A BIOLOGICALLY BASED COMPUTER MODEL

To further explore the role of synaptic interactions in the B photoreceptor cell (B cell) and the transformation of the GABA synapse, a computer model of the *Hermissenda* visual–vestibular network was developed (Werness et al., 1992). The model included only the essential visual–vestibular network, using simplifications derived from previous empirical measurements of the biological networks neurons and their connections. The responses of the network model to the same stimuli used to train the snails were analyzed and quantitative comparisons were made of animal behavior with the output from computer simulations. The close correlation between computer network responses and *Hermissenda* behavioral and electrophysiological responses (to the same training stimuli) offers strong support to the notion that the essential network modeled is indeed responsible for the associative learning behavior previously observed in the animal (Werness et al., 1993).

The results of research on *Hermissenda*, previously described, together with results from Olds et al. on the hippocampus of associatively trained rabbits (1989) and on water maze trained rats (1990), suggest that learning takes place locally, on small areas of the dendritic cell membrane in the immediate vicinity of paired (CS and UCS) incoming stimuli, areas that we refer to as "patches." In marked contrast to earlier theories of neuronal learning, it is hypothesized that learning depends on the local interaction of a multiplicity of synapses and less, if at all, on the output of the postsynaptic neuron. A non-Hebbian learning rule has been incorporated based on these neurobiological insights into an artificial neural network (ANN) that we call Dystal (dynamically stable associative learning), which has a number of advantageous properties (Alkon et al., 1990, 1991, 1993; Blackwell et al., 1992; Vogl et al., 1991, 1992, 1994).

Dystal learns to associate a CS input pattern with a UCS input pattern and demonstrates the learned association by output of the UCS pattern in response to the CS pattern alone. Just as *Hermissenda* and the rabbit have two separate inputs (vestibular and visual for *Hermissenda,* auditory and somatosensory for rabbit), Dystal has two separate input pathways, the CS and the UCS. In addition to the two input pathways, a single layer Dystal network (equivalent to a three layer feedforward network in the context of Kolmogorov's theorem, Hecht–Nielsen, 1987) consists of output elements and their patches. Each output unit (illustrated in Figure 2) receives input from a receptive field (a subset of CS inputs) and one (scaler) component of the UCS input vector via its patches. All of the patterns learned by Dystal are, collectively, stored in the entire set of patches; however, each patch individually stores only an association between a group of similar CS inputs and their associated UCS component. As illustrated, a patch is composed of: a patch vector, the running average of CS input patterns that share similar UCS values; the running average of similar UCS values; and a weight that reflects the frequency of utilization of the patch. The number of output units is equal in number to the number of components in the UCS vector. The UCS vector can be a classification vector, or of the same size as the CS input to permit pattern completion (reconstruction). Note that both the CS and the UCS inputs can be noisy; for pattern completion the UCS can, in fact, be drawn from the same population as the CS inputs.

Prior to training, no patches exist. The number and content of the patches are determined during training and are a function of the content of the training sets

and of global network parameters. The training procedure is described in Figure 3: the first paired CS-UCS presentation causes each output unit to create one patch. At each subsequent presentation, each output unit compares its CS input with its patch vectors and selects the patch whose vector is most similar. In the present implementation of Dystal, Pearson's R correlation is our measure of similarity (Irvine et al., 1994). If the stored UCS is different from the input UCS, or the CS vector is not sufficiently similar to the input CS (less than a global threshold), a new patch is created and it becomes the active patch. Otherwise, the selected patch is updated and it becomes the active patch. The output equals the similarity times the stored UCS for the most similar patch. It should be noted that all decisions required in the training process are made on the basis of information available locally at the level of the patches of each output element irrespective of whether the output element fires or not. Consequently, neither local nor global feedback is required and iterations in the learning loop (associated with nonlinear searches) are avoided. Thus, learning is monotonic and rapid. We have not encountered a problem that, despite very noisy inputs, requires more than 10 presentations of the training set to reach fully trained equilibrium; for most problems two to four presentations suffice, and the number of iterations is independent of the number of patterns in the training set, or the size of the patterns (Barbour et al., 1992). Performance is insensitive to the dominant global parameter, the patch creation threshold, and default values provide good performance on a wide variety of problems. The only difference between training and testing is that during testing the CS inputs are *not* accompanied by a UCS input. Therefore, training and testing can be intermingled as required.

Dystal can be considered a self-organizing clustering algorithm in which the number of clusters is determined dynamically as the associations among the input data are established. One of the roles of the UCS is to allow Dystal to group clusters (patch vectors) containing different information within the same class. For example, the full face image of John and the image of John's profile can both be associated with the lexical label "John." Another role of the UCS is to control the radius of the clusters formed. In sparsely populated regions of pattern space, large clusters are formed. In densely populated regions and along boundaries, small clusters are made, to closely approximate the boundary between pattern classes. These characteristics provide a facility to Dystal not available to clustering algorithms that either require the prespecification of the number of clusters to be formed, or that force each cluster to be the same size.

A complementary view of Dystal can be obtained by considering that networks whose measure is determined by a dot product use hyperplanes to divide the space defined on the feature set. The minimum number

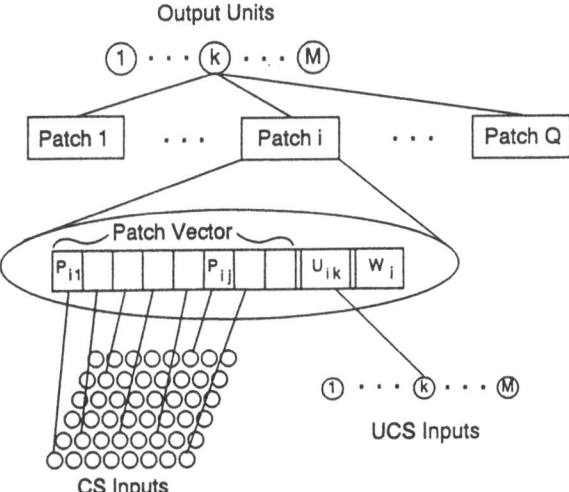

FIGURE 2. The patch structure of the Dystal output neuron. There is one output neuron for each component of the UCS vector. The patch structure for one of the patches of an output neuron is shown. In the present implementation, the CS inputs, which may be a subset (receptive field) of the entire input pattern, remain the same for each patch of an output neuron.

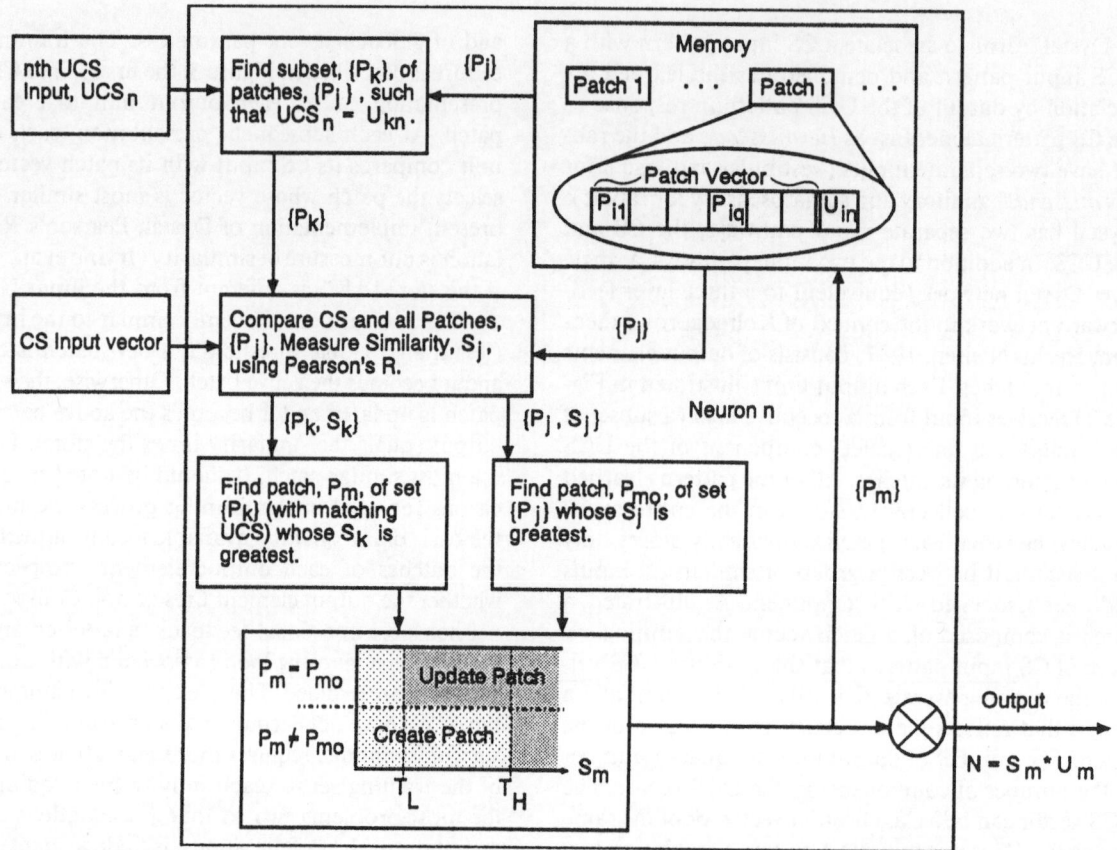

FIGURE 3. An overview of the Dystal algorithm, which is carried out independently for each output unit. The CS input pattern is compared to the patch vector of each patch on the neuron. (P_j is the patch vector of the jth patch; S_j is the similarity between the CS inputs and P_j.) The patch, m, whose patch vector is most similar to the input CS is selected. If the measure of similarity (Pearson's R in the present implementation) is sufficiently high (i.e., $>T_c$) and the stored UCS value, U_m, is close to the input UCS value, the patch is considered as matched and both the patch vector and the UCS value are updated by the use of a running average. If the similarity is too low, a new patch is created using the input CS and UCS values. The output of the neuron is the product of the stored UCS, U_m, and the similarity measure, S_m ($0 > S_m > 1$). A detailed description of the algorithm may be found in Blackwell (1992), Vogl (1992), and Alkon (1993).

of such planes that is needed to enclose the volume of feature space that defines a class attribute increases rapidly as the dimensionality of feature space increases. Dystal, by using a correlation measure (a normalized dot product), of which Pearson's R is but one example, defines domains in feature space that are areas on the surface of a unit hypersphere whose centers are located by the patch vectors. Thus, irrespective of the number of dimensions, a class attribute is represented by a single vector and an associated radius given by the allowed range of correlation (see Irvine et al., 1994) and the distance between two patches by an angle between two vectors. This measure avoids the loss in specificity with increasing dimensionality of feature space that is inherent in networks based on dot product measures.

3.1. Experimental Analyses of Dystal's Performance

Dystal has been utilized both in pattern classification (character recognition) and in pattern completion (image restoration) tasks. In both of these applications,

the ANN is remarkably insensitive to changes in its global parameters. Satisfactory performance on all three of the tasks described below can be obtained using identical global parameters; fine-tuning is accomplished by univariate adjustment of a single global parameter, the patch creation threshold.

3.1.1. Pattern Completion. Consider the problems of face recognition and reconstruction. Figure 4 shows 36 images, nine different facial expressions of each of four different persons. Each image consists of 2496 pixels (64 high by 39 wide), 256 shades of gray. The images were chipped from digitized photographs, but no preprocessing was utilized. The four left-most pictures (in Figure 4) of each of the four subjects are used as the training set and provide both CS and UCS input patterns. Therefore, for the task of image reconstruction, CS input array, UCS input array, and output array each contain 2496 elements. Each of the output elements receives input from one UCS component and a 17 × 17 pixel subfield of the CS. During training, one of the

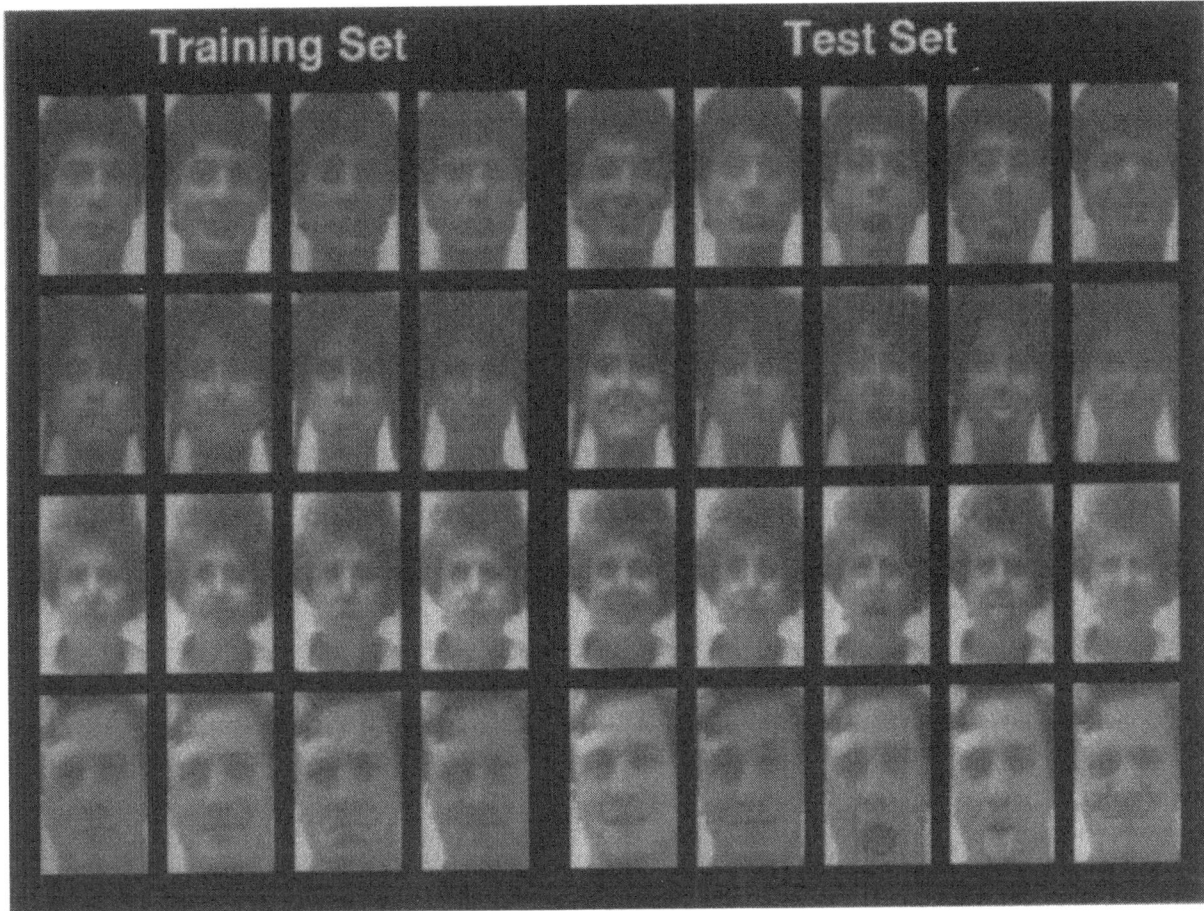

FIGURE 4. The nine images of each of four individuals used in the experiments on face recognition. The four images on the left constitute the training set; the five images on the right constitute the testing set. Note the difference in facial expressions across each row. Each image is 64 pixels high by 39 pixels wide. To achieve image completion (rather than classification) the number and arrangement of output neurons is the same as the CS input. The receptive field for each output neuron is 17 × 17 pixels.

four images of a subject is randomly chosen to be the CS and a second random choice from the set of four is used as the UCS. This is repeated for each of the four images of the four subjects. Four such sets of pairings of each subject constitute the entire training procedure that require approximately 50 min on a Silicon Graphics Personal Iris workstation. Two experiments were performed to test how well Dystal had learned each individual.

In the first experiment, the five expressions of each subject not included in the training set were presented to the trained network. The output from the network for the five images in the testing set is shown in Figure 5. Note that each output image for each individual is a composite image of the four different expressions of that individual. Note that the facial expression of the output is independent of facial expression of the input.

In the second experiment (illustrated in Figure 6), Gaussian noise was added to each of the images by adding to each pixel a random value drawn from Gaussian distributions whose variances are 0.02, 0.08, and 0.25 for the second, third, and fourth images from

the left, respectively. Figure 6 show that Dystal can reconstruct recognizable images even when the input images are so degraded by noise as to be barely recognizable as faces by human observers. We emphasize that the noise degraded input images were derived from images of facial expressions not included in the training set.

3.1.2. Classification. Pattern classification can be considered a special case of pattern completion, in which the UCS pattern is much smaller than the CS pattern. Two series of classification experiments were performed, the first using a hand-written ZIP Code compiled by the US Postal Service and the second using hand-printed Japanese Kanji characters.

Handwritten ZIP Codes. A set of segmented digits from the USPS ZIP Code data set was provided to us, in rectified format, by the staff of the ERIM Post Office project in Ann Arbor, MI. The rectified set of about 11,000 digits, scaled to fit into a 22 × 16 pixel array and rotated to a roughly vertical orientation, was divided into a 2000 digit training set (200 of each digit)

253

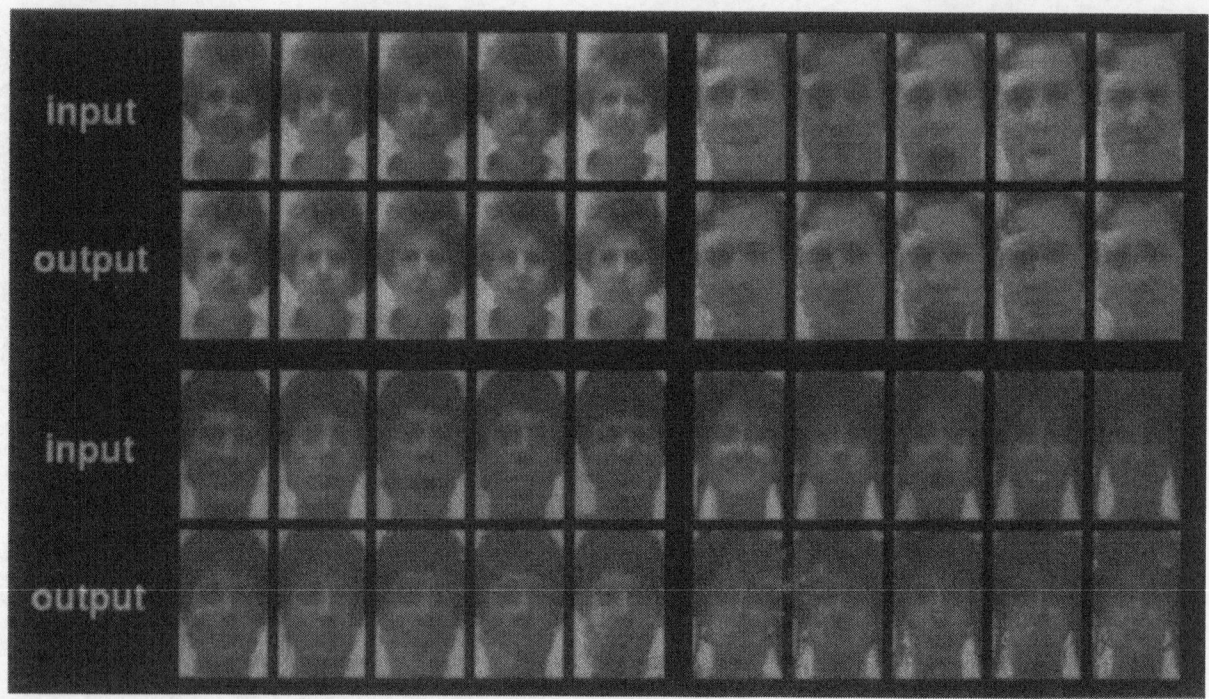

FIGURE 5. The response of the network to the 20 testing images whose expressions are different tha n those of the training set. Note that the expressions of the output do not reflect the expressions of the test images; rather, they represent a composite of the images in the training set for each individual. Neutral gray areas in the output are areas where the similarity of the input CS to the patch vector is less than the 0 (i.e., there is no correlation between the two). In that case, the output of that neuronal output element is neutral gray.

and a 8964 digit testing set. The CS input to Dystal consisted either of a randomly chosen subset of CS input pixels or of the coefficients of the first 20 principal components derived from the entirety of the training set (Hyman et al., 1991; Rubner & Schulten 1990). The UCS input is a 10 element vector all of whose elements but one are 0. The ordinal of the single element of the UCS vector whose value is unity corresponds to the ordinal number of the digit. The network was trained by presenting each member of the training set and its corresponding UCS *a single time.* During testing the input is considered classified if, and only if, one and only one of the 10 output units has an output activity over 0.5. If the ordinal number of the output unit with the activity over 0.5 matches the ordinal number of the input, the network is considered to have correctly classified the input. Testing the network with the 9000 member test set shows that the network classifies in excess of 90% of the inputs and correctly classifies 98% of those inputs that it classified. Figure 7 shows that only four presentations of the training set is required to reach an asymptotic level of performance. Using inputs preprocessed by principal component analysis resulted in a decrease in the number of unclassified inputs.

Hand-Printed Japanese Kanji. Kanji characters present a significantly more formidable challenge than digits. Not only is each character much more compli-

cated (see Figure 8) but for a practical system 3000 or more different Kanji must be classified. We review preliminary results from a Dystal based Kanji recognition system. The database consists of 956 different Kanji characters, 63×64 binary pixels in size, with 160 exemplars of each character, each exemplar being written by a different individual. The database, ETL8B2 was provided to us by the courtesy of Drs. T. Mori and K. Yamamoto of the Electrotechnical Laboratory (ETL), Tsukuba, Japan. The training set consisted of the first 20, 40, 80, or 120 exemplars of each of 20 different characters and the testing set consisted of the remaining exemplars of those characters. Figure 7 presents the classification performance of Dystal on Kanji preprocessed using the PCA preprocessor. Clearly, performance is limited by the size of the training set. When the training set contains 120 exemplars, performance on this far more complicated set of characters is comparable to that on ZIP code digits.

Because Dystal is modeled after biological associative learning, it has a local synaptic learning rule: learning does not depend on the output of the postsynaptic neuron, or on a global error term. Training does not involve nonlinear optimization. Consequently, Dystal is inherently parallel, and training time is independent of the separability of the patterns, and the number of output units. Also, training time increases linearly with the number of training patterns. To demonstrate these

254

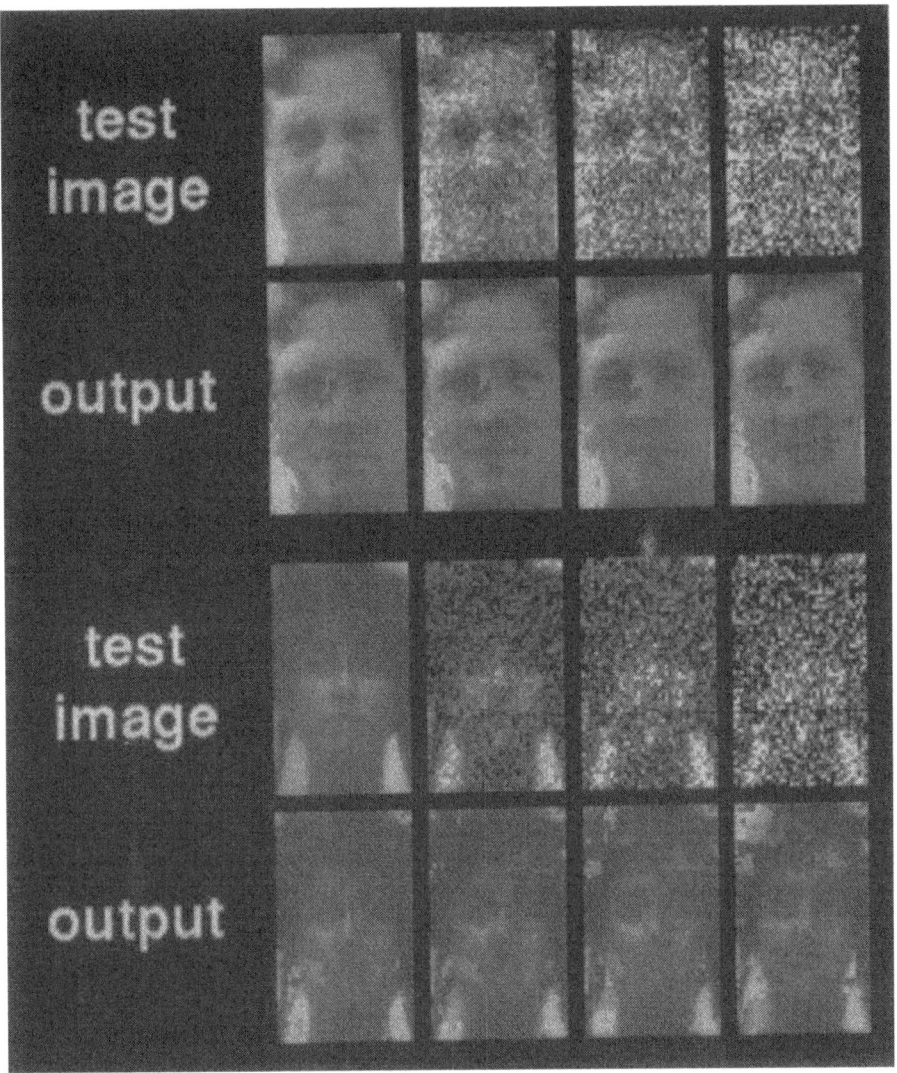

FIGURE 6. The response of the network to two of the test images with various amounts of added Gaussian noise. In each pair of rows, the top row is the same image with the amount of additive noise increasing from left to right. Note that even in the case with the most noise, the right-most images, the reconstruction is clearly recognizable despite increasing areas of neural gray (unrecognized) or other misrecognized areas in the output.

properties we compiled Dystal on a MasPar MP-1 computer, and tested Dystal using the Kanji data set. To evaluate training time as a function of separability, we increased the numbers of patches by increasing the patch creation threshold (T_c in Figure 3). An increase in the number of patches is Dystal's normal response to learning a more difficult (less separable) data set. Figure 9 shows that training time is independent of the number of patches and independent of the number of output units. Figure 10 demonstrates that, as predicted, training time increases linearly with the number of training patterns.

4. CONCLUSIONS

We have developed a description of associative memory formation and retrieval in biological systems derived from neurophysiological experiments on both invertebrate and vertebrate animals. The data can be explained by a non-Hebbian learning rule in which the inputs to a postsynaptic patch of membrane are modulated by a sign-transformable UCS synapse to store associations among inputs that occur in the correct, temporally proximate sequence.

As demonstrated by computer modeling, the specifics of the architecture of the neural network are as important as the intraneuronal mechanisms in facilitating learning and memory storage. Based on these insights, we have developed an ANN, called Dystal, that utilizes these non-Hebbian learning rules and network architecture. The performance, learning acquisition rate, and storage capacity of Dystal demonstrate the important contributions that neurobiological input can make to the design of artificial neural networks

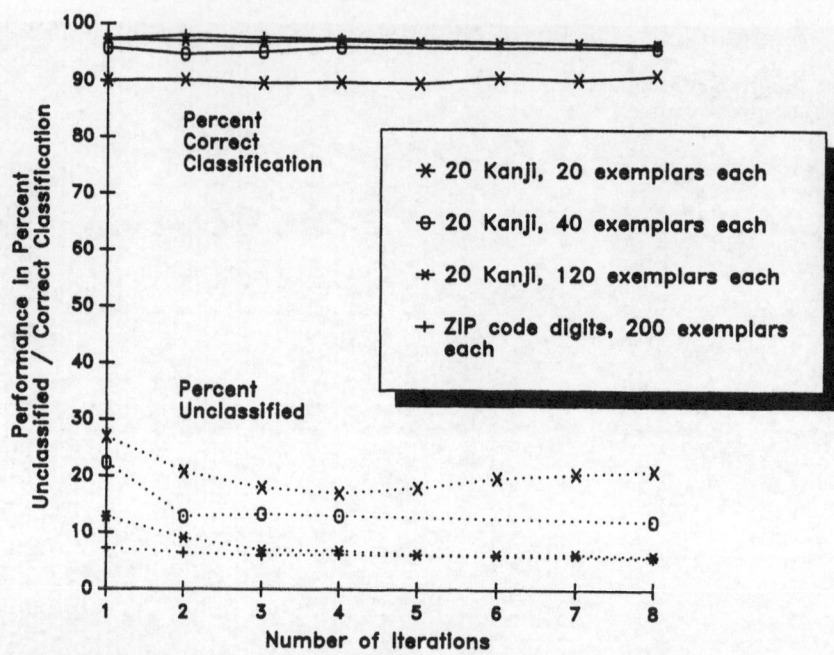

FIGURE 7. Performance of Dystal on 20 different characters from the Kanji set, and on ZIP code digits as a function of iterations through the training set and number of exemplars in the training set. The top curves show percent unclassified; the bottom curves show percent errors. Dystal is completely trained in four iterations, independent of the data set. Dystal's performance on Kanji recognition is comparable to that of digit recognition if trained on enough exemplars, even though twice as many Kanji characters as digits were used.

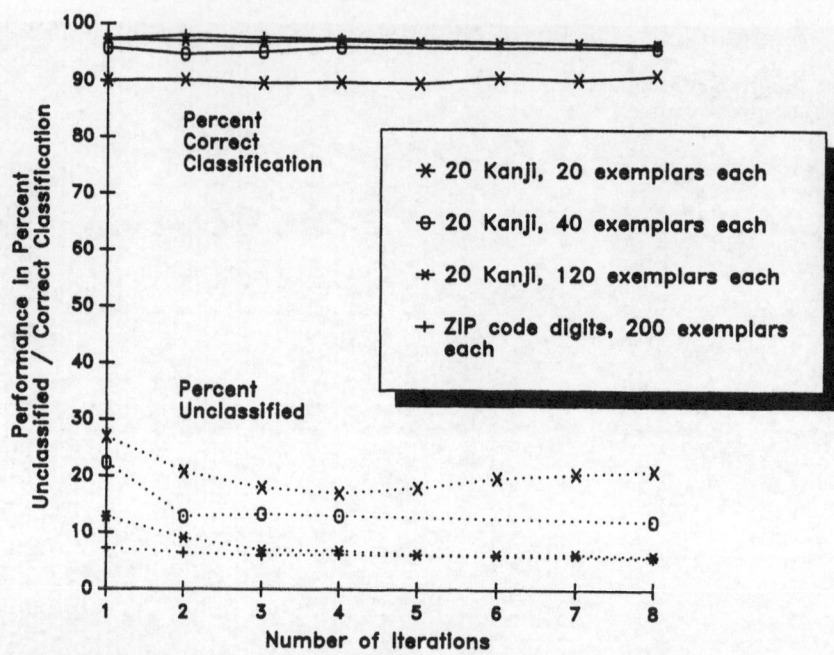

FIGURE 8. Examples of Japanese Kanji characters from the database of hand-printed Kanji. Note the variation in handwriting across the rows, which are different individual's hand printing of the same character. These examples are not chosen as extreme cases. Each character is 63 × 64 binary pixels.

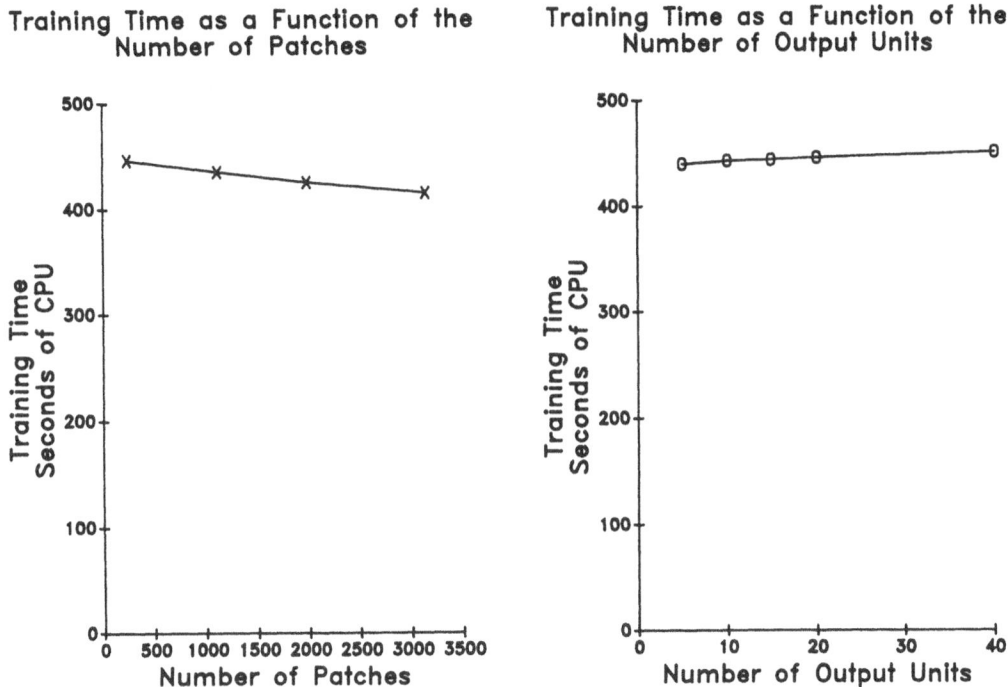

FIGURE 9. Training time as a function of patches (left graph) and output units (right graph). Training time (in CPU seconds) is listed for four iterations through the data set on a MasPar MP-1 massively parallel computer. Notice that training time is essentially constant.

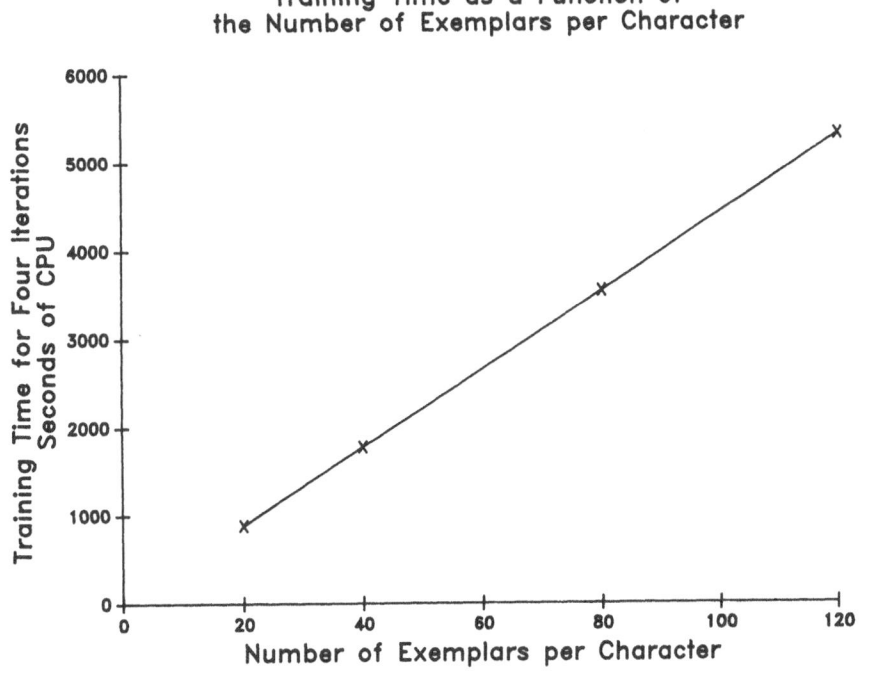

FIGURE 10. Training time as a function of training patterns on a MasPar MP-1 computer. Training time is linear with the number of training patterns.

and demonstrate that, even at our present state of knowledge, reverse engineering of principles underlying mechanisms and features of brain function is not only possible, but extremely productive. As we learn to concatenate such networks into more elaborate structures, these "third generation" neural networks [such as Dystal, Faust (Wilson, 1994), and Artmap (Carpenter et al., 1991) will demonstrate capabilities far beyond those of the currently popular second generation ANN.

REFERENCES

Alkon, D. L. (1983). Learning in a marine snail. *Scientific American*, **249**, 70–84.

Alkon, D. L. (1984). Calcium-mediated reduction of ionic currents: A biophysical memory trace. *Science*, **226**, 1037–1045.

Alkon, D. L. (1987). *Memory traces in the brain*. Cambridge, UK: Cambridge University Press.

Alkon, D. L. (1989). Memory storage and neural systems. *Scientific American*, **July**, 42–50.

Alkon, D. L., Blackwell, K. T., Barbour, G. S., Rigler, A. K., & Vogl, T. P. (1990). Pattern-recognition by an artificial network derived from biological neural systems. *Biological Cybernetics*, **62**, 363–376.

Alkon, D. L., Blackwell, K. T., Vogl, T. P., & Werness, S. A. (1993). Biological plausibility of artificial neural networks: Learning by non-Hebbian synapses. In Mohamad H. Hassoun (Ed.), *Associative neural memories*. Oxford: Oxford University Press.

Alkon, D. L., Vogl, T. P., & Blackwell, K. T. (1991). Artificial learning networks derived from biological neural systems. In Milutinovic, V. (Ed.), *Prentice Hall series on neural networks, Part IV* (pp. 24–46). Englewood Cliffs, NJ: Prentice Hall Publ.

Alkon, D. L., & Rasmussen, H. (1988). A spatial-temporal model of cell activation. *Science*, **239**, 998–1005.

Anderson, J. A. (1983). Cognitive and psychological computation with neural models. *IEEE Transactions on Systems, Man, and Cybernetics*, **SCM-13**, 799–815.

Anderson, J. A. (1986). Cognitive capabilities of a parallel system. In E. Bienenstock (q.v.) et al. (Eds.), *Disordered systems and biological organization*. Berlin: Springer–Verlag.

Bank, B., DeWeer, A., Kuzirian, A. M., Rasmussen, H., & Alkon, D. L. (1988). Classical conditioning induces long-term translocation of protein kinase C in rabbit hippocampal CA1 cells. *Proceedings of the National Academy of Sciences USA*, **85**, 1988–1992.

Bank, B., Gurd, J. W., & Chute, D. L. (1986). Decreased phosphorylation of synaptic glycoproteins following hippocampal kindling. *Brain Research*, **399**, 390–394.

Barbour, G., Blackwell, K., Busse, T., Alkon, D., & Vogl, T. (1992). Dystal: A self-organizing ANN with pattern independent training time. *Proceedings of the IJCNN '92 Baltimore*, **IV**, 814–819.

Blackwell, K. T., Vogl, T. P., Hyman, S. D., Barbour, G. S., & Alkon, D. L. (1992). A new approach to the classification of hand-written characters. *Pattern Recognition*, **25**, 655–666.

Carpenter, G. A., Grossberg, S., & Reynolds, J. R. (1991). Artmap: Supervised real-time learning and classification of nonstationary data by a self-organizing neural network. *Neural Networks*, **4**, 565–588.

Changeux, J. P., Konishi, M., & Baudry, M. (Eds.) (1987). *The neural and molecular bases of learning: Report of the Dahlem workshop on the neural and molecular bases of learning, Berlin 1985*. New York: Wiley.

Disterhoft, J. F., Coulter, D. A., & Alkon, D. L. (1986). Conditioning-specific membrane changes of rabbit hippocampal neurons mea-

sured in vitro. *Proceedings of the National Academy of Sciences USA*, **83**, 2733–2737.

Fukushima, K. (1988). Neocognitron: A hierarchical neural network capable of visual pattern recognition. *Neural Networks*, **1**, 119–130.

Grossberg, S. (1982). *Studies of mind and brain*. Holland. Dordrecht, D. Reidel Publishing Co.

Grossberg, S. (1987). Competitive learning: From interactive activation to adaptive resonance. *Cognitive Science*, **11**, 23–63.

Hebb, D. (1949). *Organization of behavior*. New York: Wiley.

Hecht-Nielsen, R. (1987). Kolmogorov's mapping neural network existence theorem. *Proceedings of the IJCNN'87*, **III**, 11–13.

Hopfield, J. J. (1982). Neural networks and physical systems with emergent collective computational abilities. *Proceedings of the National Academy of Sciences USA*, **79**, 2254–2258.

Hopfield, J. J., & Tank, D. W. (1986). Computing with neural circuits: A model. *Science*, **233**, 625–633.

Hyman, S. D., Vogl, T. P., Blackwell, K. T., Barbour, G. S., Irvine, J., & Alkon, D. L. (1991). Classification of Japanese Kanji using principal component analysis as a preprocessor to an artificial neural network. *Proceedings of the IJCNN'91*, **I**, 233–238.

Irvine, J. M., Blackwell, K. T., Alkon, D. L., & Vogl, T. P. (1994). Angular separation in neural networks. *Journal of Artificial Neural Networks*, **1**, 169–182.

Ito, E., Oka, K., Collin, C., Schreurs, B. G., Sakakibara, M., & Alkon, D. (1993). Intracellular calcium signals are enhanced for days after Pavlovian conditioning. *Journal of Neurochemistry*, **90**, 8209–8213.

Kienker, P. K., Sejnowski, T. J., Hinton, G. E., & Schumacher, L. E. (1986). Separating figure from ground with parallel network. *Perception*, **15**, 197–216.

Klopf, A. H. (1988). A neuronal model of classical conditioning. *Psychobiology*, **16**, 85–125.

Kohonen, T. (1988). *Self-organization and associative memory*. Berlin: Springer–Verlag.

Lippmann, R. P. (1987). An introduction to computing with neural nets. *IEEE Acoustics, Speech, and Signal Processing Magazine*, **4**, 4–22.

LoTurco, J. L., Coulter, D. A., & Alkon, D. L. (1988). Enhancement of synaptic potentials in rabbit CA1 pyramidal neurons following classical conditioning. *Proceedings of the National Academy of Sciences USA*, **85**, 1672–1676.

McClelland, J. L. (1985). Putting knowledge in its place: A scheme for programming parallel processing structures on the fly. *Cognitive Science*, **9**, 113–146.

McClelland, J. L., & Rumelhart, D. E. (1981). An interactive activation model of context effects in letter perception: Part 1. An account of basic findings. *Psychological Reviews*, **88**, 375–407.

Nelson, T., & Alkon, D. L. (1989). Specific protein changes during memory acquisition and storage. *BioEssays*, **10**, 75–80.

Nelson, T., Collin, C., & Alkon, D. L. (1990). Isolation of a G protein that is modified by learning and reduces potassium currents in *Hermissenda*. *Science*, **247**, 1479–1483.

Olds, J. L., Anderson, M. L., McPhie, D. L., Staten, L. D., & Alkon, D. L. (1989). Imaging memory-specific changes in the distribution of protein kinase C within the hippocampus. *Science*, **245**, 866–869.

Olds, J. L., Golski, S., McPhie, D. L., Olton, D., Mishkin, M., & Alkon, D. L. (1990). Learning—Specific changes in rat hippocampal protein kinase C distribution for two water maze discrimination tasks. *Journal of Neuroscience*, **10**, 3707–3713.

Olds, J. L., Bhalla, U. S., McPhie, D. L., Bower, J., & Alkon, D. L. (in press). Lateralization of membrane-associated PKC distribution in rat olfactory cortex: Specific to learning in the olfactory modality. *Behavioral Brain Research*.

Rubner, J., & Schulten, K. (1990). Development of feature detectors by self-organization. A network model. *Biological Cybernetics*, **62**, 193–199.

Rumelhart, D. E., McClelland, J. L. (Eds.) (1986). *Parallel distributed processing: Explorations in the microstructure of cognition.* Cambridge, MA: MIT Press.

Rumelhart, D. E., & Zipser, D. (1985). Feature discovery by competitive learning. *Cognitive Science, 9,* 75–112.

Sejnowski, T. J., & Kienker, P. K. (1986). Learning symmetry groups with hidden units: Beyond the perceptron. *Physica D (Netherlands),* **22,** 260–275.

Sejnowski, T. J., & Rosenberg, C. R. (1986). NETtalk: A parallel network that learns to read aloud. *The Johns-Hopkins University Electrical Engineering and Computer Technical Report,* JHU/EECS-86/01.

Vogl, T. P., Blackwell, K. T., Hyman, S. D., Barbour, G. S., & Alkon, D. L. (1991). Classification of hand-written digits and Japanese Kanji. *Proceedings of IJCNN '91,* **I,** 97–102.

Vogl, T. P., Blackwell, K. T., Irvine, J. M., Barbour, G. S., Hyman, S. D., & Alkon, D. L. (in press). Dystal: A neural network architecture based on biological associative learning.

Vogl, T. P., Blackwell, K. T., Barbour, G. S., & Alkon, D. L. (1992). Dynamically stable associative learning (DYSTAL): A neurobiologically based ANN and its applications. *Proceedings of SPIE's O/E Aerospace Sensing,* Technical Conference 1710, Orlando, FL, April 22–24.

Werness, S. A., Fay, S. D., Vogl, T. P., Blackwell, K. T., & Alkon, D. L. (1992). Associative learning in a model of *Hermissenda crassicornis* I. Theory. *Biological Cybernetics,* **68,** 125–133.

Werness, S. A., Fay, S. D., Vogl, T. P., Blackwell, K. T., & Alkon, D. L. (1993). Associative learning in a model of *Hermissenda crassicornis* II. Experiments. *Biological Cybernetics,* **69,** 19–28.

Wilson, C. L., Wilkinson, R. A., & Garris, M. D. (in press). Self-organizing neural network character recognition using adaptive filtering and feature extraction. In C. L. Wilson & O. M. Omidvar (Eds.), *Progress in neural networks* (Vol. III). Norwood, NJ: Ablex Publishing Co.

11.
Learning in the Brain: An Engineering Interpretation

Paul J. Werbos

Learning in the Brain: An Engineering Interpretation

Paul J. Werbos
Room 675, National Science Foundation[1]
Arlington, Virginia, USA 22230
pwerbos@nsf.gov

INTRODUCTION

Earlier in this conference, Peter Killeen argued very forcefully that we now face an historic opportunity in the field of cognitive neuroscience: by trying to understand the mathematical laws which underlie learning, in a generic way, spanning large parts of the brain, we can begin to create a precise and universal science here, with a serious similarity to physics. We face an opportunity to reinvent neuroscience, in much the same way that Newton reinvented physics.

In this paper, I will not repeat Peter's arguments. (I have made similar points myself in previous papers[1-3].) However, I would like to add one additional point: that a key to Newton's success in reinventing physics was that he had new mathematics to back up his intellectual strategy. Above all, he had a new concept -- the derivative -- which was essential to his theory of gravity, and so on.

In the development of artificial neural networks (ANNs) for engineering, we have also developed fundamentally new mathematics -- including a new type of derivative, the ordered derivative [1] -- which will be essential, in my view, to this new intellectual revolution that Peter has talked about. In this paper, I will try to give a broad overview of that mathematics, and of how I see it applying to the brain.

In a conference or book of this sort, bringing together diverse disciplines like neuroscience, psychology, physics and engineering, we often have serious problems really understanding each other, because of diverse technical jargon and notation. Therefore, for the sake of readability, I will provide the reader with an cleaned-up version of the actual conference talk, modified only to account for some recent technical work. The result may be a little less formal and a lot less precise than one might expect in an academic paper, but I hesitate to make any changes which might make it harder to read. For more precise and complete details, one may refer to [1-4].

Because the brain is far more sophisticated than anything ever built by engineers, the mathematics required for real understanding -- even just first-order understanding -- are also very sophisticated. Therefore this paper will describe the key concepts and variables, without providing the equations. Again, see the citations for equations and details, enough to make it possible to actually replicate this kind of system.

WHAT IS NEUROENGINEERING AND HOW DOES IT MATTER TO NEUROSCIENCE?

Before I describe the new mathematics, I first need to tell you where it comes from. It comes from a new field of research which the National Science Foundation (NSF) calls Neuroengineering[5].

The Neuroengineering program at NSF is part of the Engineering Directorate. When NSF first set up this program in 1987, it was not motivated by any kind of desire to

---

1.The views herein are the personal views of the author, not the views of NSF; however, as government work, the paper is in the public domain subject to proper citation.

learn more about the brain for its own sake. The program was set up as an exercise in engineering, as an effort to develop more powerful information processing technology. The goal was to understand what is really required to achieve brain-like capabilities in solving real and difficult engineering problems, <u>without</u> imposing any constraints on the mathematics and designs except for some very general constraints related to computational feasibility. In a sense, you could characterize this as abstract, general mathematical theory; however, these designs have been subjected to very tough real-world empirical tests, in proving that they can effectively control high-speed aircraft, chemical plants, cars and so on -- empirical tests which a lot of "models of learning" have never been confronted with.

More precisely, the Neuroengineering program began as an offshoot of the Lightwave Technology (LWT) program at NSF. LWT was and is one of the foremost programs in the US supporting the most advanced research in optical technology. It furthers the development and use of advanced optical fibers, lasers, holography, optical interface technology, and so on, across a wide range of engineering applications -- communication, sensing, computing, recording, etc. Years ago, several of the most advanced engineers in this field came to NSF and argued that this kind of technology could be used to generate computing systems far more powerful than conventional electronic computers. The basic idea is illustrated in Figure 1.

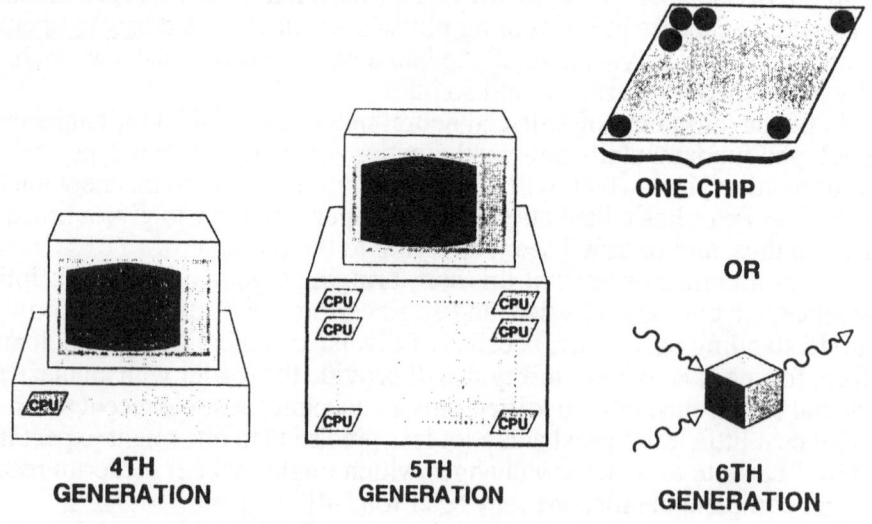

Figure 1: Three Generations of Computer Hardware

The computer on your desktop is a remarkable advance over the computers of twenty years ago. We would call it a "fourth generation" computer. The key to that computer is its Central Processing Unit (CPU), the microchip inside which does all the real substantive computing, one instruction at a time. A decade or two ago, advanced researchers pursued a new kind of computer -- the fifth generation computer, or "massively parallel processor" (MPP) or "supercomputer" -- illustrated in the middle of Figure 1. The MPP may contain hundreds or thousands of CPU chips, all working in parallel, in one single box. In theory, this permits far more computing horsepower per dollar; however, it requires a new style of computer programming, different from the one-step-at-a-time FORTRAN or C programming that most people know how to use. The U.S. government has spent many millions of dollars trying to help people learn how to use the new style of computer programming needed to exploit the power of these machines.

In the late 1980's, the optical engineers came to NSF, and claimed that they could develop a sixth generation of computing, as far beyond the MPP as the MPP is beyond the ordinary PC. Using lasers and holograms and such, they claimed they could produce a thousand to a million times more computing horsepower per dollar than the best MPP. Naturally, this was of great interest to NSF. When basic research could possibly open up a thousand-fold improvement in productivity in an industry worth hundreds of billions of dollars or more, one tends to pay attention.

Next, however, NSF went to the skeptics and asked them their opinion of this idea. The skeptics agreed that optical computing might be able to increase computing horsepower as claimed -- but only at a price. Using holograms, you can get a huge throughput of computing operations. But you have to have very simple operations at each pixel of this hologram. You have to have very simple operations performed over and over again in a stereotyped kind of way. You just can't just stick in a FORTRAN program. And then they said that most people in the world want to run FORTRAN programs, so that this kind of computing could never be more than a niche market. Yes it will give high throughput, but it will never be very interesting.

Finally, NSF went to a guy named Carver Mead, at CalTech, who is the inventor of VLSI. VLSI is the method we use to actually design microchips -- all kinds of microchips. Think about what that means: every chip in the world is due to VLSI these days, and we owe that to Carver Mead; this is an important person. And he said we can do the same kind of thing in chips that we can do in optics. We can put thousands or millions of extremely simple processing units on a single chip. But would it be limited to a niche market?

Carver Mead then pointed out that the human brain itself uses this same kind of architecture. It uses billions and billions of very simple units -- like synapses or elements of a hologram -- all working in parallel. But the human brain is not a niche machine. It seems to have a fairly general range of computing capability. Thus the human brain becomes an existence proof, to show that one can indeed develop a fairly general range of capabilities, using sixth generation computing hardware. The Neuroengineering program was set up to follow through on this existence proof, by developing the designs and programs to develop those capabilities.

Let me emphasize that in my program we are not just developing computer programs. We are developing general mathematical designs, to understand the general capabilities of anything made out of this type of architecture, which would logically include the brain. The brain was the existence proof underlying the whole thing.

WHAT IS NEUROENGINEERING?

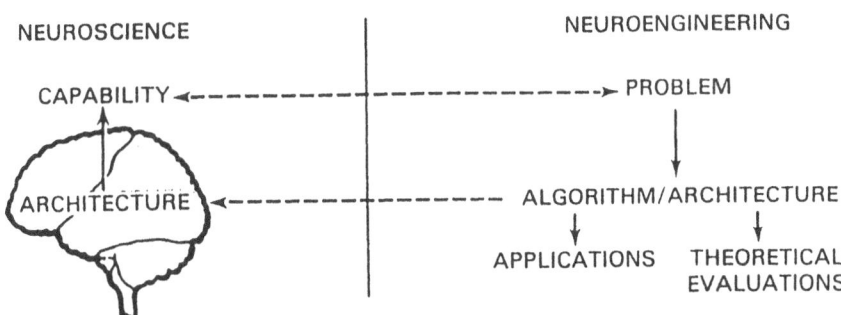

Figure 2: Neuroscience Versus Neuroengineering

Soon after NSF set up the Neuroengineering program, they called in a new person, Dr. Nick DeClaris, to look over this activity. Nick came up with a new definition of the program which sounded very different, but was really the same in practice. Figure 2 is an update of an earlier figure he produced. Nick said that there are guys over there in neuroscience, and they try to figure out the capabilities of the brain. (Maybe it's really the psychologists who try to figure out the capabilities.) They also try to figure out what are the architectures or circuits that give rise to those capabilities. On the neuroengineering side, we try to replicate these capabilities in mathematical designs that we can implement in

265

a whole lot of ways. In developing these designs, we try to use what has been learned over in neuroscience, but we also use basic principles of control theory, statistics and operations research. We try to come up with designs that will really work in solving tough engineering problems.

So that tells you how the program got started. But it doesn't tell you why I actually took the job of running it when Nick asked me to. The reason was very simple. I am interested in using this kind of engineering challenge as a test bed for developing the kind of mathematics we need to really scientifically understand brain function and behavior.

I'm really happy to hear someone else who likes the analogy to Isaac Newton. (See the chapter by Peter Killeen.) I could talk about that one for half an hour. It's a great analogy. Let me just add one little bit to it. On the one hand Newton had some very specific concepts related to dynamics and learning which are critical. But he also had some new mathematics. And I would argue that to really understand these two systems we need both. And so what I am trying to do here is develop the mathematics. The hope is that we then have a feedback loop back to understanding the brain. Then maybe what we learn in engineering can be brought back to the brain to suggest experiments, hypotheses, and even explanations of what is going on in a functional computing kind of sense. Maybe this sounds like a bit of exaggeration, but we really do have some interesting mathematics here.

Unfortunately, at many neural network conferences, these kinds of possibilities get lost because of personality conflicts and the like. We have made a lot of progress in the last few years, but there still are problems. We have people in computational neuroscience, building very precise models in one community. Their models look like neural nets; they use little circles and boxes representing differential equations, local processing and so on. We have other people using artificial neural nets to accomplish technological goals. And we even have a lot of connectionist cognitive scientists, though perhaps we need more of some other kinds of psychologists. We've got a whole bunch of connectionist cognitive scientists who have neural network kinds of models to describe certain kinds of behavior. And all of these different kinds of neural network researchers always have trouble understanding each other. They keep asking questions like: "What is this other guy doing? What is he trying to do?"

For example, there is prominent neuroengineer who is famous for some of his attacks on a famous neuron modeller. Many people believe that this is just a personality conflict, but it is not. In private, the engineer will often say," I can't understand what that guy is trying to do anyway. I've read his stuff, but it doesn't make any sense. Zillions and zillions of words and equations, but they don't really fit together, they don't add up to a functioning system. How can he go on to print more and more equations, without even knowing what the first equations really do, whether they really work? What is all this stuff supposed to mean anyway?"

In reality, what is going on here is very simple. What is going on is that there are three different validation criteria. In the computational neuroscience people are asking, "Does it fit the circuit?" In connectionist cognitive science they are asking, "Does it fit the behavior?" In our case, we are asking, "Does it work? Can it produce solutions to very challenging tasks?" But in actuality, whatever really goes on in the brain has to pass all three tests, not just one. Thus logic suggests we really need to be combining all three validation criteria. What the engineers can provide to some extent is the development of mathematics that begin to describe really general functional capabilities. This is very hard to do.

We see hundreds of learning models in computational neuroscience which can give you associative memory. And I feel a little uncomfortable saying anything negative about these models with Walter Freeman in the room because he knows that many of these models are really excellent models within the scope of what they try to do. But beyond associative memory, you don't see a lot of models that are capable of very sophisticated problem solving, because it's very hard to do. It also turns out that the classical neuron dogma inhibits this kind of modeling to an important degree; that's one problem we don't have so much on the engineering side.

The bottom line here is that we have developed some mathematics that we think might be useful to you. And we need to collaborate on that.

VARIETIES OF NEUROENGINEERING: RELEVANT VERSUS IRRELEVANT

Those of you who have heard a little bit about the neural network community may find it hard to reconcile what I just said with the kinds of papers you have actually seen out there. What I just said may have sounded a little exciting, but you will read hundreds and hundreds of papers that really seem to do some simple-minded kinds of stuff.

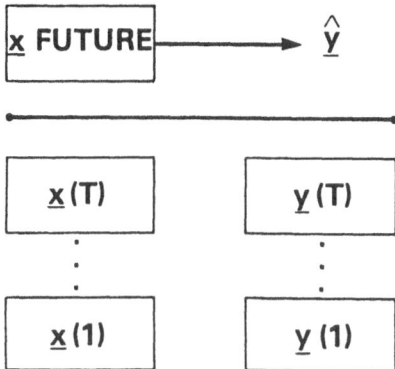

Figure 3. Supervised Learning, A Common Engineering Task

Most of the engineering applications of artificial neural nets today are applications of a very simple idea called supervised learning, shown in Figure 3. Supervised learning is a very simple idea: you have a very simple neural net; you plug in some inputs (X), which are really independent variables; you have a desired response or some target (Y); and you adapt some weights in the network similar to synapse strengths, in such a way that the actual outputs match the desired outputs, across some range of examples. You try to do all this in such a way that you would get good results in the future, when you use this network on new data. This is a simple-minded idea, but we do have systems that can implement it. These systems do have practical applications.

Nevertheless, I do get just as upset as Karl Pribram does when I read some popularized books that seem to suggest that supervised learning can explain all of what goes on in the brain. For example, Paul Churchland's new book on the mind[6] is extremely logical in some ways; however, after telling us how we are discovering lots of important new things about the mind, he then gives a whole bunch of examples all based on supervised learning. The examples are all very cognitive in nature. But there is no motivation; there is no behavior; all there is is a system to classify something correctly. And we know that the brain does more than that.

In this conference, I was very glad to hear some people (like Peter Killeen) actually say some good things about Skinner this morning. In the old days of Skinner I was very upset about his focusing only on the behavioral level, but it seems to me that the hatred of Skinner has gone to the opposite extreme where people now have models of the brain that have no motivation and no behavior. I really wonder how anyone can go about making models of intelligence that have no motivation and no behavior. It is bizarre. And I can assure you that the kind of neural net models that we are developing in engineering do not have that limitation. So maybe we are the guys who should be working with you, instead of the purely cognitive people.

To make things work in engineering we have to add a few components, above and beyond cognition. A robot that does not move is not a very useful robot. But even supervised learning by itself does have its uses. Somebody was talking earlier today about applications. There are hundreds or thousands of applications of neural nets working in industry today. [7].

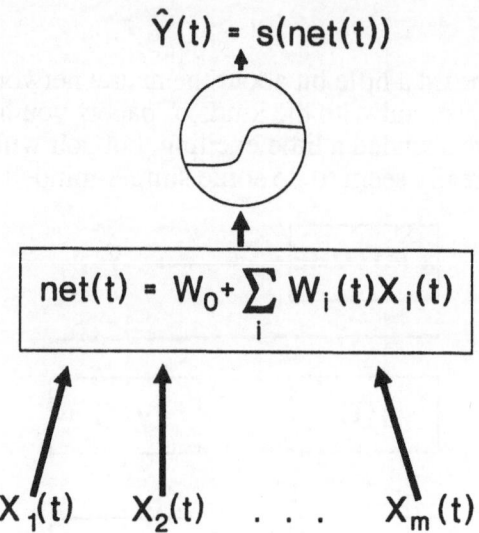

$$\hat{Y}(t) = s(net(t))$$

$$net(t) = W_0 + \sum_i W_i(t)X_i(t)$$

$$X_1(t) \quad X_2(t) \quad \ldots \quad X_m(t)$$

Figure 4. McCulloch Pitts Neuron, Modern Variant

For historical reasons, a majority of ANN applications today are based on the old McCulloch-Pitts model of the neuron, shown in Figure 4. According to this model, the voltage in the cell membrane ("net") is just a weighted sum of the inputs to the cell. The purpose of learning is simply to adjust these weights or synapse strengths. The output of the cell is a simple function ("s") of the voltage, a function whose graph is S-shaped or "sigmoidal." (For example, most people now use the hyperbolic tangent function, tanh.) Those ANN applications which are not based on the McCulloch-Pitts neuron are usually based on neuron models which are even simpler, such as radial basis functions (Gaussians) or "CMAC"[8,11].

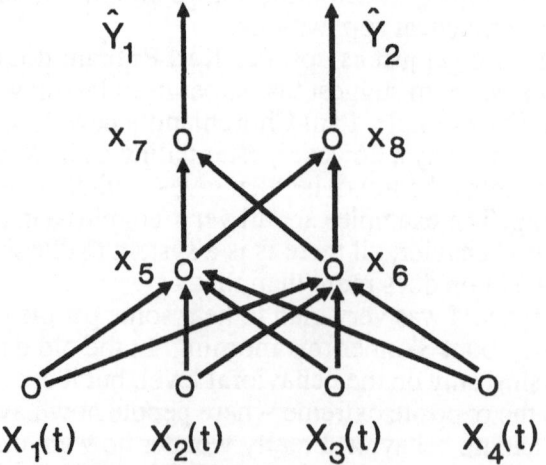

Figure 5. Three-Layered Multilayer Perceptron (MLP)

In most applications today, the McCulloch-Pitts neurons are linked together to form a "three-layered" structure, as shown in Figure 5, where the first (bottom) layer is really just the set of inputs to the network. We know that the brain is not so limited. But even this simple structure has a lot of value in engineering.

THEOREMS

1. All nets approximate nice functions
2. 4-layer MLP for tracking control (Sontag)
3. As number of inputs grow
   MLP does better (Barron)
   (SRN best; local minima)
→ **Speed versus generalization dilemma**

Figure 6. Function Approximation Theorems

It has been proven for example that a simple three layered MLP can approximate any smooth function, in an efficient way [9]. Most people in engineering today will say that is the end of the story, any smooth function. You don't need anything else. You can say that this structure is simple and dumb, but it is powerful enough to do any job. That is what people mostly think these days.

Now it turns out that this is not the whole story. Why? Because oftentimes when a system must make difficult decisions in real life, its choices cannot be described by a smooth function. Later on in this paper I will come back to why you need more sophisticated nets in the real world for functions that are not smooth; however, engineers are only just now beginning to catch up to that level of decision making. Some of us are catching up, but only very recently.

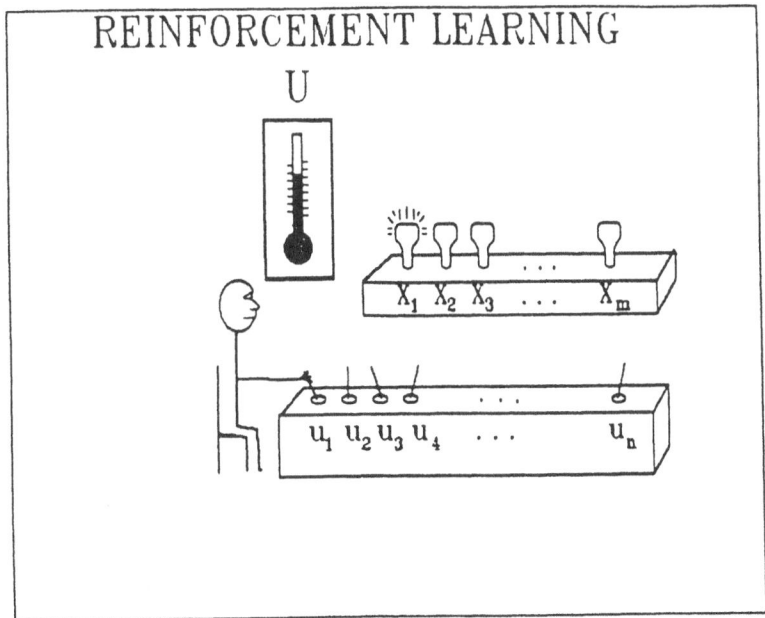

Figure 7. Reinforcement Learning: A Fundamental Task

Not everyone in the ANN field has been limited to supervised learning. Many of us have been working on a more interesting problem shown in Figure 7. This is what we call reinforcement learning.

Reinforcement learning has been a controversial idea in psychology. The reasons for this are very strange. Back in the days of Skinner, he used to say that this idea is too anthropomorphic, that it ascribes too much intelligence to human beings and other animals. Nowadays many people are saying just the opposite -- that it's not purely cognitive enough (because it has motivation in there) and that it's also too mechanistic. But in reality, it may be a good thing that we're pursuing an idea which is halfway between these two extremes [4].

In any case, the problem here for an engineer is straightforward. You've got this little person in Figure 7 who has a bunch of levers to control. We'll label them u_1 to u_n. You have a number -- u_3 or whatever -- to characterize each one of these levers. The set of n numbers forms a vector. Don't be scared by the word "vector;" it's just an ordered set of numbers; that 's all it is. Likewise, the person sees a bunch of lightbulbs labeled X_1 through X_m, representing sensory input. Finally, there is something that looks like a big thermometer in the Figure; next to the thermometer, you'll notice the big letter U which means utility (not temperature). The problem to be solved is as follows: we have to find a computer program or neural net design which can handle the job of the little person in this figure. The little person starts out knowing nothing at all about the connection between the lights, the levers and the thermometer. He must somehow learn how these things work, enough to come up with a strategy that maximizes the utility function U over the long term future.

The psychologists in the room know that we could talk for hours about just this concept right here [4]. A few quick thoughts are in order, however. This kind of reinforcement learning is not the same as self-gratification. You could think of the function U as a measure of gratification. But the problem here is more like a problem in delayed gratification. The essence of the problem is not just to maximize this in the next instant. The problem is to find a strategy over time to achieve whatever goals are built into this U; these could be very sophisticated goals.

Almost any planning or policy management problem can be put into this framework. An economist would say that this connection is very straightforward. If we simply choose U to represent net profits, then the learning task here -- to maximize profits over the long-term -- encompasses quite a lot.

Figure 7 may not be a good higher order description of the brain, but I would claim that it has been extremely productive as a good first order motivator of engineering research.

There are a few other aspects of reinforcement learning of some importance to understanding the brain.

First of all, it turns out that you can't build a really powerful reinforcement learning system if you only have one simple neural net. You need to have modules within modules within modules. And that's very exciting, because that's also what we see in the brain. Modules within modules. This is not like the AI systems where you have an arbitrary kind of hierarchy. Instead, you have a lot of modules because there are a lot of pieces that you need to do this kind of task effectively over time.

Secondly, the criticism which I usually hear from psychologists about this idea is as follows. They say, "Wait a minute. Human beings are not completely rational. They do not maximize anything in a consistent or reliable way." However, if you try to build a real engineering system that tries to learn how to do this over time, then in order to make it work you have to insert a lot of the stuff that you see in humans. You have to insert a lot of the same features that psychologists describe as the limitations of humans. For example, you have to insert exploratory behavior. If you don't have exploratory behavior the system is going to get stuck; it will be a whole lot less than optimal. So there is a lot of behavior that people do which is exploratory. Exploratory behavior is often called irrational, but you've actually got to have it if you are going to build one of these things.

Another issue is that human beings sometimes get stuck in a rut. There are many names for the ruts that humans get stuck in. Humans get stuck in less than optimal patterns of behavior. Unfortunately, the same thing happens to ANNs as well. They get stuck in things called local minima. If there were a mathematical way to avoid local minima, in all situations, then we would use it. If there were a mathematical way or a circuit way to keep the human brain from getting stuck in a rut, I think that nature would have implemented it too. But there isn't. It's just the nature of complex nonlinear systems that in the real world you've got a certain danger of falling into a local minimum, a rut. You also need a certain amount of exploratory behavior to reduce that danger.

The bottom line here is that nobody needs to worry about an engineer building a model so optimal that it is more optimal than the human brain could be. That's the last thing you have to worry about. Thus I would argue that reinforcement learning may still be a plausible first-order description of what the brain is doing, computationally. Again, I recognize that this is a strong statement, but there is a lot behind it [4,10].

NEUROCONTROL IN GENERAL

Back in 1988, when I took over the NSF program in Neuroengineering, I realized that I couldn't just focus the entire program on reinforcement learning. There would have been a lot of risks in that kind of strategy. Therefore, I decided to broaden it by emphasizing "control" -- as engineers define "control." (Some people prefer the phrase "decision and control" or "decision-making systems" or "cybernetics.")

In 1988, I defined a neurocontroller as a well defined mathematical system containing a neural network whose output is actions designed to achieve results over time[11,12]. Whatever else we can say about the brain as an information processing system, clearly its outputs are actions. And clearly the function of the brain as a whole system is to output actions. In this paper, I'm not going to talk about the larger question of how the brain relates to the mind. I have already talked about that issue at some length in some other places[4,10]. In this paper, I will limit my attention to the issue of how we can understand the brain itself as a computing device.

For the brain as a computer, control is its function. If you want to understand the components of a computer, you have to understand how they contribute to the function of the whole system. In this case, the whole system is what I would call a neurocontroller. Therefore the mathematics you need to know in order to understand the brain are in fact the mathematics of neurocontrol. That is the kind of mathematics which I am trying to develop in my program.

I knew that neurocontrol would be my emphasis right from the beginning, because of my interest in the brain. Thus the first thing I did at NSF back in 1988 was to set up the workshop which really created an independent discipline of neurocontrol. This discipline, in my view, is a subset both of neuroengineering and of control theory -- the intersection of the two fields. The book that came from our workshop back in 1990 really was the start of this now organized field called neurocontrol[11].

Because of the many successes of neurocontrol in engineering and in theory, we held another workshop jointly with McDonnell-Douglas. That workshop resulted in another book [8] which came out in 1992 which many of us now call the "Bible." That book is still the best place to go to find the core, fundamental mathematics. If you want to see all the equations, that's the place to go for now. And it's still many years ahead of what you will see in other books on intelligent control which have become rather numerous.

But let me warn you -- the Bible is not a comic book. It is not light reading. It has a very high density of fundamental ideas, many of which have only just begun to be explored. I think that my own book from 1994 [1] could be useful as an introduction to make this hard stuff a little more understandable. Basically, my book has tutorials in the back explaining what backpropagation is and what it really does. Backpropagation is a lot more general than the popularized stuff that you may have heard about. The real general form of backpropagation is described in my book, including the original papers from 1974 and 1981 when I discovered this stuff. The book can help explain the basis for designs which use backpropagation in a very sophisticated way. (Also, an abbreviated version of some of this material appears in the chapter on backpropagation in [13].)

Since 1992, there has been great progress in applying and extending these ideas. See [7] for some of the developments in neurocontrol in general. See [3] for a current overview of the more brain-like designs (and of some typographic errors in [8]).

ALTERNATIVE FORMS OF NEUROCONTROL

NEURAL NETS IN CONTROL:

1. **In Subsystems**

2. **Copy Experts**
   Supervised Control

3. **<u>Follow</u> Path, Setpoint, Ref. Model**
   Direct Inverse Control
   Neural Adaptive Control

4. **<u>Optimal</u> Control Over Time**
   Backpropagation of Utility (Direct)
   Adaptive Critics

Figure 8. Taxonomy of Neurocontrol

As part of my job, I have had to survey what has been done in engineering and in computational neuroscience that works in neurocontrol. After reading hundreds and hundreds of papers, I have found that they are all based on three basic principles, as shown in Figure 8. There are some people who are only using neural nets as subsystems of some other kind of controller; I won't count them. But there are true neural net designs that perform cloning, tracking and optimization. That's the trilogy. Those are the kinds of capabilities that we can really use in engineering.

<u>Cloning</u> means something like cloning a preexisting expert. I won't talk about that here, because that's not what the brain does. There is some kind of learning in the brain based on imitating other people, but it's nothing like the simple cloning designs that we've been using in engineering. In fact, imitative behavior in human beings depends heavily on a lot of other more fundamental capabilities which we need to understand first [4,10].

<u>Tracking</u> is the most popular form of control in engineering today. In fact, many classical control engineers think that control <u>means</u> tracking, that they are the same thing.

This is rather sad. A well-trained control engineer knows better. But a narrowly trained control specialist thinks that control means tracking.

But what do I mean by tracking? I mean something like a thermostat. There is a desired temperature, and you want to control the furnace to make the real temperature in the room track the desired setpoint. (The "setpoint" is the desired value for the variable which you are trying to control.) Or you could have a robot arm, and a desired path that you want the arm to follow. You want to control the motors so as to make the arm fit (track)the desired path.

A lot of engineering work goes into tracking. But the human brain as a whole is not a tracking machine. We don't have anyone telling us where our finger has to be every moment of the day. The essence of human intelligence and learning is that we decide where we want our finger to go. Thus tracking designs really do not make sense as a model of the brain. Nevertheless I will talk about tracking a little bit here, because some people think it could be used by part of the brain, for lower-level motor control.

Finally there is **optimization over time**. And this is the stuff that gets into reinforcement learning. This is the kind of stuff that really can be related to brain structures.

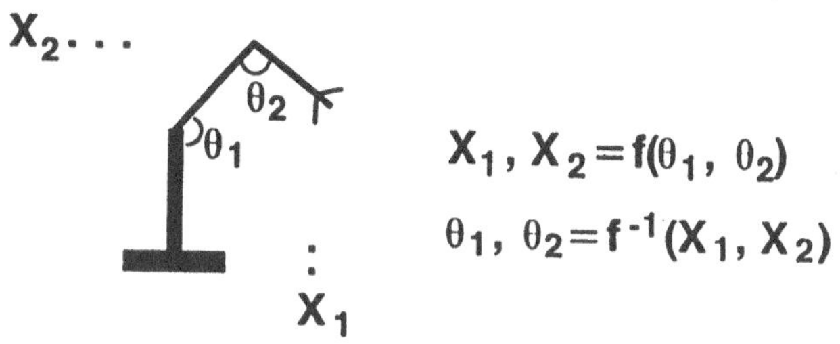

$$X_1, X_2 = f(\theta_1, \theta_2)$$
$$\theta_1, \theta_2 = f^{-1}(X_1, X_2)$$

Input X_1, X_2; Target θ_1, θ_2

Figure 9. Direct Inverse Control

Nevertheless, I promised to say something about tracking before moving on to optimization. Figure 9 gives a simple-minded example of what is called direct adaptive control -- direct tracking. The idea here is very simple: you want the robot hand to go to some point in space, defined by the coordinates x_1 and x_2. You have control over θ_1 and θ_2. You know that x_1 and x_2 are functions of θ_1 and θ_2. If the function happens to be invertible -- and that's a big assumption! -- then θ_1 and θ_2 are a function of x_1 and x_2. So what some robot people have done is as follows: they will take a robot, and flail the arm around a little bit. They will measure the x variables and the θ variables, and then they try to use simple supervised learning to learn the mapping from the x's to the θ's.

This approach does work -- up to a point. If you do it in the obvious way, you get errors of about 3% -- too much for anybody to accept in real-world robotics. If you are sophisticated, you can get the error down a lot lower[11]. There are a few robots out there that use this approach. But the approach has some real limitations. One limitation is this assumption that the function has to be invertible; among other things, this requires that the number of θ variables (degrees of freedom) has to be exactly the same as the number of x variables. The other thing is that there is no notion of minimizing pain or energy use. There have been lots of studies by people like Kawato and Uno[11], and also a lot of work by a guy named Mahoney who came from Cambridge University, who has done work on biomechanics[14]. There is lots and lots of work showing that the human arm movement system does have some kind of optimization capability.

There are lots of degrees of freedom in the human arm, and nature does not throw them out. Nature tries to exploit them to minimize pain, collision damage, whatever. The point is that direct tracking models are simply not rich enough to explain even the lowest level of arm control.

An interesting aspect of this is that there are lots of papers still out there in the biology literature talking about learning the mapping from spatial coordinates to motor coordinates. What I am saying is that this is only a metaphor. It is not a workable system. Perhaps it is useful at times in descriptive analysis, but it would be totally misleading to incorporate it into any kind of model of learning.

In actuality, in neuroengineering, most people do not use direct inverse control, even when they are trying to solve very simple tracking problems. There is another approach called <u>indirect adaptive control</u> ,where you try to solve a tracking problem by minimizing tracking error in <u>the next time period</u>. This myopic approach is now extremely popular in neuroengineering. But this approach tends to lead to instabilities in complex real-world situations (using either ANNs or classical nonneural designs). There are lots of theorems to prove that such designs are stable, but the theorems require a lot of conditions that are hard to satisfy.

Because of these instability problems, I don't think that indirect adaptive control is a plausible model of arm movement either. Furthermore, it still doesn't account for the work of Kawato and Mahoney and such, who show some kind of optimization capability <u>over time</u>. Therefore, I would claim that optimization over time is the right way to model even the lowest level of motor control.

If you look back at Figure 8, you will see that there are two forms of optimization over time which have been used in practice for reasonably large-scale problems in neuroengineering. (There are also a few brute-force approaches used on much smaller-scale problems; these are obviously not relevant here.) One of them is a direct form of optimization based entirely on backpropagation. Direct optimization over time leads to a very stable, high-performance controller. It has been used a whole lot in classical engineering and in neuroengineering both. For example, I suspect that you will see it in ANNs in some Ford cars in a couple of years. I wish I could say more, but this gets into some proprietary sensitivities. Still, I can say that it is very real. It is not hypothetical stuff.

Nevertheless, the kind of stuff that you can do in the brain is a little different from what you can do with microchips in a car. The direct form of optimization requires calculations which make no sense at all as a model of the brain. This leaves us with only one class of designs of real importance to neuroscience -- a class of designs which has sometimes been called reinforcement learning, sometimes called adaptive critics, and sometimes called approximate dynamic programming (ADP). Actually, these three terms do have different histories and meanings; in a strict sense, the designs of real relevance are those which can be described either as adaptive critics or as ADP designs[3,15].

THE BRAIN AS AN ADAPTIVE CRITIC OR ADP CONTROLLER

The kind of optimization over time that I believe must be present in the brain is a kind that I would call approximate dynamic programming (ADP). There is only one other kind of optimization over time that anybody uses (the direct approach), and that's not very brain-like. So this is the only thing we have left. But what is dynamic programming?

Dynamic programming is the classic control theory method for maximizing utility over time. Any control theorist will tell you that there is only one exact and efficient method for maximizing utility over time in a general problem and that is dynamic programming.

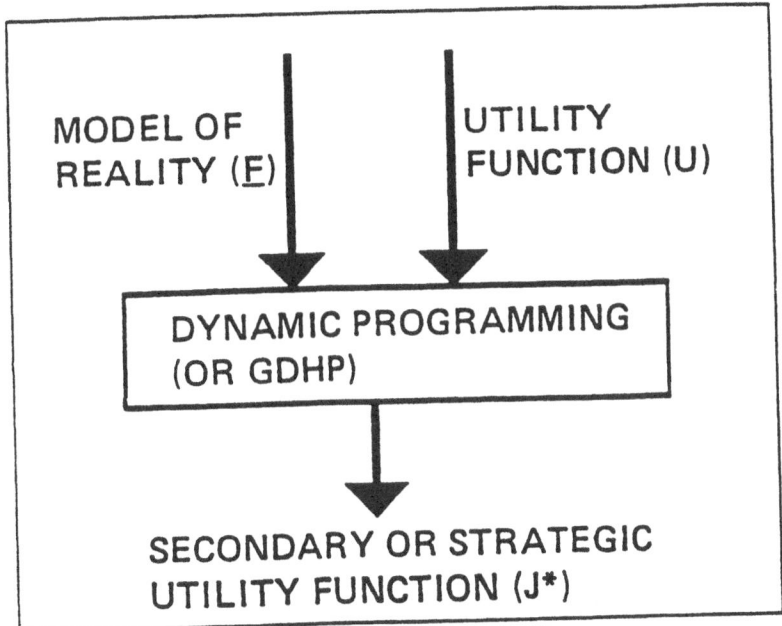

Figure 10. What Dynamic Programming Does For Us

Figure 10 illustrates the basic idea of dynamic programming. The incoming arrows represent the two things that you have to give to dynamic programming before you can use it. First, you must give it the basic utility function U. In other words, you must tell it what you want to maximize over the long-term future. This is like a primary reinforcement signal, in psychology. Second, you have to feed it a stochastic model of your environment. And then it comes up with another function that I like to call a strategic utility function, J.

The basic theorem in dynamic programming is that this J function will always exist if you have a complete state model. Maximizing J in the short term will give you the strategy which maximizes U in the long term. Thus dynamic programming translates a difficult problem in planning or optimization over time into a much more straightforward problem in short term maximization.

If dynamic programming can solve any optimization problem over time, and account for all kinds of noise and random disturbance, then why don't we use it all the time? The real answer is very simple: it costs too much to implement in most practical applications. It requires too many calculations. To run dynamic programming on a large problem is too expensive. It just won't work. But there is a solution to that problem, called approximation.

In Approximate Dynamic Programming (ADP), we build a neural net or a model to approximate this function J. Thus instead of considering all possible functions J, we do what you do if you are an economist building a prediction model. You build a structure with some parameters in it and you try to adapt the parameters to make it work. You specify a model or a network with weights in it, and you try to adapt the weights to make this a good approximation to J. A neural network which does that is called a Critic network. And if it adapts over time, if it learns, we call it an adaptive critic. So right now in engineering we have almost three synonyms. Approximate dynamic programming, adaptive critics, and reinforcement learning -- those are almost the same thing.

Based on all of this logic, I would conjecture that the human brain itself must essentially be an adaptive critic system. At first glance, this may sound pretty weird. How could there be dynamic programming going on inside the brain? What would this idea mean in terms of folk psychology, our everyday experience of what it feels like to be human? Like Karl Pribram, I would agree that a good model of the brain should fit with our personal experience of how the brain really works. That's part of the empirical data. We don't want to ignore it. So does this theory make sense in terms of folk psychology? I will argue that it does.

Domain	Intrinsic Utility (U)	Strategic Utility (J)
Chess	Win/Lose	Queen = 9 points, etc.
Business Theory	Current Profit Cash Flow	Present Value of Strategic Assets (Performance Measures)
Human Thought	Pleasure/Pain Hunger	Hope/Fear Reaction to Job Loss
Behavioral Psychology	Primary Reinforcement	Secondary Reinforcement
Artificial Intelligence	Utility Function	Static Position Evaluator (Simon) Evaluation Function (Hayes-Roth)
Government Finance	National Values, Long-Term Goals	Cost/Benefit Measures
Physics	Lagrangian	Action Function

Figure 11. Examples of J and U

I would like to give you a few examples of where this J versus U duality comes in, in different kinds of intelligent behavior.

Those of you who have followed artificial intelligence (AI) or chess playing probably are aware that in computer chess the basic goal, the U, is to win the game, and not to lose it. This is in computer chess, not in real chess, in computer chess. But there is a little heuristic they teach beginners. They teach you that a queen is worth 9 points, a castle is worth 5, and so on. You can compute this kind of score on every move. This score has nothing to do with the rules of the game. But people have learned that if you maximize your score in the short term, that's the way to win in the long term.

When you get to be a good chess player, you learn to make a more accurate evaluation of how well you are doing. For example, you learn to account for the value of controlling the center of the board, regardless of how many pieces you have. Studies suggest that the very best chess players are people who do really sophisticated stuff, a really high quality strategic analysis of how good their position is one move ahead. Those are the studies I've seen. So basically, this evaluation score is like a J function. It's a measure of how well you are doing.

Figure 11 also lists a few other examples. In animal learning, U is like primary reinforcement, the inborn kind of stuff. It reminds me of the hypothalamus and the epithalamus. And J is like secondary reinforcement, the learned stuff, learned reinforcers. U is like pleasure or pain, an automatic kind of response, while J is like hope and fear. And in a way all of this fancy theory is just saying hey, I think hope and fear is hard-wired into the brain. We respond to hopes and fears from day one. Hopes and fears drive everything we do and learn.

Based on some things that Peter Killeen said, I'm going to stick my neck out and mention one more analogy that I usually don't talk about because it gets into some difficult and complex subjects. It turns out that this stuff also matches physics. In fact, the Bellman equation we use in dynamic programming is exactly what is called the Hamilton-Jacobi equation in physics. If you read Bryson and Ho[16], they even call it the Hamilton-Jacobi-Bellman equation. In physics, they would say that the universe is maximizing a Lagrangian function instead of calling it a utility function; thus they use the letter L instead of the letter U, but it's the same equation. And it turns out that our J refers to something they call "action." And the things we call "forces" in physics turn out to be the gradient of the J function[17,18].

If I were just a bit more precise about describing my theory, I would actually say that there are circuits in the brain that are computing what you would call forces. But there are two major levels of learning in this theory. There is the level of learning where you change your behavior in response to your hopes and your fears, in response to emotional forces. But then there is also the level where the forces themselves change because of secondary reinforcement. And these two levels of learning work together.

This mathematics also has some interesting implications for motor control. In recent years, Hogan and his collaborators have claimed that they can best describe motor behavior by using concepts like force fields... which are fully consistent with what I am discussing here! Hogan's people have argued very strenuously with Kawato's group about the idea of optimization; however, the mathematics of forces and the mathematics of optimization turn out to be the same underneath if you pursue it to this level.

Finally, I would like to say just a little bit more about the intuition behind U and J. If there were an economist in the room, he might well say, "Wait a minute. There is utility and there is price; these are different things. These functions U and J are global measures of how happy you are. But what if you want to know what is the value of a specific object? For example, what is the market value of a peanut?" An economist would say that this is an easy question. The value of a product is equal to its marginal utility. The marginal utility refers to the increase in your U function which would result if you had one extra peanut. It is the derivative of U with respect to peanut consumption. Thus the derivatives of U represent values. The derivatives of J are what give you market values and forces. Thus values are the derivatives here, and we have some adaptive critic designs where the network outputs the derivatives, the values, rather than the raw quantities U and J.

ADAPTIVE CRITIC DESIGNS: SIMPLE VERSUS BRAIN-LIKE, FREUD AND BIOLOGICAL EVIDENCE

When I say that I think the brain is an adaptive critic system, this does not tell you exactly how the learning works. Just as there are lots and lots of ANN designs in general, so too are there lots and lots of adaptive critic designs. I like to think of these designs as forming a kind of "ladder," rising up from the simplest and most popular designs, which are easy to implement, through to more complex and more powerful designs, ultimately including the human brain itself. The designs now used in engineering can be classified as level zero up to level five [3].

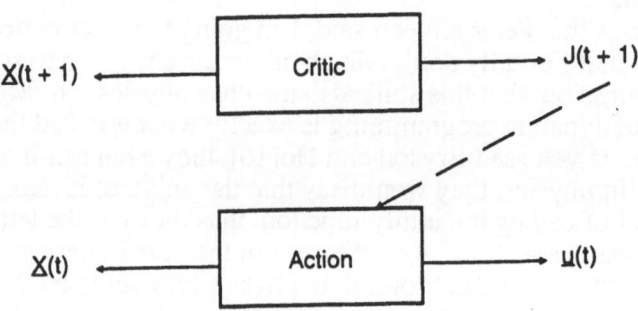

Figure 12. The Barto-Sutton-Anderson Adaptive Critic System (Level 1)

The most popular design of all, at present, is the Barto-Sutton-Anderson (BSA) design, shown in Figure 12. Strictly speaking, this design is more popular in computer science than in engineering. It was first published back in 1983[19]. Barto and Sutton have written many papers on this kind of design, showing how it can implement theories of animal learning like the Rescorla-Wagner theory and so on[8]. In fact, the animal psychologist Harry Klopf and the engineer Bernie Widrow really developed a lot of the ideas which went into this design. It was actually Bernie Widrow who coined the word "Critic," and implemented the first ANN adaptive critic system[20].

In any event, the BSA design is very simple, as you can see in Figure 12. There is one network -- the Action network -- which really does the control. It inputs the sensor data $\underline{X}$ and it outputs the actions $\underline{u}$. Then the Critic network gives a kind of gross reward and punishment to the Action network. So the Action net does the real work, and the job of the Critic is just to help train the Action net.

There are convergence theorems for this kind of design. But there is also a problem. It only works on very small systems. It works very well on small systems. And Barto would be quick to add that the world's best backgammon player is based on this kind of system (with some special features added[21]). Backgammon is not entirely a small problem, but it is small in one respect: at each move, it only requires a choice between a few choices of action, only a few action variables. The reason why this design doesn't work well on truly large problems is that the feedback from the teacher to the Action net is very limited. It's just one gross scalar measure.

Suppose that you are a student trying to learn, say, a hundred numbers. You write down a hundred numbers, and you know they are probably wrong and you give them to the teacher. And the teacher looks and says, "No good." You do it again. "No good." How long will it take for you to find the right hundred numbers? But suppose instead that the teacher told you, for each number, "Make that bigger; make this smaller; this is really important, turn this up." Then it might be possible for you to find the numbers. So the point is this: if there are a lot of weights, if there are a lot of action variables, then the scalar kind of feedback won't work very well. What you really need to have here is feedback to each action variable, indicating which way to adjust it.

Strictly speaking, of course, this design will still converge for large problems -- theoretically. The practical problem is that the speed of learning or convergence gets to be slower and slower as problems get more and more complex. For middle-sized problems (about 10 variables) involving continuous variables in engineering, everyone I know who has tried this method says that it is unacceptable. There are even a few engineers who extrapolate too far and say that "reinforcement learning is slow in general." But those engineers should wake up to the fact that there are other reinforcement learning designs available here.

LIMITS OF SIMPLE CRITIC:

- ## $\underline{X}$(t) Versus $\underline{R}$(t)

- ## U_o and Tantrums

- ## Multicolinearity (Exams, Cause-and-Effect, Blue Shirt)

Figure 13. Limitations of the BSA Design

There are some other limitations with the BSA design, listed in Figure 13. The most important limitation has to do with $\underline{X}$ versus $\underline{R}$ -- two concepts or vectors which merit a lot of explanation.

You may recall from Figure 7 that I use the letter $\underline{X}$ to represent the external sensory data. By contrast, $\underline{R}$ represents something more like an internal representation of external reality. Engineers would call it an "estimated state vector." Intuitively, it could also be seen as a kind of short-term memory or working memory (as per Pribram and Goldman-Rakic).

It's kind of fun to have to make this distinction here, because the exact same issue was discussed at length, in a different context, in other talks in this conference. People from psychology discussed how important it is to consider our internal representation of reality, as opposed to our current sensory experience. Is it really important to think about such internal representations?

Well, from the engineer's point of view, to make these systems work, you really have to have a representation of reality. It turns out that all of the theorems for dynamic programming require that you have what is technically called a Markov Model or a state space model of the environment you are trying to influence or control. In practice, what this means is that you can't just use sensor input data. You have to reconstruct an estimated state vector. You have to build up a representation of the external world. And I like to use the letter $\underline{R}$ to represent the reconstructed representation of reality through recurrent networks. Thus the biggest problem with the BSA design is that we really need a way to build up that kind of representation and feed it into the network.

Given time, I could say more about some of the other points in Figure 13. I should note that there is some recent research in control theory which argues that our estimated state vector should not be based purely on a cognitive, value-free model of the world; instead, to get the right solution to the control problem, we need to use some kind of value-weighted model or procedure [22]. This fits in very well with the adaptive critic approach, and with our knowledge of how salience measures from the limbic system (a Critic) influence our learning of representations in the neocortex.

From the viewpoint of animal learning, Grossberg has criticized the BSA model severely and justifiably. He has argued that the lack of an expectations system makes this model fundamentally unable to address the huge literature on classical or Pavlovian conditioning, which shows how animals change their expectations through learning. The need for an expectations system leads us naturally up to the next design.

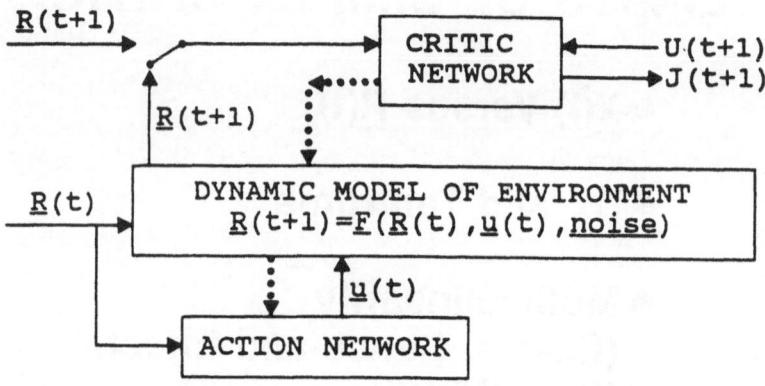

Figure 14. THE HDP/BAC Design (Level 3 of the Ladder)

Way back in 1977, before Barto, Sutton and Anderson, I came up with another design that was based on three networks[23]. I call this design Heuristic Dynamic Programming (HDP) with a Backpropagated Adaptive Critic (BAC).

If you compare Figure 14 with Figure 12, you can see that we now need a third network -- a Model network. The Model network serves as a kind of predictive model of the external world. It also serves to build up a representation of reality to use in making predictions. You could also think of it as an expectations system. So now we have three networks, all of which should be adapted concurrently in real time, if we really want to imitate the brain.

But how is this adaptation done? In particular, how can we adapt the Model network, the new part of this design? In actuality, engineers have spent a lot of time on this kind of issue. They call it the "system identification task." Even before ANNs were used, engineers spent decades building up a complex theory of how to do system identification for dynamic systems[16]. I can't summarize that literature in ten minutes, but I can assure you that we have a lot of technology in engineering for adapting networks to describe or characterize the world, to build up state space models and all that kind of stuff[8].

Where does this idea of a Model network fit in with neuroscience? For those of us who attend the Radford conferences, there is an obvious answer. There was a talk here last year by Nicolelis and Chapin[24] describing some new experiments on the thalamus. More precisely, they described how different cells in the thalamus respond to different ways of stimulating the whiskers of a rat. They showed how some cells in the thalamus tend to act as advance predictors of other cells, the cells which register the primary sensory events. Then they described experiments in which they used lesions to make the predictor cells into bad predictors. After learning, however, the predictor cells would somehow learn to use different inputs, and learn a way to become good predictor cells again. This strongly supports the theory that the underlying learning mechanism here is one which tries to minimize prediction errors. Nicolelis and Chapin, and Pribram's group at Radford, were beginning to undertake new experiments, last year, to strengthen these results, by changing the correlations between different inputs coming into the rat (rather than using lesions), to test the ability of this system to learn a new model. This kind of work is extremely exciting. It ties directly into what we have been doing in engineering.

Strictly speaking, our neuroengineering designs for system identification have some aspects which sound rather strange, at first. We need one subsystem which predicts $\underline{X}$ (i.e., these cells in the thalamus, supported by inputs from layer VI of the neocortex). We need a subsystem which reconstructs $\underline{R}$ (the neocortex, especially layer V). But to adapt this kind of system, we require some kind of clocked control, and an alternation between a phase of forward calculation when real predictions are made, and a backward phase when adaptation takes place. (See the discussions of Time-Lagged Recurrent Networks in chapters 10 and 13 of [8].)

At a recent meeting on Capitol Hill, I had a chance to ask Barry Richmond of NIH what his recent experiments show about lag times and temporal coding in the neocortex. This work was not yet final and not yet published; therefore, the indications for now are very tentative. Tentative or not, however, they seem extremely exciting and extremely important.

Richmond's group has done new studies involving synchronization in the cortex. This is not the kind of synchronization that neuroscientists talk about when they discuss epilepsy; it's not the kind of synchronization where all the cells fire at once. Rather, it's the kind of synchronization engineers would think about, where all the cells send a meaningful signal at the same time -- a signal which may be an on signal or an off signal. Richmond describes it as a kind of "window" in time. He said that he found that the usual 100-millisecond-or-so sampling time of the neocortex actually contains only a 30-40 millisecond "window" for the forward calculations which generate the output of the neocortex. There is another 30-40 millisecond window of active calculations which somehow do not lead to a change in outputs, which seems relatively mysterious. Richmond speculated -- with some idea of how to test this further -- that this mysterious second window is the cycle which leads to adaptation. If so, then the neocortex may well share exactly those features of our engineering designs which modellers have been most troubled by.

Coming back to Figure 14, however, it's not enough for us to explain how the Model network is adapted. We also need to specify the learning rules used to adapt the Critic network and the Action network, in order to complete our mathematical design or model. The way we adapt the Action network is by calculating the derivatives of J, by propagating these derivatives back on through the Model network, and then using those derivatives to adapt the Action network. The backwards broken arrows in Figure 14 represent this backwards flow of derivative calculations. I don't have time to explain the Critic today.

Immediately you may ask,"A backwards flow of information? Where does this come from and what sense does it make?" Well, this is a form of backpropagation. Backpropagation, in its simplest form, is used in the vast majority of ANN applications today. But the form of backpropagation shown in Figure 14 is not the simplest form of backpropagation. It is not error backpropagation. It is not supervised learning. It is something else.

The form of backpropagation used in Figure 14 is the original first form of backpropagation, which I developed well before my well-known 1974 thesis[1,25]. The idea really came from Sigmund Freud. To develop backpropagation, all I did was to translate an idea from Sigmund Freud into mathematics. So anyone who says that Freud doesn't have applications should learn about this causal link.

Freud did not start out his career by diagnosing hysterical patients. He started out by trying to understand the dynamics of learning in the brain. He started out with an idea of neurodynamics which he returned to again in the later part of his life[26,27]. Karl Pribram has a book which describes this much better than I will in 5 minutes, but I will now give my 5 minute summary of what Karl's book says on this subject.

Freud said, "Look -- first of all, human behavior is controlled by emotions." "That's very obvious," he thought. When I read what some people write in the AI field today, I wonder if they have ever seen a human being; however, it seemed very obvious to Freud at least that emotions are dominant in human behavior. And it seemed very clear that we place emotional loadings on objects in our environment. We like this; we don't like that; Freud called this phenomenon "cathexis," an emotional charge. I would say that we place a value on a variable, where he would say that we place a charge on an object. But these are just the same idea expressed in different words.

Freud then asked, "Where does emotional charge come from? How does it work?" He said,"Well, first of all, it's clear that we have to learn something about cause and effect in our lives. So let's say, for example, that we learn that object A causes B. We learn to associate A with B. We see A at one time followed by B later. But how is that represented in the brain?" Freud said that there must be a cell representing A and a cell representing B. He proposed that a forward causal association would be represented by a connection from A to B somehow, with a strength W representing a synapse strength, a connection strength. Now if A causes B with strength W, then if you place a value on B you should place a value on A. If A causes B and you want B, then you should want A. I don't see any way you can avoid that. No matter how fancy you get, you can't develop a system that learns to do complicated strategies unless it can learn that A causes B. And it has to exploit the fact that if A causes B, then if you want B, then you want A. You have to have a mechanism that does that. I don't see any way you can avoid that in engineering or in any other way.

So what is the mechanism here? Freud went on to reason: "If A causes B with strength W, then there must be a flow of cathexis or emotional charge from B back to A. That flow," he said,"must be proportional to the cathexis on B and to the strength of the association." This is something that I could write down as a mathematical equation. And in fact, this is the basic equation of backpropagation. All I did was to write down the equation, dress it up a little and prove that it is a theorem. It is simply an equation for calculating derivatives (values), and it makes perfect mathematical sense as such.

If you were an audience of engineers, I might not even mention Freud. I would say,"Look at these equations. This is a new form of the chain rule. It works. You can use it in a lot of different ways in practical applications." (In fact, I really had to do this, in order to get my Ph.D. from Harvard[1].) But it really comes from Freud.

Back in the 1970's, when I first proposed the HDP/BAC design and some further improvements, I found it hard to generate a lot of interest. The papers I published back then were very hard papers. And I didn't have a feeling for how to simplify and explain this kind of stuff. But just since November 1993, the engineers have finally caught up with this kind of design[3]. Just in the last two years, people have gone ahead and implemented adaptive critic designs which have at least these three basic components -- a Model, a Critic and an Action net -- where you really use the Model to adapt the whole system. One of the people who has done it is Rob Santiago, an undergrtaduate student who is attending this conference[28]. There is also a student named Daniil Prokhorov at Texas Tech, a recent Russian immigrant, and a guy named Balakrishnan from India, and several others. They have all demonstrated that this kind of design gives you more accurate control in difficult simulated engineering problems than anything else that exists.

Balakrishnan has done it with missile interception. Would you want to bet that people have spent money on how to do missile interception? Balakrishnan had worked with McDonnell-Douglas, and knew the existing methods. He tried a benchmark test, a very simple simulation benchmark test. He tested ten of the standard methods against one of these three-net kinds of critics, and he was able to reduce error by an order of magnitude on missile interception.

Prokhorov, Santiago and Wunsch have studied two difficult benchmark problems taken from [11]: a bioreactor problem and the autolander problem.

Many of the biologists in the room may already know what a bioreactor is. It is little vessel you use to grow cells in. You use it to grow cells which produce some kind of useful chemical product. (I can remember a meeting at NSF where someone described this as a totally new technology. It was amusing when Dr. Heineken corrected him.) The problem here is that cells are nasty little creatures. If you try to use conventional control, the whole thing blows up, becomes unstable, at least for the system described in [11]. But if you use optimizing neural net control, it is possible to stabilize and optimize this thing. Likewise, the autolander problem in [11] was suggested by NASA Ames. It is a very difficult problem of automatically landing an airplane on a short runway. The problem that Prokhorov et al had was that a very simple adaptive critic could solve the original problem easily. Thus they made the problem harder; they multiplied the random wind shear by a factor of four or ten. They shortened the runway by a factor of four. And at that point even the conventional adaptive critics were crashing every time. The conventional classical controllers were also crashing every time. And at least the brain-like stuff could come in 80% of the time. Thus they achieved much higher performance in noisy nonlinear problems with this kind of architecture.

More recently, in late 1995, Wunsch and Prokhorov have reported the first successful implementations of a level 5 adaptive critic system which, as expected, performed better than the level 3 system. Prokhorov has also done some work with Feldkamp, Puskorius and others at Ford. See [3] for an update on these kinds of engineering applications, and for an updated discussion of the plausibility of backpropagation in the brain.

FROM NEUROENGINEERING TO THE BRAIN: NEXT STEPS

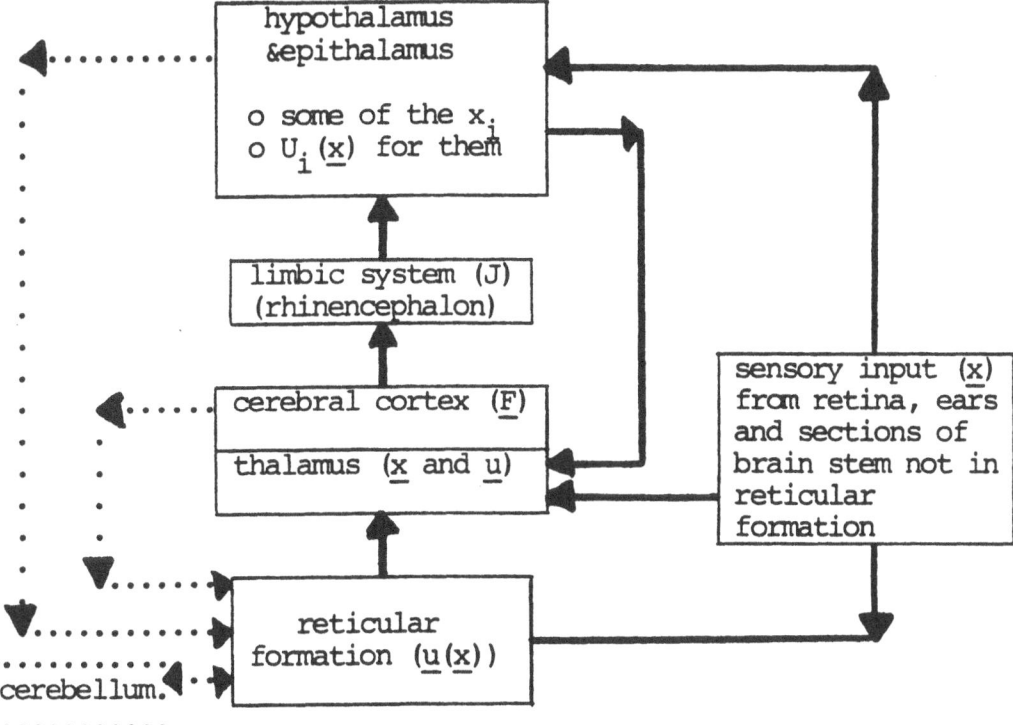

Figure 15. A Strawman Theory of Learning in the Brain
(as a Level 5 System[29])

In 1987 I published something I was a little embarrassed by [29] -- a strawman theory or model of learning in the brain. I was embarrassed because I knew that the real brain is a lot more complicated. But over time, after I have seen what other people have published in computational neuroscience, and after I've seen what the problems are in that literature, I'm not quite as embarrassed as I was in the past.

The basic idea is shown in Figure 15. In this picture, the hypothalamus and the epithalamus provide the raw utility function U. The limbic system calculates the J function discussed at length above. In other words, the limbic system acts as the emotional system of the brain. It amazes me how many modelers think that the limbic system is just another memory store; it's phenomenal! And of course Karl Pribram will tell them better[30]. There is work due to Olds and Papez going back for decades showing the importance of the limbic system in generating secondary reinforcement signals. And then we have some system identification going on in the cerebral-thalamic system, as discussed above. I know that the cerebral cortex has other functions as well -- i.e. that learning in the neocortex is based on the sum of several sources of feedback, not just prediction errors -- but system identification appears to be the primary function. And then, down at the bottom of the figure, we have some Action or motor circuits.

Even in 1987, however, I recognized that the cerebellum does not entirely fit this simple picture. This was a nice first cut model, but it took a long time before I began to understand the role of the cerebellum here [8,31].

Earlier in this talk, I mentioned how simple ANNs can approximate any function. I mentioned that there are beautiful theorems, particularly by a guy in Yale[9], saying that you can approximate smooth functions very efficiently with a simple feedforward net. But what if it's not a smooth function? There is a guy named Sontag at Rutgers[32] who has studied the problem of tracking control. He has asked what kind of networks do you need to solve a tracking problem, where the response pattern you need is not always a smooth function. Sontag found out that a 4 layer feedforward net with just 2 hidden layers can do well enough in simple tracking control. One hidden layer is not enough, but with two hidden layers you can do OK on tracking control. But then it turns out for really tough problems, you need something I call a simultaneous recurrent net (SRN[2,33,34]) That is something that has been implemented really seriously just in the last 6 months. And it has not even been published yet. But there has been some theoretical work on it before.

A key feature of these SRNs is that they are very expensive in a certain sense. They take a long time to settle down. You can't just plug in the inputs, and then read out the outputs a millisecond later. You've got to plug in the inputs, and then let the thing settle down, and that takes a little bit of time. But when you do fast motor control you want maximum speed; you want 100 Hz or 200 Hz. What can you do?

For tracking control -- or for lower-level control in general -- a two-hidden-layer feedforward net is good enough. It turns out that the cerebellum, this relatively lower level part of the brain, is basically a feedforward network with two hidden layers[34]. You've got a granule layer, a Purkinje cell layer, and then your output layer is actually the cerebellar nucleus and the vestibular nucleus together. Those two nuclei together really form the output layer. You need a lot of neurons to make this kind of feedforward net work, but there really are a lot of neurons in the granule layer. This leads up to the picture in Figure 16, discussed in [31].

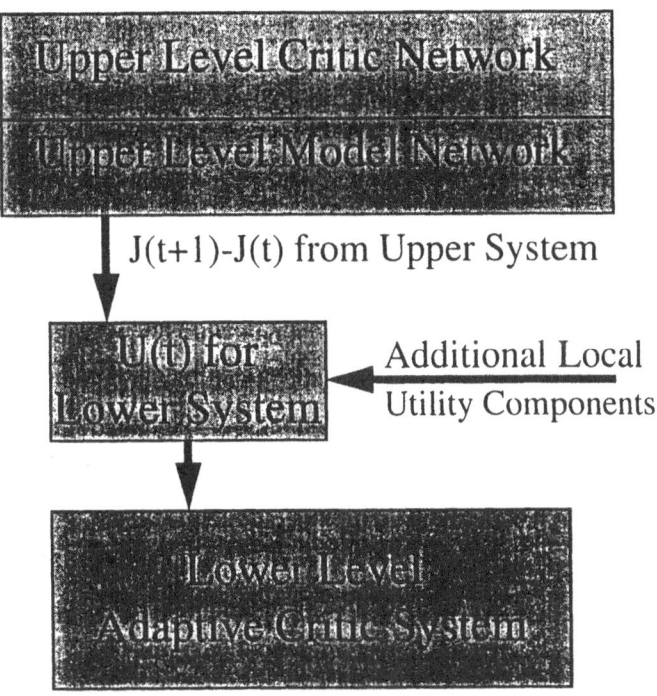

Figure 16. The Brain As Two Entire Adaptive Critic Systems, as "Two Brains in One"

The basic idea here is that we have not one brain but two brains. We have two entire adaptive critic control systems, an upper system and a lower system. The upper system is like the system shown in Figure 15, with the limbic system acting as a Critic and the neocortex as a Model. The upper system, made up of SRN components, requires a long computational cycle but has the ability to solve very difficult problems. The lower level system uses feedforward networks, primarily, to achieve fast operation at the cost of less sophisticated planning abilities. The lower-level system clearly includes the cerebellum as an Action network, and the inferior olive as the Critic network to train that Action network. The values or forces calculated in the upper system, the delta J from upstairs, becomes the U for the lower system. The upstairs J function becomes the downstairs U function. This is one way to chain a pair of controllers in a master-slave kind of arrangement.
(More precisely, the U(t) which the lower system tries to maximize may be <u>defined</u> as something like the upstairs J(t+1) - J(t) plus a local downstairs U(t) component calculated at a higher sampling rate; the actual feedback may involve derivatives of all these quantities. The local U(t) might include terms like finger pain and so on.)

By the way, when I mention all this to social psychologists, they go on to say, "Aha! Perhaps the U function in your upper brain depends in turn on social feedback, so that your individual brain is like a cerebellum of the collective intelligence." There may be some degree of truth in that.

In any event, the basic idea here is that we have two entirely different systems coupled together in one brain. There is real reason to believe the inferior olive acts as a critic here, but there are also some crucial experiments that have never been done. The next most critical experiment, in my view, is to demonstrate that the inferior olive is capable of learning. (After that comes an understanding of the learning equations, in effect, starting from experiments suggested in [31].) To do this, you could culture some olive cells <u>together with</u> some Purkinje cells and maybe some spinal cells, and then start studying plasticity in those olive cells. Nobody is doing those experiments, and it's kind of scary why they aren't. But this is an example where engineering can motivate exploring where people haven't looked yet. So maybe we need more collaboration to stimulate this kind of stuff.

By the way, there is another aspect of the lower control system, discussed in chapter 13 of [8], which could also be tested further. My theory here is that the cerebellum builds up its reconstruction of reality, **R**, using an approach rather different from what has been used before in engineering. Instead of using system identification, it simply uses time-lagged recurrence in the Purkinje layer. This requires the existence of what I call an Error Critic embedded in that layer, perhaps involving the basket cells. This approach would not allow such rapid learning as the conventional approach allows, in theory, but it does allow very fast operation -- the critical issue in the lower system. Again, there is a lot of empirical work needed to test out these kinds of ideas. Even as I write this, however, Ford is testing out the possibility of using this arrangement in automotive control.

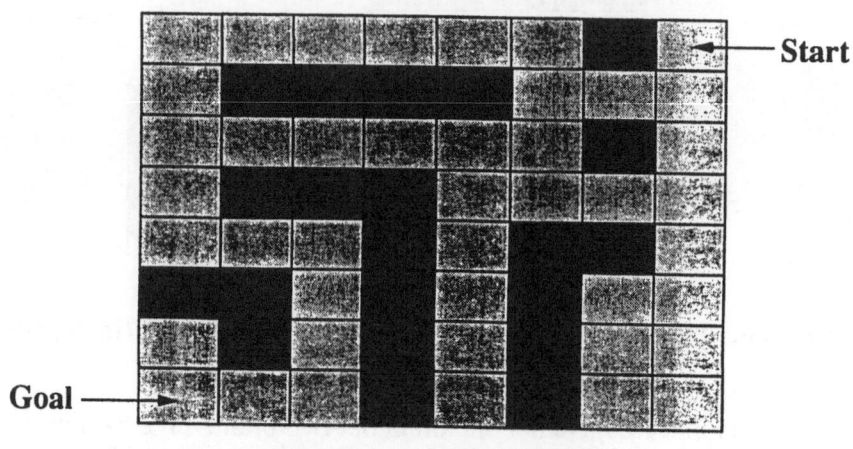

Input entire grid
Output next move

Figure 17. Example of a Task Requiring an SRN [34]

Figure 17 illustrates an example of a mapping problem that you can't solve with a feedforward net[33,34]. Actually, this problem is similar to some other problems that Minsky discussed back in 1969 that the world forgot.

A group of Frenchmen posed this particular problem. It's a problem in robot navigation: you want to go from the start to the goal by the fastest possible route. Now when most computer scientists study this problem, they say, "Here is a maze, and I'll train my neural net to learn the maze." But that's not the way human beings do it. If we see a new room with new obstacles in it, we don't bounce through the room a hundred times in order to learn the room. If you do it like that, you are going to smash a lot of machinery. In real-world robot navigation, you want to build a robot that looks at the room and sees the pathway through. So the task is this: the robot has to learn to see the room and respond with a path for that room.

When these Frenchmen set up this problem, they set up a whole bunch of mazes at random. They asked the system to learn the mapping for which the input vector is the pixels (i.e., they see a picture of the room from overhead), and the output is the desired behavior. It turns out a feedforward net can't learn that. It cannot even represent the mapping accurately when you hard-wire the weights. But in our recent work, we have shown that an SRN can represent this mapping exactly [34], for an arbitrarily large maze.

I might add that the structure of the SRN appears to include the kind of nets that Grossberg and Hopfield have talked about. But the kind of learning that they have used will not produce this kind of mapping. The kind of Hebbian learning they have used will only produce associative memory. And to solve robot navigation, that's not a memory task, that's an analytical kind of task. And the only way I know to adapt a network to be able to solve that kind of tough mapping problem is with some form of backpropagation.

In [8], I suggested that we might use simultaneous backpropagation [2,33], a method which I first proposed and implemented back in 1981, closely related to some of the later work of Pineda and Almeida. However, as I re-examine some of the observations of Walter Freeman regarding "searching behavior" in the olfactory system, and as I consider some very complex learning problems, I begin to worry that the highest levels of intelligence may instead require the use of an Error Critic design [8, ch.13]. (In essence, the Error Critic design can provide a real-time approximation to the methods proposed by Rumelhart, Hinton and Williams for adapting associative-memory recurrent networks[36].). Complex as they are, these kinds of network should open the door to solving very complex planning and scheduling problems in industry, problems that could not be solved with simpler ANNs.

THREE BRAINS IN ONE?

Finally, over the past four years -- thanks in part to the Radford conferences -- I have begun to realize the need to go beyond even the two-brain model discussed above.

The biggest hole in the two-brain model is the basal ganglia. At present, I am beginning to think that the basal ganglia really form a third entire brain, or, more precisely, a third entire adaptive critic control system. The evidence here comes from many sources, both from neuroscience and from engineering, discussed further in [3].

On the biological side, for example, James Houk claims that the substantia nigra pars compacta, the dopamine system in the basal ganglia, serves as an adaptive critic.[37]. He has shown how the learning mechanisms in that organ seem to follow the mathematical rules that we have derived for how to adapt a Critic network. In fact, he claims that this is the best empirical evidence we have for a Critic network anywhere in the brain.

On the engineering side, there are many limitations in the two-brain model, which a third major system can remedy. For example, there is the issue of discrete versus continuous variables. In neuroengineering, we have mainly been dealing with continuous variables. At the highest level of human learning, we are dealing with emotions, which are also continuous variables. At the lowest level, the important variables are again mainly continuous. But in-between the highest level and the lowest level, there is a kind of AI 1/0 world that we also have to live in. Some of Peter Killeen's examples today involve these discrete choice kind of things, these 1/0 kind of things. Furthermore, the idea of sending reinforcement back to the representation as opposed to the output fits beautifully with some of the possibilities on the engineering side. This may be an area where three-way collaborations between psychology, neuroscience and engineering will be critical.

Another key role for the basal ganglia would involve the problem of temporal chunking, which relates to the effective-foresight problem I mentioned in [11]. If we postulate that the basal ganglia basically evolve a "menu" of action schemata, we may use them to address both the discrete choice problem and the chunking problem, together. Each schema would have to include its own "membership function" or "confidence measure," perhaps adapted in a way similar to what is used with mixture-of-experts systems. It would also have to involve its own local critic, able to translate directly from goals at the end of a task to values at the start. It turns out that the development of such a local critic can be equivalent to the development of a local forecasting system, which tries to predict the result of the task directly from its initial state, with appropriate constraints on the structure.

An interesting problem in designing such a system is the choice between digital all-or-nothing choice versus fuzzy choice. Even when there are are strict, mutually exclusive choices to be made between action schemata, a fuzzy system may work better, for a variety of reasons, involving flexibility, learnability, and the ability to learn to do two things at once and so on. Nevertheless, the choice of which schema to "turn on", to what degree, is related to the problem of discrete choice. It suggests the need for explicit decision-making cells, perhaps in layer V of neocortex, with a certain kind of stochastic aspect, analogous to the "temperature" mechanism discussed by Dan Levine and others. To create an efficient, functional neural net embodying this idea, one can use either the Stochastic Encoder/Decoder/Predictor (SEDP) architecture [8,ch.13] or some recent extensions of that architecture. In the SEDP architecture, there is an interesting duality between "predicted R_i" and "estimated R_i" variables, which might possibly be reflected in the duality between calculations in the apical dendrites of giant pyramid cells and calculations in the cell body. This architecture may also make it natural to train layer V as a "dual-use" structure, making decisions and reconstructing reality at the same time, and learning based on the sum of feedbacks from both activities. Alternatively, one might ascribe the decision-making to the striatum itself, albeit still with feedback that affects learning in layer V.

Some of these details may be elaborated on further in [3] and in a patent disclosure anticipated before the publication of this book. A very interesting test problem for some of these designs would be the Chinese/Japanese game of Wei Chi or Go, which combines elements of large-scale continuous decision-making with local discrete choice and exploration, plus a need for some of the special tricks in [34].

REFERENCES

1. P.Werbos, *The Roots of Backpropagation: From Ordered Derivatives to Neural Networks and Political Forecasting*, Wiley, 1994.
2. P. Werbos, The brain as a neurocontroller: New hypotheses and new experimental possibilities. In K.Pribram, ed., *Origins: Brain and Self-Organization*, Erlbaum, 1994.
3. P. Werbos, Intelligent control: Recent progress towards more brain-like designs, *Proc. IEEE*, submitted and accepted for special issue, E.Gelenbe ed., to appear in 1996.
4. P. Werbos, Optimization: A foundation for understanding consciousness. In D.Levine & W. Elsberry (eds) *Optimality in Biological and Artificial Networks?*, Erlbaum, 1996.
5. Go to website www.nsf.gov; then click on Engineering, then ECS, then neural nets.
6. P. Churchland, *The Engine of Reason, the Seat of the Soul*, MIT Press, 1995.
7. E.Fiesler and R. Beale, eds, *Handbook of Neural Computation*, Oxford U. Press and IOP, 1996
8. D.White and D.Sofge, eds., *Handbook of Intelligent Control*, Van Nostrand, 1992
9. A.R.Barron, Universal approximation bounds for superpositions of a sigmoidal function *IEEE Trans. Info. Theory* **39**(3) 930-945, 1993
10. R. Cytowic and P.Werbos, forthcoming book
11. W.T.Miller, R.Sutton & P.Werbos (eds), *Neural Networks for Control*, MIT Press, 1990, now in paperback.
12. P.Werbos, Backpropagation and neurocontrol: a review and prospectus. In *Proceedings of the International Joint Conference on Neural Networks*, IEEE, 1989
13. P.Werbos, Backpropagation, in M.Arbib (ed) *Handbook of Brain Theory and Neural Networks*, MIT Press, 1995.
14. R.A.Mahoney, A stochastic control model of human target-directed movements. In J.Vossoughi, ed., *Biomedical Engineering: Recent Developments* (Proc. 13th Southern Biomedical Eng. Conf.), Engineering Research Center, University of the District of Columbia, 1994. See also his *Human Target-Directed Posture Control*, Ph.D. thesis, Engineering Dept., University of Cambridge, UK, 1993.

15. P.Werbos, The cytoskeleton: why it may be crucial to human learning and neurocontrol, *Nanobiology*, Vol. 1, No. 1, 1992

16. A.Bryson & Y.C.Ho, *Applied Optimal Control*, Ginn, 1969.

17. F.Mandl, *Introduction to Quantum Field Theory*, Wiley, 1959.

18. V.G.Makhankov, Yu.P.Rybakov and V.I.Sanyuk, *The Skyrme Model: Fundamentals, Methods, Applications*, Springer-Verlag (800-777-4643), 1993

19. A.Barto, R.Sutton and C.Anderson, Neuronlike adaptive elements that can solve difficult learning control problems, *IEEE Trans. SMC*, Vol. 13, No.5, 1983, p.834-846.

20. B.Widrow, N.Gupta & S.Maitra, Punish/reward: learning with a Critic in adaptive threshold systems, *IEEE Trans. SMC*, 1973, Vol. 5, p.455-465

21. G.J.Tesauro, Practical issues in temporal difference learning. *Machine Learning*, 1992, 8:p.257-277..

22. J.S.Baras and N.S.Patel, Information state for robust control of set-valued discrete time systems, *Proc. 34th Conf. Decision and Control (CDC)*, IEEE, 1995. p.2302.

23. P.Werbos, Advanced forecasting for global crisis warning and models of intelligence, *General Systems Yearbook*, 1977 issue. Strictly speaking, this design also appeared in more detail in my 1972 Ph.D. thesis proposal to Harvard U. -- which was rejected as being too complex -- and was presaged by the discussion in 1968 [25].

24 M.Nicolelis, C.Lin, D.Woodward & J.Chapin, Induction of immediate spatiotemporal changes in thalamic networks by peripheral block of ascending cutaneous information, *Nature*, Vol.361, 11 Feb. 1993, p.533-536.

25 P.Werbos, The elements of intelligence. Cybernetica (Namur), No.3, 1968

26 D.Yankelovich and W.Bartlett, *Ego and Instinct: The Psychoanalytic View of Human Nature* - Revised, Vintage, 1971

27 K.Pribram and M.Gill, *Freud's Project Reassessed*, Basic Books, 1976.

28 D.Prokhorov, R.Santiago & D.Wunsch, Adaptive critic designs: a case study for neurocontrol, *Neural Networks*, Vol.8, No.9, 1995.

29 P.Werbos, Building and understanding adaptive systems: a statistical/numerical approach to factory automation and brain research, *IEEE Transactions on Systems, Man and Cybernetics*, Vol. 17, No. 1, 1987.

30. K.Pribram, *Brain and Perception: Holonomy and Structure in Figural Processing*. Erlbaum, 1991.

31 P.Werbos and A.Pellionisz, Neurocontrol and neurobiology: new developments and connections. In *Proceedings of the International Joint Conference on Neural Networks (Baltimore)*. IEEE: New York, 1992.

32. E.D.Sontag, Feedback stabilization using two-hidden-layer nets, *IEEE Trans. Neural Networks*, Vol. 3, No.6, 1992.

33. P.Werbos, Supervised learning: can it escape its local minimum, *WCNN93 Proceedings*, Erlbaum, 1993. Reprinted in V. Roychowdhury et al (eds), *Theoretical Advances in Neural Computation and Learning*, Kluwer, 1994

34. X.Pang and P.Werbos, New type of neural network learns to navigate any maze. *Proc. IEEE Conf. Systems, Man and Cybernetics (Beijing)*, IEEE, 1996.

35. W.Nauta & M.Feirtag, *Fundamental Neuro-anatomy*, W.H.Freeman, 1986

36. D.Rumelhart, G.Hinton and R.Williams, Learning internal representations by error propagation. In D.Rumelhart and J.McClelland, *Parallel Distributed Processing*. Vol.1, MIT Press, 1986.

37 J.Houk, J.Davis & D.Beiser (eds), *Models of Information Processing in the Basal Ganglia*, MIT Press, 1995

The Neural Systems Level: Process

12.
Morphogenesis and Mental Process

Jason W. Brown

Morphogenesis and mental process

JASON W. BROWN
New York University Medical Center, Department of Neurology

Abstract

Parcellation and heterochrony (neoteny) reflect the pattern and rate of a growth mechanism in morphogenesis. Structure (morphology) and function (behavior) are staged realizations of morphogenetic process. This process continues into adult cognition in the actualization of the mind/brain state. Parcellation obtains in the pruning of cells and connections in early growth, whereas inhibition obtains in a relatively stable morphology with constraints on context : item transforms in microgeny. Selective retardation in process (neoteny) leads to growth at earlier (juvenile) phases. This accounts for the specification of the language areas and elaboration at preliminary phases in mind — for example, dominance, introspection, and creativity.

. . . we create distinctions, then
Deem that our puny boundaries are things
Which we perceive, and not which we have
made.

> *Wordsworth*
> *Prelude II:221*

In the past, efforts were made to show that developmental abnormalities could be interpreted as an attenuation or deviance of normal maturation (e.g., Werner, 1948). These accounts were based for the most part on a temporal or diachronic perspective that gradually fell out of favor with the advent of cognitive science. This was partly a result of work that was critical of simple correlative accounts, but there was also a strong bias against this mode of explanation that resulted from the synchronic perspective typical of cognitive psychology. The specification of systems through growth was incompatible with the concept of modules for various functions conceived as genetically specified organs. In this view, pathology, whether developmental or acquired, represents the perturbation of a system (i.e., a program) that is prewired and no more than fine-tuned by the environment. The development of the brain is treated as an unimportant detail that mediates between the innately delivered program and the mature function. This way of thinking is clearly linked to artificial intelligence models of the mind, where the pattern and direction of growth in the brain are considered no less irrelevant to the final organization of cognition than the mode of construction of a computer to the software.

In recent years, however, there has been a revival of interest in the relationship between normal and pathological development. This field, termed *developmental psychopathology* (e.g., Cicchetti, 1984, 1993), has led to a rethinking of normal and aberrant development in terms of common underlying mechanisms. The emphasis on timing as a key to the understanding of deviant behavior and the life-span approach to growth and pathology are central to this account. These principles have also been an important feature of the microgenetic theory of adult pathology and its relation to evolutionary and developmental growth trends (Brown, 1988; Brown & Jaffe, 1975). This theory holds that an understanding of the nature of the symptom or error rather

Address correspondence and reprint requests to: Jason W. Brown, 66 East 79th Street, New York, NY 10021.

than the deficit is of critical importance if a common basis of growth and pathology is to be uncovered. But what precisely are the mechanisms that growth and pathology have in common? This paper is an attempt to specify some of the morphogenetic patterns that may account for behavior in the normal state and in cases of developmental or acquired pathology.

Morphogenesis

Ontogeny is a bridge from the genetic code to the phenotype, whereas phylogeny is the sum, or average, of the ontogenies of a given line. Ontogeny covers a period from conception to adult structure with no clear boundary except death at the end of a developmental sequence. At some relatively stable point in this sequence, an organ is judged to express a phenotype. The phenotype is usually assumed to be *directly* generated by the genome. For example, in brain organization, there is assumed to be a correlation between the genetic code and the specialized adaptations of the mature organism. Ontogeny is the process through which this correlation is achieved.

The extent to which morphology is specified in the genome is uncertain. Presumably, the genetic code contains a set of instructions for the processes—or events that lead to the processes—that will realize the code in structure. Such processes as are specified by the genetic code are still a long way from phenotypic structure. Neurobiologists recognize this problem and focus, therefore, on the epigenetic functional relations or algorithms that translate code to structure, not the manner in which genes produce nerve cells or circuits (Stent, 1981). However, our knowledge of relations underlying the morphogenetic process, the transition from the genetic code to the developed organism, is still very incomplete.

From the standpoint of mental development, the onset of sensation and learning at birth creates a natural testing ground for the study of genetic specificity. The newborn is prepared for many complex behaviors through a genetic endowment that is relatively uncontaminated by learning. Yet the innate capacities of the mind/brain of the newborn are difficult to specify and continue to be a source of lively debate (see Carey & Gelman, 1991). Speculations range from innate rules, ideas, and/or mental processes to the more conservative notion of constraints on action and perception.

The problem with studies of innateness that begin with the newborn is the focus on behavior as an interaction of a delivered morphology with experience. Circuits in the brain are taken to be the outcome of a prenatal process of growth when, in fact, growth continues on into late life. Similarly, brain function is assumed to be determined by the pattern of electrical activity that maturing circuits generate (i.e., the output of the structure) when the pattern of activity, that is behavior, is an expression of sustained growth. The cleavage between growth and function, or structure and process, results in a neglect of the formative history of the mind/brain, in evolution or in utero. The goal of development is a machine that can be instructed or that can realize a functional program. In this view, the fetal segment of ontogeny is less informative than the confrontation of the innate and the acquired in the earliest period of postnatal life.

In contrast, suppose that development is morphology (i.e., that morphology is an artificial slice through development with behavior its four-dimensional structure) (Brown, 1991; cf. Striedter & Northcutt, 1991). From this standpoint, the newborn is not a starting point to study the innate determinants of language and behavior but, rather, is a phase in a life-span process. Onto- (morpho)-genetic process leading to the mind of the newborn also lays down function after birth. The dichotomy between the innate and the acquired is orthogonal to the *nature* of mental activity, and this activity is independent of birth as a pivotal event.

With this perspective, an early phase in development deposits morphology, and a late stage deposits function. A common process elaborates both morphology and function, with function being the iteration of growth through the morphology. There have been prior speculations along these lines (e.g., the ideas of Loeb [1907] and Goldscheider [1906], that ". . . configurations experienced in perception might derive from excitation in the brain resembling the 'force lines' that determine form during embryogenesis" (cited in Pribram, 1991, p. 25). A relationship has been suggested between "ontogenetic sculpting" and mechanisms of learning (Thatcher, 1992b) and information representation in the brain (Malsburg & Singer, 1988). Tucker (1992; see also Brown, 1990) commented that physical growth in the brain is psychological growth. The problem is to specify the growth process and determine whether or not this process is related to processes underlying cognition in the adult.

There have been attempts to define the morphogenetic process with greater precision. Goodwin (1982) argued that developing organisms have "an extensive range of morphological potential, describable in terms of probabalistic fields which collapse . . . into specific morphologies reflecting the particular conditions, internal and external, which act upon them" (p. 52). According to Goodwin, the generative principles that account for the progression from whole to part recur so that a taxonomy of biological form can be achieved through a hierarchic ordering of the transformations. Katz (1983) proposed that "ontogenetic buffer mechanisms" mediate the transition to functional brain architecture. These mechanisms include exuberant growth with specificity through parcellation or pruning of connections, possibly by competitive interaction (Edelman, 1987), and heterochrony, a variation in the timing of developmental process.

It is a thesis of this paper that processes in the development of the brain recur as processes in cognition, that development and cognition — ontogeny and microgeny — are different ways of looking at the same process. Two lines of evidence for this thesis are explored: the relation of parcellation in development to specification in cognitive

processing, and heterochrony as a theory of developmental abnormality in relation to pathological symptoms in mature organisms.

Parcellation

Morphology

Parcellation is the pruning of exuberant connections in the growth of the brain as a way of achieving specificity in mature brain structure. According to parcellation theory, the connectivity of the brain is accomplished, at least in part, through a loss of connections. Indeed, most structures in the vertebrate brain have a larger number of neurons during development than in adulthood. The decrease in synapses is even more striking. In macaque neocortex there is a loss of over 2 trillion synapses by the 5th year of life (Rakic, 1989, 1992).

Ebbeson (1984) has written that most, if not all, systems go through phases of diffuse projections that later become more restricted, presumably by the degeneration of selected axonal branches or the loss of selected neurons. The finding of initial proliferation and later elimination in the progression from the general to the specific has been described mainly in the study of sensory systems, where exuberant growth in juveniles with loss of cortical connections (neurons and synapses) in adults is a characteristic feature. There is also evidence for parcellation in the growth of callosal fibers (Innocenti, 1984); connections are initially diffuse and abundant and then become specified through elimination. In studies of cerebral dominance, hemispheric asymmetries are related to callosal thickness and may reflect pruning rather than accentuated growth (Witelson, 1990). Even cytoarchitectonic specificity has been attributed to the gradual connectivity of initially homogeneous neocortex (Creutzfeldt, 1977). Innately driven process determines the connectivity but so does experience. Early visual deprivation in animals can prevent foveal specificity and lead to a more ancestral state of diffuse or ambient perception

(Ebbeson, 1984). Experience enhances the specification through constraints on emerging form (Brown, 1988).

At some point in morphogenesis, presumably after most anatomical connections are established, the parcellation effects that produced the structure of the brain give way to parcellation-like effects that characterize processing within this structure. The transition is from the *elimination* of connections in development to the *inhibition* of connections in maturity. The question is, Is specification in neuronal development only analogous to later processes of differentiation through inhibition, or are development and behavior manifestations of a common process?

Physiology

There are numerous examples of *physiological* "parcellation." Many years ago, Coghill (1964) proposed that partial patterns (e.g., local graded reflexes) individuate out of global patterns (e.g., mass reflexes) and that the earlier state can reappear in pathological conditions. Thus, spasticity with central nervous system injury represents a recurrence of the excitatory patterns of earlier stages.

Another example of physiological specification is the diffuse to focal gradient in dominance establishment (e.g., regional specification of the language zones) (Brown, 1977; Semmes, 1968). Evoked potentials show a gradual restriction to the cortical site of the stimulated modality with recurrence of the global pattern in cases of brain damage. The progressive specification could be attributed to synaptic pruning, but the recurrence of the generalized pattern with pathology implies disinhibition. Wall (1988) suggested that some forms of recovery after brain injury may reflect disinhibition of latent synapses and reenlargement of receptive fields.

Such instances of progressive specificity in postnatal development can occur through either the active inhibition of synapses or their elimination, or both. There is a transition from a predominantly morphological pattern to one that is predominantly physio-

logical. The specificity achieved through elimination in the development of the brain continues like the selectivity achieved through inhibition in the further development of functional systems. Lateral or surround inhibition (and figural contrast) accomplishes the specificity in process that parcellation accomplishes in growth. The implication (see later) is that the putative modular organization of the mind/brain *results* from a process in which constraints are applied at successive phases in the differentiation of form (Karmiloff-Smith, 1991), whether in morphology or behavior.

Growth and process

Our concepts of morphology and process are shaped by implicit beliefs about the relation of structure to function (i.e., that function is the output of structure). However, in a temporal context, organic structure is neither the source nor residue of process but the momentary appearance that process takes on. In growth, process generates morphology. In a developed morphology, process generates function. The question is, Is there a common process that elaborates structure and function depending on the stage in growth?

Put differently, Is the process of nerve growth and connectivity the basis of the process that the connectivity instantiates in function? If the meaning of growth is restricted, say, to alterations in synaptic protein, and if the concept of mental process is confined to the output of a population of cells, there will be a gulf between the two activities, even if synaptic growth is a determinant of the activity pattern. However, if the concept of growth is expanded from a cellular event to a population dynamic, and if mental processes are conceived as the configural properties of this population, then it becomes possible to map one process to the other.

Cognition

Parallels in the development of structure and process extend to the activity underlying cognition in the adult. The emergence of elements through parcellation and the development of specificity through inhibition correspond with the analysis of cognitive wholes through context : item transformations.

The description of the cognitive process as a treelike series of nested context : item shifts is a core feature of microgenetic theory. Selection of items occurs through the actualization of elements within manifolds with each tier serving as a ground for another specification. The specification occurs through constraints on emerging content. The constraints are both intrinsic, on the unfolding mental representation, and extrinsic, in the shaping by sensation of perceptual representations. The account of the process of actualization as one that is driven by intrinsic and extrinsic constraints corresponds with the idea of physiological inhibition guiding the extraction of featural detail. Essentially, elements are contrasts. In this theory, details are neither modeled nor assembled but, rather, exposed, sculpted, or realized through a suppression of alternative routes. A few examples from studies of language, action, and perception in adults and children follow.

There is a line of thinking in which features are not the elements of object construction but "emergent characteristics of form" (Pribram, 1991). In children, phonemes are not concatenated into words but develop as emergents (Best, 1991). Gestalt recognition precedes feature analysis in studies of object (face) identification. Word categories are established early in lexical acquisition with derivation of specific words out of categorial representations. The acquisition of word-meaning in children shifts from representations based on characteristic features to those based on defining features (Keil, 1991), a process similar to the analysis of wholes or the articulation of concepts from the primitive to the scientific.

In aphasia, disorders of word-meaning show a zeroing-in on lexical targets in the process of word-finding. Error patterns confirm a specification from wide to narrow semantic relatedness as the word is finally selected. The emergence of the skeletal frame of the word evokes a specification of phonemes out of phonological gestalts as word sounds fill in the "slots" of abstract lexical representations (Brown, 1988).

In action, infants show progressive differentiation beginning with proximal and axial movements and continuing to the specification of distal grasp and manipulation (Trevarthen, 1984). In frame : content theory, a postural to analytic shift underlies the evolution of manual asymmetry (MacNeilage, Studdert-Kennedy, & Lindblom, 1987). Digital movement is elicited out of background axial and proximal motor systems (Brown, 1988). The proximal setting provides the context in which the distal movement develops (Goldberg, 1985).

"Blindsight" phenomena (Bard, 1905; Bender & Krieger, 1951; Brown, 1977) reveal substantial perceptual ability in extrastriate regions and suggest that patterns are analyzed into their constituent features. The many instances of implicit perception (e.g., Reber, 1992) reflect holistic or contextual properties of stimuli *subsequently* analyzed into conscious or explicit perception. Holistic and analytic are in a relation of precedence, a shift from context to item, or whole to part, not simply the product of asymmetric or parallel brain systems (Brown, 1988).

Vision in the newborn is controlled by subcortex with cortically mediated behavior coming into play during the first few postnatal months. This shift coincides with increasing selectivity and awareness of spatial detail, discrimination, and orientation. In some respects (e.g., fixation), the behavior of a 1-month-old is similar to that of an adult with striate lesions (Van Sluyters et al., 1987). The process corresponding to this shift has been described as an emergence of adult patterns of connectivity through the refinement of an initially diffuse set of connections.

Summary

The generative principles of developmental growth include a process of proliferation and elimination — parcellation — in the establishment of connectivity in the brain. This principle is akin to the evolutionary process of adaptation in which unfit exemplars are pruned by the environment. A process similar to that of parcellation is described in physiological studies of neuronal populations. This process is a shift from a diffuse to a focal organization in which inhibition plays the role in function that elimination played in growth. This process recurs in context : item shifts in cognition (Brown, 1990).

According to microgenetic theory, the multiple levels (phases) of mind are generated through a progressive context : item shift that retraces patterns in phyloontogeny; that is, one operation lays down multiple levels (phases) rather than multiple operations acting on separate contents. Content is specified through a phased transformation at successive moments in the mind/brain state, which itself is cyclic and recurrent (Brown, 1991). There are points of contact with evolutionary models (e.g., Goodwin's [1984] description of limb formation in which elements are specified along a proximodistal gradient as solutions in a single periodic generative process). A context : item shift is an actualization process, not a sequence of self-replications. Still, there is a resemblance between this process and fractal geometry (MacLean, 1991; Robertson, 1991; Vandervert, 1988, 1990).

Heterochrony

Heterochrony is a temporal disparity in the development of organ systems. The timing of development is uneven. The change in rate can affect a process that is focal and delimited or one that is widespread and pervasive with the result that organs develop at different rates. Both retardation and acceleration occur. The result is that adult and juvenile features are loosely bound in covariant sets (Gould, 1982). In addition to the rate of process, the timing of onset and offset is important.

Neoteny is a form of heterochrony in which a retardation of development prolongs the duration of a juvenile stage. Neoteny can be of adaptive value, when juvenile features survive in the adult as an escape from the more rigid specializations of adult structure. Gould (1977) has argued that neoteny is an important mechanism in human evolution. Some human features reminiscent of a juvenile primate include a more upright posture and flat face, high brain to body weight, absence of brow ridges, thin skull bones, central position of the foramen magnum, and reduced body hair. The juvenilization of morphology accompanies a retention of physiological features associated with juvenile ancestral stages. For example, increasing brain size over the primates is associated with prolongation of sexual immaturity, from 2 years in the lemur to 7 years in the great apes to about 14 years in humans (Bonner & Horn, 1982).

Heterochrony can lead to successful adaptations, but it can also lead to developmental abnormalities. Serres (1860, in Gould, 1977) argued that when different parts of the fetus develop at different rates, "monstrosities" (monstres par défaut ou excès) can arise if certain parts lag behind and retain, at birth, the character of a lower animal. In his account, Serres combined a recapitulation argument with differences in the timing of development of different organs. The recapitulation is not for "lower animals" but embryonic stages, while the maladaptations that result from altered timing may be subtle and qualitative, not just omission and excess.

One implication of the theory of heterochrony is that a rate change early in development should have more generalized effects than one late in development. Another implication is that an altered timing of brief duration that leads to anomaly rather than truncation should not exclude a normal *subsequent* development, with the resultant deformation carried into future stages as a signature of the altered phase. For example, delayed closure of the cranial sutures is a neotenous feature that permits the rapid expansion of a similarly neotenous brain. Closure occurs but at a later phase in development. In this example, the delay is associated with other features that identify the phase in ontogeny where the retardation began. The focal change and the context of the change identify the phase. The question is whether or not an error from a disparity in the timing of developmental process is comparable to a symptom of brain damage, that is, whether or not early (developmental) and late (cognitive) errors can be explained by heterochronic change.

Efforts to establish a correspondence between the maturational sequence and the pattern of errors in adult decomposition have floundered on the misconception that disorders of adult cognition should recapture stages in acquisition, the so-called regression hypothesis. This approach, at least in the work of Hughlings Jackson or Roman Jakobsen, has been disconfirmed by the finding of weak correlations between, say, grammatical errors in aphasics and stages in the acquisition of grammar (Gleason, 1978). It is true that pathology does not unpeel the acquisitional sequence, but this obscures the deeper truth that mental process reiterates developmental *trends*, not facts; that is, the commonality of development and cognition—morphogenesis and microgenesis—is in process, not content.

Heterochronic Principles of Error Analysis

Lesions and errors

The principle effect of a brain lesion is to retard process, not to destroy function. The effect is not an ablation of elements in a circuit board but, rather, a change in a configuration that is a type of traveling wave. There is evidence for such "waves" in studies of brain development (Thatcher, 1992a). The lesion is comparable to an obstruction in a river (Figure 1). The obstruction impedes and delays flow but does not interrupt it. The retarded segment persists as a local disparity or dyssynchrony in relation to the cross-section of the stream (process) in which it is embedded. Unlike a river that leaves the obstruction behind, mental process is recurrent, so the obstruction is encountered anew in each traversal.

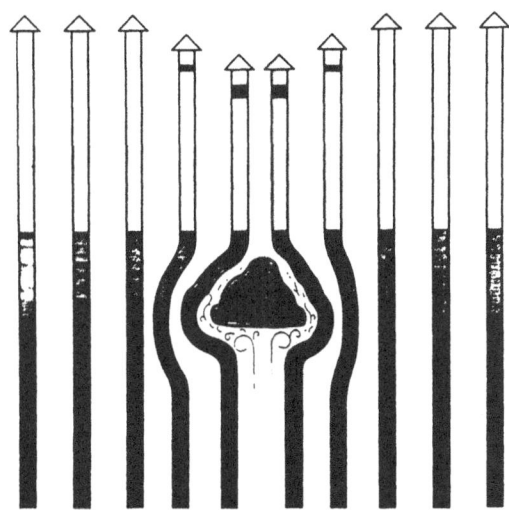

Figure 1. The lesion delays flow in a segment of process. The retarded (neotenous) segment is the symptom; the unobstructed flow parallel to the segment is the normal context in which the symptom occurs (see discussion in Brown, 1988, pp. 12–14).

The nature of the error is determined by the stream (component) involved (e.g., action, perception, language) and the extent to which that segment of the stream has developed prior to the disruption. The error depends on the dynamic of flow and differs slightly for each traversal. The segment corresponding to the error is only approximated by the error type. The error identifies the damaged segment. Conversely, error pattern, together with the location of damage, provides a basis for reconstructing the normal sequence.

Errors reflect perturbations as well as "obstructions." In a river, eddy currents develop around an obstruction. A lag in mental process may produce local eddies around the damage point. Such eddies and whirlpools have been described mathematically in Hopfield simulations (Hoffman, 1987) that link errors to normal processes. There might be analogous conditions in development, where a local delay impacts on regional systems not directly involved by the defect. Still, a perturbation effect should be comparable to a lesion effect (i.e., retardation [or acceleration] of process).

A disparity in the timing of a process that advances like a wave front should lead to a

301

local delay (or acceleration) that is "out of sync" with concurrent streams of process. The local delay is the error, and the concurrent streams establish the normal context within which the error occurs. An (adult) error is abnormal because it is out of context with ongoing behavior. This is the meaning of abnormality. If a person says *chair* when he or she should say *table*, the error involves word choice but also the fit between the word and the context in which it is used. Hallucination while dreaming is normal but pathological while awake. The dislocation between a focal deviance and a normal surround is the basis for the pathological symptom. A few examples of error types illustrate this concept (see Brown, 1988, for a discussion of the full range of neuropsychological symptoms from this point of view).

Aphasic errors are preliminary normal stages. For example, a substitution such as *chair* for *table* reflects a phase of equivalence for either item. The error reflects the semantic category prior to the individuation of the correct item. The deviation reflects an attenuation (juvenilization) of the process of lexical realization. In this example, parcellation and neoteny combine in the symptomatology. Incomplete fractionation is a result of a local neoteny in process. The delay produces an incomplete specification with the naming error revealing the context prior to the intended item. Even a correct name (*table*) can be an error, when a holophrastic noun does not achieve the referential or denotative specificity of a fully individuated word. The name is used to label objects that are not usually incorporated within its semantic field. There are similarities with early language learning, where nouns are used for objects in a class (e.g., *dog* for many different animals).

In amnestic cases, the inability to revive an event reflects the retarded activation of that event in the stream of ongoing mentation. The retardation leads to omission or substitution errors according to the depth of the delay. An amnestic gap points to the survival of a preliminary (unrealized) segment into mature (end-stage) cognition.

That the event is still active we know from intrusions (delayed recurrence), implicit memory, cuing and contextual facilitation, affective correlates, personality growth, and the potential for recovery.

Patients with perceptual disorders can have errors of object meaning (e.g., the [perceptual] misidentification of a knife as a fork, but the object [knife] is seen [or drawn] properly). The error in such cases reveals the underlying object concept, while subsequent analysis of form is unaffected. In other cases, form is involved (e.g., a knife perceived as a stick), with preserved object meaning. Object detail does not individuate, and objects are perceived on the basis of size, shape relations, and so forth. In both instances, an error of object concepts or object form, a background phase is carried into end-stage cognition. Put differently, a focal "juvenilization" of process persists as a symptom of the retarded segment.

In sum, pathological symptoms are focal attenuations in action, perception, or their derivations into language, at various depths in the mental state. The focal delay is a brief neoteny of a covert phase that survives in behavior. The preservation (normal rate) of process parallel to the focus establishes the normalcy of behavior apart from the altered segment. Intact processing prior to the delay (i.e., the depth of the lesion) establishes the error type within the component. Because the process laying down the mental state is reiterated, and the direction of processing is obligatory, the involved segment is traversed in every iteration.

Errors: Developmental and acquired

The relation between developmental deficits in children and acquired deficits in adults is complex. As mentioned, in adults the deviance is in relation to the context. The delay occurs at a segment in a mature system. The maturity influences the error because the error samples the context. In contrast, errors in young children reflect a still-forming context. The shift from context to item occurs—context is critical in learning — but because the context is limited the distribution of errors is reduced. This is why errors are more predictable in children than in adults. Because the context is impoverished, the deviance has to be sought for in the relation to what is age appropriate in other children or to better performance in another (intact) domain of function. An error will respect the structure of a system, or a primary component of a system, at the point of the delay. In young children, the juvenilization tends to affect a system or component as a whole.

Consider, for example, the developmental language disorders. In such cases, the disturbance is usually generalized at first and resolves into a disorder of production or comprehension. This can be interpreted as delayed maturation of language generally (i.e., the linguistic derivation of action and perception), with a residual deficit in one of these domains (action or perception), whichever is most delayed (Sahlén, 1991). Considerable evidence indicates that motor or perceptual deficits play an important role in the etiology of developmental language disorders, with little support for a disturbance in a language acquisition device (Bishop, 1992). A disparity in the rate of development could affect either (or both) action and perception or different phases or epochs in the development of these components. This could give the heterogeneity that has been associated inter alia with age and severity (i.e., the rate of language development or the timing of the onset of the delay). In any event, symptoms still reflect a general or local alteration in the *rate* of development.

Typically, errors in developmental language disorders are abnormal only by virtue of being inappropriate for a child of that age. Simplifications are the rule. Performance tends to be characterized in terms of norms (e.g., a 3-year-old at the stage of an 18-month-old). Errors that violate rules of normal development, other than timing, are rare and controversial. Contextual cues are less effective in children than in adults. These observations suggest that the difference between the developmental and the ac-

quired reflects neotenous change in the initial formation of a system as opposed to a segmental delay in the reinstantiation of a system already formed.

For many, the acquisition and disturbance of syntax is a critical issue, even if perceptual deficits explain most syntactic errors (Bishop, 1992). Syntax involves the extraction of relations between levels in mental structure. This raises the question of whether or not context : item shifts at successive moments in the unfolding of the mental state — as transitions across nested or embedded units — are process-equivalents of the core rules of syntax.

Double dissociation

A process can be retarded at one segment without an effect on a subsequent segment. For example, an alteration of timing in development gives syndactyly or webbing of the digits but spares the ensuing stage of nailbed formation. The defect labels a segment of process but the process continues to develop. Naturally, a severe disruption can abort development (or mentation). But with focal alterations, ensuing phases may be normal though postponed.

The specificity to a segment in process, in development or cognition, relates to the concept of "double dissociation." This concept is important in (adult) neuropsychological study because it seems to show that functions are interactive and modular. An example would be a separate impairment of phonetic and lexical reading. If a lesion separately disrupts phonetic and lexical reading, these performances are presumably independent (i.e., one operation is not contingent on the other). Obviously, a double dissociation pertains to functions that are cognitively "close." Limb paralysis and reading dissociate, but this is not of interest. The dissociation suggests interactive systems and seems to refute the idea of serial processing for the involved symptoms (i.e., if A goes to B, damage to A should prevent the occurrence of B). However, the concept of a symptom as a local delay is compatible with a dissociation of serial elements.

303

Thus, in the example of lexical and phonetic reading, if phonetic reading derives or develops from lexical reading, disruption of the prior segment would give errors through local retardation. The patient reads the word *horse* as, say, *zebra*. Lexical realization is attenuated; the selected item falls within the word category. The symptom is a persistent "juvenile" (preliminary) feature. The successive phase of phonetic encoding would then occur on a deviant lexical form. In this example, the disruption (lag) displays a segment of process as an incomplete transform without affecting a subsequent phase. The subsequent phase is spared because the deviant content on which it develops is premature, not "abnormal."

In children with developmental language disorders, the dissociation is less emphatic; for example, lexical-semantic disorders tend to be accompanied by disorders of phonology though isolated phonological disorders occur. This is due to the lack of sufficient context for errors of derailment to provide a substrate for ensuing transformations, and the graded nature of development, so that preliminary phases must be accomplished before subsequent stages can begin. A prolonged delay puts ensuing stages at risk. Indeed, the linkage between deficits in children is greater evidence for their continuity than is the specificity of deficits in adults for their autonomy.

Heterochrony and creativity

Heterochrony has been discussed in relation to pathology but is an important feature of normal cognition. In microgenetic theory, a reminiscence (memory image) is an attenuated perception; inner speech is a preliminary utterance (Brown, 1988). Metaphor, imagery, and creative thinking are elaborations at submerged phases in mentation. A preliminary interior phase of perception or language survives into consciousness. More precisely, self-awareness and mental content (i.e., the "space" of introspection) owes to a juvenilization in microgenesis with anticipatory stages coming to the fore as final

forms. The brief delay permits elaboration of segments that are normally analyzed into behavior.

Neoteny is an important mechanism in the evolution of the human mind/brain. The persistence of earlier features accounts not only for the dramatic expansion of the brain but also for the emergence of regions unique to humans (i.e., the language areas) within the "integration" neocortex (Figure 2), a "growth plane" preliminary to the "sensorimotor" cortex in forebrain evolution (Sanides, 1975). The language areas arise as preliminary phases to mediate an interior cognition that is also preliminary. Self and introspection are not appended to the repertoire of cognition but are accentuations of preliminary phases in mental process (Brown, 1991). Neoteny is an evolutionary and cognitive mechanism that generates novelty through delay at earlier form-building stages.

Summary

The principle of heterochrony (neoteny) can be applied to the analysis of "lesion" effects in developing and mature brain. Symptoms reflect a local or generalized delay in process. The major change appears to be retardation, though acceleration is theoretically possible. The delay results in an incomplete transition through the damaged segment with preliminary (juvenile) features carried to subsequent phases. Such phases are postponed but may occur normally on deviant (attenuated) forms. This occurs more commonly with acquired disorders in adults than with developmental disorders in children. On this basis, the significance of double dissociation for the imputation of functional autonomy is rejected.

The onset of a microgenetic series (mental state) is presumed to be driven by a pulselike activation (pacemaker) that could vary as to pulse frequency. Offset can be truncated, as in sleep and dream or in coma and hallucination. Acquired errors in adults represent a dyssynchrony between the local retardation and the contextual surround.

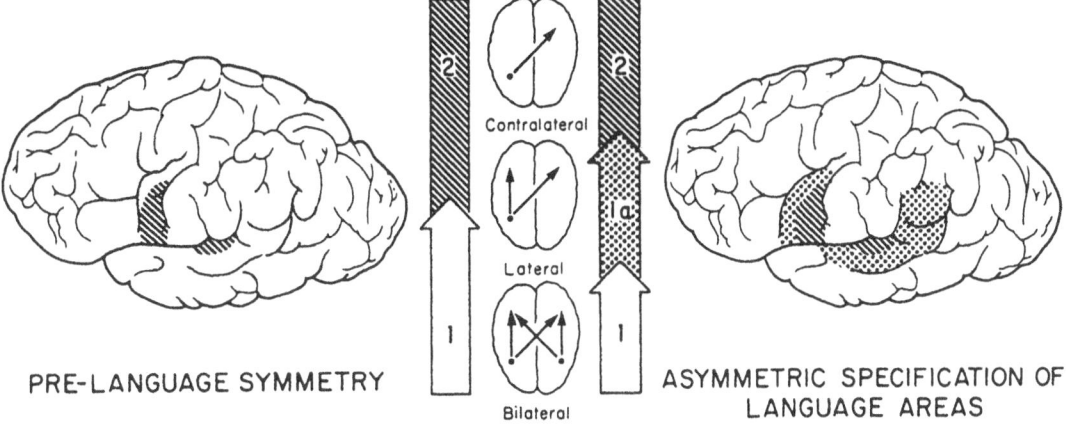

PRE-LANGUAGE SYMMETRY

ASYMMETRIC SPECIFICATION OF
LANGUAGE AREAS

Figure 2. The emergence of the Wernicke and Broca zones within the "integration" neocortex represents neotenous expansion in brain development corresponding to the development of cerebral asymmetry through parcellation. Neoteny and parcellation combine in the emergence of asymmetric language representation. This phase is interposed between bilateral (limbic) and contralateral (motor cortex) representation (see discussion in Brown, 1988, pp. 149–151).

Errors reflect context at normal preliminary phases that unduly persist. In developmental language disorders, there is a retardation of the fine derivation of perception and/or action into the complementary language components. Contextual frames for errors are weakly established and opportunities for derailment reduced, so error "tokens" show less variance than in adults.

Retardation at "juvenile" phases in mental process permits expansion at subsurface levels. The elaboration of preliminary phases in the development of acts and objects populates the mind with intrapersonal content, creating an intrapersonal space of introspection prior to the exteriorization of object space. Awareness, self, choice, and imagery are due to the postponement (retardation) of mental process at earlier phases. The ontogeny of human brain and mentality reflects neotenous change in microgenetic process.

Conclusion

This paper has examined the relevance of two morphogenetic phenomena, parcellation and heterochrony, to processes underlying mature cognition. It is argued that the specification of elements in growth establishes a physiological dynamic that contin-

ues as the specification of contents in mind. Growth is not an open-ended sequence but a reiterated dynamic. The parcellation that is reiterated in growth continues in whole : part or context : item shifts in adult mentation.[1] A process that is stretched out (but is, in reality, *repeated*) over ontogeny lays down a "track" for the momentary unfolding (microgenesis) of the mental state.

In development, structure is deposited as a milestone of change. The slow development of structure creates an historical perspective that is difficult to resolve with the idea of growth as a resurgent process. Yet growth can be interpreted as a cyclic (Thatcher, 1992a, 1992b) or reiterated dynamic. Cell migration begins in the periventricular region with the deeper layers of neocortex established prior to the superficial ones. Rhythmic patterns of depth-to-surface layering occur through successive waves of proliferation, migration, and pruning. After the morphology has been established, these patterns persist as "force-lines," which determine the pattern of physiological activity. The pattern of growth and physiological activity continues into microgenetic process,

---

1. The reverse process (i.e., item to context or part to whole, is a more complex problem relating to emergence and the anisotropy of time.

which deposits the depth to surface "layers" of mental structure.

Specification and heterochrony are related as the pattern of a process to its rate. Retardation leads to incomplete specification at a given segment, while progressive arborization (context : item transforms) requires that every branch (?fractal) issuing from a context remain *in phase* with other parallel branches (i.e., that the rate of specification is uniform across concurrent segments).

As the saying goes, timing is everything. The cognitive process (microgeny) can be considered a speeded-up ontogeny. But what aspect of ontogeny is speeded up? Microgeny does not *retrace* ontogeny; that would require progressive delay in the duration of recapitulation or continuous acceleration to accommodate new layers of acquisition. It is not stages that recur but the process leading to the stages. *The recapitulation is for the process, not its actualized elements.*

The timing of process is critical. The notion of time is bound up with the nature of process. Acceleration, retardation, and recapitulation are temporal concepts. Indeed, the shift from a context to an item is like the analysis of a category. A category is like a duration (Brown, 1991). Duration and categorization are abstract entities with vague boundaries and indistinct centers. The passage from a category to an instance in that category is comparable to the isolation of instants in a duration of time (Brown, 1991). Exemplars in categories are virtual, like instants in a duration. Both categorization and duration are fundamental properties of mind. The nature of time and duration are crucial problems for any theory of physical growth (morphogenesis) and mental process (microgenesis).

The idea that ontogeny retraces embryonic stages in evolution implies that deep phyletic time is collapsed in developmental or life-span process. In fact, phylogeny is not an extended sequence but a reconstruction of a series of ontogenies. Ontogenetic process is the primary reality with phylogeny an extrapolation from the ontogenies of a given line. What, then, is the time of an ontogeny? Is it the life span of an organism, a day, a year, a century? Alfred North Whitehead thought the duration of a thing (e.g., atom, object) was the minimal period for the thing to be what it is. I would argue that the duration of an ontogeny is the minimal duration of the process that sustains it and that this process is replicated throughout the life of the organism. In other words, a process of some duration is repeated — a minimal unit of growth (later, of mind) — with ontogeny the sum of the repetitions of these units. With each iteration, the organism changes. The sequence of change *appears* to be the effect of a life-span process, when actually the sequence is the pattern of change the process lays down.

References

Bard, L. (1905). De la persistence des sensations lumineuses dans le champ aveugle des hémianopsiques. *Semaine Médicale, 25,* 253–255.

Bender, M., & Krieger, H. (1951). Visual function in perimetrically blind fields. *Archives of Neurology and Psychiatry, 65,* 72–99.

Best, C. (1991). The emergence of native-language phonological influences in infants: A perceptual assimilation model. *Haskins Laboratory Status Report on Speech Perception, SR 107/108,* 1–30.

Bishop, E. (1992). The underlying nature of specific language impairment. *Journal of Child Psychology and Psychiatry, 33,* 3–66.

Bonner, J., & Horn, H. (1982). Selection for size, shape and developmental timing. In J. Bonner (Ed.), *Evolution and development* (pp. 259–276). Berlin: Springer-Verlag.

Brown, J. W. (1977). *Mind, brain and consciousness.* New York: Academic Press.

Brown, J. W. (1988). *Life of the mind.* Hillsdale, NJ: Erlbaum.

Brown, J. W. (1990). Overview. In A. Scheibel & A. Wechsler (Eds.), *Neurobiology of higher functions* (pp. 357–365). New York: Guilford Press.

Brown, J. W. (1991). *Self and process.* New York: Springer-Verlag.

Brown, J. W., & Jaffe, J. (1975). Hypothesis on cerebral dominance. *Neuropsychologia, 26,* 183–189.

Carey, S., & Gelman, R. (Eds.). (1991). *The epigenesis of mind: Essays on biology and cognition.* Hillsdale, NJ: Erlbaum.

Cicchetti, D. (1984). The emergence of developmental psychopathology. *Child Development, 55,* 1–7.

Cicchetti, D. (1993). Developmental psychopathology: Reactions, reflections, projections. *Developmental Review, 13*, 471–502.

Coghill, G. (1964). *Anatomy and the problem of behavior*. New York: Hafner.

Creutzfeldt, O. (1977). Generality of the functional structure of the neocortex. *Naturwissenschaften, 64*, 507–517.

Ebbeson, S. (1984). Evolution and ontogeny of neural circuits. *Behavioral and Brain Sciences, 7*, 321–366.

Edelman, G. (1987). *Neural Darwinism*. New York: Basic Books.

Gleason, J. (1978). The acquisition and dissolution of the English inflectional system. In A. Carramazza & E. Zurif (Eds.), *Language acquisition and language breakdown* (pp. 109–120). Baltimore, MD: Johns Hopkins University Press.

Goldberg, G. (1985). Supplementary motor area structure and function: Review and hypotheses. *Behavioral Brain Sciences, 8*, 567–616.

Goodwin, B. (1982). Development and evolution. *Journal of Theoretical Biology, 97*, 43–55.

Goodwin, B. (1984). Changing from an evolutionary to a generative paradigm in biology. In J. Pollard (Ed.), *Evolutionary theory: Paths into the future* (pp. 99–120). New York: Wiley.

Gould, S. (1977). *Ontogeny and phylogeny*. Cambridge: Harvard University Press.

Gould, S. (1982). Change in developmental timing as a mechanism of macroevolution. In J. Bonner (Ed.), *Evolution and development* (pp. 333–346). Berlin: Springer-Verlag.

Hoffman, R. (1987). Computer simulations of neural information processing and the schizophrenia–mania dichotomy. *Archives of General Psychiatry, 44*, 178–188.

Innocenti, G. (1984). Commentary. *Behavioral Brain Sciences, 7*, 340–341.

Karmiloff-Smith, A. (1991). Beyond modularity: Innate constraints and developmental change. In S. Carey & R. Gelman (Eds.), *The epigenesis of mind: Essays on biology and cognition* (pp. 171–198). Hillsdale, NJ: Erlbaum.

Katz, M. (1983). Ontophyletics: Studying evolution beyond the genome. *Perspectives in Biology and Medicine, 26*, 323–333.

Keil, F. (1991). The emergence of theoretical beliefs as constraints on concepts. In S. Carey & R. Gelman (Eds.), *The epigenesis of mind: Essays on biology and cognition* (pp. 237–256). Hillsdale, NJ: Erlbaum.

MacLean, P. (1991). Neofrontocerebellar evolution in regard to computation and prediction: Some fractal aspects of microgenesis. In R. Hanlon (Ed.), *Cognitive microgenesis: A neuropsychological perspective* (pp. 3–31). New York: Springer-Verlag.

MacNeilage, P., Studdert-Kennedy, M., & Lindblom, B. (1987). Primate handedness reconsidered. *Behavioral Brain Sciences, 10*, 247–303.

Malsburg, C. von der, & Singer, W. (1988). Principles of cortical network organization. In P. Rakic & W. Singer (Eds.), *Neurobiology of neocortex*. New York: Wiley.

Pribram, K. (1991). *Brain and perception*. Hillsdale, NJ: Erlbaum.

Rakic, P. (1989). Competitive interactions during neuronal and synaptic development. In A. Galaburda (Ed.), *From reading to neurons* (pp. 443–462). Cambridge: MIT Press.

Rakic, P. (1992). *Developmental origin of cortical diversity*. Schmitt lecture, Rockefeller University, New York.

Reber, A. (1992). The cognitive unconscious: An evolutionary perspective. *Consciousness and Cognition, 1*.

Robertson, D. (1991). Feedback theory and Darwinian evolution. *Journal of Theoretical Biology, 152*, 469–484.

Sahlén, B. (1991). *From depth to surface: A case study approach to severe developmental language disorders*. Studies in Logopedics and Phoniatrics No. 1, Lund University, Lund, Sweden.

Sanides, F. (1975). Comparative neurology of the temporal lobe in primates including man with reference to speech. *Brain and Language, 2*, 396–419.

Semmes, J. (1968). Hemispheric specialization: A possible clue to mechanism. *Neuropsychologia, 6*, 11–26.

Stent, G. (1981). Strength and weakness of the genetic approach to the development of the nervous system. *Annual Review of Neuroscience, 4*, 163–194.

Striedter, G., & Northcutt, R. (1991). Biological hierarchies and the concept of homology. *Brain, Behavior and Evolution, 38*, 177–189.

Thatcher, R. (1992a). Are rhythms of human cerebral development "traveling waves"? *Behavioral Brain Sciences, 14*, 575.

Thatcher, R. (1992b). Cyclic cortical reorganization: Origins of human cognitive development. In G. Dawson & K. Fischer (Eds.), *Human behavior and brain development*.

Trevarthen, C. (1984). Emotions in infancy. In K. Scherer & P. Ekman (Eds.), *Approaches to emotion*. Hillsdale, NJ: Erlbaum.

Tucker, D. M. (1992). Developing emotions and cortical networks. In M. Gunnar & C. Nelson (Eds.), *Minnesota Symposium on Child Psychology: Vol. 24. Developmental Behavioral Neuroscience* (pp. 75–128). Hillsdale, NJ: Erlbaum.

Vandervert, L. (1988, 1990). Systems thinking and a proposal for a neurological positivism. *Systems Research, 5*, 313–321, *7*, 1–17.

Van Sluyters, R., Atkinson, J., Banks, M., Held, R., Hoffmann, K., & Shatz, C. (1987). The development of vision and visual perception. In L. Spillman & J. Werner (Eds.), *Visual perception: The neurophysiological foundations*. New York: Academic Press.

Wall, P. (1988). Recruitment of ineffective synapses after injury. In S. Waxman (Ed.), *Functional recovery in neurological disease* (pp. 387–400). New York: Raven Press.

Werner, H. (1948). *Comparative psychology of mental development*. New York: International Universities Press.

Witelson, S. (1990). Structural correlates of cognition in the human brain. In A. Scheibel & A. Wechsler (Eds.), *Neurobiology of higher functions* (pp. 167–184). New York: Guilford Press.

13.

Topographically Different Regional Networks Impose Structural Limitations on Both Sexes in Early Postnatal Development

Harriett W. Hanlon

Topographically Different Regional Networks Impose Structural Limitations on Both Sexes in Early Postnatal Development

Harriet W. Hanlon

Abstract

Principal components analysis of 56 time series of mean EEG coherence from all brain regions of 224 girls and 284 boys reveals no commonality in regional patterns of cerebral cortex growth from birth to age 16 years. Coherence time series are oscillating waves that represent change over time in EEG wave similarity at two spatially separated electrode sites. Coherence oscillations are thought to mirror the increasing complexity of gray matter structures, particularly overproduction and pruning of synapses at sites compared (local networks). The 56 local networks examined unfold in cycles of growth and pruning that are sexually dimorphic in time, rate, amplitude and flexibility. Bifurcations occur in all local networks around ages 6 and 11 years leading to a partitioning of the first 16 years of postnatal development into early, middle and late periods. It is hypothesized that the bifurcations are points of structural limitation orchestrated by the DNA to shift developmental focus in each sex in preparation for the next life cycle stage. The 11 regional networks determined for each sex represent in-phase growth that accounts for 95 percent of variance in local networks, represents all brain regions in topographically different groupings and has complementary regional networks that reach the adult stage or are retained at the juvenile stage. Models for quantitative and qualitative differences in the sexes' complementary signatures are discussed.

... "Homo sapiens did not appear on the earth, just a geologic second ago, because evolutionary theory predicts such an outcome based on themes of progress and increasing neural complexity. Humans arose, rather, as a fortuitous and contingent outcome of thousands of linked events, any one of which could have occurred differently and sent history on an alternative pathway that would not have led to consciousness. ..."

Stephen Jay Gould
Scientific American, October, 1994

This paper will demonstrate that the development of the human brain involves an integration of isolated local networks into efficient regional networks with distributed processing. Based on derived measures of electroencephalogram (EEG) readings for 508 girls and boys for the postnatal period from birth to age 16 years, a stage theory will be proposed that describes the sequence of steps involved in the transformation of isolated local networks into globally interconnected regional networks. This developmental process for networks will be evaluated in terms of gender differences.

Although evidence of sex differences in central tendency, variability and number of high scorers on verbal and spatial tasks has been documented (Maccoby & Jacklin, 1974; Feingold, 1992; Kimura, 1992; Hedges & Nowell, 1995), until very recently there was insufficient neurological evidence to support the assertion of any structural differences in the sexes' cerebral cortices. A growing body of developmental evidence, however, suggests that sexual dimorphism exists, begins early in fetal development and continues throughout the life cycle.

Anatomical studies. At 13 weeks' gestation, the entire right cerebral hemisphere is more developed in males and the left prefrontal cortex is more developed in females; the male brain also appears more asymmetrical than the female brain (de Lacoste, 1985; 1991). Around age 2 years the male brain begins growing at a faster rate than the female; by age 6 years, with height controlled, the male brain is 12 percent larger than the female, a weight difference maintained for the rest of the lifespan (Swaab, 1984). By mean age 9 for a sample of 15 female and 15 male children, Caviness (1994) using magnetic resonance imaging found "exuberant and delayed maturation" in volumetric growth of structures during childhood. Structures with greater volume in childhood when compared to adult structures of the same sex include the female pallidum (112 percent), the male caudate nuclei (115 percent), the male putaminal nuclei (108 percent) and the male amygdala (115 percent); structures with less volume in childhood for both sexes include cerebral central white matter (84 percent), the hippocampus (86 percent) and the brainstem (89 percent). Caviness found that mean female volume for neocortex, diencephalon and cerebral central white matter is proportionately larger than male volume for these structures in children aged 6 to 11 years. His findings support sexual dimorphism in developing brain structures during the middle childhood period.

Physiological studies. Functional neuroimaging studies are expensive, invasive and limited in their use in early development to young children who are ill. Studies of adults (Naylor, 1991; Gur, 1995; Shaywitz, 1995) indicate gender differences in brain regions at rest or activated for the same cognitive task. However, whether used for adults or children, functional neuroimaging techniques do not offer the degree of resolution necessary to track gray matter changes at the synaptic level. In contrast, neurometric EEG techniques, refined over the past 20 years, may offer the window needed for accurately localizing these structural changes in cortical gray matter with the added benefit that the techniques are inexpensive and noninvasive. The algorithms for spectral analysis, wavelength analysis and pattern recognition used with EEG signals were developed with a view toward evaluating anatomical integrity, developmental maturation and the mediation of sensory, perceptual and cognitive processes associated with brain growth (Harmony, 1984). EEG coherence, one of many derived EEG neuromeasures, is a cross- correlation of the amplitudes of the independent wave components of the EEGs at paired sites. Its purpose is to measure wave similarity at two scalp locations (Nunez, 1981, pages 303-320; Lopez da Silva, 1989; Duffy, 1994). Over long periods of time, change in EEG coherence is thought to primarily reflect changes in the gray matter at the paired sites (Thatcher, 1986; Tucker, 1986). Neurophysiological mechanisms responsible for change in the numbers and/or strength of cortico-cortical connections include, but may not be limited to the following changes in the gray matter: synaptogenesis, pruning of synaptic connections, axonal and dendritic sprouting, expansion of existing synaptic terminals and changes in strength of neurotransmitter secretions during presynaptic and postsynaptic exchanges (Changeaux, 1984; Kandel, 1985; Thatcher, 1992).

Recent morphometric visual cortex studies by Huttenlocher (1990; 1994) have increased the support for synaptic pruning and maturation of retained synapses as the primary change processes in human cortex development, rather than change in number of neurons or axonal and dendritic expansions of neurons. He found no loss of neurons from the 28th week of gestation to age 71 years (oldest subject), no increase in dendritic length between 21 months and age 7 years (oldest subject), but prolonged postnatal changes in the synaptic net. Overproduction of synapses in the visual cortex reached its peak at 8 months and was 42 percent greater than the adult synaptic level reached through synapse elimination by age 10 years (1994). Hence, as our knowledge of the developmental changes in the synaptic net have increased, it has become clear that EEG coherence is the measure of choice to evaluate how brain networks are changing over time (Thatcher, 1983-1994; Hanlon, 1994-1995).

EEG coherence has been used to:
- to confirm onset of cortical connections for visual memory in infants (Fox & Bell, 1990);
- to identify gender patterns in brain regions favored by college-age high performers on the graduate record exam (Corsi-Cabrera, 1989);
- to classify school-age children into IQ groupings from below-average to gifted (Thatcher, 1983);
- to compare rate and timing differences in growth of the two cerebral hemispheres in children, adolescents and young adults (Thatcher, 1987);
- to determine gender differences in local and regional cortical networks during the first 16 years of postnatal development (Hanlon, 1994; 1995).

The Hanlon mean coherence studies reported in this paper used these well-verified EEG methods of tracking development to establish that the sexes have qualitatively and quantitatively different developmental signatures. All studies involve analyses of selected portions of the female and male normed data base for EEG coherence collected by R. W. Thatcher from 1979 to 1986 at the Applied Neuroscience Institute, University of Maryland, Eastern Shore. The analyses do not include any environmental dimension for subject or time point. Consequently, the gender differences evident are more likely to be general phenotypical traits of the specie. The studies are limited by small samples for time points beyond 16 years.

Methods

Time series of mean EEG coherence and time series of standard deviation (sd) of mean EEG coherence were constructed for 76 consecutive time point samples of female subjects aged 0.2 year to 24.1 years and 82 time point samples of male subjects aged 0.2 year to 23.4 years. Each time series represents a local network. The oscillations in the time series estimate change in cortico-cortical connectivity and associated gray matter at paired brain regions separated by 6 to 28 centimeters.

Appendix 1 lists variable names for the 64 electrode-pair sites evaluated, the biological names of the two regions being compared and the distance between the two sites (centimeters). Appendix 2 consists of flow charts that summarize methods used in constructing the coherence time series in the primary study.

Principal components analysis with Varimax rotation determines regional networks consisting of local networks time-linked in patterns of growth. All principal components analyses were on unsmoothed mean coherence time series.

Participants

The primary study (Hanlon, 1994) was restricted to the first 16 years of postnatal development. It was based on a total of 508 children, 224 females and 284 males, selected from a pool of subjects recruited by Thatcher for studies related to cognitive functioning of normal children. Table 1 lists statistics for age, race and handedness for each sex. Subjects in the female and male samples have a mean full scale IQ of 109, represent all socioeconomic groups and have no history of neurological disorders or abnormalities in the prenatal, perinatal and postnatal development periods. Normed criteria replicated the stringent criteria established by Matousek and Peterson in thefirst normed absolute power EEG data base collected on female and male subjects, aged from 1 to 21 years, in 1973.

Table 1

Demographics of Female and Male Subjects

Sex	Age Range In Years	Handedness Right	Left/Mixed	Race (American) European	African	Asian
Female	0.29-16.22	87%	13%	65.5%	31.4%	3.1%
Male	0.19-16.08	80%	20%	73.9%	23.2%	2.9%

Data

For each child, 1 minute of eyes-closed EEG was recorded at a digitization rate of 100 hz in a no-task condition from 19 channels (international 10/20 system of electrode placement). EEG coherence was computed for all pairwise combinations of 16 channels in the standard frequency bands (Thatcher, 1983). EEG coherence is equivalent to the absolute value of Pearson's r correlation determined in real space. Graphically, it is the magnitude of the resultant vector in the complex plane, determined by correlating the frequency components in phase-locked EEG waves at the paired sites. It ranges in value from 0 to 1 and is most frequently expressed as a percent. See Thatcher (1983) for a more comprehensive discussion of algorithms for EEG coherence used in this study. All Hanlon studies examine EEG coherence data from the theta frequency band (3.5 to 7.0 hz).

Unit of Analysis

Female and male subjects were placed in age-appropriate samples for time points at 3-month intervals, beginning with sample 1 (birth to 0.25 year) and ending with sample 64 (16.0 years to 16.25 years). Replicating Thatcher's method (1992) for increasing the frequency resolution of coherence oscillations of growth and pruning across the 16-year developmental period, sliding averages were computed for each time point, using 1-year epochs and .25-year increments. Independent subjects appear in at most four time points and each new time point has at most a 25 percent increase in new subjects. For example, coherence data for a 9.78-year-old male subject was included in time point samples for 9.00 years (mean age 9.483), 9.25 years (mean age 9.781), 9.50 years (mean age 9.975) and 9.75 years (mean age 10.173). Median sample size for female time points was 16 subjects, with a range of 3 to 22; median sample size for male time points was 19.5 subjects, with a range of 4 to 33. Since comparison of gender cortico-cortical development required closely matched time points (mean ages), 8 poorly matched time points were discarded, reducing the time series to 56 well-matched points.

Time series of sample means and variance were constructed for 28 intrahemisphere electrode-pair sites in each hemisphere. Time series were also constructed for 8 homologous regions in each hemisphere, for a total of 64 time series. Time series were smoothed and first derivative time series were constructed for each of the 64 electrode-pair sites, using a least-square procedure of Savitzky and Golay (1964) which includes a quadratic convolution that prevents degradation of sharp peaks.

Growth (pruning) spurts were defined as positive (negative) peaks in the velocity time series equal to or greater than one sd change from the baseline mean of the mean coherence time series.

Using velocity time series brings two benefits:
   1) identifies postnatal ages where mean coherence is experiencing its maximum rate of increase (decrease);
   2) shifts the mean axis of coherence to 0 where gender timing, rate and amplitude differences in the growth and pruning peaks can more easily be seen and compared.

Local Network Analyses
   Gender comparisons of local networks for the same brain regions include:
   1) visual plots of time series;
   2) comparison of baseline difference;
   3) comparison of amplitude difference;
   4) comparison of high energy cyclic components;
   5) counts of growth (pruning) peaks 1 SD above (below) the mean baseline;
   6) topography of growth and pruning peaks at 56 time points;
   7) comparison of plasticity of unfolding structures;
   8) comparison of wave flow;
   9) global correlation measure of similarity.

Regional Network Analyses
   Gender comparisons of regional networks for the 56 local networks include:
   1) comparison of topography across anterior-posterior spatial gradient;
   2) defining a sequential process of development of four stages common to both sexes;
   3) comparison of networks reaching adult stage and networks retained at the juvenile stage.

Results

<u>Local Networks</u>

    <u>Genders have similar shifts in developmental process</u>. Mean coherence development experiences discontinuities when the power driving the oscillating growth and pruning wave switches to a different set of frequencies. For each sex, local networks in all brain regions experience these discontinuities, with most occurring around ages 6 and 10 years. These ontogenetically orchestrated shifts (bifurcations) divide cortical development into early, middle and late periods, each with defining characteristics dependent on sex and hemisphere. The shifts, separating the distinct growth periods, appear as cusps when the rapidly changing mean coherence and its velocity are plotted against time in a 3-dimensional phase plot. Figure 1 contrasts female and male development of connectivity between right temporal lobe sites T4 and T6. Frequency shifts or cusps can be seen in the female velocity plot and 3-dimensional phase plot around ages 7 and 14 years (Figure 1A) and in the male plots around ages 6 and 12 years (Figure 1B). Sexual dimorphism is apparent in the timing, rate and amplitude of the growth and pruning peaks in this local network.

    <u>Genders have similar wave flow across spatial gradients in each hemisphere</u>. Mean coherence development consists of postnatal oscillating waves that flow across lateral-medial and anterior-posterior gradients in each hemisphere in both sexes. Neurogenetic gradients are similar to those found in mice and rats (Smart, 1983; Bayer & Altman, 1991). Decomposing the wave flow into anterior-posterior and lateral-medial dimensions is best illustrated using a fixed reference point. Figure 2 shows female and male growth spurt wave flow at left-hemisphere frontal pole site Fp1 decomposed into lateral-to-medial clockwise rotations and anterior-to-posterior expansions. Figure 2A shows a 3-year cycle for the female beginning at F1T3 at age 2 years; Figure 2B shows a 3-year cycle for the male beginning at F1F7 at age 3.8 years.

    <u>Genders have similar patterns of oscillating variance around the time series means.</u> In both sexes, rather than a constant band of variance around the means, standard deviations for the mean coherence time series are oscillating waves with variance increasing at some time points and decreasing at others. Variance at each sample time point reflects the influence of internal and external effects that modify the genetic expression of the networks for each child in the sample.

    Velocity plots and cross correlations at 0 phase shift were used to compare coherence means and coherence standard deviations for each of the sexes' 56 local networks. Plotting the rate of change in mean coherence against the rate of change in the standard deviation gives a picture of how these two measures covary. A negative correlation indicates variance is greater during the synaptic pruning phase. A positive correlation indicates variance is greater during the synaptic growth phase.

A fundamental assumption of this paper is that networks with negative correlations have greater plasticity with which to accommodate new growth and respond to requirements of the environment, while networks with positive correlations have reduced plasticity because of previous commitments essential for survival.

Across all time point samples, flexible networks (negative correlations) have minimal variance during the growth phase and maximal variance during the pruning phase. In these networks synapses seem to be activated by the genetic program for broad functional purposes, while the cortico-cortical connections pruned vary among the individuals in each sample. For example, it is assumed that auditory center synapses are activated for recognizing all possible human sounds, but connections are pruned for sounds not heard by the individual. Hence, individuals in the time point samples vary more in choice of synapses pruned than in how they accommodate new growth.

In contrast, networks with positive correlations are assumed to have lost some of their plasticity in meeting prior functional requirements. Across all time point samples, less flexible networks have maximal variance during the growth phase and minimal variance during the pruning phase. In these networks, it is assumed that individuals vary in accommodating new synaptic growth because the existing functional purpose of the network places structural limitations for both growth and pruning on the individual. For example, it is assumed in association areas for integrating visual inputs with tactile inputs, that individuals would have similar limitations in pruning synaptic connections between the primary sensory regions and the association region, but would vary in the complexity of the internal connection pattern established within the association region. Hence, the individuals in the time point samples for these networks vary in the choices they make to accommodate new growth, but behave similarly in the restrictions the functioning network imposes on pruning choices of connections to primary sensory cortices.

Correlations of first derivative time series of means and standard deviations for each sex's 56 local networks are listed in Appendix 3. There is sexual dimorphism in the local networks, but, in both sexes, the correlations show:
- the majority of sites separated by 6 cm correlate negatively; i.e., variance is least during the synaptic overproduction phase and greatest during the pruning phase;
- the majority of sites separated by 24 to 28 cm correlate positively; i.e., variance is greatest during the synaptic overproduction phase and least during the pruning phase;
- the majority of sites representing spatially separated sites of intermediate distance have

time series that correlate negatively and positively for portions of development, with the shift from negative to positive occurring most often at a bifurcation.

Figure 3 shows velocity plots for each sex for sites in both hemispheres. Figure 3A shows sites where each sex has its highest negative correlation--Fp2F8 for the female and Fp2F4 for the male; Figure 3B shows sites where each sex has its highest positive correlation--O1T3 for the female and F7T5 for the male; and Figure 3C shows site O1F7 where there is sexual dimorphism in the velocity patterns.

Genders have differences in mean coherence baselines, amplitudes of growth and pruning spurts and energy in cycles of growth and pruning supporting the amplitude differences. Baselines in mean coherence plots are lower in the female than the male in 42 of the 56 local networks while amplitudes for growth and pruning spurts about those baselines are greater in 47 of the 56 networks. The differences are obvious even upon visual inspection and are summarized in Table 2. Note the 6 to 1 ratio in the right hemisphere of female to male mean baselines and in both hemispheres of female to male amplitudes peaks. While not as dramatic, the 2 to 1 ratio for the left hemisphere baselines is still an indicator of female complexity for many functions verses male specialization for selected functions. See Appendix 4 for specific baseline and amplitude difference for each local network.

Lower coherence is associated with greater differentiation of gray matter in the neural circuits; i.e., lesser connectivity between the neural circuits at the two more distant sites and greater connectivity within or to nearby columnar matrices. Greater amplitude in growth and pruning peaks (variance about baseline) indicates greater complexity in the connectivity patterns made and reflects more high energy frequency components in the developmental time series; e.g., an electrode-pair site with high energy in 2- and 3-year cycles will have its greatest peak in multiples of the 6-year cycle. Development in local network F8T4 illustrates these differences (Figure 4).

Table 2
Sex Differences in Time Series' Baselines and in Peak Amplitudes

	Mean Baseline Lower*		Variance in Peaks Greater**	
Hemisphere	Female	Male	Female	Male
Left	18	10	24	4
Right	24	4	23	5

* For 56 local networks, baseline range is from -4.113 at F8T4 (female lower) to 1.756 at F7T3 (male lower), mean of -0.944; i.e., female baselines are about 1% lower than male baselines.

** For 56 local networks, variance ratios range from 0.442 at O2T4 (1/.442 = 2.262 or 162% greater variance in male peaks) to 8.346 at T3T5 (735% greater variance in female peaks), with mean of 1.873; i.e., female growth and pruning peaks have amplitudes that are about 88% greater than male peaks.

Gender differences in mean coherence development between right hemisphere lateral-frontal site F8 and anterior-temporal site T4 are shown in Figure 4A. The female baseline is 4.113 percent lower than the male; female variance about that baseline is 286 percent greater (variance ratio of 3.856).

Figure 4B shows velocity plots for F8T4 time series shown in Figure 4A. A limitation of velocity plots is that "percent change in mean coherence over time" does not reflect actual change; e.g., the velocity peak of 4 percent in female mean coherence at age 2 years (Figure 4B) represents a mean coherence change from 38 to 66 percent (Figure 4A) that takes place during the 2-year growth spurt beginning around 1 year of age.

Female neural networks have more high energy components. The greater amplitude in the female growth and pruning peaks can be examined in the frequency domain to discern which cycles in the spectra are the high energy components. Fourier analysis increases precision over time domain analysis because of independence of spectra frequencies over which energy is distributed. The inverse of frequency gives the length of time needed to complete one growth and pruning cycle; i.e., it gives the period. Recall that there is energy in every period from 6 months to 16 years. However, Fourier analyses for the 56 female and male time series show the major high energy components are in periods that range from 2 to 5 years in length.

Analysis of the sexes' 56 intrahemisphere time series shows very low power in most frequencies, with one or two components having most of the developmental energy. Table 3 summarizes the range of mean power distribution for all components and the range of power for the few high energy frequencies.

The female's left temporal lobe has a greater number of high energy growth cycles than any other local network. Complexity of the female's left temporal lobe is particularly striking; e.g., T3T5 has 5 high amplitude cycles; C3T5 has 4; F7T5 and F3T3 have 3; and Fp1T3, Fp1T5, F3T5, T3C3, T3P3 and T3O1 have 2. Figure 5 shows Fourier analysis of the female's T3T5 time series which includes high-amplitude components for 2-year, 2.3-year, 4-year, 8-year and 16-year periods. Another major assumption of this paper is that the 2.3-year fundamental is a singular network; the 4-year fundamental is a complex network with elaborate, highly correlated subcycles at age 2, 8 and 16 years.

Table 3

Mean Power and High Energy Components in Sexes' Time Series

	Power in Developmental Frequencies (Range of Means for 56 time series, uv^2)	
Sex	All Components	High Energy Components
Female	.002-.024	.100-.460*
Male	.001-.006	.100-.178**

* Female periods for 1 cycle range from 1.8 years to 16 years; amplitudes of components range from .316 to .678 uv.

** Male periods for 1 cycle range from 2.0 years to 8 years; amplitudes of components range from .316 to .433 uv.

In the time domain, the different high-energy growth cycles of T3T5 culminate in a time series plot of high amplitude growth and pruning peaks, with a most dramatic conjunction at age 14 years when mean coherence changes from 47.5 percent to 89.3 percent over a 3-year period. The first derivative shows this growth spurt as a 7 percent velocity peak (see Figure 6).

Figure 7 shows distribution of the high energy development cycles in female and male brains. Female development has 55 high energy cycles, of which 36 are at 18 electrode-pair sites in the left hemisphere and 19 are at 15 electrode-pair sites in the right (Figure 7A). Females have 15 sites with multiple high energy development cycles, 11 in the left hemisphere and 4 in the right. Median developmental rate for the 55 high energy cycles is 2.7 years. Male development has 18 high energy cycles, 9 at 9 electrode-pair sites in the left hemisphere and 9 at 7 electrode-pair sites in the right (Figure 7B). Males have two sites--Fp2F8 and Fp2T4, each of which has two high energy cycles. Median development rate for these 18 cycles is 3.2 years.

Other analyses offer additional verification of baseline mean and variance ratio differences. Statistical analysis of all paired mean coherence time series reveals gender baselines are significantly different (Female Mean Baseline - Male Mean Baseline; Wilcoxon nonparametric T = -4.005; p < .0000), as are the variance ratios about those baselines (Female Variance/Male Variance; Fischer nonparametric B = 5.078; p < .0000).

Several other methods verify differences in mean coherence baselines and variance ratios. Box plots of each sex's time series show females have lower means, wider interquartile spread and more extreme low and high values at most sites. If the outliers (growth and pruning spurts) are excluded, the box plot distributions appear Gaussian for both sexes.

Hotelling's T^2, which gives a multivariate statistical assessment primarily of amplitude differences between the sexes for all 64 intra- and inter-hemisphere mean coherence time series, gives an F statistic of 8.120, df=64,47, p < .0000. Twenty-seven of the univariate F tests are statistically significant with p-values < .05; of these paired sites, 18 intrahemisphere and 2 interhemisphere variables have p-values < .01. The homologous regions of greatest gender statistical difference involve frontal lobe sites F3F4, C3C4 and F7F8. This contrasts with the regions of greatest timing difference determined with the cross-correlogram; i.e., T5T6 (6 years), P3P4 (5.75 years) and Fp1Fp2 (2.75 years).

Genders differ in time and rate of growth and pruning cycles in 56 local networks. Sexual dimorphism in the timing of growth and pruning peaks for the 56 local networks is shown in Figure 8. Figures 8A, 8B and 8C show sexual dimorphism in growth and pruning peaks at three consecutive time points in early, middle and late development, respectively.

The number of growth and pruning peaks and the number of 3-month periods where the growth or pruning spurt exceeds 1 sd were determined for all local networks. Length of most growth spurts is less than one year, but each sex has sites with prolonged plateau periods of growth; e.g., females have a 21-month growth period at Fp1F7 that begins at age 2.0 years and males have 18-month growth periods at F4F8, C4F8 and F4C4 that begin at age 13.4 years. Length of pruning spurts shows a similar pattern; e.g., females have a 24-month pruning period at F4T4 that begins at age 5.5 years and males have an 18-month pruning period at C4P4 that begins at age 8.3 years. Summary totals for peaks and spurts for each hemisphere are given in Table 4.

Table 4
Gender Patterns in Growth and Pruning Spurts across Development

| | No. of Spurts* | | | | | No. of 3-Mo. Periods in Spurt | | | |
| | Growth | | Pruning | | | Growth | | Pruning | |
Hemisphere	F	M	F	M		F	M	F	M
Left	84	89	79	89		242	232	205	231
Right	77	66	82	69		226	221	237	220
Both	161	155	161	158		468	453	442	451

* synaptogenetic peak 1 sd above baseline mean of time series; synaptoreduction peak 1 sd below baseline mean of time series.

Intersex difference in counts. While whole-brain totals for growth and pruning peaks 1 sd above the mean are similar in the sexes, differences are apparent in the hemispheres. Females have 11 more growth spurts and 13 more pruning spurts in the right hemisphere than males, while males have 5 more growth spurts and 10 more pruning spurts in the left hemisphere than females. When length of developmental spurt is measured by number of 3-month periods during which the peak is 1 sd above the mean coherence baseline, several gender differences exist. The total number of 3-month periods of growth and pruning is dissimilar in the sexes, with females having 15 more 3-month periods of growth than males and males having 9 more 3-month periods of pruning than females. Both sexes have more 3-month periods of growth in the left hemisphere, but pruning differences are complementary. Males have 26 more 3-month periods of pruning than females in the left hemisphere, while females have 17 more 3-month periods of pruning than males in the right. As with the peaks, pruning spurts are greater for each sex in the hemisphere where development lags in the first half of prenatal development (de Lacoste, 1986; 1991).

Intrasex difference in counts. Across hemispheres, the total number of growth and pruning peaks is asymmetric in the male, almost symmetric in the female. Males have 23 more growth peaks and 20 more pruning peaks in the left hemisphere than in the right; females have 7 more growth peaks in the left and 3 more pruning peaks in the right.

When length of developmental spurt is measured by number of 3-month periods, growth and pruning are asymmetric in both sexes; females favor the left hemisphere in growth, but the right hemisphere in pruning; males favor the left hemisphere for both growth and pruning. Females have 16 more 3-month periods of growth in the left than the right, 32 more 3-month periods of pruning in the right than the left; males have 11 more 3-month periods of growth and pruning in the left than the right.

Gender timing of growth and pruning peaks is weakly correlated at 0 phase shift. Visual comparison of mean coherence plots by gender suggests the sexes have differences in time of growth and pruning peaks in all local networks, but greater dissimilarity in connectivity patterns in local networks involving connections between medial and lateral sites than between medial and medial sites. At zero-phase shift, left hemisphere correlations of 28 local networks have a median of +.208, with a range from -.495 to +.618; correlations of 28 right hemisphere local networks have a median of +.182, with a range from -.648 to +.648. A value of +1 indicates in-phase growth and pruning spurts with the same amplitudes; -1 indicates growth patterns are 180 degrees out-of-phase with the same amplitudes.

In cross-correlograms, the time shift required for the best correlation of the sexes' developmental time series for each local network is determined. Table 5 summarizes the phase shift required for the greatest positive correlation of the gender's time series. The positive correlation was selected because it is the best match of growth spurts. It was assumed that learning readiness issues would be apparent if either a significant phase-shift or low correlation existed between the sexes.

Table 5

Comparison of Gender's 56 Local Networks Phase-Shifted Match

	Number of Sites for Specified Brain Regions			
	(median phase shift in years; median correlation)			
Phase Shift* For Best Match Of Time Series	Medial To Medial	Medial To Lateral	Lateral To Lateral	Total
0 to 1 year	16 (½;.424)	11 (0;.517)	1 (0;.601)	28 (¼;.467)
1.25 to 6.75 years	4 (4;.450)	19 (4½;.443)	5 (4;.424)	28 (4.38;.450)
Total	20 (½;.439)	30 (2.38;.465)	6 (3¼;.483)	56 (1.13;.450)

* Absolute value of lag or lead in phase-shift needed for greatest positive correlation.

Gender phase-shifts required for best match are greatest in networks involving older brain systems; i.e., at sites involving temporal lobes and ventrolateral cortex, with phase shifts greatest at right hemisphere anterior-temporal site T4 and left hemisphere lateral-frontal site F7. Brain networks involving newer brain systems seem to be at a juvenile stage of development in both sexes at age 16 years.

Regional Networks

Sexes have different time table for development of same brain regions. Principal components analyses of time series of unsmoothed EEG coherence data were restricted to eigenvalues explaining at least one percent of the variance. In a combined intrahemisphere analysis of 112 local networks (56 for each sex), 18 components, 9 for each sex, explain 94 percent of the variance in human mean coherence development.

The factors satisfy Thurstone's simple structure criteria; i.e., "variables should have large loadings on as few of the factors as possible and low or zero loadings on the remaining factors (Guilford, 1954)." The negligible cross-loading by the sexes supports the poor correlation of the sexes' local networks at zero phase shift. Of the 2,016 loadings on the 18 components, 7.4 percent load greater than .400 with the majority of loadings near 0. See Table 6 for distribution of the loadings.

Sexes have same number of regional networks. Principal components analyses conducted on each sex's time series separately maximize the regional network patterns and remove any confounding effects from the other sex's data. Results show that each sex has 11 regional networks (components), representing all brain regions and explaining 89 percent of variance in female development and 90 percent in male development. When analyses are conducted on early, middle and late time series subsets that bracket the bifurcations at age 6 and 11 years, 95 percent of variance is explained in each sex's development.

Each regional network consists of a limited number of local networks with high loadings on the component. The female has 17 local networks with loadings > .800 on 9 of the 11 regional networks; marker variables for the other two networks have loadings of .799 and .774. The male has 19 local networks with loadings > .800 on 8 of the 11 regional networks; marker variables for the other three networks have loadings of .708, .695 and .673.

Table 6

Distribution of Loadings on Sexes' Regional Network Factors

Minimal Cross-Loading on Combined-Sex Factor Analysis Of 112 Intrahemisphere Local Networks, 56 for Each Sex   [2,016 Loadings for 112 Variables (X) on 18 Factors]

Loading on Factors	Number			Cumulative	
	Male	Female	Both	Percent	Percent
X > .900	3	1	0	0.198%	0.198%
.800 < X < .899	14	10	0	1.190%	1.388%
.700 < X < .799	10	15	0	1.240%	2.628%
.600 < X < .699	9	7	0	0.794%	3.422%
.500 < X < .599	16	12	2*	1.488%	4.910%
.400 < X < .499	16	24	11	2.523%	7.433%
Total X > .400:	68	69	13	7.433%	-
Total X < .399		1,866		92.567%	100%

* Two male local networks load on female regional networks; examination of time series shows one network is out-of-phase and the other network is in-phase during middle development.

Forty-two female local networks load on only one regional network across the 16 years of development, with 1 loading below .600; 5 local networks cross-load, with all shifting from one regional network to another at the age 6 years bifurcation. Female development includes 9 local networks (16 percent) not strongly aligned with any regional network*.

Forty-three male local networks load on only one regional network across the 16 years of development, with 2 loading below .600; 11 local networks cross-load, with 9 shifting from one regional network to another at the bifurcation around age 6 years and 2 shifting at the bifurcation around age 10 years. Male development includes 2 local networks (4 percent) not strongly aligned with any regional network.

Sexes' regional networks show four stages across first 16 years of development. The sequence of developmental stages that each regional networks follows can be seen by examining

---

* Nonaligned networks seem to be immature networks since they are in brain regions known to develop later.

in-phase and out-of-phase behaviors of local network velocity time series during early, middle and late development periods. Local network mean coherence baseline behaviors also aid in confirming that a regional network has shifted to a higher stage. Baseline behaviors include: dramatic shifts (up or down) at a bifurcation, a constant slope up or down during in-phase development segments and a steady baseline during out-of-phase development segments.

The stage theory proposed, as diagrammed in Figure 9, assists in understanding the gender differences in regional networks for the same brain regions. The sexes differ in which brain regions experience accelerated or slowed growth in prenatal and early postnatal development. The consistency with which accelerated and slowed growth and pruning cycles are orchestrated across the 16-year postnatal period in the different brain regions suggests a DNA input that differs in the sexes.

Gould's theories (1977; 1982) of neoteny (delayed or slow growth for greater complexity) and progenesis (accelerated or fast growth for specie preservation) provide a useful model for understanding the complementary timing differences in the sexes' development of the same brain regions.

In the stages listed below, Stage 4 represents the beginning of an iterative process that continues for the remainder of the life cycle, a process of in-phase growth and pruning of synapses in the regional networks followed by out-of-phase globalization; i.e., accomodation and assimilation of changes throughout the synaptic net.

Stage 1 - Juvenile state: involves local networks becoming operational after conception; i.e, axons have reached target locations. A defining characteristic of this undifferentiated stage is out-of-phase growth of all components that will form a regional network at a later period. It is assumed that local networks are in a "stand-by" position for some future function, with excessive number of synapses in an immature state. Each sex has many local networks that have completed this stage in utero and a few local networks possibly still at this stage at age 16 years. The sexes have complementary development in networks that are in the juvenile stage at birth.

Stage 2 - Formation stage: involves two or more local networks establishing a regional network by developing in tandem; i.e., regions sharing cortico-cortical changes in the gray matter are assumed to be a functioning neural network. A defining characteristic of this stage is high-amplitude in-phase growth and pruning across one or more developmental periods. Each sex has a few regional networks functioning in this manner at birth; others are formed after the bifurcation around age 6 years that marks the end of the early development period. The sexes have complementary development in networks that are in the formation stage at birth.

Stage 3 - Globalization stage: involves the newly formed regional network differentiating connections within and to other brain regions; i.e., development is moving the regional networks toward increased globalization. A defining characteristic is out-of-phase, low-amplitude growth and pruning. A few networks in both sexes enter Stage 3 after the bifurcation at age 6 years; many more enter this stage after the bifurcation at age 10 years.

Stage 4 - Adult stage: involves all local networks in the regional network returning to in-phase, high-amplitude growth and pruning after completing Stage 3; i.e., a resculpting of the networks that enables the individual to function at the more complex level of the adult. Psychometric evidence and neurological studies were considered in determining that Stage 4 is equivalent to young adulthood (Witelson, 1976; Goldman, 1974; Piaget, 1985; Kimura, 1984; 1992).

Figure 10 shows head diagrams for the sexes' 11 regional networks. Stage of development and local networks in each regional network are specified. Regional networks show symmetry across the left and right hemisphere if they have topographies that involve homologous local networks, are at the same stage of development and show similar time, rate and amplitude changes in the growth and pruning cycles. Representative time series are shown in later figures.

Figure 10A shows anterior regional networks, where female development involves fewer, more asymmetric networks (greater lateralization of function). Figure 10B shows posterior regional networks, where male development involves fewer, more asymmetric networks (greater lateralization of function).

The female favors symmetry in timing of development of regional networks at posterior sites with only the bilateral frontal pole network having symmetry in the anterior region; the male favors symmetry in networks at anterior sites. Within regional networks in the same hemisphere, females accelerate growth in local networks across anterior-posterior spatial gradients, while males accelerate growth across posterior-anterior spatial gradients. While network topographies differ, the sexes share some brain regions of slowed growth; e.g., the right- and left-hemisphere parietal association areas.

Sexes' regional networks are complementary in accelerated growth. The focus of networks that are accelerated to the adult stage by age 16 years is very complementary. Female networks favor left hemisphere regions involved in language processing, verbal and sequential memory and sensorimotor integration and bilateral regions involved in broad executive decision making skills. Male networks favor left and right hemisphere regions for visual and spatial processing, right hemisphere nonverbal memory and bilateral frontal pole and prefrontal medial regions for planning and decision making related to movement and targeting. Figure 11 show topographies of regional networks reaching Stage 4 and associated local network time series.

<u>Sexual dimorphism in language and spatial processing networks.</u> Time series for gender differences in regional networks involved in left-hemisphere language processing (Figure 12) and right hemisphere spatial processing (Figure 13) contrasts one sex's accelerated development with the other sex's slowed development.

Figure 12 contrasts the female's one accelerated language processing network with the male's two later developing networks. The female network involves 16 local networks in the left hemisphere; the two male networks involve 4 local networks in the posterior regional network and 6 in the anterior local network. Broca's and Wernicke's regions are developed at the same time in the female and at different times in the male.

Figures 12A and 12B compare first derivative time series for three common local networks from anterior regions for each sex, while Figures 12C and 12D are for local networks from posterior regions. Bifurcations for the time series occur around ages 8 and 12 years in the female's language network (Figures 12A and 12C), around ages 7 and 11 in the male's posterior language network (Figure 12D) and around age 8 in the male's anterior language network (Figure 12B).

The female's accelerated regional network integrates many other left hemisphere functions with the language networks. The male networks are much more focal, with Broca's and Wernicke's areas separated, possibly indicating less robust functional connectivity. Within the female regional network, anterior speech networks F3T3, F7T3 and C3T3 and posterior language formation networks P3T3, T3T5 and C3T5 complete Stage 1 before 6 months (start of study), complete Stage 2 around age 8 years, complete Stage 3 around 12 years, and are in the first cycle of Stage 4 as the study ends. Within the male regional networks, the anterior networks F3T3, F7T3 and C3T3 complete Stage 1 around age 8 years and Stage 2 around age 16 years. The male's posterior networks P3T3, T3T5 and C3T5 complete Stage 1 around age 7 years, complete Stage 2 around age 11 years and are in Stage 3 as the study ends. As with most male regional networks, posterior local networks have accelerated development over anterior networks. Table 7 summarizes the stages and synchrony-asynchrony patterns in the language processing networks.

Table 7

Sexual Dimorphism in Stages of Language Processing Networks

Brain Region and Local Networks	Synchrony-Asynchrony Growth Pattern (Stage)			
		Developmental Period		
	Sex	Early	Middle	Late
Anterior	Female	In-phase (Stage 2)	Out-of-phase (Stage 3)	In-phase (Stage 4)
F7T3, C3T3 & F3T3	Male	Out-of-phase (Stage 1)	In-phase (Stage 2)	In-phase (Stage 2)
Posterior	Female	In-phase (Stage 2)	Out-of-phase (Stage 3)	In-phase (Stage 4)
C3T5, P3T3 & T3T5	Male	Out-of-phase (Stage 1)	In-phase (Stage 2)	Out-of-phase (Stage 3)

Figure 13 contrasts the female's four slow developing posterior right hemisphere spatial processing networks with the male's one accelerated regional network. The female networks involve 8 local networks with two networks having bilateral components in the posterior left hemisphere (O1T5 and O1P3). The male network involves 8 local networks with O1T5, the only bilateral component in the posterior left. Nonverbal memory, spatial processing and integration of vision with somatosensory networks develop at the same time in the male, while these functions develop separately and over a different time schedule in the female.

Figure 13 compares first derivative time series for T4T6 and O2T4; i.e., cortico-cortical connections between nonverbal memory centers and the occipital center. Bifurcations for the time series occur around ages 7 and 14 years in the female's spatial network (Figure 13A) and around ages 6 and 11 in the male's network (Figure 13B).

The female local network for nonverbal memory and sensory integration with vision completes Stage 1 around age 7 years, Stage 2 around age 12 years, and is in Stage 3 as the study ends (Figure 13A). Within the male regional network, all local networks complete Stage 1 before 6 months (start of study), complete Stage 2 around 5 years, complete Stage 3 around age 12 years, and are in the first cycle of Stage 4 as the study ends (Figure 13B). Table 8 summarizes the stages and synchrony-asynchrony patterns in the spatial processing networks.

Table 8

Sexual Dimorphism in Stages of Spatial Processing Networks

Brain Region and Local Networks	Synchrony-Asynchrony Growth Pattern (Stage)			
		Developmental Period		
	Sex	Early	Middle	Late
Posterior Right	Female	In-phase (Stage 1)	Out-of-phase (Stage 2)	In-phase (Stage 3)
O2T4 & T4T6	Male	Out-of-phase (Stage 2)	In-phase (Stage 3)	In-phase (Stage 4)

Summary

Experimental Results

Mean coherence measurements are able clearly to confirm, bound, localize and quantify the stages and the cyclic nature of development and to estimate gender differences in cortico-cortical connections relevant to cognitive functioning. The validity of the results rests on the following assumptions:

● Mean EEG coherence is a window on the synaptic processing changes by which the neural circuits increase in complexity;

● The bifurcations are evidence of genetic input to theprocess of growth and pruning that shapes the local networks across the life cycle; the discontinuities in the wave flow indicate that structural limitations are imposed on the individual to insure that he/she prepares for a new stage of life;

● Gender determines organization of local networks into regional networks with different topographies and different developmental time tables.

Appendix 5 summarizes the primary gender results in table form with references to parallel research by other scientists.

Potential Inferences

Yin-yang is a term defined in Taoist philosophy, a belief system that emphasizes a meditative life close to nature. The symbol used to summarize the core ideas in Taoism contains geometric representations of the two basic forces of the universe--yin (female) and yang (male)-- in total balance and unity, yet completely complementary.

World Book Encyclopedia, 1976

Human Self-Organizing Responses in a Structurally-Limited System. The development of the human brain seems to be a marriage of an individual's "feed forward" and "feed backward" self-organizing responses to genetically initiated inputs. The response is influenced by gender and the unique sexual hormone distribution system related to femaleness and maleness. In the data time frame of my studies, genetic inputs occur around ages 6, 10 and 20 years, leading to bifurcations in the local networks. With stage of development retarded dependent on gender, the local network enters either a period of disequilibrium or a period of globalization.

Feed forward systems[*]. The disequilibrium periods lead to an increase in synaptic maturation; these growth spurts lead to modification of the system in a feed-forward mode. The individual self-organizes to restore equilibrium between output behaviors possible and input structural capabilities; it is possible to describe the process in terms of chaotic behavior[**]. Females and males attempt to retain output behaviors that have historically been useful for their survival; however, the new structural capabilities in the context of a guiding culture permit exchange of old output behaviors for similar, but more complex, new behaviors. The individual modifies the structure by favoring some synapses over others; this structural modification eventually leads to modification of behavioral outputs.

Gender and sex hormones add a layer of complexity to the feed forward processes for which Marion Diamond's work (1988) with rats provides some insight. Building on her work, I suggest the following may be happening. Gender may have a greater role than sex hormones in

---

[*] Feed forward mechanisms in the genetic program lead to bifurcations in network sculpting, forcing the individual in the context of her/his environment to shape all brain networks for the next stage of development.

[**] In a preliminary study on female networks T3T5 and T4T6, a sequence of 4 attractors was found that paralleled the stages defined in this paper. T3T5 was exhibiting the 4th attractor at age 16 years; T4T6 was exhibiting the 3rd attractor at age 24. T3T5 and T4T6 were selected because the former has accelerated development and the latter has retarded development.

structurally limiting the networks in early development to complementary roles needed forpropagating the specie, but the level of sex hormones will influence the continuum of female/male structures and behaviors. In rats, Diamond found that some gender-specific structures were retained even when the gonads were removed early in development. For example, male sex organs removed at birth led to the male rat retaining development of occipital cortices like males but developing frontal and somatosensory cortices like female rats. In a similar manner, female organs removed at birth led to the female rat retaining development of frontal and somatosensory cortices like females but developing motor cortices like male rats. Diamond found gonads removed at later time periods had less effect on the brain structures. Hence, variance in the levels of sex hormones, particularly early in development, may lead to the complex continuum of overlapping female and male behaviors that we see in humans.

Diamond found that as female rats age they develop sites favored by male rats in early development (right posterior networks) while male rats do not develop sites favored by female rats until a much later period. In humans, a similar phenomenon seems to occur, but with both sexes shifting developmental focus around 8 years to brain regions favored prenatally and early in development by the other sex. The female has growth bursts in the right hemisphere posterior spatial region, bilateral occipital networks and bilateral long-connection networks, while the male has growth bursts in the left hemisphere anterior language and memory regions and in bilateral frontal decision-making centers.

While one benefit of neoteny is to tailor the late developing networks to better fit the environment, the disadvantage of focal networks favored by the sexes in late development is the potential for dysfunction if any portion of the system is injured. Kimura's studies on aphasia and apraxia (1984) show males, with two networks, have poorer outcomes after left hemisphere strokes than females. It would seem predictable that females, with four networks, would have poorer outcomes to injuries to right hemisphere spatial networks than males.

Feed backward systems. Feed backward mechanisms are most effective when the response time of the system being modified is very slow. Mean EEG coherence time series show long time intervals, on the order of 2 to 16 years, in the length of resculpting cycles in the local networks. It is assumed that this time is needed for regional network consolidation of learned programs, and, presumably, for adjustments in interconnections to other networks to eliminate redundancy of function and to establish efficient routes for neurons to communicate with each other; i.e., to increase globalization.

As we observe the process of the feed forward and feed backward mechanisms over long periods of time, the products of self-organization become apparent. In a manner similar to the way in which the Hubble telescope studies the universe, we now can observe boundary patterns for self-organization in neural networks, whether specified by gender or by time. The Taoist symbol nicely summarizes the mysteries of intrasex and intersex relationships in the human specie.

Personal Observations

I think it is important that we not define human behavior by the extreme tails of the female/male continuum. Categories like "hunter" and "gatherer" imply separate activities for the sexes, with no overlap, rather than the integrated activities that will be necessary for the twenty-first century. We need to give humans permission to reach cognitive and affective potentials at every stage of the life cycle; to develop their maleness and femaleness rather than restricting them to one aspect of their nature. Our genetic programs evolved very slowly to optimize our successful navigation. We now can develop an accurate taxonomy of behaviors linked to gender brain stages to guarantee a less stressful, more enjoyable passage for all individuals. An underlying theme is that females and males are constantly preparing for the future in a complementary manner; they need to understand these changes to better support each other while providing the needed companionship for optimal specie survival. As Marian Diamond (1989) says, "four brains are better than two."

Future research

Until very recently, studies of children's performance focused on the cognitive products of thinking, with the assumption that girls and boys had the same neural circuits with which to perform age-linked tasks. Given the prime finding of this study that the genders differ in both neurological structures of the regional networks and the timing of their development, this assumption need to be re-examined.

How the sexes, with their complementary developments, navigate interpersonal relationships, particularly at life cycle bifurcations where cognitive and affective abilities shift rather dramatically, needs to be examined. Longitudinal studies are needed to determine whether changes experienced by humans throughout the life cycle are the result of a biological catalyst.

Acknowledgments

I am indebted to Dr. Robert Thatcher for the care and precision with which he collected the data base, to Dr. John Dockery for his encouragement and support in development of the regional network stage theory, to Dr. Jerry Cline for his continued direction and guidance regarding postdoctoral positions and to Robert Hanlon for his continued support of all my work.

REFERENCES AND NOTES

Bayer S. A. & Altman, J. (1991). Summary and Conclusions. Chapter 17 in <u>Neocortical development</u>, 216-223. New York, NY: Raven Press.

Caviness, V. (1994). Chief of Pediatric Neurology, Massachusetts General Hospital & Joseph & Rose Kennedy Professor of Child Neurology & Mental Retardation, Harvard Medical School, personal communication (1994): for 30 female and male children, with mean and median age of 9 years and with a range of 6 to 11 years, volumetric measurements determined: the neocortex and diencephalon are proportionately larger in the female; structures of cerebral central white matter, caudate and putaminal nucleus, pallidum and hippocampus are proportionately larger in the female; amygdala of male child 's brain is 115% of adult size; only the pallidum (right > left) and putamen (left > right) show asymmetries greater than 5% and when the asymmetries were examined by gender, asymmetries were present only in the female pallidum (right 8% > left) and caudate (left 4% > right).

Changeux, J.-P., Heidmann, T. and Patte, P. (1984). Learning by selection, <u>The biology of learning</u>, P. Marler & H.S.Terrance (Eds.). New York, NY: Springer-Verlag.

Corsi-Cabrera, M., Herrera, P. & Malvido, M. (1989). Correlation between EEG and cognitive abilities: sex differences. <u>International Journal of Neuroscience, 45</u>, 133-141.

de Lacoste, M.-C. & Horvath, D. S. (1985). Sex differences in the development of morphological asymmetries in human fetuses (abstract). <u>American Journal Physical Anthropology, 66</u>, 163.

de Lacoste, M.-C., Horvath, D.S. & Woodward, D.J. (1991). Possible sex differences in the developing human fetal brain. <u>Journal of Clinical and Experimental Neuropsychology, 13(6)</u>, 831-846.

Diamond, M. C. (1988). <u>Enriching heredity, the impact of the environment on the anatomy of the brain.</u> New York, NY: The Free Press.

Diamond, M. C. (1989). Sex and the cerebral cortex. <u>Biological Psychiatry</u>, 823-825.

Duffy, F. (1994). The role of quantified electroencephalography in psychological research. In <u>Human behavior and the developing brain</u>, G. Dawson & K.W. Fischer (Eds.). New York, NY: Guilford Publications.

Epstein, H. T. (1974). Phrenoblysis: special brain and mind growth periods. I. Human brain and skull development. <u>Developmental Psychobiology, 7(3)</u>, 207-216.

Feingold, A. (1992). Sex differences in variability in intellectual abilities: a new look at an old controversy. <u>Review of Educational Research, 62(1)</u>, 61-84.

Fox, N. A. & Bell, M. A. (1990). Electrophysiological indices of frontal lobe development. In <u>The development and neural bases of higher cognitive functions</u>, A. Diamond (Ed.). New York, NY: The New York Academy of Sciences.

Goldman, P. S. et al. (1974). Sex-dependent behavioral effects of cerebral cortical lesions in the developing rhesus monkey. <u>Science, 186</u>, 540-542.

Gould, S. J. (1977). <u>Ontogeny and phylogeny</u>, Cambridge, MA: Harvard University Press.

Gould, S. J. (1982). Change in developmental timing as a mechanism of macroevolution. <u>Evolution and development</u>, J.T. Bonner (Ed.).

Gould, S. J. (1994). The evolution of life on earth. <u>Scientific American</u>, 85-91.

Guilford, J. P. (1954). <u>Psychometric methods</u>, New York, NY: McGraw-Hill Book Co.

Gur, R. C. et al. (1982). Sex and handedness differences in cerebral blood flow during rest and cognitive activity. Science, 217, 659-661.

Gur, R. C., et al (1995). Sex differences in regional cerebral glucose metabolism during a resting state. Science 267, 528-531.

Hanlon, H. W. (1994). Differences in female and male development of the human cerebral cortex from birth to age 16, U.M.I. Dissertation Information Service, #9507750.

Hanlon, H. W. (1995). Evolution of sexual differences as reflected in local and regional cerebral cortex networks.

Abstract distributed at Washington Evolutionary System Society meeting.

Harmony, T. (1984). Functional neuroscience, vol.3, neurometric assessment of brain dysfunction in neurological patients. Hillsdale, NJ: Lawrence Erlbaum Associates.

Hedges, L. V. & A. Nowell (1995). Sex differences in mental test scores, variability and numbers of high-scoring individuals. Science, 269, 41-45.

Hudspeth, W. J. & Pribram, K. H. (1990). Stages of brain and cognitive maturation. Journal of Educational Psychology, 82(4), 881-884.

Huttenlocher, P. R. (1990). Morphometric study of human cerebral cortex development. Neuropsychologia, 28(6), 517-527.

Huttenlocher, P. R. (1994). Synaptogenesis in Human Cerebral Cortex. In Human behavior and the developing brain, G. Dawson & K. W. Fischer (Eds.). New York, NY: Guilford Publications.

Kandel, E. R. (1985). Synapse formation, trophic interactions between neurons, and the development of behavior. In Principles of neural science, Kandel, E. R. & Schwartz, J. H. (Eds.). New York, NY: Elsevier Science Publishers.

Kimura, D. & Harshman, R. A. (1984). Sex differences in brain organization for verbal and non-verbal functions. In Progress in Brain Research, 61, DeVries, G.J. (Ed.). Amsterdam: Elsevier Science Publishers.

Kimura, D. (1992). Sex differences in the brain. Scientific American (September), 119-125.
Lopez da Silva, F. et al. (1989). Brain topography, 2.

Maccoby, E. E. & Jacklin, C. N. (1974). The psychology of sex differences. Stanford, CA: Stanford University Press.

Matousek, M. & Petersen, I. (1973). Frequency analysis of the EEG in normal children and adolescents. In Automation of clinical electroencephalography, P. Kellaway & I. Petersen (Eds.). New York, NY: Raven Press.

Naylor, C. (1991). Personal communication In Brain Sex: the real difference between men and women, Moir A. & Jessel, D., pg. 195. New York, NY: Carol Publishing Group.

Nunez, P. L. (1981). Electric fields of the brain: the neurophysics of EEG. New York, NY: Oxford University Press.

Piaget, J. (1985). The equilibration of cognitive structures, the central problem of intellectual development. Chicago, IL: The University of Chicago Press.

Savitzky, A. & Golay, M. J. E. (1964). Smoothing and differentiation of data by simplified least square procedures. Analytical Chemistry, 36, 1627-1639.

Shaywitz et al. (1995). Sex differences in the functional organization of the brain for language. Nature, 373, 607-609.

Smart, I. H. M. (1983). Three dimensional growth of the mouse isocortex. Journal of Anatomy, 137(4), 683-694.

Swaab, D. (1984). Sexual differentiation of the human brain, a historical perspective. Progress in Brain Research, 61, De Vries et al (Eds.), 361-374.

Thatcher, R. W. (1983). Hemispheric EEG asymmetries related to cognitive functioning in children. In: Cognitive processing in the right hemisphere, Perecman, E. (Ed), New York, NY: Academic Press.

Thatcher, R. W., Krause, P. J. & Hrybyk, M. (1986). Cortico-cortical associations and EEG coherence: a two-compartmental model. Electroencephalography and clinical neurophysiology, 64.

Thatcher, R. W., Walker, R. A. & Giudice, S. (1987). Human cerebral hemispheres develop at different rates and ages.
Science, 236, 1110-1114.

Thatcher, R. W. (1991). Maturation of the human frontal lobes: physiological evidence for staging. Developmental Neuropsychology, 7(3), 397-419.

Thatcher, R. W. (1992). Cyclic cortical reorganization during early childhood. Brain and Cognition, 20, 24-50.

Thatcher, R. W. (1994). Cyclic cortical reorganization: origins of human cognitive development. In Human behavior and the developing brain, G. Dawson & K. W. Fischer (Eds.).
New York, NY: Guilford Publications.

Tucker, D. M., Roth D. L. & Bair T. B. (1986). Functional connections among cortical regions: topography of EEG Coherence. Electroencephalography and clinical neurophysiology, 63.

Witelson, S. F. (1976). Sex and the single hemisphere: specialization of the right hemisphere for spatial processing. Science, 193, 425-427.

World Book Encyclopedia (1976 edition).

Appendix 1
Electrode Sites: Abbreviation and Descriptor for Variables

Fp1 and Fp2: frontal lobe pole (area near forehead)
F3 and F4: dorsal-medial (top-middle), frontal lobe
F7 and F8: ventral-lateral (bottom-outside), frontal lobe
C3 and C4: medial-central, frontal lobe (posterior)
T3 and T4: anterior (front area), temporal lobe
T5 and T6: posterior (back area), temporal lobe
P3 and P4: medial-posterior, parietal lobe
O1 and O2: occipital lobe pole (back of head)

Variable (Electrode-Pair) Left Hemisphere	Distance between Two Sites: Right Hemisphere	(in centimeters)
Fp1F3	Fp2F4	6 cm
Fp1F7	Fp2F8	7 cm
Fp1C3	Fp2C4	12 cm
Fp1T3	Fp2T4	14 cm
Fp1P3	Fp2P4	18 cm
Fp1T5	Fp2T6	21 cm
Fp1O1	Fp2O2	24 cm
F3F7	F4F8	6 cm
F3C3	F4C4	6 cm
F3T3	F4T4	10 cm
F3P3	F4P4	12 cm
F3T5	F4T6	14 cm
F3O1	F4O2	18 cm
C3F7	C4F8	10 cm
C3T3	C4T4	6 cm
C3P3	C4P4	6 cm
C3T5	C4T6	10 cm
C3O1	C4O2	12 cm
P3F7	P4F8	14 cm
P3T3	P4T4	7 cm
P3T5	P4T6	6 cm
P3O1	P4O2	6 cm
O1F7	O2F8	21 cm
O1T3	O2T4	14 cm
O1T5	O2T6	7 cm
F7T3	F8T4	7 cm
F7T5	F8T6	14 cm
T3T5	T4T6	7 cm

338

Appendix 2

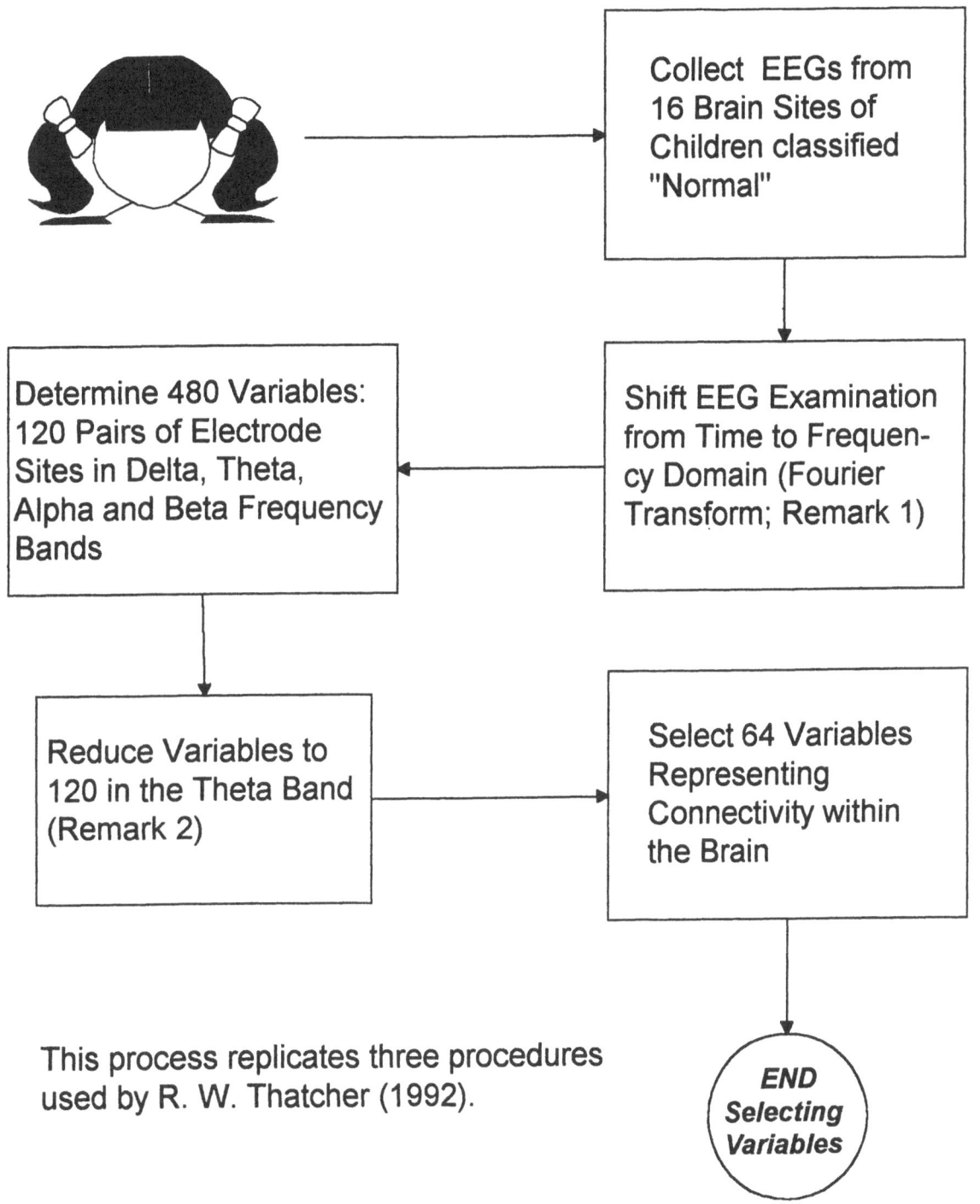

Appendix 2, Figure 1: Reduction Procedure for Selection of Variables

Remarks for Appendix 2, Figure 1

1. Frequency bands, including the center frequencies (Fc) and one-half power values (B) were: Delta (0.5 to 3.5 Hz; Fc=2.0 Hz and B=1.0 Hz); Theta (3.5 to 7.0 Hz; Fc=4.25 Hz and B=3.5 Hz); Alpha (7.0 to 13.0 Hz; Fc=9.0 Hz and B=6.0 Hz); and Beta (13.0 to 22.0 Hz; Fc=19.0 Hz and B=14.0 Hz). Modern filter theory was used which results in minimal distortion of the wave of interest (Thatcher, 1987).

2. The Theta band contains all frequencies from 3.5 to 7.0 Hz. The sampling rate of 100 Hz permits undistorted reproduction of frequencies as high as 50 Hz (Nyquist criterion). If the EEG signal were to be a perfect sinusoid in the Theta band, then the Fourier transform would yield one term in the series with the frequency of that sinusoid. Since the EEG is the summation of sinusoids at many different frequencies, this trivial case would not occur. For example, an EEG might have power in the Theta band at 2 frequencies or at 10 frequencies, depending on the signal detected at the electrode site. Coherence is a cross-correlation of the power of all component frequencies within this spectrum.

    Thatcher (1991) analyzed the data in the four frequency bands, without identifying the variables by sex, and determined the factor structure to be the same in each band for the 64 variables considered here. He found 9 to 11 factors in each band.

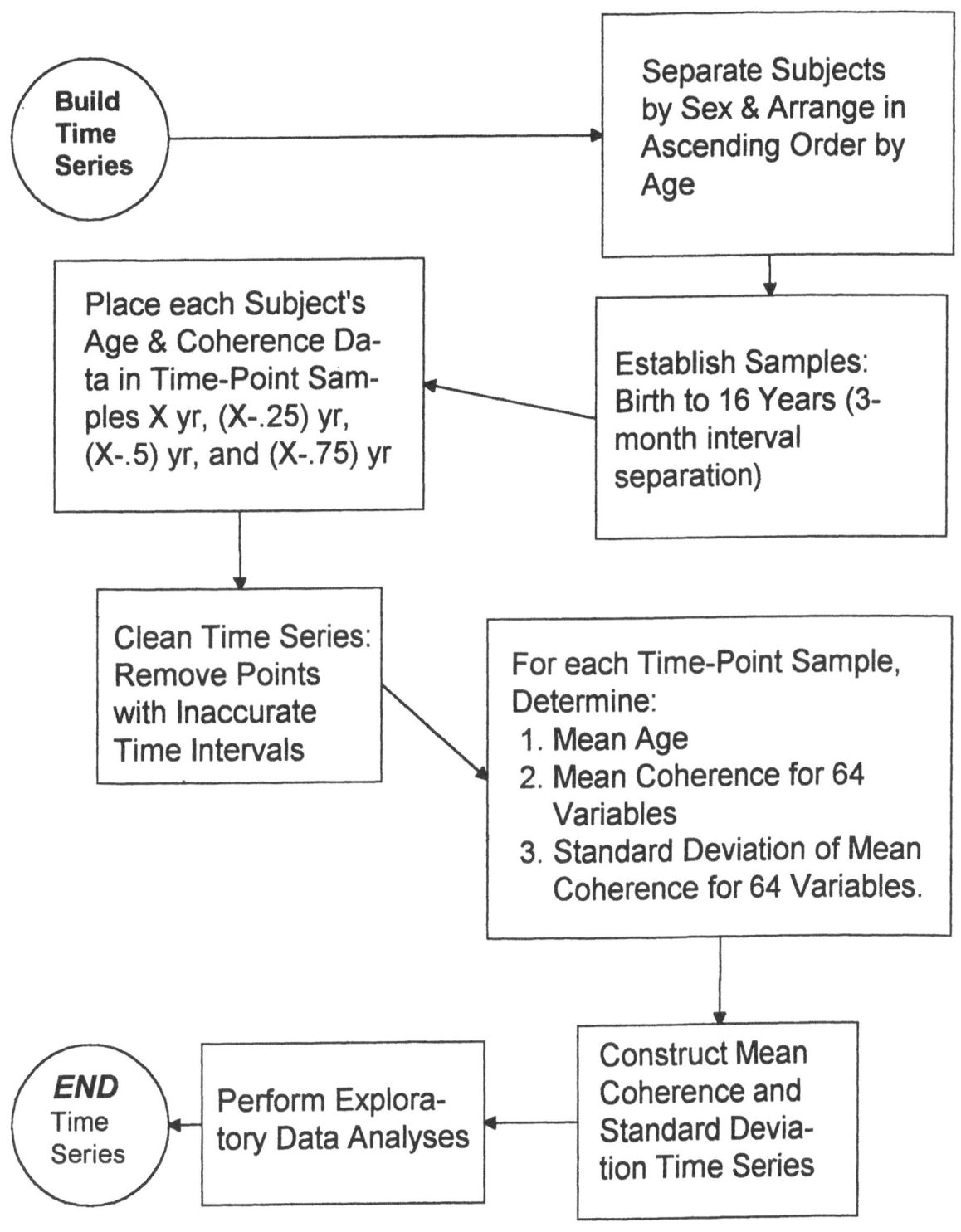

Appendix 2, Figure 2: Samples for Time Points in Each Sex's Development

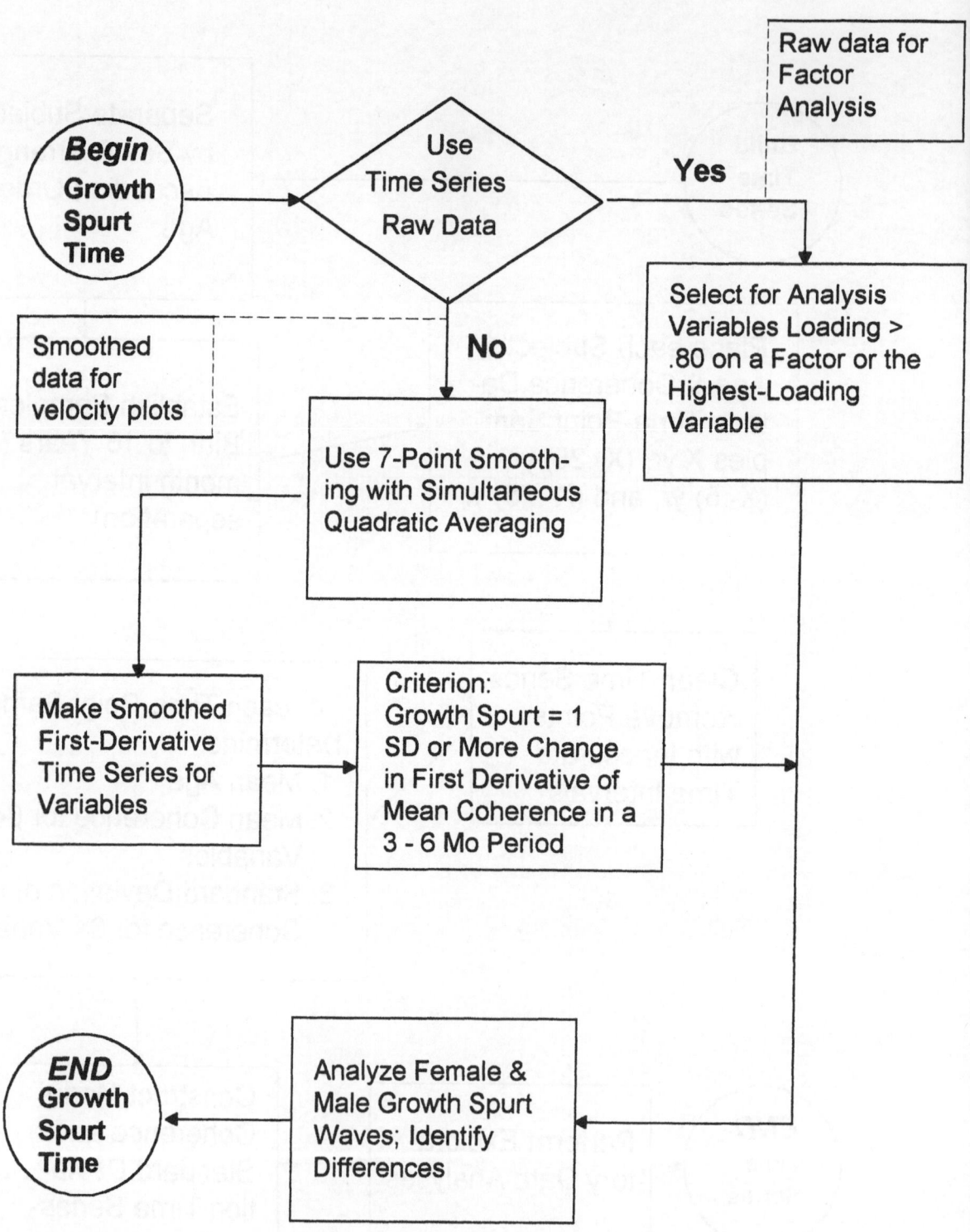

Begin Growth Spurt Time

Use Time Series Raw Data

Yes

Raw data for Factor Analysis

Smoothed data for velocity plots

No

Select for Analysis Variables Loading > .80 on a Factor or the Highest-Loading Variable

Use 7-Point Smoothing with Simultaneous Quadratic Averaging

Make Smoothed First-Derivative Time Series for Variables

Criterion:
Growth Spurt = 1 SD or More Change in First Derivative of Mean Coherence in a 3 - 6 Mo Period

END Growth Spurt Time

Analyze Female & Male Growth Spurt Waves; Identify Differences

Appencix 2, Figure 3: Select Variables, Prepare Time Series for Analysis

Appendix 3

Correlations: Female Mean Coherence and Variance Time Series

Time Series (Electrode-Pair)	Correlation At 0 Phase Shift	Distance Between Electrode Pairs
F1F3	-.827	6cm
F2F4	-.865	6cm
F1F7	-.693	6cm
F2F8	-.913	6cm
F3C3	-.612	6cm
F4T4	-.676	6cm
F8T4	-.672	6cm
C4F8	-.645	6cm
C4T4	-.790	6cm
P3T5	-.631	6cm
P3O1	-.604	6cm
O1T5	-.757	6cm
F1O1	+.672	24cm
F1T3	+.661	18cm
F1T5	+.643	21cm
F7T5	+.732	12cm
F4O2	+.621	18cm
P3F7	+.696	18cm
O1T3	+.822	18cm
O2T4	+.726	18cm
O2F8	+.778	18cm

5 electrode-pair sites with correlations between -.599 and +.599 at 0 phase shift:F1C3:+.131;
1P3:+.554; F2C4:-.173; F2P4:+.327; F2O2:+.396; F2T4:-.141; F2T6:+.523; F3F7:-.557; F3T3:-.165;
3T5:+.577; F3P3:-378; F3O1:+.518; F4F8:-.457; F4C4:-.271; F4P4:-.194; F4T6:+.508;F7T3:-.109;
8T6:+.469; C3F7:-.029; C3T3:-.205; C3T5:-.005; C3P3:-.506; C3O1:+.020; C4P4:-.520; C4T6:-.206;
4O2:+.066; P3T3:-.067; P4F8:+.316; P4T4:-.539; P4T6:-.478; P4O2:-.393; O1F7:+.491; O2T6:-.219;
3T5: +.038; T4T6:-.199

Correlations: Male Mean Coherence and Variance Time Series

Time Series (Electrode-Pair)	Correlation At 0 Phase Shift	Distance Between Electrode Pairs
F1F3	-.723	6cm
F2F4	-.857	6cm
F1F7	-.669	6cm
F2F8	-.840	6cm
F3F7	-.799	6cm
F3C3	-.619	6cm
C3T3	-.662	6cm
C3P3	-.694	6cm
C4P4	-.811	6cm
P3T5	-.611	6cm
P4T6	-.770	6cm
P3O1	-.725	6cm
P4O2	-.628	6cm
T3T5	-.608	6cm
F1T5	+.646	21cm
F3T5	+.718	18cm
F3O1	+.804	18cm
F7T5	+.824	12cm
F8T6	+.606	12cm
O1F7	+.772	18cm
O2F8	+.748	18cm

<u>35 electrode-pair sites with correlations between -.599 and +.599 at 0 phase shift:</u>F1C3:+.253; F1T3:+.018; F1P3:+.513; F1O1:+.582; F2C4:-.109; F2T4:+.089; F2P4:+.298; F2T6:+.347; F2O2:+.187; F3T3:-.279; F3P3:-.069; F4F8:-.579; F4C4:-.577; F4T4:-.408; F4P4:-.329; F4T6:+.167;F4O2:+.322; F7T3:-.583; F8T4:-.112; C3F7:-.378; C3T5:-.261; C3O1:-.497; C4F8:-.292; C4T4:-.355; C4T6:-.456; C4O2:-.442; P3F7:+.382; P3T3:-.124; P4F8:+.031; P4T4:-.293; O1T3:-.078; O1T5:-.588; O2T4:+.301; O2T6:-.063; T4T6:-.159

Appendix 4

Difference Scores and Variance Ratios for Females and Males

Variable	Mean Difference (Female-Male)	Median Difference (Female-Male)	Variance Ratio (Female/Male)
Fp1F3	-1.099	-0.769	2.155
Fp1C3	-2.407	-3.317	2.277
Fp1P3	-0.129	-1.972	1.442
Fp1O1	0.309	-0.004	1.857
Fp1F7	0.077	1.211	2.153
Fp1T3	-0.245	-0.055	3.153
Fp1T5	3.019	2.540	3.232
Fp2F4	-0.397	-0.834	0.833
Fp2C4	-2.917	-3.574	1.366
Fp2P4	-1.081	-1.353	1.034
Fp2O2	-0.062	-0.325	1.635
Fp2F8	-0.251	-0.714	1.159
Fp2T4	-1.941	-3.362	1.380
Fp2T6	-0.151	-0.172	1.784
F3C3	-2.472	-2.660	1.129
F3P3	-1.701	-1.736	1.040
F3O1	-0.417	-0.070	1.138
F3F7	0.309	0.258	0.629
F3T3	-1.407	-1.262	2.838
F3T5	1.536	1.605	1.911
F4C4	-3.014	-3.282	1.041
F4P4	-2.853	-2.276	1.716
F4O2	-0.833	-0.731	1.800
F4F8	-0.365	0.135	1.048
F4T4	-2.996	-2.756	2.068
F4T6	0.433	0.301	1.221
C3P3	-1.790	-1.344	2.658
C3O1	-0.413	-0.662	1.614
C3F7	-2.320	-3.695	0.562
C3T3	-0.133	0.257	1.825
C3T5	1.167	1.354	5.740
C4P4	-1.891	-0.331	3.456
C4O2	0.039	0.936	1.643
C4F8	-3.101	-3.056	1.992

Appendix 4 (continued)

Variable	Mean Difference (Female-Male)	Median Difference (Female-Male)	Variance Ratio (Female/Male)
C4T4	-2.264	-2.236	3.200
C4T6	1.458	2.001	1.121
P3O1	-0.768	-0.635	0.863
P3F7	-1.691	-1.812	1.195
P3T3	-0.090	-0.261	3.262
P3T5	0.775	0.424	1.645
P4O2	-1.278	-1.241	1.138
P4F8	-2.785	-2.377	1.545
P4T4	-3.072	-3.681	1.273
P4T6	0.219	0.134	0.795
O1F7	-0.213	-0.709	1.127
O1T3	0.605	-0.311	2.730
O1T5	-2.168	-1.767	3.770
O2F8	-0.085	-0.594	1.078
O2T4	-3.012	-2.737	0.442
O2T6	-2.834	-2.948	1.150
F7T3	-2.525	-2.884	0.898
F7T5	1.756	0.644	2.351
F8T4	-4.113	-3.601	3.856
F8T6	-0.254	-0.259	0.945
T3T5	0.393	0.254	8.346
T4T6	-1.342	-2.558	0.645

MEAN -0.944 MEAN -0.938 MEAN 1.873

Wilcoxon Large Sample Approximation for Distribution-Free Signed Rank Test for mean differences (T=-4.005; p<.0000).

Fisher Large Sample Approximation for Distribution-Free Sign Test for variance differences (B=5.078; p<.0000).
****

Appendix 5

Gender Difference/Similarity in Mean Coherence Development

Hanlon Results	Parallel Supporting Research
BASIC HUMAN DEVELOPMENTAL PROCESSES	
1. <u>In both sexes.</u> Fourier analysis of mean coherence time series shows phase shifts in all frequencies of the spectrum. The net effect is a discontinuity or bifurcation that appears as a cusp in the 3-dimensional phase portraits. <u>Bifurcations appear at differ times in the sexes with respect to a specific local network</u>. In both sexes, bifurcations occur around ages 4-8 and 9-12 years, dividing development into early, middle and late periods.	1a. R. W. Thatcher (1992, 1994): anatomical growth cycles of mean coherence are marked by phase transitions around age 6 and 10 years (human studies). 1b. J. Piaget (1985): stage theory describes qualitative rather than quantitative differences between sensorimotor, preoperational, concrete operational and formal operational stages (human studies).
2. <u>In both sexes, mean coherence is a traveling wave.</u> The spatial gradients include lateral-to-medial clockwise and counter-clockwise rotations and anterior-to-posterior expansions and contractions. The <u>sexes differ in rate of wave flow</u> within a hemisphere <u>and in spatial gradient of flow</u>; e.g., one sex may have a clockwise rotation while the other sex has a counterclockwise across the same brain regions.	2a. R.W. Thatcher (1992): growth cycles have a period of approximately 2 to 4 years and involve both a rostral-caudal expansion and contraction as well as a lateral-to-medial rotation (human studies). 2b. I.H.M. Smart (1983); S.A. Bayer & J. Altman (1991): same developmental wave pattern observed in mice and rats during fetal development.

3. <u>In both sexes, variance of mean coherence is heterogeneous.</u> Relationship between 56 mean and variance velocity time series has correlation ranging from -.9 to +.9. See Appendix 3.
 	
    It is hypothesized that plasticity in the network decreases as correlation goes from -1 to +1. Both sexes have many sites, particularly those separated by short distances, with negative correlations in the late development period.
 	
    The sexes have sites at which the correlations differ. O1F7 shows limited plasticity in the male's from birth (r=.772), while female network's plasticity is not reduced until the bifurcation at age 6 (r=.491). Male O1F7 is in a regional network at Stage 2 at birth; while female O1F7 is at the juvenile Stage 1 at birth, with Stage 2 beginning around age 8 years.
	3a. Developmental plasticity studies indicate brain functions in humans can be preserved dependent on time of insult and whether brain region injured or improperly stimulated has synaptic plasticity. Most notably, strabismic amblyopia vision studies show this condition can be reversed if diagnosed during the period from birth to age 4 years when the visual cortex has great plasticity (Huttenlocher, 1990).
	
	3b. Goldman (1974). Male rhesus monkeys with orbital prefrontal lesions were impaired on behavioral tests at 2.5 months of age whereas deficits were not detected in females with comparable lesions made at birth until 15 to 18 months of age.
	
	    Note: rhesus monkeys may begin to have deficits when they establish regional network. The female monkey may have been able to use alternative juvenile networks to complete the functional task until structural limits were imposed by the genetic program around 15 months.

GENDER DIFFERENCES IN LOCAL NETWORKS	
4. <u>Temporal pattern of growth and pruning waves</u> that estimates changes in cortico-cortical networks across the first 16 years of development <u>differs between sexes</u> in all local networks.	4a. No comparable neurological human study. 4b. P. Goldman (1974): found female and male chimps differ in timing of development of connections between the occipital and orbital frontal lobe.
5. The total <u>number of growth and pruning spurts</u> is similar in the sexes, but there are intrasex and intersex differences with respect to the left and right hemispheres.	5. No parallel study.

6. <u>Baselines of female mean coherence time series are lower</u> than male in 42 of 56 intrahemisphere local networks. Mean differences across the 56 sites indicate female baselines are 1 percent lower.
 	
    Lower coherence supports proportionately greater density of gray matter in the female and greater density of white matter in the male.
	6a. R. C. Gur (1982):
	
	    Intrasex: females have proportionately more gray matter; males have proportionately more white matter.
	
	    Intersex: females have more gray matter than males.
	
	6b. V.Caviness (1994): females have proportionately more white and gray matter than males at mean age 9 years.
	
	6c. M. C. Diamond (1988): limited study shows human females with more glial cells per neuron in right and left parietal association areas
	than males, indicating more evolved brain regions with metabolically demanding nerve cells.
	    Differences in humans parallel differences found in rats.

7. <u>Variance of female mean coherence time series is greater</u> than male in 47 of 56 intrahemisphere local networks. Variance ratios indicate growth and pruning peaks are 88 percent greater in the female.	7. No parallel study in humans.

8. <u>Components in mean coherence spectrum</u> (Fourier analysis of time series) <u>vary in period</u> from 6 months to 16 years, with components having highest power being cycles between 2 and 5 years in length.
 	
    Females have more local networks with multiple high energy components (15 compared to 2 in the male); T3T5 has the most with 5. Females have 55 high energy components, males have 18.
 	
    In the time domain, the dominant cycles contribute to the amplitudes of the growth and pruning peaks being greater.
	8a. H.T.Epstein (1974): skull measurement studies showed growth spurts and plateaus in brain and mind development (human studies).
	
	8b. Hudspeth & Pribram (1990): determined single maturation trajectories for four brain regions using the normed EEG data published by Matousek and Petersen in 1973. Curves show growth spurts and plateaus at five time periods that coincide with time frame specified by Piaget for his stage theory (human studies).
	
	8c. R. W. Thatcher (1992, 1994): observed 2-year growth cycles in EEG mean coherence nested within 4-year interhemispheric rotations (human studies).

GENDER DIFFERENCES IN REGIONAL NETWORKS	
9. The sexes have 11 <u>regional networks</u> that <u>differ in topography, rate of development and asymmetry relationships</u> across left/right and anterior/posterior spatial gradients. About 95 percent of variance across the 16 years of the study is accounted for in each sexes' development when time series are analyzed using principal components analyses across early, middle and late development periods defined by bifurcations.	9a. R. W. Thatcher (1991): 9 to 11 factors determined for each of the four frequency bands for mean EEG coherence time series examined (human studies). Anatomical patterns of regional networks are similar independent of frequency bands.

10. <u>Sexual dimorphism exists in developmental time needed for regional networks</u> to pass from juvenile stage of undifferentiated synaptic connectivity to adult stage of globally complex interconnected networks.

Females have 2 networks and males have 4 that reach adult functional level by age 16 years. The greater number of networks with juvenilized morphology in the female supports Gould's theory of progenesis since female sexual maturation is more precocious than males. However, the networks of the female that are accelerated are those needed to make judgments on remembered, pragmatic events both in making decisions for the present and planning for the future; these networks would be most beneficial for making good judgments with regard to caring for and protecting children.

The male networks, on the other hand, seem to define survival in terms of nobly protecting the specie since they favor speed in movement, complex abilities to focus on targets and planning strategies that focus more on the future, with little concern for past events since left hemisphere memory systems for them are very juvenile.

Examination of time series for the sexes for 16 to 24 years of age indicates that the most juvenilized regional networks in both sexes are in Stage 4 by age 24; i.e., the female is working on higher-level thinking networks and the male is working on judgment and memory systems.

10a. S. Gould (1977): evolutionary ideas of <u>progenesis</u> and <u>neoteny</u>, where the former favors precocious sexual maturation with juvenilized morphology as a secondary consequence and where the latter represents fine tuning of the organs for immediate ecological conditions.

10b. M. Diamond (1988).

On female rats: "In contrast to the male, the development of the female cortex follows a different course..The frontal cortex is quite well developed at birth, growing only by 2% in the week from 7 to 14 days..reaching a peak by 18 days...The somatosensory was also more developed at birth than the male ...it may be that the female cortex is more highly developed at birth to ensure a better start for the reproduction of the species."

On male rats: "All regions grow equally rapidly after birth until somewhere between 26 and 41 days of age, when they begin a gentle but steady decline for the remainder of their lives..In the newborn male rat the right cerebral cortex is thicker than the left and this pattern persists into advanced maturity."

On enriched environments: in the female, all regions thickened, with somatosensory the most; in the male, the occipital changes more than any other area... Males have a more responsive visual cortex, while females have a more responsive general sensory cortex."

11. Asymmetry differences in accelerated regional networks:
females favor left and anterior regions; males favor right and posterior regions.

Female development is accelerated in lateral and medial frontal pole executive networks for decisions related to present, past and future events and in left hemisphere networks involved with verbal memory, somatosensory integration, language formation and speech processing.

In a complementary manner, male development is accelerated in frontal lobe executive networks for planning related to movement and visual targeting, bilateral occipital lobes and right hemisphere spatial association and nonverbal memory centers.

Note: a relevant study on rats by M.Diamond (1988) addresses the role of sex gene verses sex hormone in shaping the networks. When the ovaries are removed from the female at birth, the only difference is in the medial motor cortex. We can hypothesize that the female's executive decisions will now involve movement. When the gonads are removed from the male, the differences are more dramatic. The frontal cortex and left temporal and somatosensory cortex now develop like the female.

Kimura (1992) found females verbal skill increased during high estrogen portion of monthly cycle, while math skills increased during low estrogen portions.

11a. M. C. Diamond (1988): found similar asymmetry in rats. The male rat has a thicker right cortex from birth to old age (900 days) at which time the statistically significant difference ceases. The female rat has a thicker left cortex at birth, but not one that is statistically significant. By 180 days all posterior regions in her right hemisphere are thicker than the left, with area 39 becoming statistically significant by 390 days.

11b. D. Kimura (1984): aphasia and apraxia studies indicate females and males each have brain regions of greater asymmetry.

11c. D. Kimura (1992): developed psychometric tests to challenge posterior networks that favor female accelerated language and sensorimotor networks (gatherer) and male accelerated visual and spatial processing networks (hunter). Results show effect-size differences that exceeded 0.25 sd favored the sex with the accelerated network.

11d. S. Witelson (1976). Girls do not use the right temporal-occipital brain regions to process tactile-visual material until age 12 years, using other strategies until then; after age 12 they use only the right; males reach adult functioning in this right-hemisphere network at age 6 years. This study supports females developing posterior right hemispheres later than males.

Figure Caption

Figure 1 - Bifurcations in local network T4T6 development: Right hemisphere temporal lobe velocity plots for mean coherence and 3-dimensional phase portraits of mean coherence, mean coherence velocity and time. Frequency shifts and growth and pruning peaks at least one standard deviation above or below the time series baseline are labeled.

Figure 1A - Bifurcations in female T4T6 development are accompanied by frequency shifts, which show as cusps in the phase portraits at ages 7 and 14 years. Growth peaks in mean coherence are at ages 8.5, 11.2 and 14.8 years; pruning peaks are at ages 10.0 and 13.0 years.

Figure 1B - Bifurcations in male T4T6 development are accompanied by frequency shifts, which show as cusps in the phase portraits at ages 6 and 12 years. Growth peaks in mean coherence are at ages 4.2 and 14.0 years, with the early growth peak part of a long growth plateau; pruning is minimal with one peak at age 2.0 years.

Figure 2 - Spatial gradients of oscillating mean coherence waves for each sex's left-hemisphere frontal-pole networks. The wave flow across Fp1 local networks can be decomposed into a lateral-medial clockwise rotation or into an anterior-posterior expansion.

Figure 2A - Female waveflow decomposes into a 3-year lateral-medial clockwise rotation and a 1.5-year anterior-posterior expansion. The clockwise rotation begins with a 4-percent velocity peak in F1T3 at age 2.0 years, followed by peaks for F1F7 at 2.3 years, F1F3 at 2.7 years, F1C3 and F1P3 at 2.8 years, F1O1 at 3.6 years, F1T5 at 3.8 years and again at F1T3 at 5.0 years. The expansion begins at F1F7 at age 2.3 years and ends at F1T5 at age 3.8 years.

Figure 2B - Male waveflow decomposes into a 3-year lateral-medial clockwise rotation and a 2.4 year anterior-posterior expansion. The clockwise rotation begins with a 1-percent velocity peak in F1F7 at age 3.8 years, followed by peaks for F1F3 at 4.8 years, F1C3 at 5.2 years, F1P3 at 5.8 years, F1T5 at 6.2 years and again at F1F7 at 6.8 years. The expansion begins at F1F7 at age 3.8 years and ends at F1T5 at age 6.2 years.

Figure 3 - Correlation of mean coherence and variance coherence time series estimates flexibility of local networks for change. The most flexible local networks have mean and variance time series with large negative correlations and time series plots 180 degrees out-of-phase for early, middle and late development periods. The least flexible local networks have mean and variance time series with large positive correlations and time series plots in-phase for early, middle and late development periods. Sexual dimorphism is present in all local networks.

Figure 3A - Local networks with highest negative correlation for each sex involve right hemisphere frontal lobe sites. Female network F2F8, correlation of -.913, has mean and variance time series 180 degrees out-of-phase from birth to age 14 years. Male network F2F4, correlation of -.857, has time series 180 degrees out-of-phase from birth to age 12 years.

Figure 3B - Local networks with highest positive correlation for each sex involve left hemisphere temporal sites. Female network O1T3, correlation of +.822, has mean and variance time series in-phase from birth to age 16 years except for the time period from 11 to 13 years. Male network F7T5, correlation of +.824, has mean and variance time series in-phase except for the middle development period from ages 7 to 11 years.

Figure 3C - Sexual dimorphism in flexibility of local network O1F7. Female network O1F7, correlation of +.491, has mean and variance time series out-of-phase to the bifurcation at age 8 years and in-phase to the study's end at 16 years. Male network O1F7, correlation of +.772, is in-phase from birth to 16 years.

Figure 4 - Sexual dimorphism in right-hemisphere frontal-lobe local network F8T4.

Figure 4A - F8T4 mean coherence plots: female mean coherence baseline is 4.1 percent lower than the male baseline. Variance in amplitudes of growth and pruning peaks is 286 percent greater in the female than in the male.

Figure 4B - F8T4 mean coherence velocity plots: rate and timing of maximum growth and pruning spurts differ in sexes. Females have maximum growth peaks at ages 2, 4 and 8 years; males have maximum growth peaks at ages 9 and 14 years. Females have maximum pruning peaks at ages 7 and 15 years; males have maximum pruning peaks at 12.5 years. Female growth and pruning cycles are in periods approximately 4 years in length while male cycles are in periods approximately 2 years in length.

Figure 5 - Fourier analysis of female raw data mean coherence time series T3T5. Amplitude cutoff for high energy frequencies is a magnitude of 0.310 uv (power of 0.1 uv^2 in 16-year period). In decreasing order, frequencies (cases) 5, 8, 3, 2 and 9 satisfy this criterion. With period equal to 16 years (sixty-four 3-month periods) divided by case (number - 1), the high energy frequencies are equivalent to the following growth and pruning cycles in the time domain: 4 years, 2.3 years, 8 years, 16 years and 2 years.

Figure 6 - Female left hemisphere temporal lobe network T3T5 mean coherence and velocity plots.
Figure 6A - Mean coherence time series plot with mean baseline for time series of 54.1 percent.
Figure 6B - Velocity plot for mean coherence. Time points where growth increases at a rate greater than 1 percent are: 2 years, 5.5 years, 7.5 years, 11 years and 14.5 years; time points where pruning increases at a rate greater than 1 percent are: 3.8 years, 8 to 10 years and 12 years.

Figure 7 - Local networks with one or more high energy development cycles. High energy frequencies are markers for increased complexity in gray matter connectivity and show as high amplitude mean coherence growth and pruning cycles in the time domain. For both sexes, medial sites have less complex development of gray matter across the first 16 years of development.
Figure 7A - Female 55 high energy cycles. Left hemisphere: 13 local networks have no high energy cycles; 3 have one, 7 have two, F3T3 and F7T5 have three, C3T5 has four and T3T5 has five high energy cycles. Right hemisphere: 13 local networks have no high energy cycles; 11 have one and 4 have two high energy cycles. All multiple high-energy local networks except C3P3 and Fp2F8 involve the temporal lobes.
Figure 7B - Male 18 high energy cycles. Left hemisphere: 19 local networks have no high energy cycles; 9 have one high energy cycle. Right hemisphere: 21 local networks have no high energy cycles; 7 have one and 2 have two high energy cycles. Male local networks involving left hemisphere site F7 and right hemisphere site Fp2 are the only regions with more complex gray matter development.

Figure 8 - Comparison of gender growth and pruning peaks across early, middle and late development periods. When all time points from birth to age 16 years are examined the cycles of growth and pruning exhibit many interesting patterns; e.g., both sexes have growth between sites separated by short distances prior to growth between sites separated by long distances in both hemispheres during early development.
Figure 8A - Early development period from birth to age 6 years. Most notable: female bilateral growth at age 1.9 years; male bilateral pruning at age 1.7 years.
Figure 8B - Middle development period from age 6 years to age 11 years. Most notable: female right hemisphere pruning at age 10.0 years; male pruning and growth at parietal site P4 at age 9.5 years.
Figure 8C - Late development period from age 11 years to age 16 years. Most notable: female growth in long connections for medial local networks at 14.1 years; male right hemisphere growth at 13.9 years and left hemisphere pruning at 14.4 years.

Figure 9 - Regional networks, stages of development (estimated by mean EEG coherence time series). Stage 1 development involves local networks not yet designated for an ontogenetic regional network functioning as spares. Synapses in these juvenile networks represent a pool of synapses that may be used for other purposes until required to join their permanent network. Stage 2 development involves local networks forming a regional network for some functional purpose. All local networks in the newly formed regional network experience cycles of synaptic growth and pruning at the same time. Stage 3 development involves regional networks refining intra- and inter-network connections, a globalization of the synaptic net, with out-of-phase growth and pruning in the local networks. Stage 4 development involves the regional network being resculpted, with in-phase synaptic growth and pruning. Stage 4 is considered to be at the adult functional level. Stage 5 development would be another iteration of globalization. Stage 6 development would involve another DNA initiated period of in-phase synaptic growth and pruning. In the Hanlon studies, the accelerated regional networks are in Stage 6 as the study ends (age 24 years). Thatcher has evidence for continuation of this resculpting process beyond age 24 years (personal communication, 1994).

Figure 10 - Regional networks formed by local networks differ in sexes. Forty-seven female local networks and 54 male local networks are aligned with the 11 depicted regional networks by the late development period (ages 11 to 16 years). Regional networks are at different stages of development, with females having two and males having four at the most complex Stage 4. For cortico-cortical connections involving long distances, those primarily involving the occipital or posterior temporal sites were classified as "posterior networks," while those involving parietal and anterior temporal sites were classified as "anterior networks." Local networks with solid lines had greater correlation with the regional network than local networks with dashed lines. In most cases the local networks with dashed lines joined the regional network later.
Figure 10A - Four female and six male anterior regional networks. The female's dominant left-hemisphere network was placed with the anterior networks because of limited connectivity to the occipital lobe. Stage of development indicates females have more asymmetric development of the frontal lobes than males.
Figure 10B - Seven female and five male posterior regional networks. Stage of development indicates males have more asymmetric development of the occipital lobes than females.

Figure 11 - Topographies of sexes' regional networks reaching Stage 4 and time series plots for associated local networks.
Time series for female's two networks are in Figures 11A and 12A. Time series for male's four networks are in Figures 11B, 11C, 11D and 13B.
Figure 11A - Female bilateral frontal pole executive decision making networks. Local networks, involved in bilateral decision making for past, present and future events, are functioning as a unit from birth. In contrast to the male, the female's frontal pole connections to the motor cortices are in the female's left regional network for language, somatosensory processing and associated memory and in the right frontal network for nonverbal memory.
Figure 11B - Male bilateral frontal pole executive decision making networks. Local networks, involved in bilateral decision making for present and future, are functioning as a unit from birth. Frontal pole connections to the right lateral-frontal region, which involves decision making for nonverbal events in the past, is in-phase with this network until late development. After age 11 years, Fp2F8 seems to align its development with local network Fp1F7, which involves decision making for past verbal and sequential memory events. Neither of these lateral- frontal networks seem to reach Stage 4 until the mid-twenties.
Figure 11C - Male left-hemisphere long-connection network for decision making related to visual targets. This is the male's dominant left hemisphere network, is functioning as a unit from birth and involves decision making for present, past and future events. This network would support the abilities useful to the "hunter."
Figure 11D - Male right-hemisphere long-connection network for decision making related to visual targets. This network has a different time schedule for development than the left-hemisphere long-connection network and may have a different functional purpose.

Figure 12 - Sexes' left-hemisphere regional networks for language processing. Broca and Wernicke's regions develop at the same time in the female and at different times in the male. The female's regional network involves 16 of the 28 left-hemisphere local networks; the male's two regional networks involve 10 of the 28 left-hemisphere local networks. The female network reaches Stage 4; the male posterior network reaches Stage 3; and the male anterior network reaches Stage 2.

Figures 12A and 12C - Female anterior local networks F7T3, F3T3 and C3T3 (Figure 12A) and posterior local networks T3T5, P3T3 and C3T5 (Figure 12C). This network would support the association memory capabilities and somatosensory skills useful to the "gatherer." The anterior local networks in the female's regional network have a longer Stage 3 middle development period than the posterior local networks, possibly indicating more elaborate interconnections to other networks. The 16 local networks begin the adult Stage 4 between ages 11 and 13 years.

Figure 12B - Male anterior local networks F7T3, F3T3 and C3T3 remain at the juvenile Stage 1 from birth until the bifurcation around 8 years of age when the regional formation Stage 2 begins and continues to end of the study data at age 16 years.

Figure 12D - Male posterior local networks T3T5, P3T3 and C3T5 are at the juvenile Stage 1 from birth until the bifurcation around 6 years of age, at the formation Stage 2 from 6 years to the bifurcation around 11 years of age and at the globalization Stage 3 from 11 years to the end of study data at age 16 years.

Figure 13 - Sexes' right hemisphere spatial processing networks.

The female's four regional networks have very different development patterns, with all reaching Stage 3 by age 16 years except for the bilateral network connecting the occipital and parietal regions which remains at Stage 2. The male's one regional network begins the adult Stage 4 around 12 years of age. Figure 13A - Female local networks T4T6 and O2T4 form a regional network linking nonverbal memory centers in the temporal lobe with vision centers in the occipital lobe. This network is in juvenile Stage 1 until the bifurcation around age 8 years, is in formation Stage 2 from age 8 years to the bifurcation around age 12 years and is in the globalization Stage 3 to the study's end at age 16 years.

Figure 13B - The male's one regional network includes nine local networks, including T4T6 and O2T4. The local networks are in formation Stage 2 to the bifurcation around age 5 years, in the globalization Stage 3 from age 5 years to the bifurcation around 12 years of age and in the adult Stage 4 until the study's end at age 16 years.

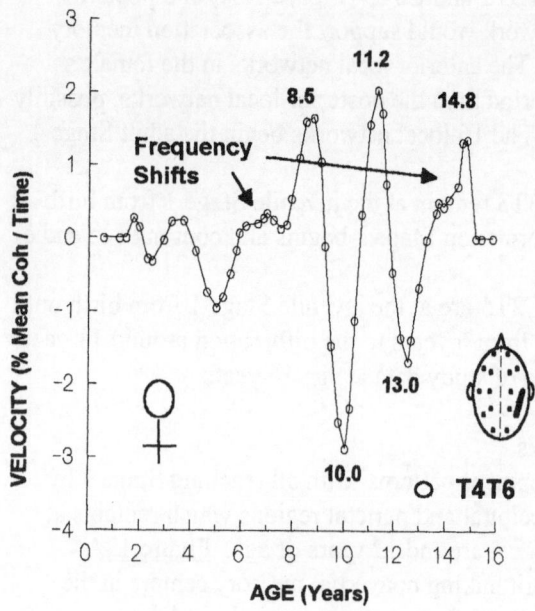

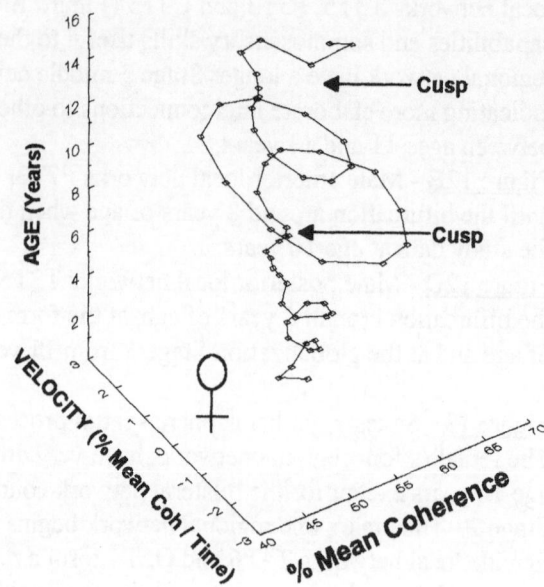

Figure 1A

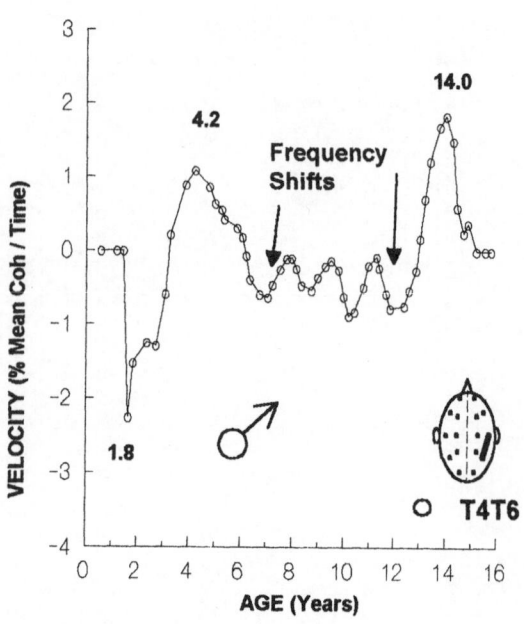

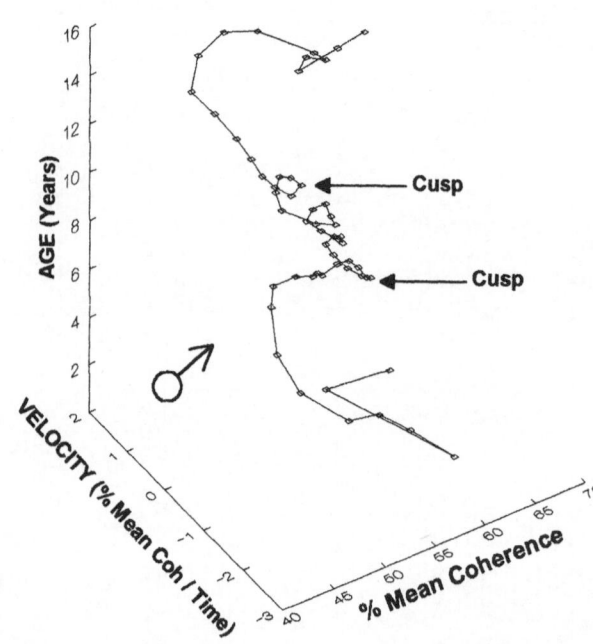

Figure 1B

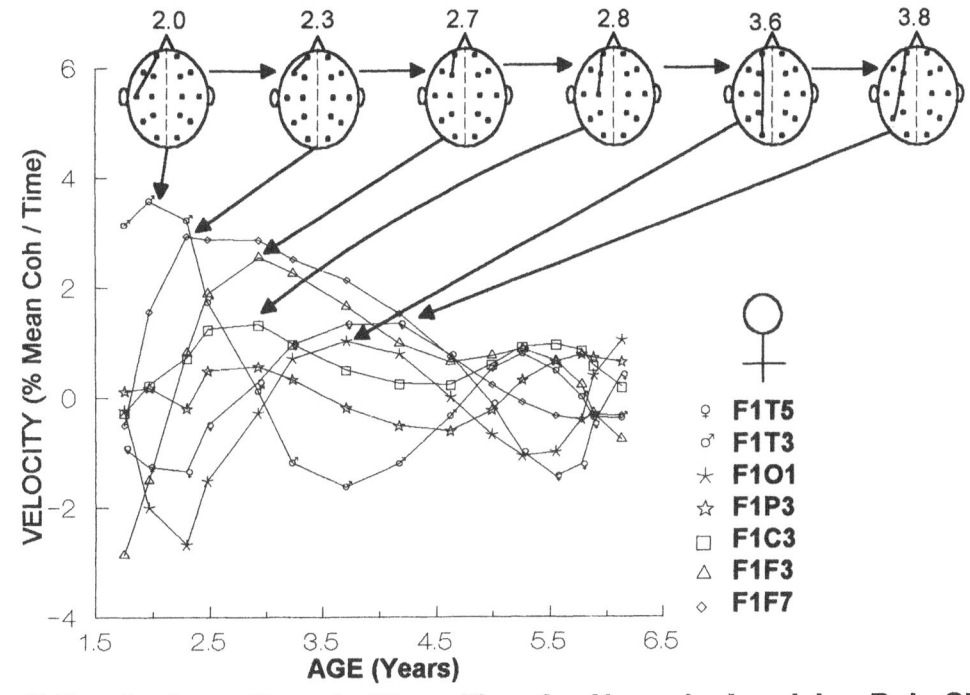

Left Hemisphere, Female Wave Flow for Neworks Involving Pole Site Fp1

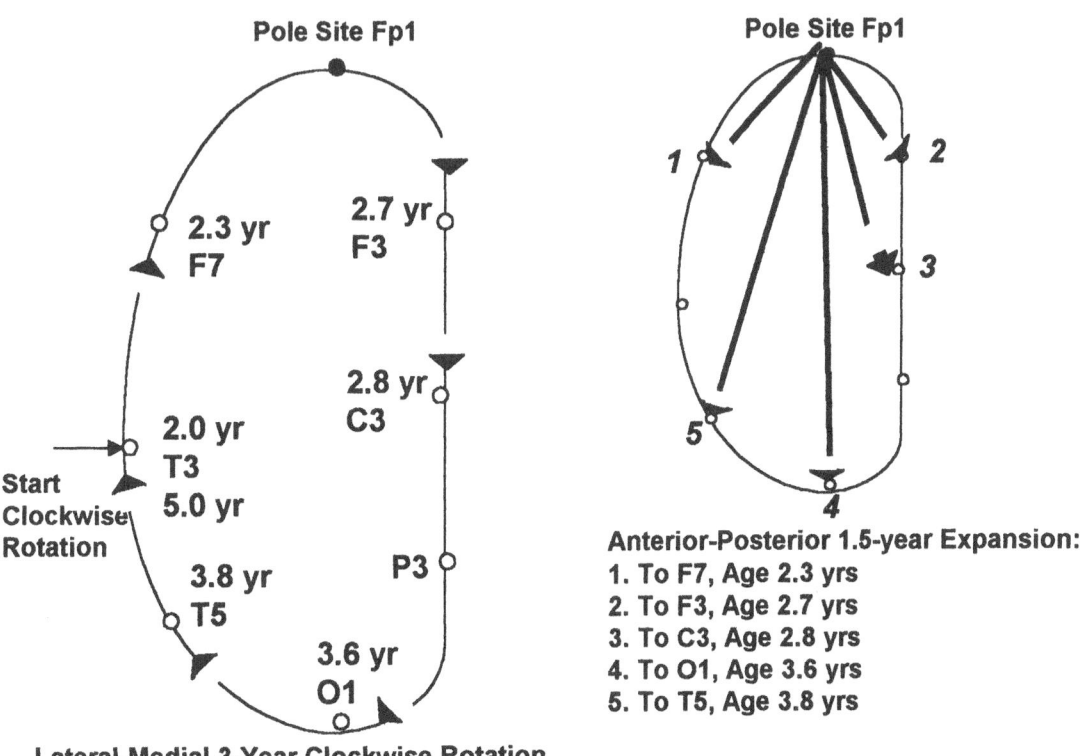

Figure 2A

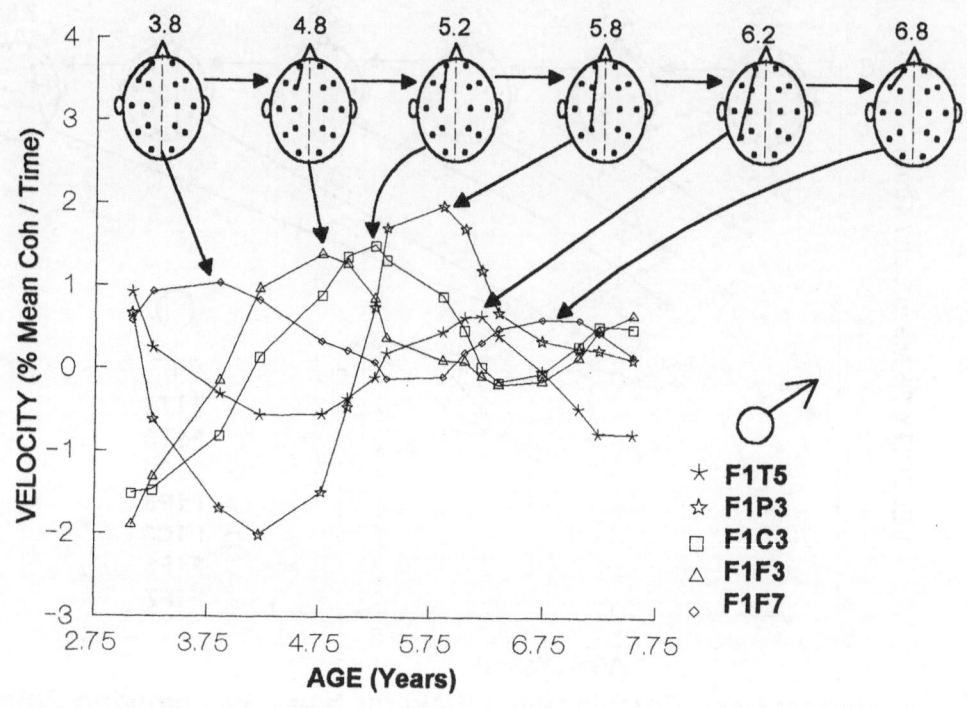

Left Hemisphere, Male Wave Flow for Networks Involving Pole Site Fp1

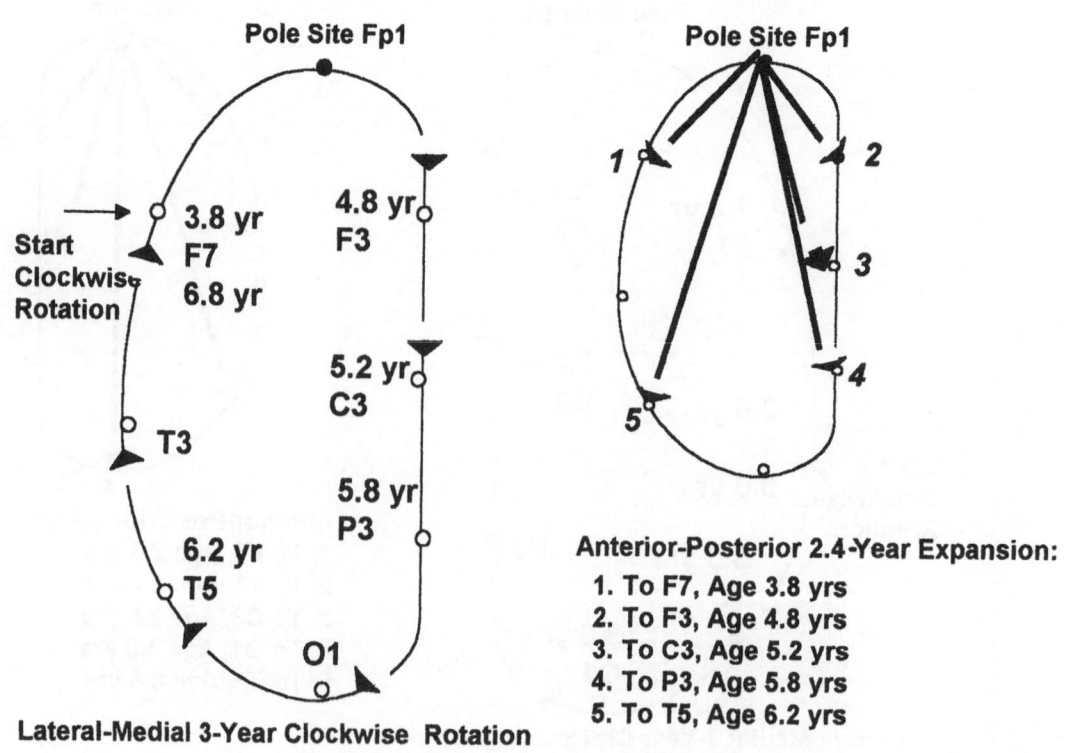

Figure 2B

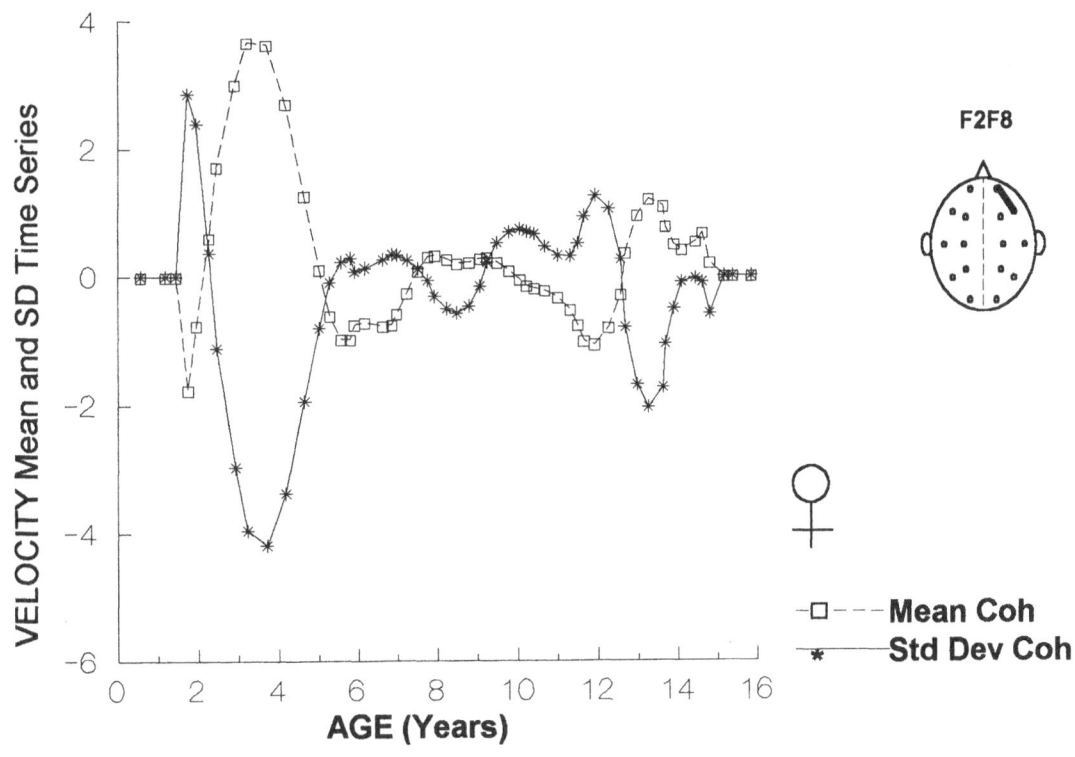

Female F2F8, Largest Negative Correlation (-.913)

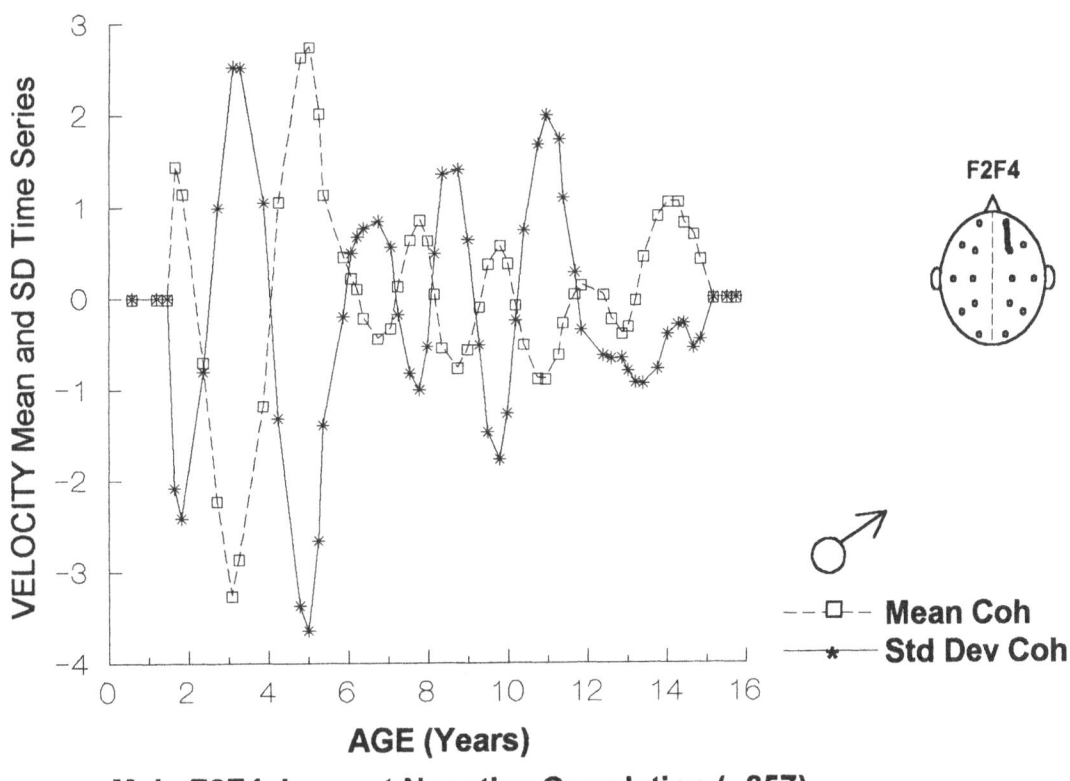

Male F2F4, Largest Negative Correlation (-.857)

Figure 3A

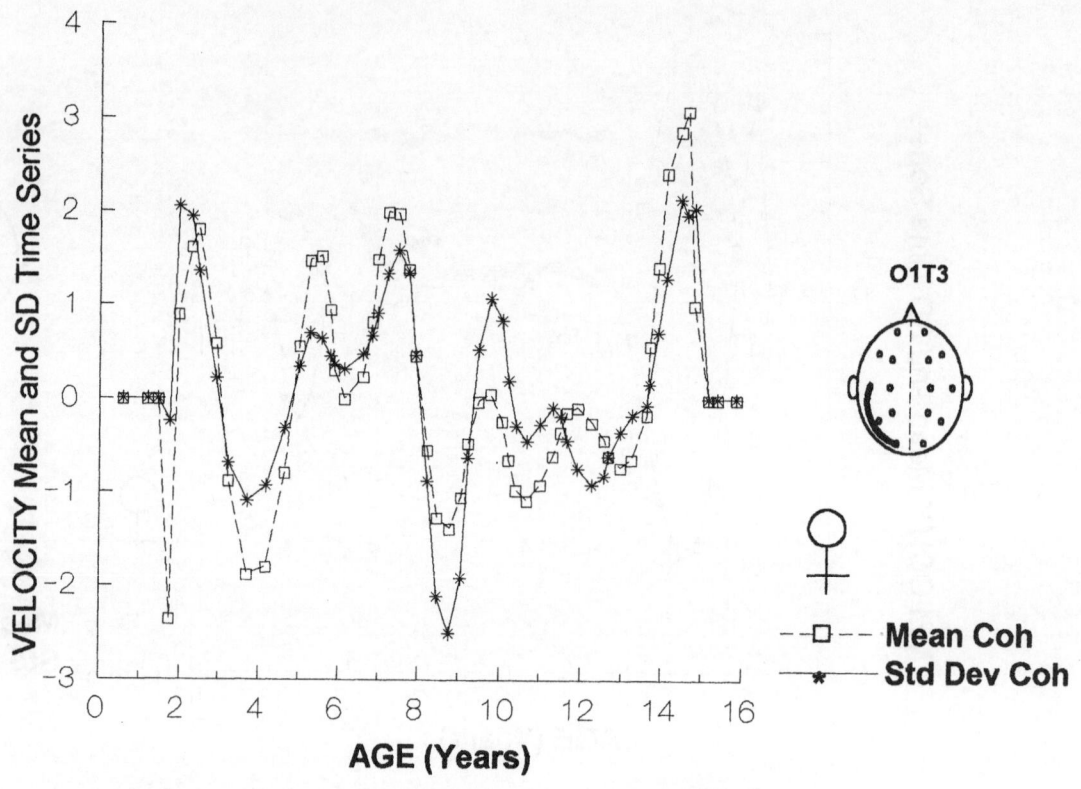

Female O1T3, Largest Positive Correlation (.822)

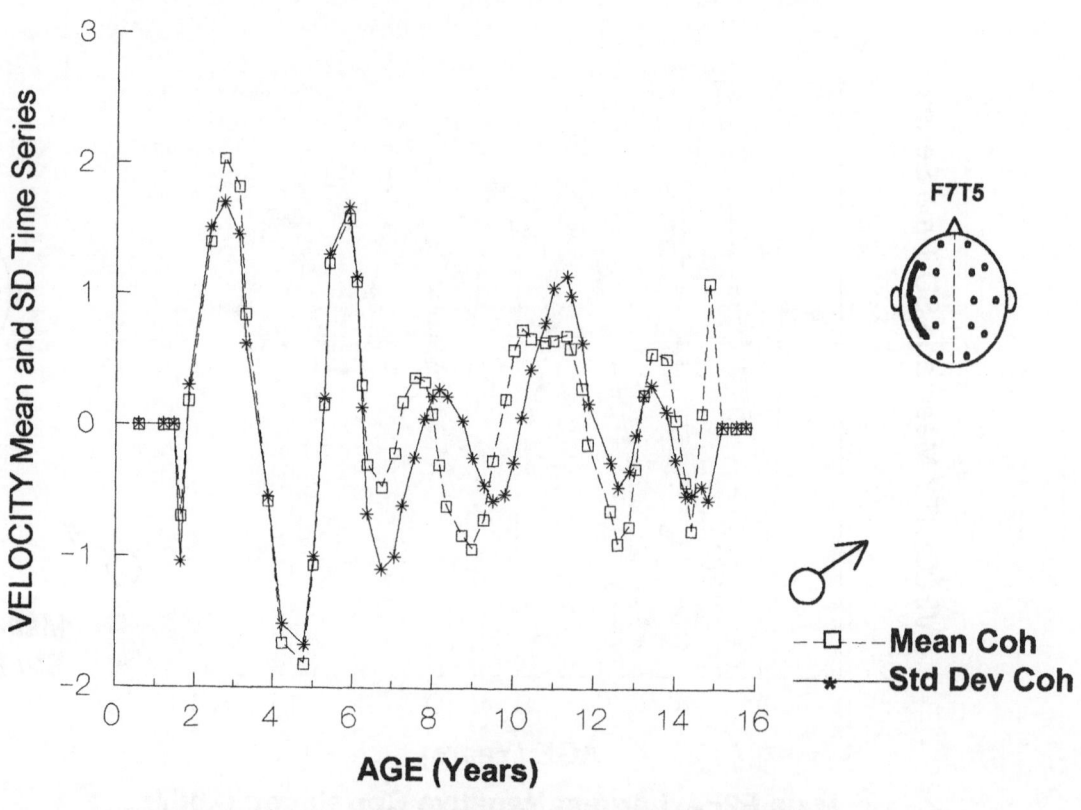

Male F7T5, Largest Positive Correlation (.824)

Figure 3B

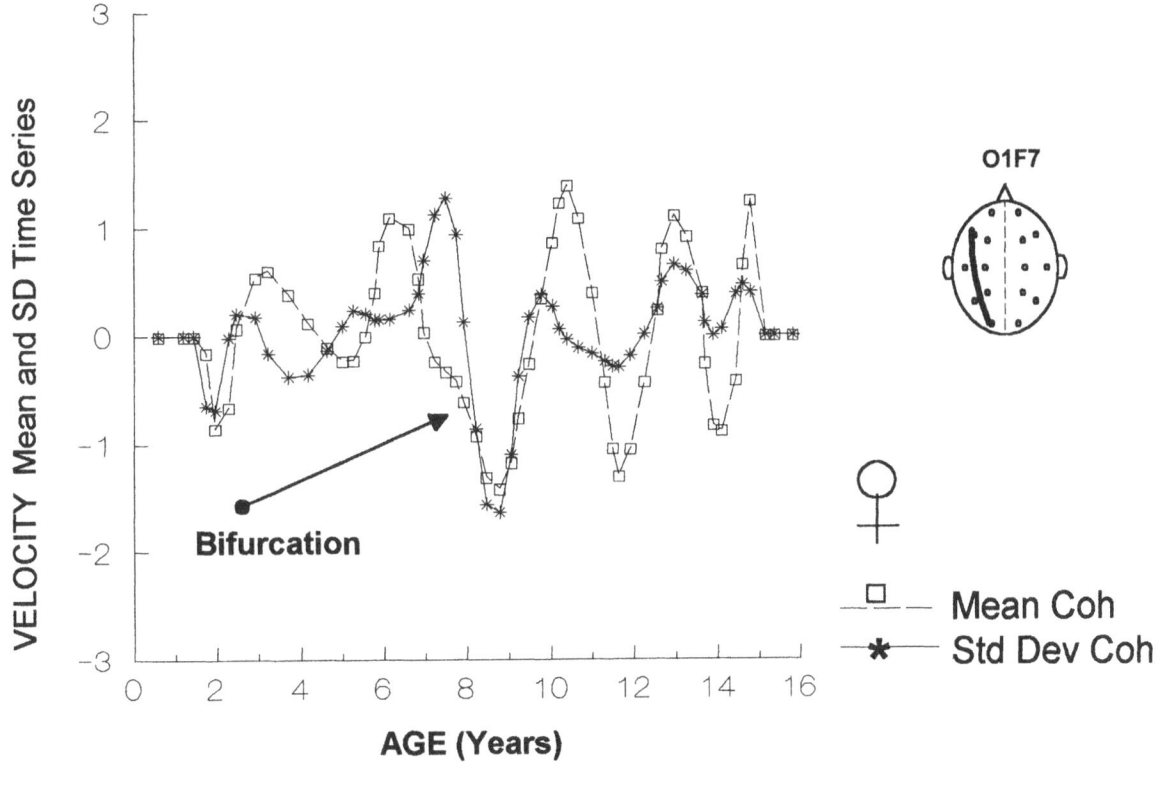

Female O1F7, Correlation +.491

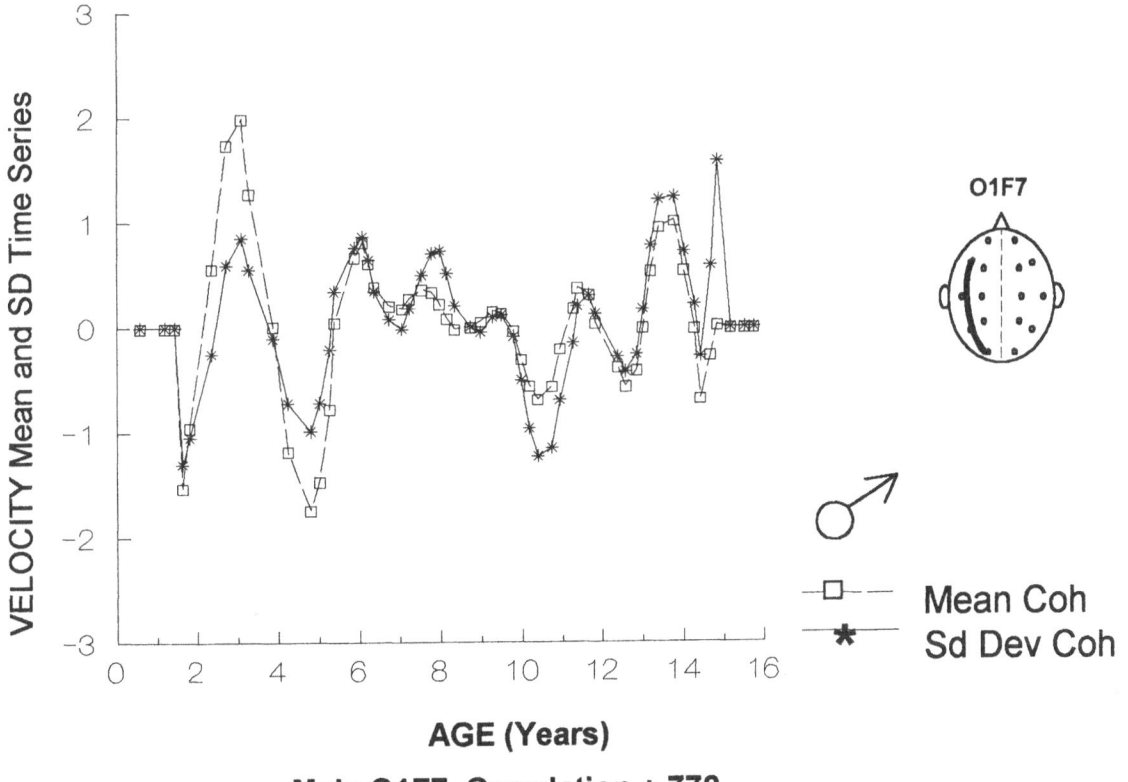

Male O1F7, Correlation +.772

Figure 3C

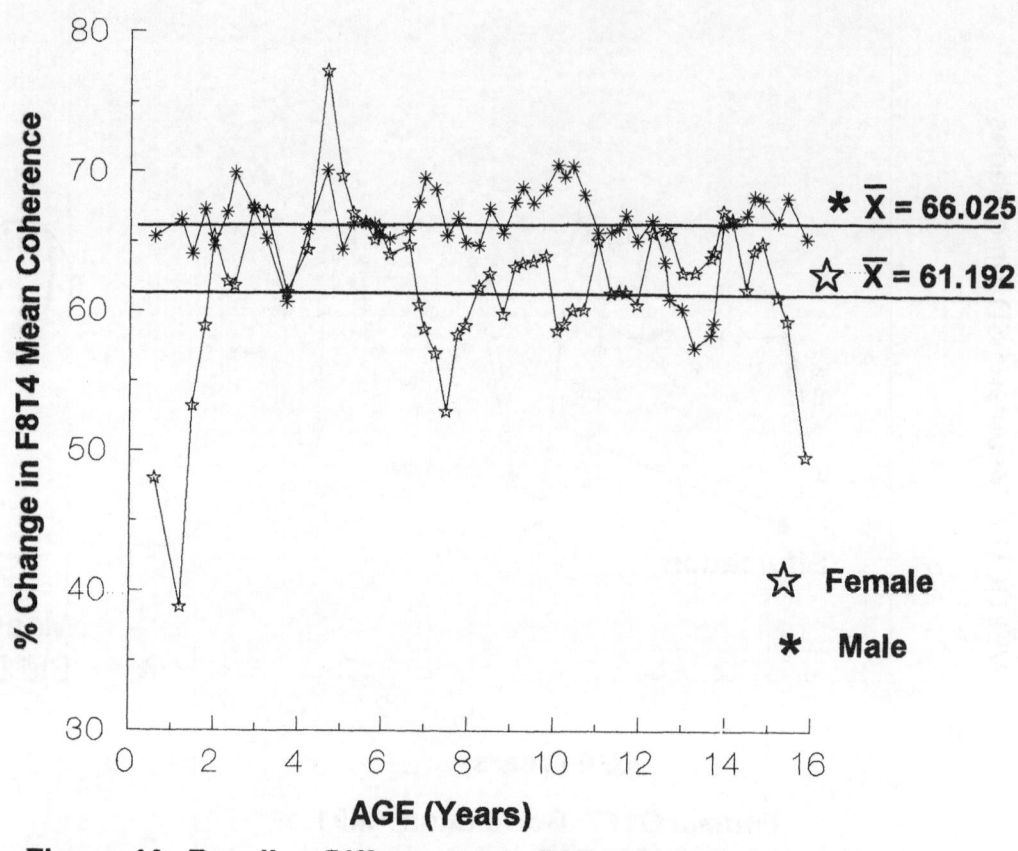

Figure 4A: Baseline Differences in F8T4 Mean Coherence

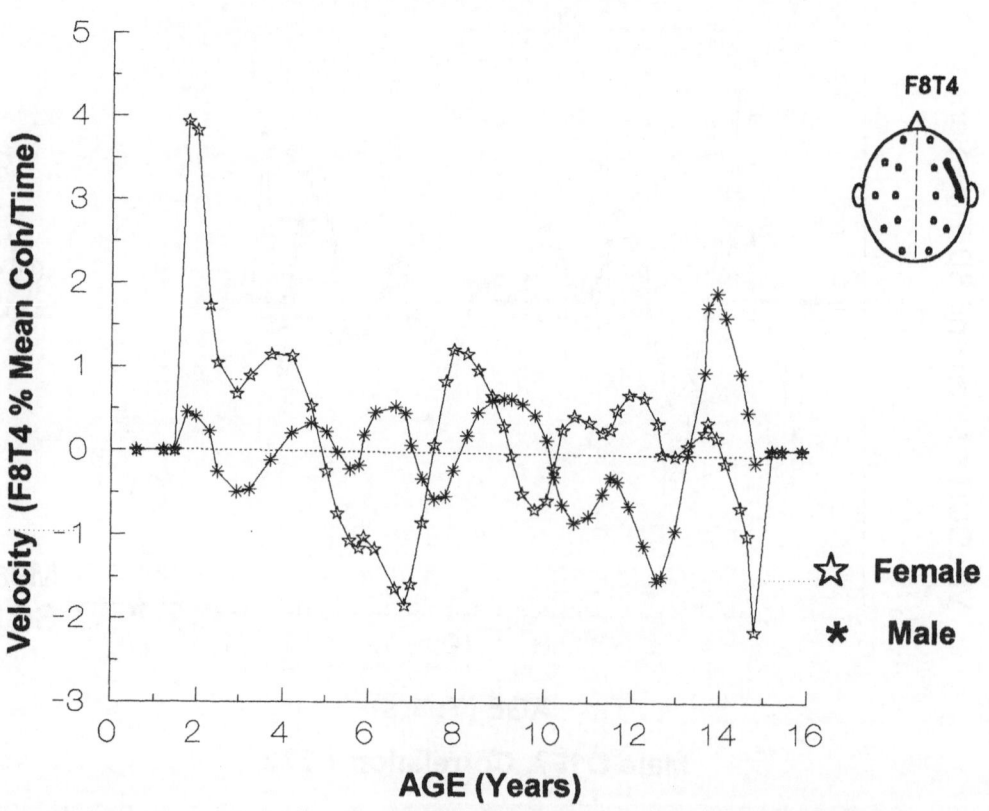

Figure 4B: Rate and Timing Differences in F8T4 Velocity Plots

Power Distribution by Frequency

Sequence Plot of Time Series

Case	Magnitude in uv	Range 0.005_____0.315_____0.678
1	0.488	constant term
2	0.376	xxxxxxxxxxxxxxxxxxxxxxxxxxxxxxxx16 yr. cycle
3	0.403	xxxxxxxxxxxxxxxxxxxxxxxxxxxxxxxxxx8 yr.
4	0.273	xxxxxxxxxxxxxxxxxxxx
5	0.678	xxxxxxxxxxxxxxxxxxxxxxxxxxxxxxxxxxxxxxxxxxxxx4 yr. 6 0.253

xxxxxxxxxxxxxxx   7   0.225   xxxxxxxxxxxxxx

Case	Magnitude in uv	Range
8	0.623	xxxxxxxxxxxxxxxxxxxxxxxxxxxxxxxxxxxxxxxxx2.3 yr.
9	0.314	xxxxxxxxxxxxxxxxxxxxxxxx2 yr.
10	0.273	xxxxxxxxxxxxxxxxxxxx
11	0.154	xxxxxxxxxxx
12	0.205	xxxxxxxxxxxxxx
13	0.112	xxxxxxxx
14	0.061	xxxx
15	0.096	xxxxxxx
16	0.113	xxxxxxxx
17	0.079	xxxxxx energy in 1 yr. cycle
18	0.030	xx
19	0.033	xx
20	0.053	xxx
21	0.052	xxx
22	0.039	xx
23	0.008	x
24	0.021	xx
25	0.035	xx energy in 9 month cycle
26	0.040	xx
27	0.028	xx
28	0.008	x
29	0.015	x
30	0.036	xx
31	0.044	xx
32	0.030	xx
33	0.030	xx energy in 6 month cycle

Figure 5 - Fourier Analysis of Female T3T5 Mean Coherence Time Series

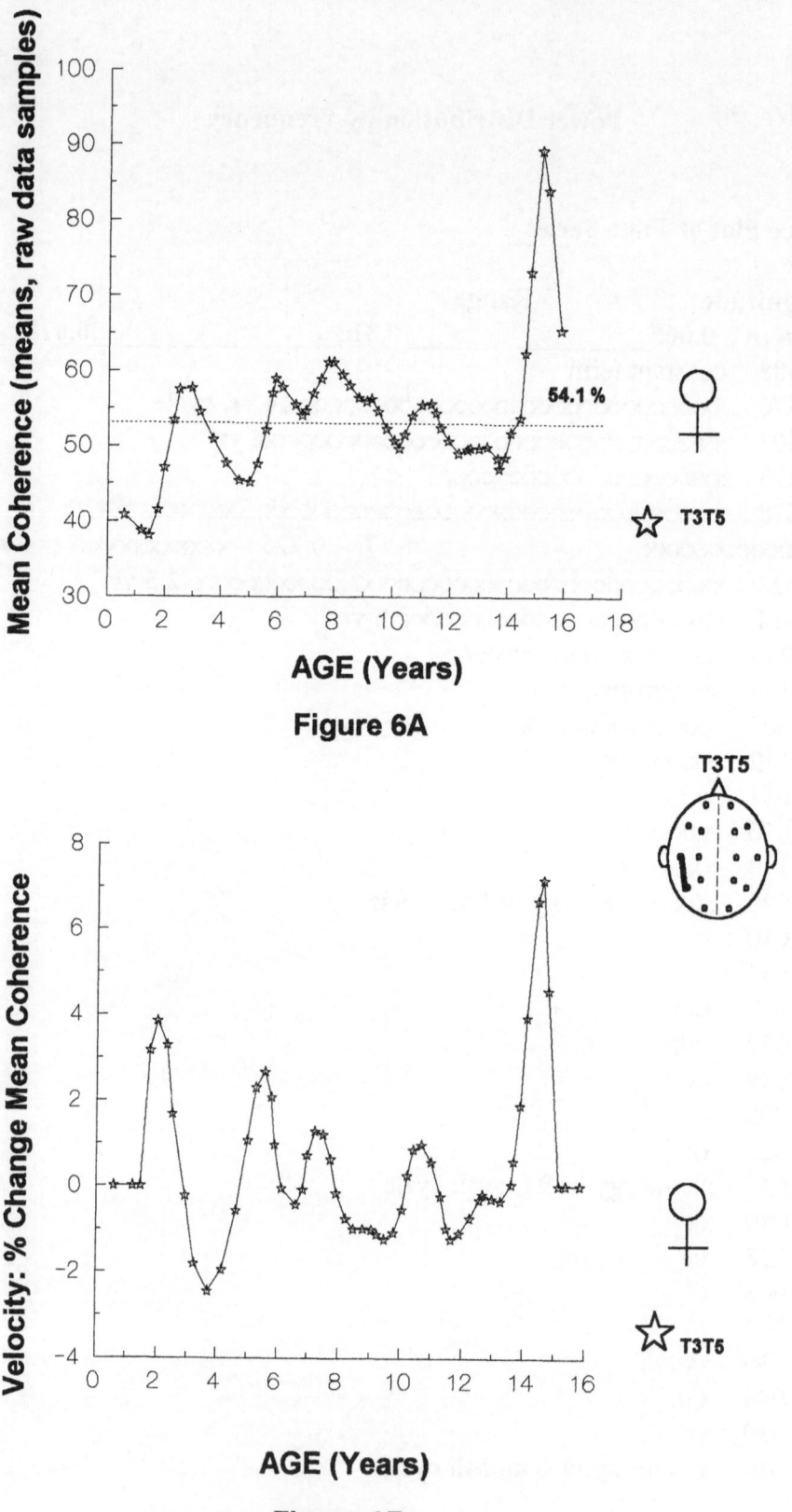

Figure 6A

Figure 6B

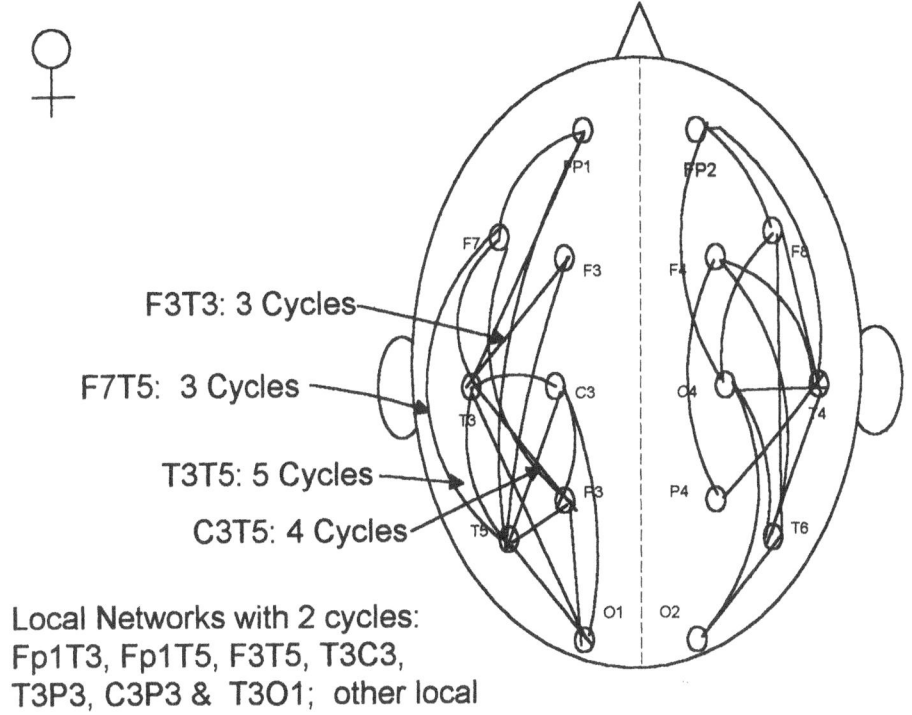

Electrode Pairs with 2 cycles: Fp2F8, Fp2T4, T4P4 & T4T6; other local networks have energy primarily in one high energy cycle.

F3T3: 3 Cycles

F7T5: 3 Cycles

T3T5: 5 Cycles

C3T5: 4 Cycles

Local Networks with 2 cycles: Fp1T3, Fp1T5, F3T5, T3C3, T3P3, C3P3 & T3O1; other local networks have energy primarily in one high energy cycle.

Figure 7A

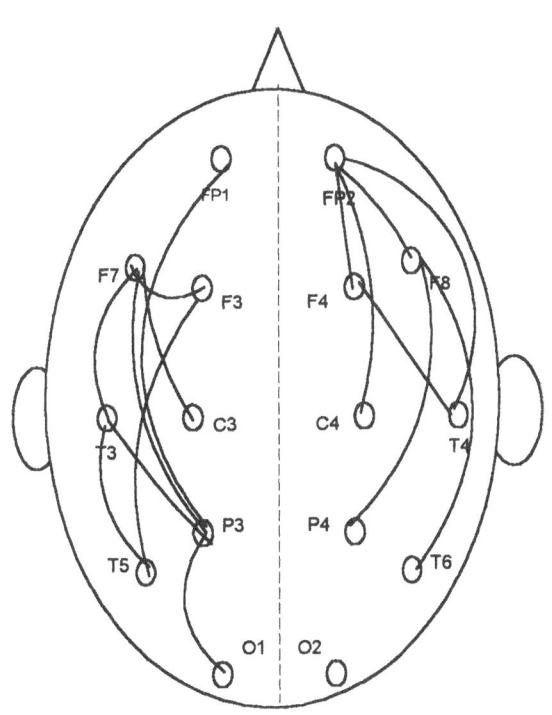

Fp2F8 & Fp2T4 have 2 cycles; other local networks have energy primarily in one high energy cycle.

Figure 7B

Sexually Dimorphic Topography: Mean EEG Coherence Growth and Pruning Peaks

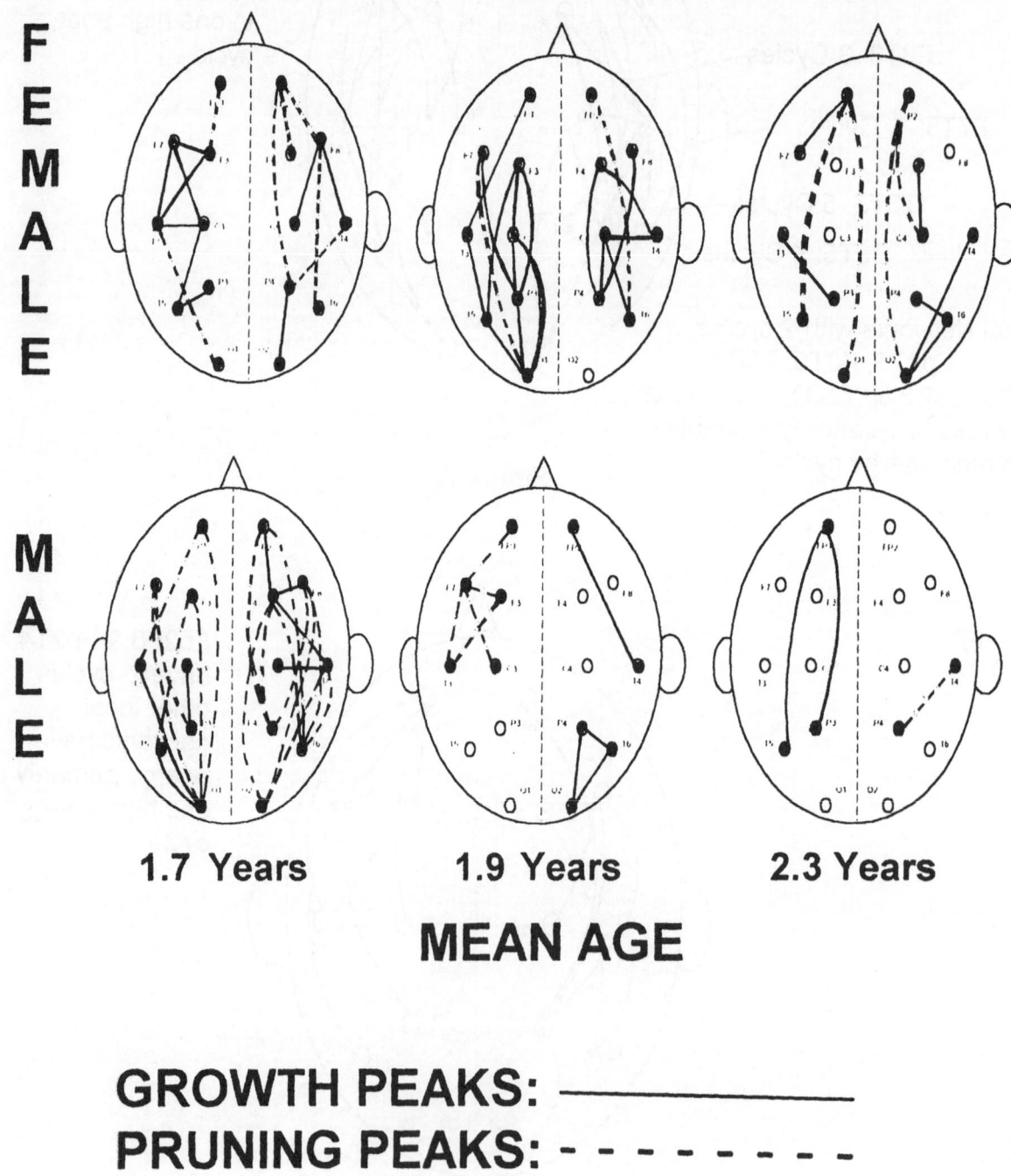

MEAN AGE

GROWTH PEAKS: ——————
PRUNING PEAKS: - - - - - - - -

Figure 8A - Sexes' Growth and Pruning Peaks in Early Development

Sexually Dimorphic Topography:
Mean EEG Coherence
Growth and Pruning Peaks

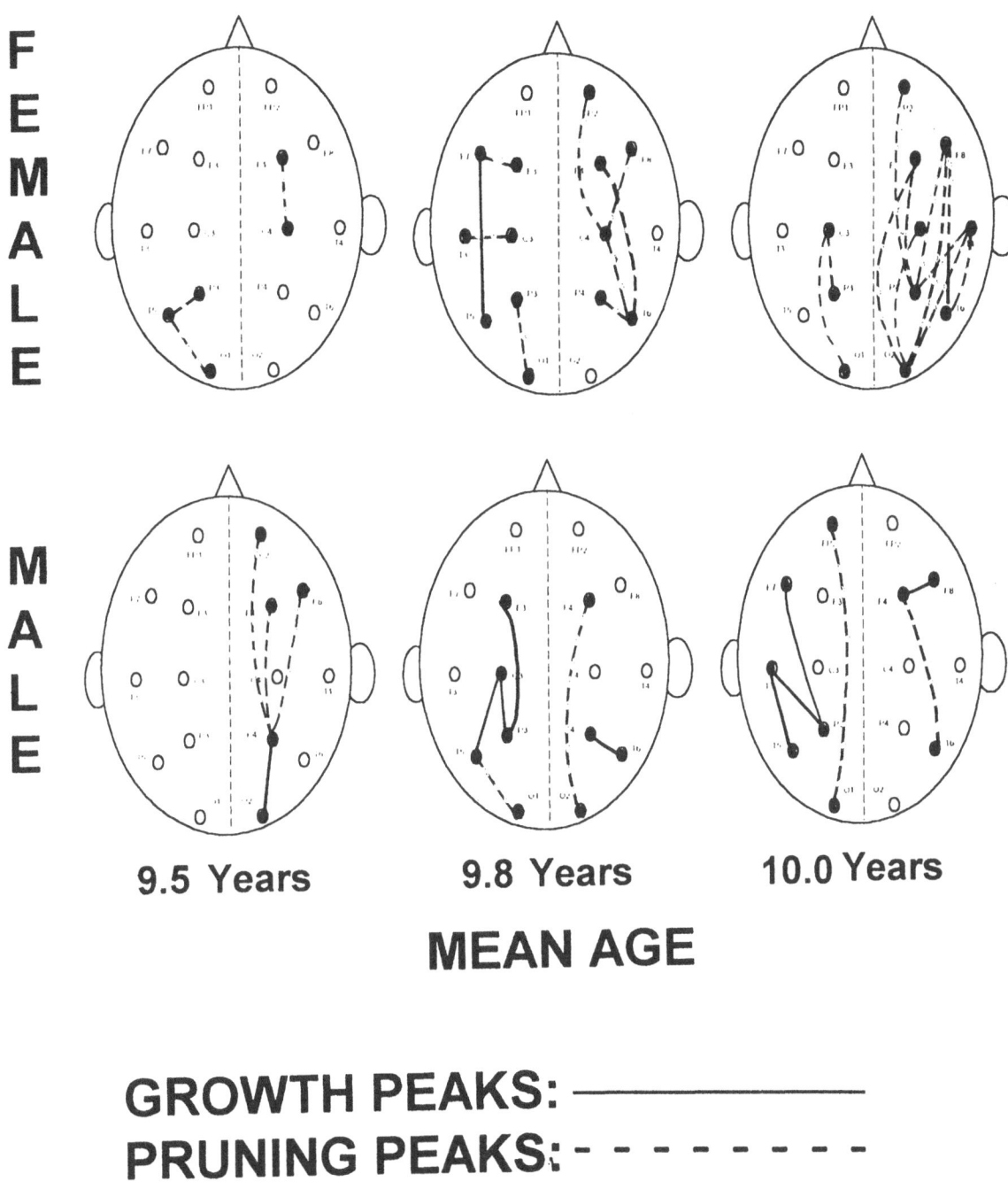

GROWTH PEAKS: ——————

PRUNING PEAKS: - - - - - - -

Figure 8B - Sexes' Growth and Pruning Peaks in Middle Development

Sexually Dimorphic Topography:
Mean EEG Coherence
Growth and Pruning Peaks

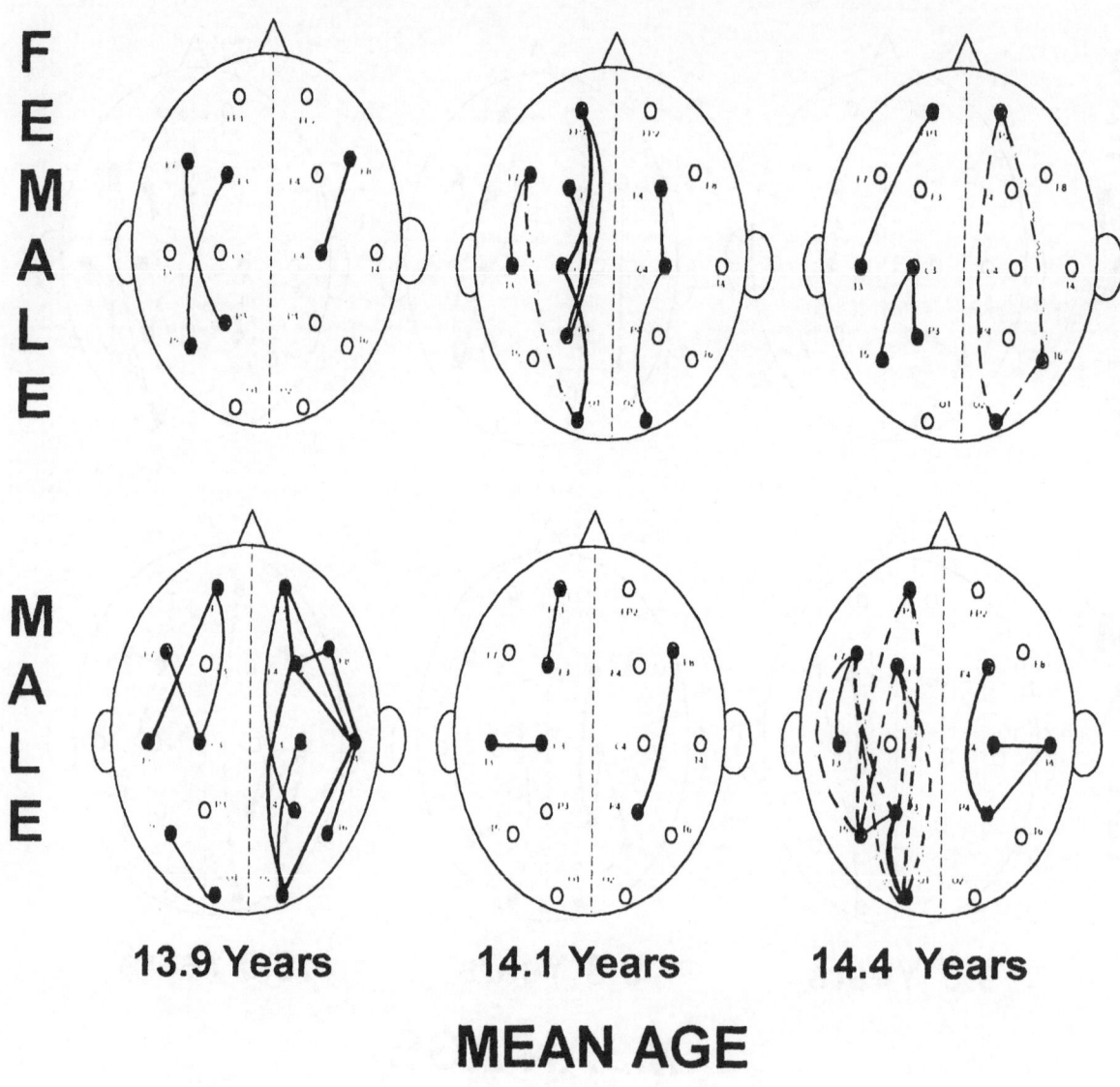

GROWTH PEAKS: —————

PRUNING PEAKS: - - - - - - -

Figure 8C - Sexes' Growth and Pruning Peaks in Late Development

Regional Network Stages of Development

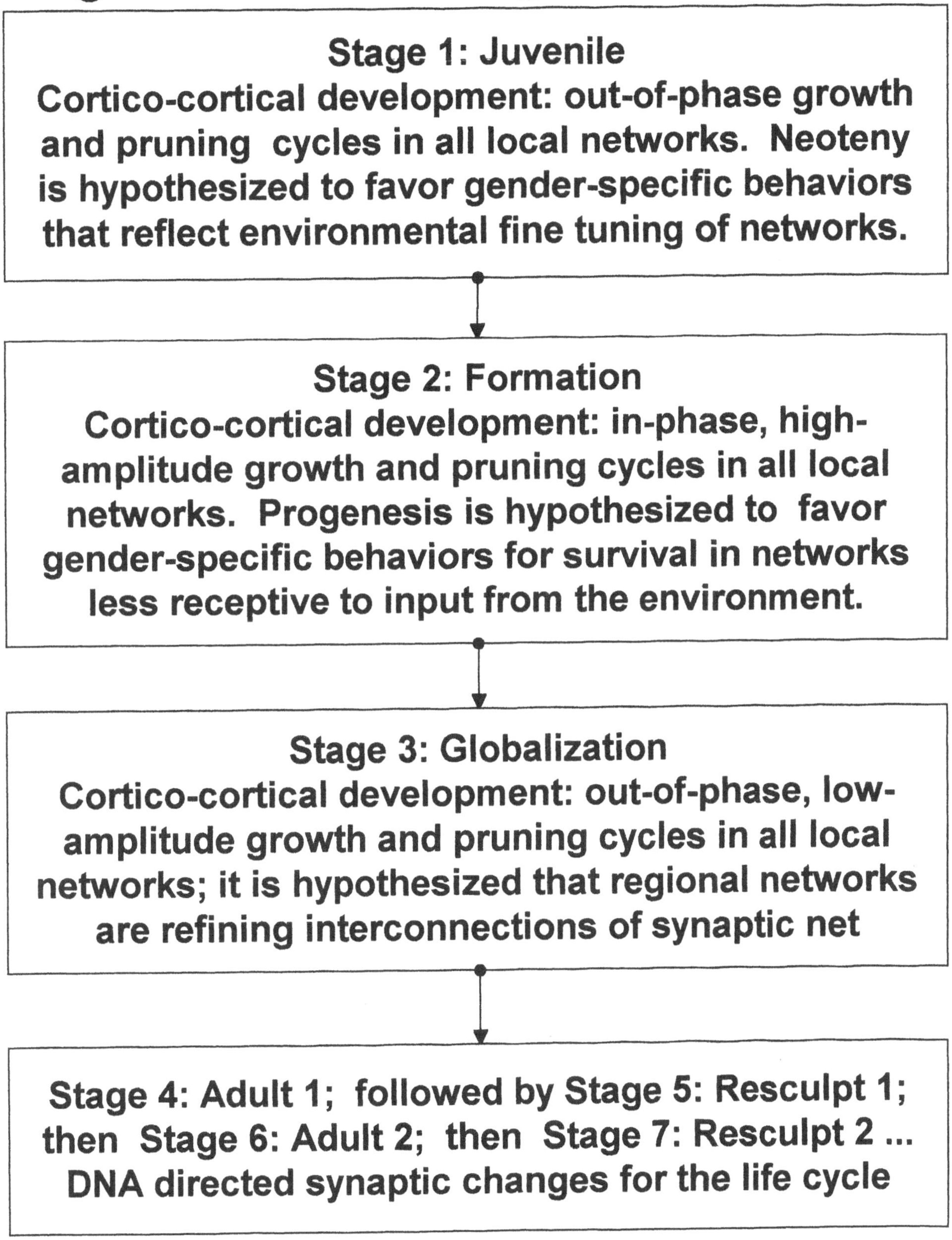

Stage 1: Juvenile
Cortico-cortical development: out-of-phase growth and pruning cycles in all local networks. Neoteny is hypothesized to favor gender-specific behaviors that reflect environmental fine tuning of networks.

Stage 2: Formation
Cortico-cortical development: in-phase, high-amplitude growth and pruning cycles in all local networks. Progenesis is hypothesized to favor gender-specific behaviors for survival in networks less receptive to input from the environment.

Stage 3: Globalization
Cortico-cortical development: out-of-phase, low-amplitude growth and pruning cycles in all local networks; it is hypothesized that regional networks are refining interconnections of synaptic net

Stage 4: Adult 1; followed by Stage 5: Resculpt 1; then Stage 6: Adult 2; then Stage 7: Resculpt 2 ... DNA directed synaptic changes for the life cycle

Figure 9

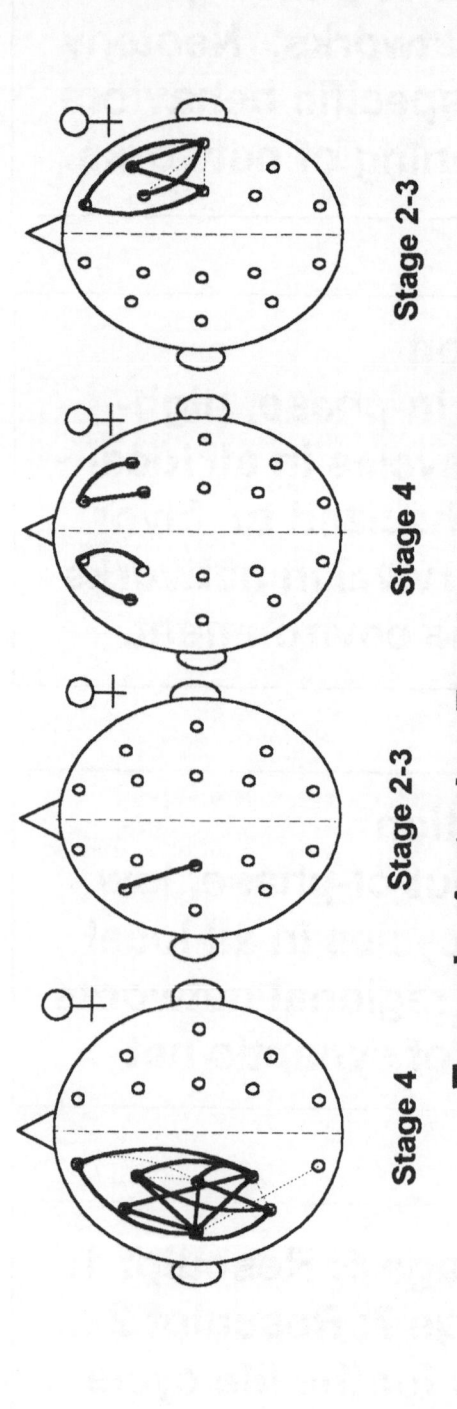

Female Anterior Regional Networks

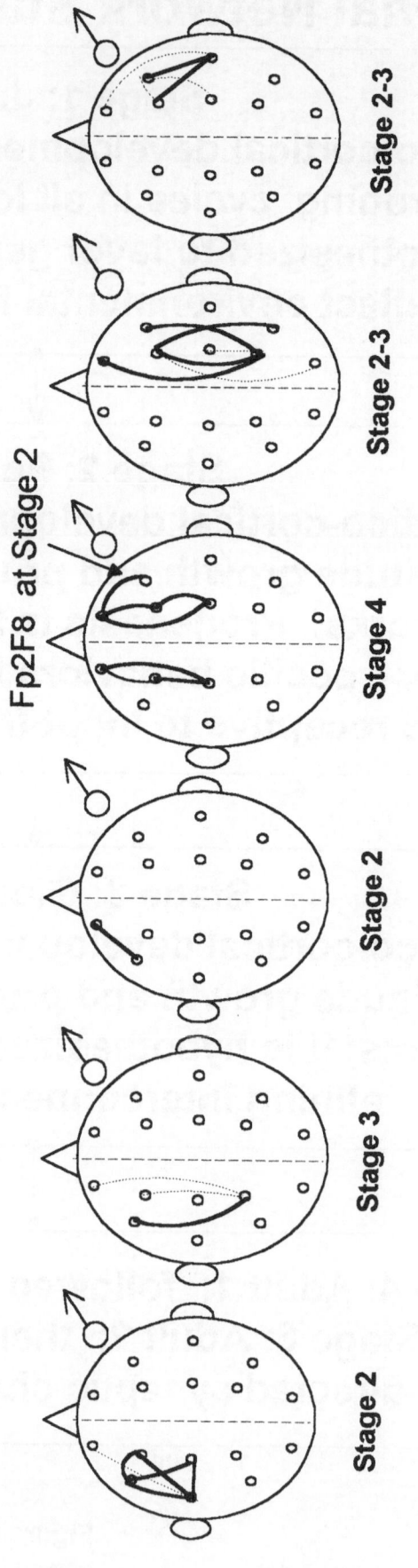

Male Anterior Regional Networks

Figure 10A

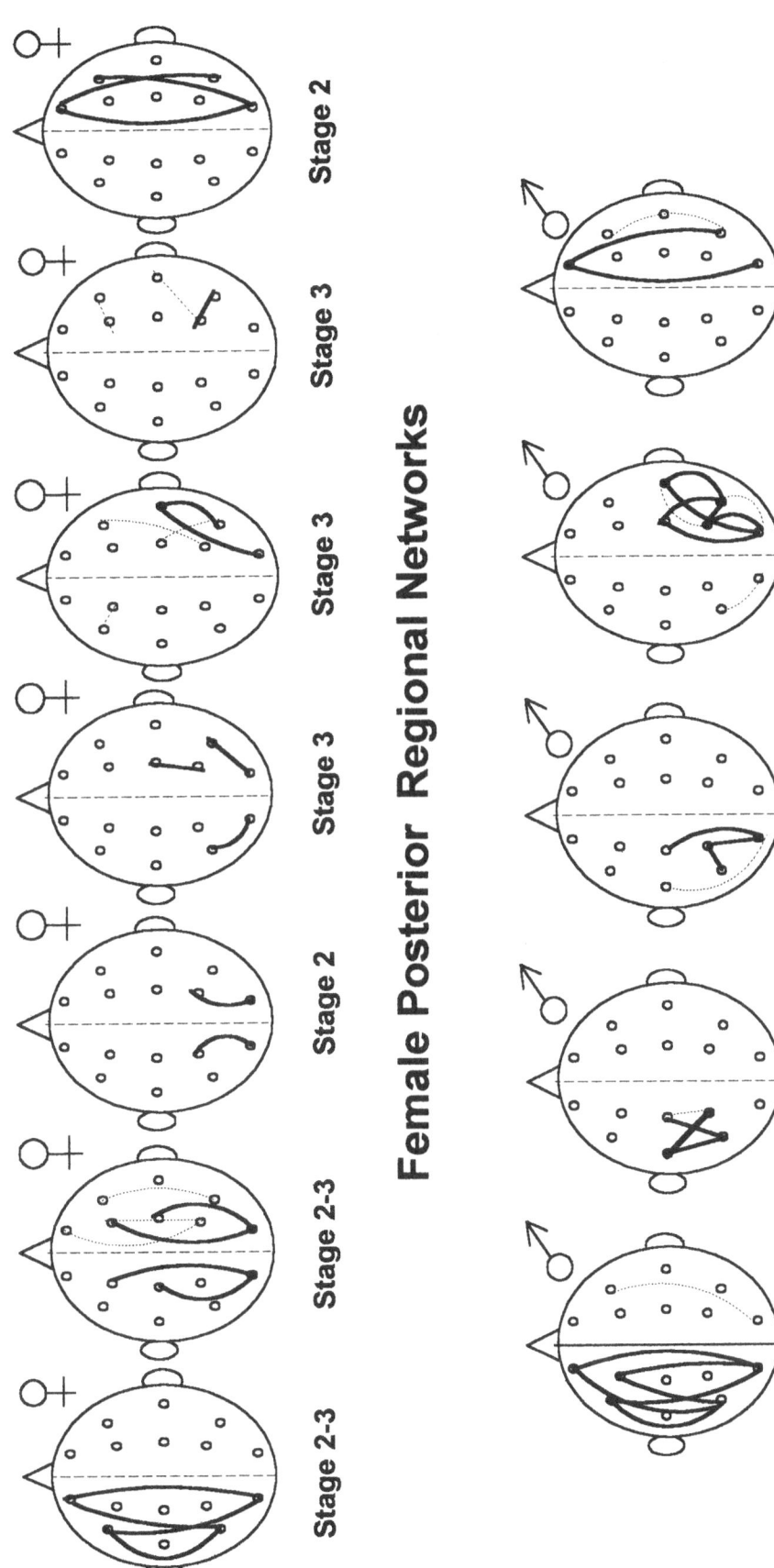

Female Posterior Regional Networks

Male Posterior Regional Networks

Figure 10B

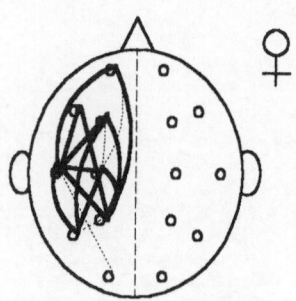

See Figure 12A for time series plots for local networks in female language processing regional network

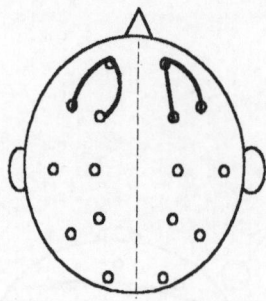

See Figure 11A for time series plots for local networks in female executive decision making regional network

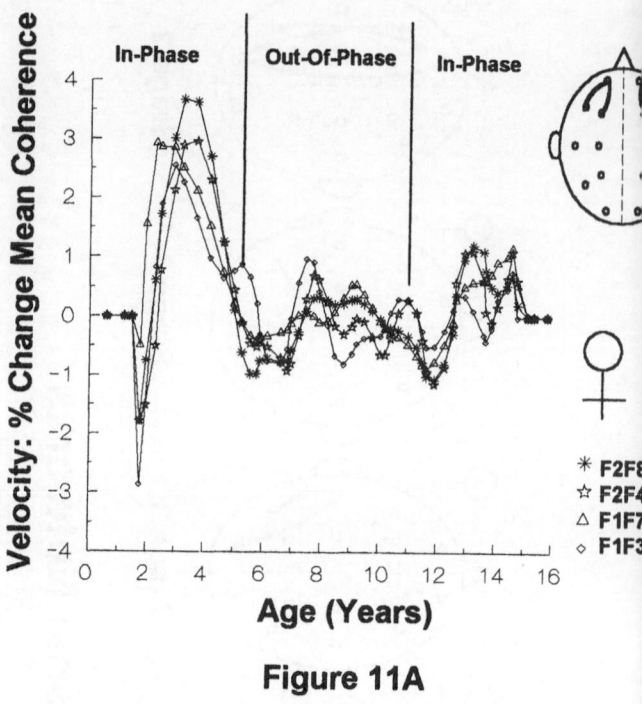

Figure 11A

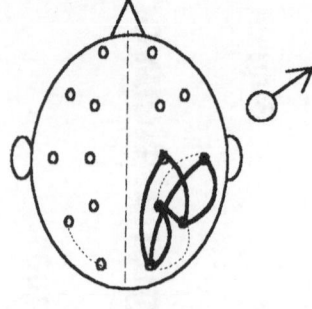

See Figure 13B for time series plots for local networks in male spatial processing regional network

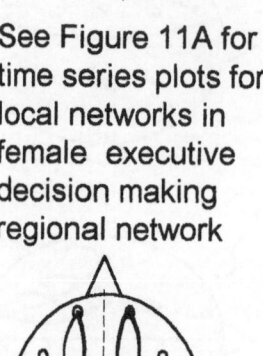

See Figure 11B for time series plots for local networks in male executive decision making regional network

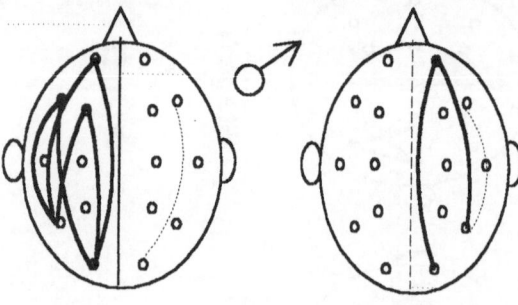

See Figure 11C, time series plots for local networks in male left hemisphere visual processing regional network and Figure 11D, time series plots for local networks in male right hemisphere visual processing regional network

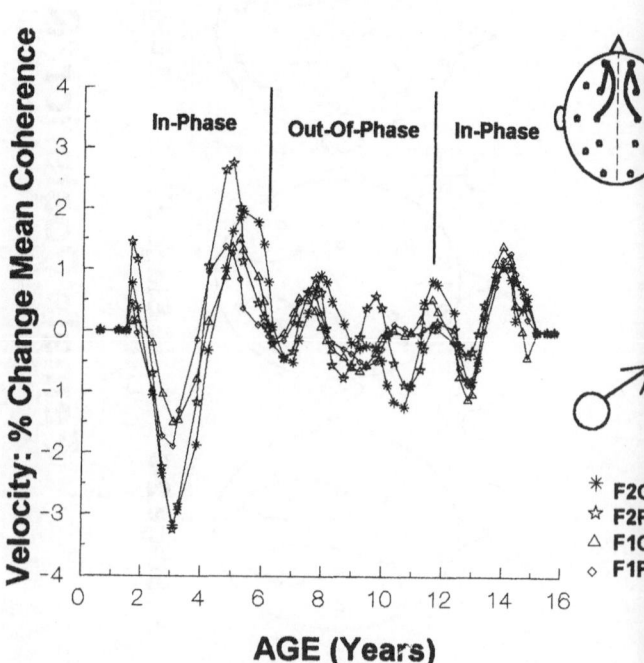

Figure 11B

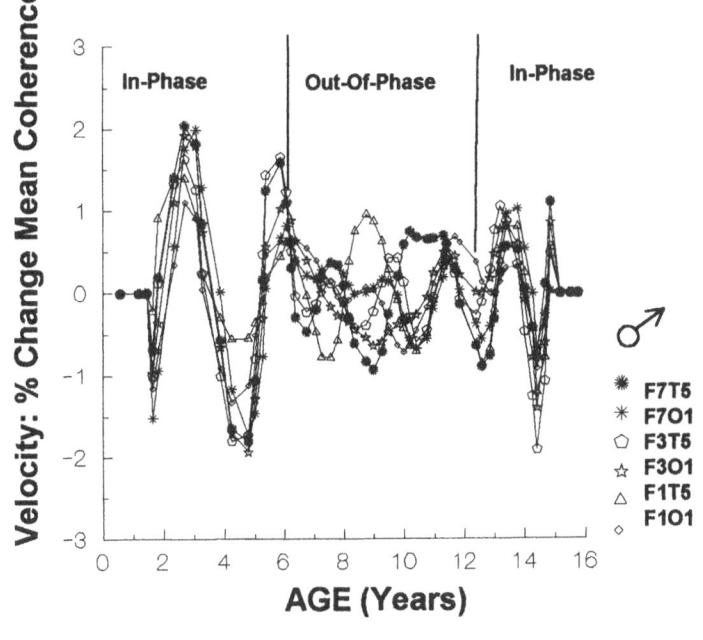

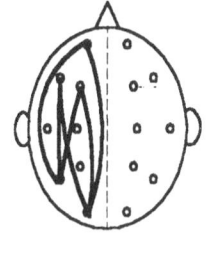

Figure 11C

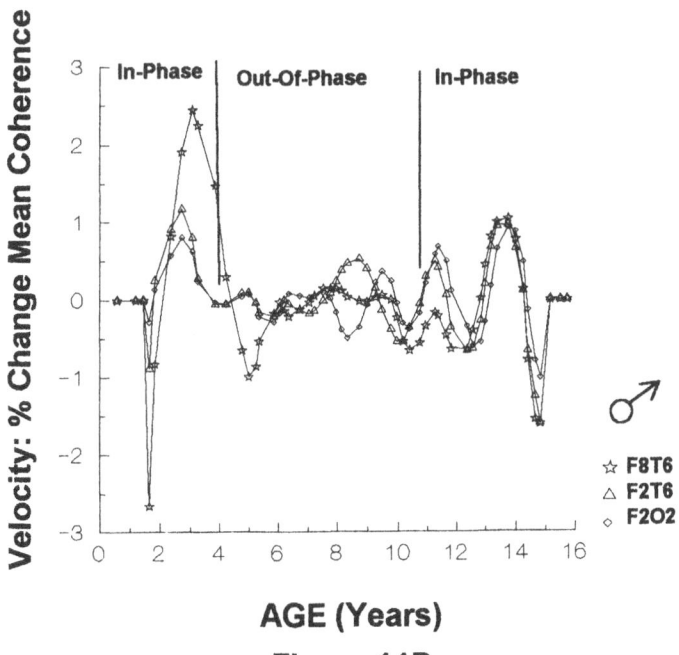

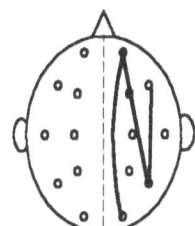

Figure 11D

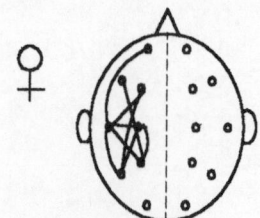

Female: Broca and Wernicke Regions are in one primary language processing network

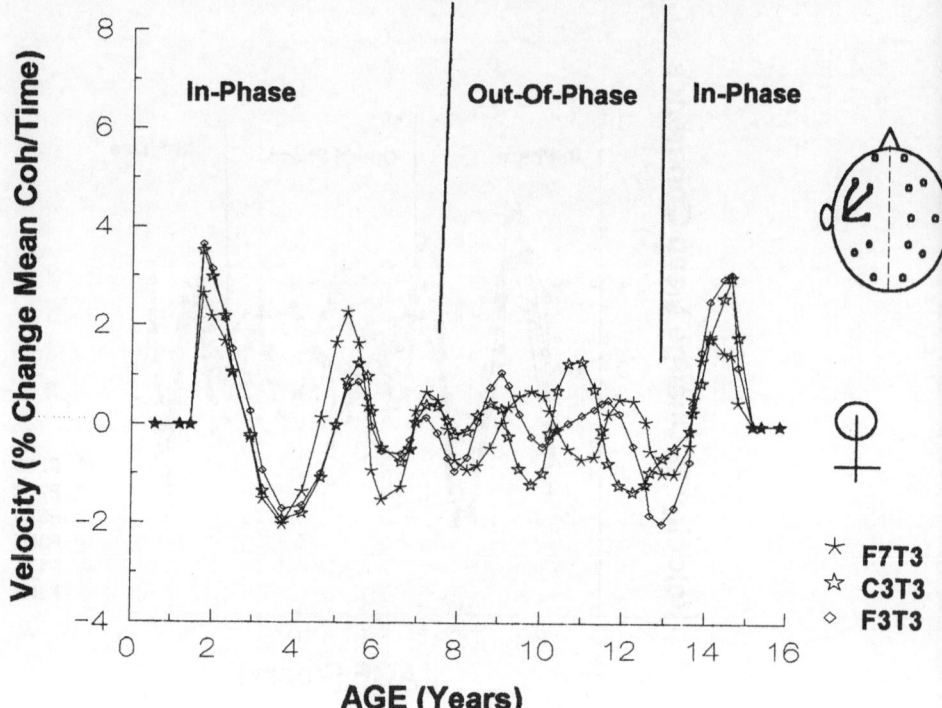

In-Phase   Out-Of-Phase   In-Phase

Velocity (% Change Mean Coh/Time)

AGE (Years)

* F7T3
☆ C3T3
◇ F3T3

Figure 12A

Male: Broca and Wernicke Regions are in two different language processing networks

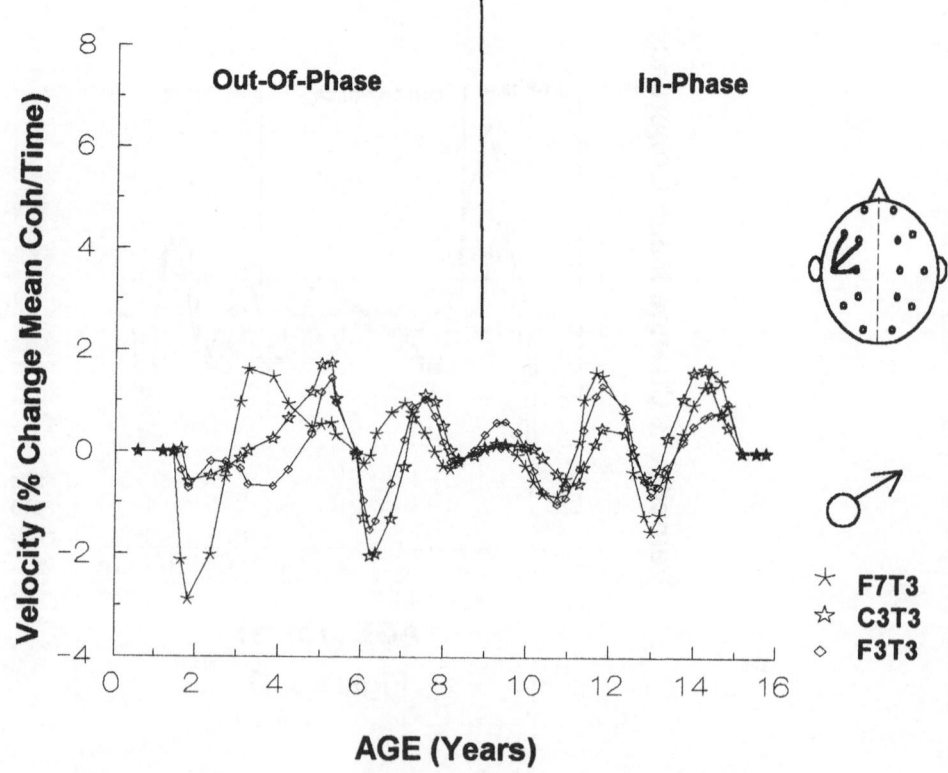

Out-Of-Phase   In-Phase

Velocity (% Change Mean Coh/Time)

AGE (Years)

* F7T3
☆ C3T3
◇ F3T3

Figure 12B

374

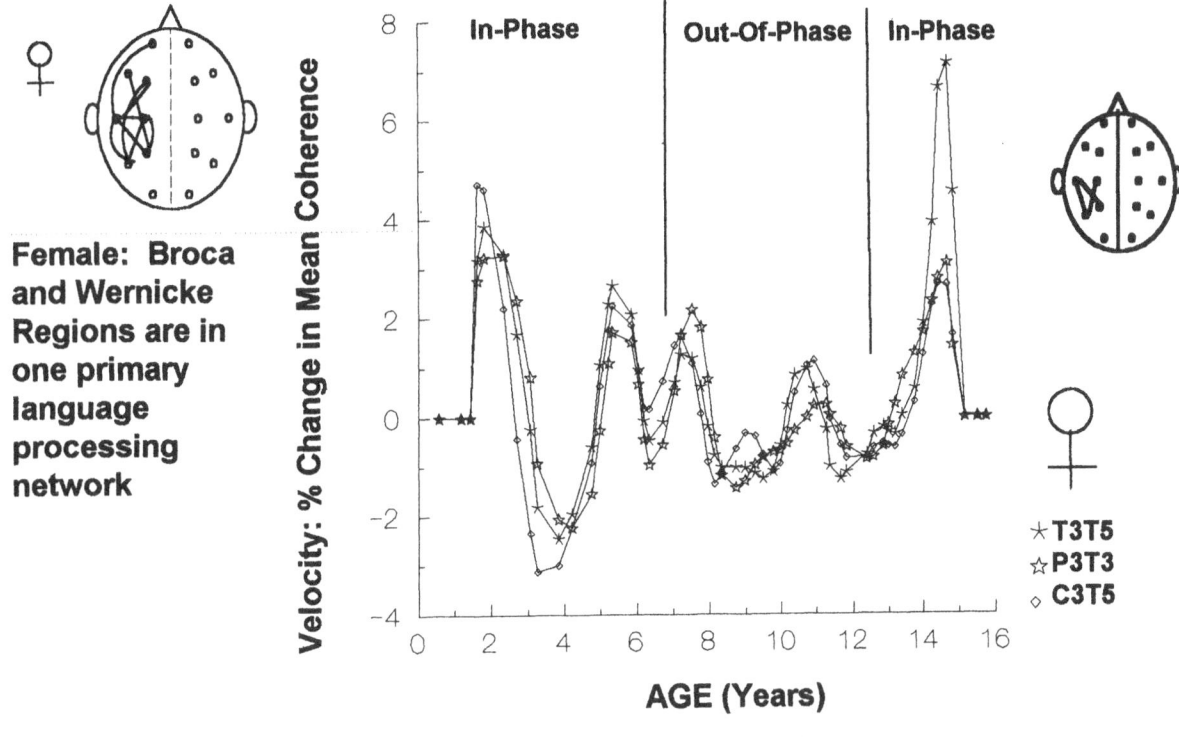

Female: Broca and Wernicke Regions are in one primary language processing network

Figure 12C

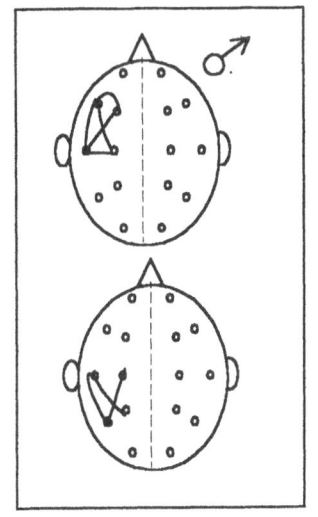

Male: Broca and Wernicke Regions are in two different language processing networks

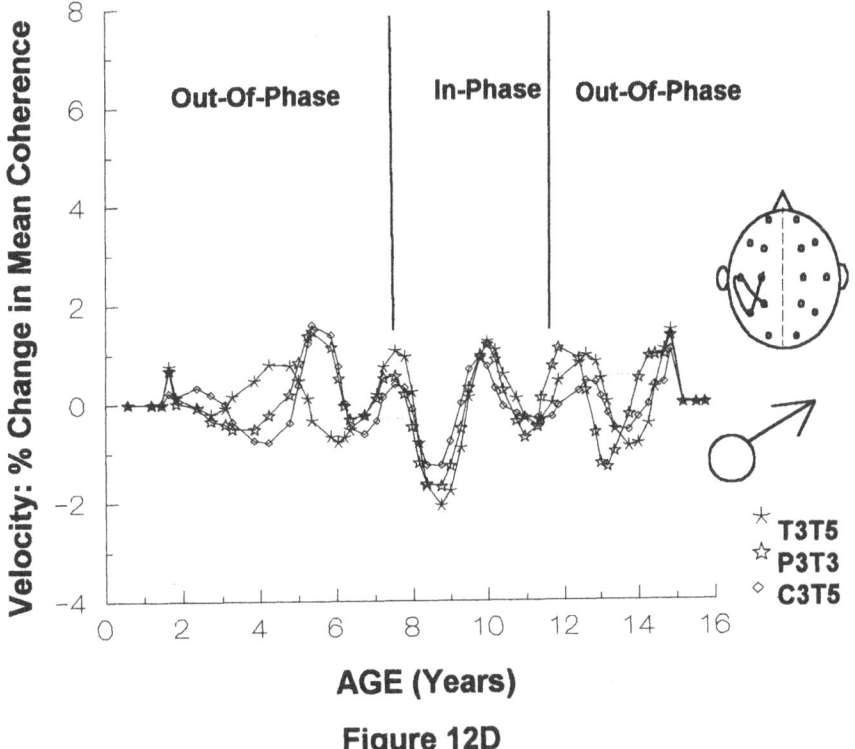

Figure 12D

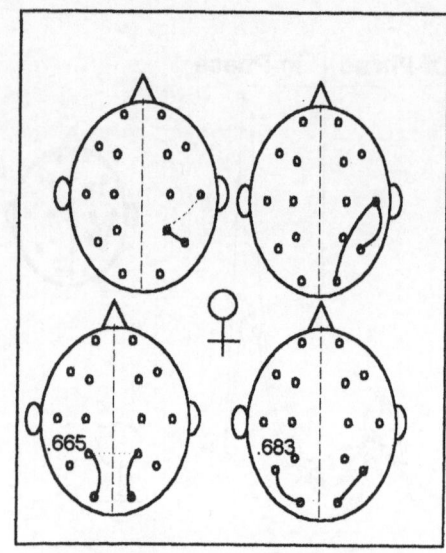

Female: Spatial processing in four regional networks

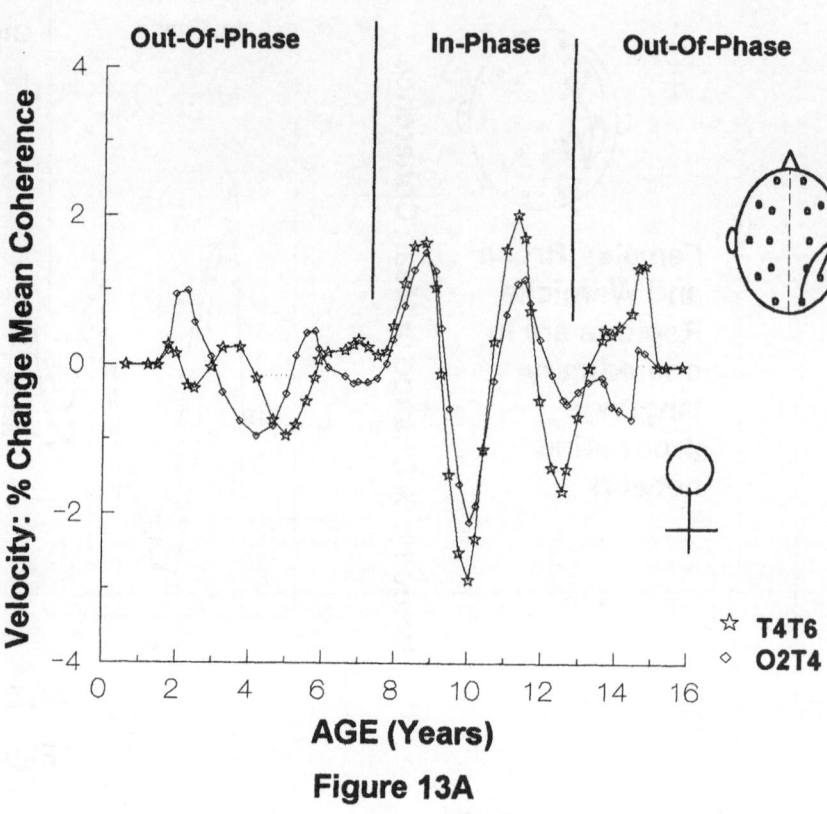

Figure 13A

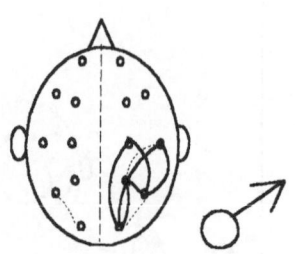

Male: Spatial processing in one regional network

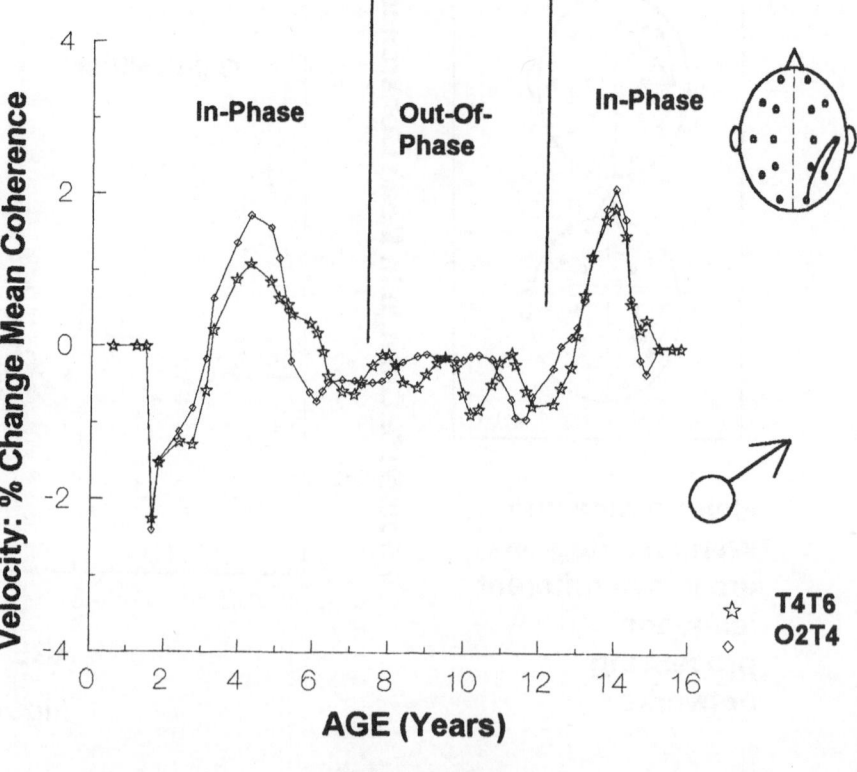

Figure 13B

14.

Brain Regions Associated With Retrieval of Structurally Coherent Visual Information

Daniel L. Schacter, Eric Reiman, Anne Uecker, Michael R. Polster, Lang Sheng Yun & Lynn A. Cooper

Brain regions associated with retrieval of structurally coherent visual information

Daniel L. Schacter, Eric Reiman, Anne Uecker, Michael R. Polster,
Lang Sheng Yun & Lynn A. Cooper

Brain regions associated with retrieval of structurally coherent visual information

Daniel L. Schacter*, Eric Reiman††, Anne Uecker§,
Michael R. Polster‖, Lang Sheng Yun‡¶
& Lynn A. Cooper #

* Department of Psychology, Harvard University, 33 Kirkland Street,
Cambridge, Massachusetts 02138, USA
† Department of Psychiatry, University of Arizona, Tucson,
Arizona 85721, USA
‡ Positron Emission Tomography Center, Good Samaritan Regional
Medical Center, Phoenix, Arizona 85006, USA
§ Division of Neural Systems, Memory and Aging, University of
Arizona, Tucson, Arizona 85721, USA
‖ Department of Psychology, University of Victoria at Wellington,
Wellington, New Zealand
¶ Department of Computer Science and Engineering, Arizona State
University, Tempe, Arizona 85287, USA
Department of Psychology, Columbia University, New York,
New York 10027, USA

An object's global, three-dimensional structure may be represented by a specialized brain system involving regions of inferior temporal cortex[1-3]. This system's role in object representation can be understood by experiments in which people study drawings of novel objects with possible or impossible three-dimensional structures, and later make either possible/impossible object decisions or old/new recognition decisions about briefly flashed studied and non-studied objects. Although object decisions about possible objects are facilitated by prior study, there is no corresponding facilitation for impossible objects, thereby implicating a system that is specifically involved in the representation of structurally coherent visual objects[4]. Here we show, by positron emission tomography (PET), that increases in blood flow in inferior temporal regions are associated with object decisions about possible but not impossible objects, and that there are increases in the vicinity of the hippocampal formation associated with episodic recognition of possible objects.

Ten 31-slice PET images of regional cerebral blood flow in 16 healthy females were obtained using the ECAT 951/31 scanner (Siemens, Knoxville, Tennessee), 45 mCi intravenous bolus injections of $H_2{}^{15}O$, and 60-s scans as they viewed 50-ms presentations of individual possible or impossible objects. During the first two scans, subjects pressed a button with their index finger when an object disappeared from the screen, providing no-decision baseline images that controlled for visual stimulation and movement. Subjects then studied a series of 20 possible and 20 impossible objects in the absence of a scan. During the next four scans, subjects made decisions about whether the objects were possible or impossible during four separate blocks of 20 old possible, 20 new possible, 20 old impossible, and 20 new impossible objects. The final four scans used the same arrangement of blocks, but subjects made old/new recognition memory decisions (Fig. 1). Despite procedural modifications necessitated by the demands of PET, behavioural data concerning object decision and recognition performance are in close agreement with previous results (Table 1). For each subtraction described below, automated algorithms were used to align the 10 PET images from each subject, spatially transform them into the coordinates of a standard brain atlas, control for variations in whole brain measurements, compute normalized t-score maps of significant increases in regional blood flow (maximal t-score > 2.58, $P < 0.005$, uncorrected for multiple comparisons), and superimpose the maps onto the subjects' spatially transformed and averaged magnetic resonance images[5-8].

To examine blood flow increases associated with object decisions about old and new, structurally coherent and structurally incoherent objects, the no-decision baseline scan was subtracted from the new and old object decision scans for possible and impossible objects, respectively. Object decisions about possible objects were associated with significant blood flow increases in the inferior temporal gyrus and the adjacent fusiform (occipito-temporal) gyrus (Fig. 2), bilaterally for old objects and on the right for new objects. In contrast, object decisions about old and new impossible objects were not associated with significant blood flow increases in the inferior temporal or fusiform gyri; the only exception was a trend ($P < 0.05$) in the right inferior temporal region in the new-object decision condition. To examine blood flow increases associated with changes in object decisions about structurally coherent objects as a result of studying them during the tool/support encoding task (priming effects; see Fig. 1, Table 1), the new possible-object decision scan was subtracted from

FIG. 1 Examples of objects used in the experiment are shown on the left. Possible objects are structures that could exist in the three-dimensional world. Impossible objects contain local edge or surface violations and hence could not be constructed as connected, three-dimensional solids. The overall experimental design is displayed on the right. Ten separate scans were used, each involving computerized presentation of 20 consecutive objects according to the following sequence: (1) a cross-hair fixation point for 250 ms; (2) a blank interval of 100 ms; (3) an object at central fixation for 50 ms followed by a key-press response; and (4) an inter-trial delay of 2,500 ms. The no-decision baseline scans (1 and 2) served as a control for visual exposure to the objects and actuation of a key-press response. Subjects next engaged in an encoding or study task, during which no scanning occurred. They saw 40 objects (20 possible and 20 impossible), presented individually in a random sequence, for 5 s each. Subjects indicated whether each object would best be used as a tool or for support. Previous research has shown that the tool/support judgement is one of several encoding tasks that produce significant effects of study, known as priming effects, on the object decision task[29]. Object decisions (OD) were assessed in the next four scans. During two of these scans (3 and 4; new possible and new impossible), subjects made decisions of possible/impossible about objects that had not appeared previously at any point in the experiment; during the other two scans (5 and 6; old possible and old impossible), they made the same decisions about objects that had appeared during the encoding task. Finally, episodic recognition (RN) was assessed by using a similar arrangement of four scans, except that subjects indicated whether or not they remembered seeing an object during the encoding task. The new possible and new impossible scans (7 and 8) each used objects

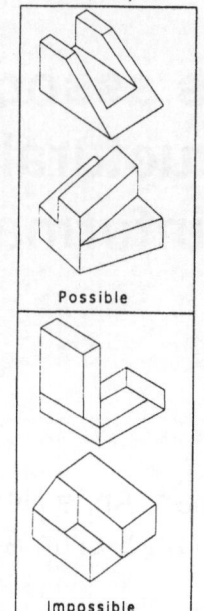

Sample objects

Possible

Impossible

Experimental Design

Task	Scan	Objects	Condition
Keypress To offset object	1 2	Possible impossible	No-decision baseline
Tool-support judgement	- - -	Possible and impossible	Encoding task
Object decision Possible or impossible structure	3 4 5 6	New possible New impossible Old possible Old impossible	New OD New OD Old OD Old OD
Recognition Previously studied or new object	7 8 9 10	New possible New impossible Old possible Old impossible	New RN New RN Old RN Old RN

that had not appeared previously in any phase of the experiment, whereas the old possible and old impossible scans (9 and 10) used objects that had appeared during the encoding task and during scans 5 or 6 of the object decision task. Assignment of objects to conditions and order of scans within conditions was completely counterbalanced for 12 subjects and partly counterbalanced for 4 subjects.

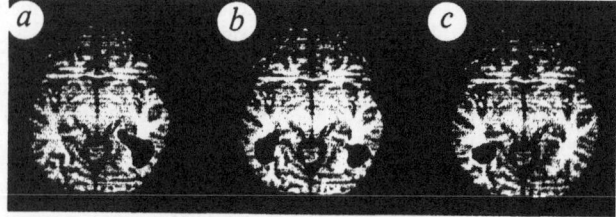

FIG. 2 For possible visual objects: a, new-object decision minus no-decision baseline; b, old-object decision minus no-decision baseline; and c, old-object decision minus new-object decision subtractions revealed significant blood flow increases in the vicinity of right, bilateral, and left inferior temporal and fusiform gyri, respectively. For each subtraction, automated algorithms were used to characterize significant increases in regional blood flow (those with a maximal normalized t-value > 2.58, $P < 0.005$, uncorrected for multiple comparisons[6-8]). For a–c, normalized t-value maps were superimposed onto a magnetic resonance image which was transformed into the coordinates of a brain atlas[5,7], volume rendered, and resected posterior to a coronal plane through the anterior commissure to reveal inferior temporal blood flow increases in a horizontal section 8 mm inferior to a plane through the anterior and posterior commissures. Increases in inferior temporal blood flow are shown in red, green and blue, which correspond to normalized t-values greater than 2.58, 2.33 and 1.64, and uncorrected probabilities less than 0.005, 0.01 and 0.05, respectively. Atlas coordinates and t-values of these and other significant blood flow increases are available from the author on request.

FIG. 3 For possible visual objects: a, new-recognition memory minus no-decision baseline; b, old-recognition memory minus no-decision baseline; and c, old-recognition memory minus new-recognition memory subtractions revealed significant blood flow increases in the vicinity of the left, bilateral and right hippocampus, respectively. For each subtraction, automated algorithms were used to characterize significant increases in regional blood flow (those with a maximal normalized t-value > 2.58, $P < 0.005$, uncorrected for multiple comparisons[6-8]). For a–c, normalized t-value maps were superimposed onto a magnetic resonance image which was transformed into the coordinates of a brain atlas[5,7], volume rendered, and resected posterior to a coronal plane through the anterior commissure to reveal hippocampal blood flow increases in a horizontal section 8 mm inferior to this plane in b and c. Increases in hippocampal blood flow (and additional increases in the vicinity of the parahippocampal gyrus and midbrain in b) are shown in red, green and blue, which correspond to normalized t-values greater

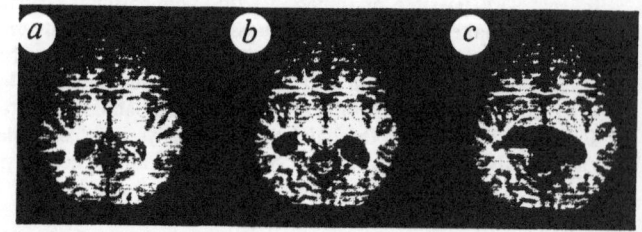

than 2.58, 2.33 and 1.64, and uncorrected probabilities less than 0.005, 0.01 and 0.05, respectively. Atlas coordinates and t-values of these and other significant blood flow increases are available from the author on request.

the old possible-object decision scan. This comparison revealed significant blood flow increases in the left inferior temporal/fusiform region (Fig. 2). When the same subtraction was performed for impossible objects, there were no significant blood flow increases in the inferior temporal and fusiform gyri. Although these findings suggest an association between priming of possible objects and left inferior temporal/fusiform blood flow increases, they could also be due to more possible but not more impossible objects being perceived as structurally coherent in the old-object decision condition than in the new-object decision condition.

Post hoc analyses were performed to characterize more directly those blood flow increases that were preferentially related to possible versus impossible objects, and those that were preferentially related to the left versus right hemisphere. Right inferior temporal and fusiform blood flow increases during object decision for possible objects were greater than those in the left hemisphere ($P < 0.005$), and were also greater than those in the right hemisphere during object decision for impossible objects ($P < 0.05$). Left inferior temporal and fusiform blood flow increases associated with studying possible objects were greater than those in the right hemisphere ($P < 0.005$), and were also greater than those in the left hemisphere during the same comparison for impossible objects ($P < 0.05$).

These findings are consistent with our proposal that a system involving inferior temporal regions is involved in the representation of a visual object's global, three-dimensional structure. Studies with non-human primates implicate inferior temporal regions in computing size-invariant representations of global object structure[1,3,9], and studies with human volunteers show that effects of prior study on object decisions about possible objects are size invariant[10] and depend on encoding of global object structure[4]. Several kinds of evidence implicate the fusiform gyri in high-level object recognition[11,12] and global or holistic visual representation[13-15].

For impossible objects only, both the old-object decision and new-object decision minus no-decision baseline comparisons revealed significant blood flow increases in the vicinity of the left hippocampal formation and the pulvinar bilaterally. The hippocampal formation has been linked previously with encoding of novel or unexpected stimuli[16-19] and the pulvinar has been implicated in controlling attention to visually salient stimuli[20]. Accordingly, these increases could reflect encoding of the novel or unusual features of impossible objects.

To examine blood flow increases associated with the recognition task, each no-decision baseline (possible or impossible) was subtracted from the corresponding old or new recognition scan, respectively (Fig. 3). For both possible and impossible objects, the new recognition minus no-decision baseline comparison revealed significant blood flow increases in the vicinity of the hippocampal formation bilaterally, predominantly on the left. The old-recognition minus baseline comparisons also revealed significant blood flow increases in the vicinity of the hippocampal formation, bilaterally for possible objects and predominantly on the left for impossible objects. For impossible objects only, these comparisons again revealed significant blood flow increases in the vicinity of the pulvinar bilaterally. Post hoc comparisons indicate that blood flow increases in the left hippocampal formation during the new-object decision, old-object decision, new recognition, and old recognition minus no-decision baseline comparisons for impossible objects were greater than those in the right ($P < 0.05$), and were also greater than those in the left during the same comparisons for possible objects ($P < 0.05$).

To examine blood flow changes specifically associated with memory for the appearance of an object in the study list, here referred to as episodic memory, the new-recognition condition was subtracted from the old-recognition condition (Fig. 3). For possible objects, this comparison revealed blood flow increases in the vicinity of the right hippocampal formation, left parahippocampal and fusiform gyri, the midbrain, and left dorsolat-

Task	Possible objects		Impossible objects	
	Old	New	Old	New
Object decision	0.67	0.56	0.63	0.58
Old/new recognition	0.82	0.33	0.68	0.36

TABLE 1 Object decision and recognition performance

For the object decision test, the table entries refer to the proportion of correct object decisions for old objects, which appeared previously during the encoding task, and the proportion of correct object decisions for new objects, which had not appeared during the encoding task. For possible objects, subjects made more correct decisions about studied than non-studied objects ($t(15) = 3.01$, $P < 0.005$), whereas, for impossible objects, significant differences between studied and non-studied objects were not observed ($t(15) < 1$). Nearly identical results were obtained in a purely behavioural pilot study using identical test procedures, carried out before the PET study. These findings replicate previously published results from a variety of conditions in numerous experiments[2,4,10,29]. Overall levels of object decision accuracy for new and old objects are also closely comparable to data from previously published behavioural studies. For the yes/no recognition test, the table entries refer to the proportion of 'old' responses made to old objects (the hit rate) and the proportion of 'old' responses to new objects (the false-alarm rate). To compare recognition accuracy for possible and impossible objects, we computed a corrected recognition score by subtracting the false-alarm rate from the hit rate. The corrected recognition score for possible objects (0.49) was significantly higher than the corrected recognition score for impossible objects (0.32), ($t(15) = 4.58$, $P < 0.001$). Nearly identical results were obtained in the pilot study using a separate set of subjects and the same test procedures. These findings replicate results from numerous previously published experiments[2,4,10,29]. The overall levels of 'old' and 'new' responses are also similar to published results from the same experiments. That we replicated both the quantitative and qualitative findings of previous research is important because the demands of PET scanning required that we present old possible, new possible, old impossible, and new impossible objects in separate blocks. This is a substantial departure from the standard practice of intermixing objects from different conditions, but our behavioural data indicate that task performance is indistinguishable from performance under standard experimental conditions.

eral prefrontal cortex (Brodmann areas 10, 44–46). For impossible objects, for which there was less accurate episodic memory (Table 1), the same comparisons revealed no significant blood flow increases. Post hoc comparisons indicate that the right hippocampal/parahippocampal blood flow increases during episodic memory for possible objects were greater than those in the left ($P < 0.05$), and were greater than those in the right during the same comparisons for impossible objects ($P < 0.05$).

Our results are consistent with evidence supporting involvement of the hippocampal formation[21-24] and dorsolateral prefrontal cortex[24-27] in episodic memory. Failure to detect significant blood flow increases in association with episodic recognition of impossible objects could indicate that recognition of these objects was not sufficiently robust, that the right hippocampal region is less involved in recognition of impossible than of possible objects, or that right hippocampal activation depends on both episodic memory and other factors that are not yet well understood[27].

This study and others have activated the hippocampal formation, but several previous memory experiments failed to do so[25-27]. Although the exact reasons for the contrasting results remain elusive[27,28], we suspect that the novelty of our objects is relevant. We suggest that the left hippocampal/parahippocampal region participated in some aspect of response to these novel stimuli (such as object encoding or identification of a mismatch between internal expectations and novel stimuli), whereas the right hippocampal/parahippocampal region was involved in episodic retrieval of the objects.

Our study attempted to indicate brain regions that are specifically involved in object decision and recognition tasks that we have previously studied behaviourally, so we presented possible or impossible objects in all conditions. It is unknown whether simply looking at possible or impossible objects, as opposed to a condition in which non-object stimuli are presented, produces differential blood flow increases for possible and impossible objects. A study using additional controls is necessary to determine how the blood flow changes reported here are related to those involved in passive perception of possible and impossible objects.

Additional significant blood flow changes will be described in a subsequent report. For instance, for both possible and impossible objects, the new-object decision, old-object decision, new recognition and old recognition minus no-decision baseline comparisons all revealed increases in the vicinity of dorsolateral prefrontal cortex (Brodmann areas 10, 44–46), which could be involved in the executive operations associated with making a decision.

Our study provides direct neuroanatomical evidence from the normal human brain that inferior temporal and fusiform gyri are selectively involved in computing global representations of structurally coherent three-dimensional objects, and extends our knowledge of the neural systems involved in episodic memory for visual objects. □

Received 28 October 1994; accepted 28 June 1995.

1. Miyashita, Y. A. Rev. Neurosci. 16, 245–269 (1993).
2. Schacter, D. L., Cooper, L. A., Tharan, M. & Rubens, A. B. J. cogn. Neurosci. 3, 118–131 (1991).
3. Tanaka, K. Science 262, 685–688 (1993).
4. Schacter, D. L., Cooper, L. A. & Delaney, S. M. J. exp. Psychol. 119, 5–24 (1990).
5. Collins, D. L., Neelin, P., Peters, T. M. & Evans, A. C. J. comput. assist. Tomogr. 18, 192–204 (1994).
6. Friston, K. J. et al. J. cerebr. Blood Flow Metab. 10, 458–466 (1991).
7. Talairach, J. & Tournoux, P. Co-planar Stereotaxic Atlas of the Human Brain (Thieme Medical, New York, 1988).
8. Mintun, M. A. & Lee, K. S. J. nucl. Med. 31, 816 (1991).
9. Gross, C. G. Phil. Trans. R. Soc. B335, 245–246 (1992).
10. Cooper, L. A., Schacter, D. L., Ballesteros, S. & Moore, C. J. exp. Psychol., Learn. Mem. Cogn. 18, 43–57 (1992).
11. Kosslyn, S. M. et al. Brain (in the press).
12. Sergent, J., Ohta, S. & MacDonald, B. Brain 115, 15–36 (1992).
13. Allison, T. et al. J. Neurophysiol. 71, 821–825 (1994).
14. Damasio, A. R., Damasio, H. & VanHoesen, G. W. Neurology 32, 331–341 (1982).
15. Haxby, J. V. et al. J. Neurosci. 14, 6336–6353 (1994).
16. Gray, J. A. The Neuropsychology of Anxiety (Oxford Univ. Press, 1982).
17. Stern, C. et al. Soc. Neurosci. Abstr. 20, 530 (1994).
18. Tulving, E. et al. NeuroReport 5, 2525–2528 (1994).
19. Vinograda, O. S. The Hippocampus Vol. 2 (eds Isaacson, R. L. & Pribram, K. H.) (Plenum, New York, 1978).
20. Robinson, D. L. & Petersen, S. E. Trends Neurosci. 15, 127–132 (1992).
21. Frackowiak, R. S. J. Trends Neurosci. 17, 109–115 (1994).
22. Grasby, P. M. et al. Neurosci. Lett. 163, 185–188 (1993).
23. Jones-Gotman, M. et al. Soc. Neurosci. Abstr. 19, 1002 (1993).
24. Squire, L. R. et al. Proc. natn. Acad. Sci. U.S.A. 89, 1837–1841 (1992).
25. Shallice, T. et al. Nature 368, 633–635 (1994).
26. Tulving, E. et al. Proc. natn. Acad. Sci. U.S.A. 91, 2012–2015 (1994).
27. Buckner, R. et al. J. Neurosci. 15, 12–29 (1995).
28. Buckner, R. & Tulving, E. Handbook of Neuropsychology (eds Boller, F. & Grafman, J.) 439–466 (Elsevier, Amsterdam, 1995).
29. Schacter, D. L. & Cooper, L. A. J. exp. Psychol., Learn. Mem. Cogn. 19, 995–1009 (1993).

ACKNOWLEDGEMENTS. We thank D. Bandy, N. Blocher, A. Evans, M. Mintun, K. Nelson and K. Zemanick for their assistance. This work was supported by grants from the McDonnell-Pew Program in Cognitive Neuroscience, the Robert S. Flinn Foundation, and the NIMH.

REGION	BRODMANN AREAS		OOD-BL ATLAS COORDINATES				NOD-BL ATLAS COORDINATES				OOD-NOD ATLAS COORDINATES			
			x	y	z	t-sc	x	y	z	t-sc	x	y	z	t-sc
inferior temporal and fusiform gyrus	21, 37	p o s s i b l e	-46 / 40	-44 / -48	-8 / -12	3.00 / 3.53	42	-48	-8	3.58	-42	-48	-8	2.86
hippocampus														
prefrontal	10, 44-46		-40 / 32	26 / 22	16 / 20	4.14 / 3.40	-38 / 30	26 / 20	16 / 20	3.63 / 4.26				
inferior temporal and fusiform gyrus	21, 37	i m p o s s i b l e												
hippocampus			-30	-30	-8	3.13	-28	-40	4	4.14				
prefrontal	10, 44-46		-36 / 36	30 / 20	16 / 20	3.79 / 4.39	-26 / 34	36 / 22	16 / 20	4.17 / 4.0				

REGION	BRODMANN AREAS		ORN-BL ATLAS COORDINATES				NRN-BL ATLAS COORDINATES				ORN-NRN ATLAS COORDINATES			
			x	y	z	t-sc	x	y	z	t-sc	x	y	z	t-sc
inferior temporal and fusiform gyrus	21, 37	P o s s i b l e												
hippocampus			-32 / 24	-28 / -34	-4 / -4	2.62 / 3.51	-22	-38	4	3.21	24	-30	-8	3.54
prefrontal	10, 44-46		-24 / 34	22 / 22	16 / 16	5.02 / 4.80	30	24	16	3.70	-26	32	12	3.22
inferior temporal and fusiform gyrus	21, 37	i m p o s s i b l e												
hippocampus			-28 / 28	-38 / -36	4 / 0	5.07 / 3.03	-30 / 24	-34 / -36	0 / 4	4.16 / 2.68				
prefrontal	10, 44-46		-24 / 32	38 / 18	16 / 20	5.04 / 5.29	-28 / 22	40 / 40	4 / 4	4.93 / 5.1				

Talairach and Tournoux (1988) coordinate values from <u>Brain Regions Associated with Retrieval of Structurally Coherent Visual Information</u>, Schacter et. al, 1995.
OOD = old object decision, NOD = new object decision, ORN = old recognition, NRN = nre recognition, and BL = baseline.

15.

Conscious Recollection and the Human Hippocampal Formation: Evidence from Positron Emmission Tomography

Daniel L. Schacter, Nathaniel M. Alpert, Cary R. Savage, Scott L. Rauch, and Marilyn S. Albert

Conscious recollection and the human hippocampal formation: Evidence from positron emission tomography

(memory/visual cortex/hippocampus/frontal lobes)

Daniel L. Schacter*†, Nathaniel M. Alpert‡, Cary R. Savage‡§, Scott L. Rauch‡§, and Marilyn S. Albert§¶

*Department of Psychology, Harvard University, 33 Kirkland Street, Cambridge, MA 02138; and Departments of ‡Radiology, §Psychiatry, and ¶Neurology, Massachusetts General Hospital and Harvard Medical School, Boston, MA 02114

Communicated by Larry R. Squire, University of California, San Diego, CA, September 11, 1995

ABSTRACT We used positron emission tomography (PET) to examine the role of the hippocampal formation in implicit and explicit memory. Human volunteers studied a list of familiar words, and then they either provided the first word that came to mind in response to three-letter cues (implicit memory) or tried to recall studied words in response to the same cues (explicit memory). There was no evidence of hippocampal activation in association with implicit memory. However, priming effects on the implicit memory test were associated with decreased activity in extrastriate visual cortex. On the explicit memory test, subjects recalled many target words in one condition and recalled few words in a second condition, despite trying to remember them. Comparisons between the two conditions showed that blood-flow increases in the hippocampal formation are specifically associated with the conscious recollection of studied words, whereas blood-flow increases in frontal regions are associated with efforts to retrieve target words. Our results help to clarify some puzzles concerning the role of the hippocampal formation in human memory.

Understanding the role of the hippocampal formation in learning and memory constitutes an enduring problem in cognitive neuroscience. Studies of brain-damaged amnesic patients implicate the hippocampal formation in explicit or conscious memory for past events. By contrast, the hippocampal formation is thought to be uninvolved in a nonconscious or implicit form of memory known as priming (1–4). Yet previous attempts to test these ideas directly by studying the normal human brain with positron emission tomography (PET) have yielded inconclusive results.

In an early PET study by Squire et al. (5), subjects studied a list of familiar words (e.g., GARNISH) and were then tested with three-letter word stems (e.g., GAR__). When subjects were instructed to provide a word from the study list on a cued recall test (explicit memory), there were significant blood flow increases in the vicinity of the right hippocampal formation compared with a baseline condition in which subjects responded to stems of nonstudied words. In a separate scan conducted in the same experimental session, subjects were instructed to complete stems of previously studied words with the first word that comes to mind (implicit memory), and a priming effect was observed: subjects preferentially completed the stems with words from the study list. Compared with the baseline condition, priming was associated with decreased blood flow in extrastriate occipital cortex and increased blood flow in the right hippocampus/parahippocampal gyrus. Because amnesic patients with hippocampal damage show intact priming effects (6–8), the former finding is consistent with the idea that such effects are mediated by brain systems outside the

hippocampal formation. But the latter finding is inconsistent with this idea.

However, performance in the priming condition may have been "contaminated" by some form of explicit memory (9): subjects may have intentionally or unintentionally remembered the primed words (5). If such contamination accounts for hippocampal activation in the priming condition, then it should be possible to abolish hippocampal activation by eliminating explicit retrieval. Yet several PET experiments have failed to find hippocampal activation even in association with explicit retrieval (10–13). Most critically, Buckner et al. (14) reported a follow-up of the Squire et al. experiment in which subjects were given three-letter word beginnings and attempted to remember words that had been studied previously either in the auditory modality or in a different typographic case. Buckner et al. observed no evidence of hippocampal activations in either condition (14). Because subjects were attempting to remember target items in both the different-modality and different-case conditions, the absence of blood flow changes in the hippocampal formation suggests that trying to retrieve a past event is not sufficient to activate the hippocampus. Hippocampal activation may be more closely related to some aspect of the actual recollection of an event. By contrast, Buckner et al. (14) found that areas in prefrontal cortex showed blood flow increases in both the different-case and different-modality conditions, thus raising the possibility that frontal activations, which have been observed frequently in PET studies of explicit retrieval (5, 10, 12–16), are related to the effort involved in trying to remember recently studied items (11).

To test these hypotheses, we performed a priming experiment in which we attempted to eliminate conscious recollection and an explicit memory experiment in which we attempted to separate out the effort to recall an event from the actual recollection of it.

METHODS

Experimental Procedure. In the priming experiment, subjects studied target words in a way that ensured that they would later have poor explicit memory for them (6). Specifically, subjects performed a shallow, nonsemantic study task that requires them to indicate the number of T-junctions in a word. After the subjects had studied 24 familiar words (20 target plus 4 nontested buffers), PET scans were carried out while subjects responded to three-letter word stems, with separate blocks of stems for studied words (priming) and nonstudied words (baseline). During each 1-min scan, subjects were instructed to respond with the first word that came to mind and to do their best to complete each stem. We refer to the nonscanned study task, the scanned priming condition, and the scanned baseline

---

Abbreviation: PET, positron emission tomography.
†To whom reprint requests should be addressed.

condition as a "study-test unit." The volunteers were then given two additional study-test units, thus yielding a total of three scans for primed words and three scans for baseline words. Order of conditions and items assigned to conditions was counterbalanced across subjects.

For the explicit memory experiment, one condition was designed to yield high levels of explicit recall and the other was designed to yield low levels of explicit recall. We accomplished this by manipulating how subjects studied a series of target words. Forty-eight different words (40 targets plus 8 buffers) were shown for 5 sec each; no scanning was performed during this study phase. The High Recall condition consisted of 20 target words that were presented four times each, with presentations distributed randomly throughout the list; each time one of these words appeared, subjects made a semantic judgment (they counted the number of meanings associated with each word). We reasoned that on a later memory test, subjects would easily recollect many of these words. The Low Recall condition consisted of 20 words that were presented only once; subjects made a nonsemantic judgment about each word (the T-junction counting task used in our first experiment). We reasoned that on a later test, subjects would recall few of these words despite trying hard to do so.

Two separate 90-sec blocks of three-letter stems, separated by a 10-min rest or study period, were presented on a computer monitor. One block contained stems that could be completed with the High Recall words and the other contained stems that could be completed with the Low Recall words; subjects were instructed to try to remember a study-list word that fit each stem. If they could not recall a study-list target, they were told to guess. Subjects were allowed up to 5 sec to respond to each stem. Immediately after their response, the next stem appeared. After subjects completed the two test scans, two further study-test units were administered. Prior to the first study-test unit and after the third, subjects performed the baseline task used in the previous experiment, in which they completed stems of nonstudied words with the first word that came to mind. Order of conditions and items assigned to conditions was completely counterbalanced across subjects.

Subjects. Six healthy male and two healthy female volunteers (mean age = 19.6 yr) participated in the priming experiment; five healthy male and three healthy female volunteers (mean age = 20.5 yr) participated in the explicit memory experiment. All subjects were screened to rule out the presence of medical, psychiatric, or neurological disorders.

PET Scanning. A gantry held the computer monitor, tilted so that the screen was readily visible from within the PET camera. PET data were acquired while subjects inhaled oxygen-15-labeled carbon dioxide ($[^{15}O]CO_2$) for 1 min. Each scan proceeded as follows: (*i*) subjects were reminded of the instructions to lie still, breathe normally, and to perform either the stem completion or cued recall task; (*ii*) the PET camera was started at time zero, and continued acquiring data for 90 sec; (*iii*) the stem completion (or cued recall) task started at time zero, preceded by four buffer items, and continued until completion of the block of 24 trials (all subjects required >90 sec to finish the stem completion or cued recall tasks); (*iv*) the final 60 sec of PET camera data acquisition (i.e., time 30 to 90 sec) coincided with the 60-sec period of active tracer inhalation; (*v*) at the end of this period, PET data acquisition and radiolabeled gas flow were terminated; (*vi*) following a 10-min tracer-washout period, the next scan was performed, until the series of eight scans was completed.

The PET facilities and procedures were very similar to those previously described (e.g., refs. 17 and 18). A General Electric–Scanditronix (Uppsala) model PC4096 15-slice whole-body tomograph was used (19). An individually molded thermoplastic face mask (True Scan, Annapolis, MD) was used to minimize head motion. Transmission measurements were made by using an orbiting pin source.

All brain images were corrected for interscan movement, by realignment with respect to the first scan, prior to further image processing. An automated motion-correction algorithm was employed (after ref. 20). Motion-corrected PET brain images were then transformed to the standard Talairach coordinate system (21) as previously described (e.g., refs. 17 and 22). Blood flow images were normalized to 50 ml/min per 100 g and were rescaled and smoothed with a 20-mm Gaussian filter.

Once the transformations of the PET data were performed and the data were expressed in stereotaxic space, statistical parametric maps (SPMs) were created. Each SPM was inspected for regions of activation with Z scores ≥ 3.00 for unplanned comparisons ($P < 0.001$, uncorrected for multiple comparisons), and >2.58 ($P < 0.005$) for planned comparisons involving the hippocampal formation, prefrontal cortex, and extrastriate occipital cortex.

RESULTS

Analysis of behavioral data from the priming experiment revealed that a significantly larger percentage of stems was completed with study-list words in the priming condition than in the baseline condition [30% vs. 17%; $F(1, 7) = 41.81$, $P < 0.0001$]. The magnitude of priming is comparable to similar effects obtained in conditions where explicit memory has been effectively eliminated (23, 24), but it is much smaller than the priming effect reported in the PET study of Squire *et al.* (5), reflecting the explicit contamination that likely occurred in that experiment. Analysis of priming effects separately for each study-test unit revealed nearly identical levels of priming in the first, second, and third test blocks ($F < 1$), providing additional evidence that subjects did not engage in intentional retrieval strategies, which would have inflated priming in later test blocks.

To examine relevant changes in regional cerebral blood flow, data from the three study-test units were combined to yield a single baseline condition and a single priming condition. When we compared these two conditions, we found that priming was associated with significant blood flow decreases in bilateral extrastriate occipital cortex (Brodmann area 19; Table 1/Fig. 1). The decrease on the right was in approximately the same location as in the previous study, whereas the decrease on the left had a more superior focus. By contrast, there were no significant blood flow changes in the vicinity of the hippocampal formation (maximum Z score = 0.33 for available points z axis = -12 to $+4$). In addition to the predicted blood flow changes in the extrastriate regions, we also observed other significant ($Z > 3.0$) decreases and increases in association with priming that will be discussed in a separate report. Baseline minus priming [decreases]: right insular cortex (39, -26, 0), right thalamus (7, -30, 4), right putamen (16, 2, 8), right motor/premotor cortex (61, -8, 28), and right parietal cortex (area 7; 30, -55, 52). Priming minus baseline [increases]: left prefrontal cortex (area 47; -39, 30, -8), left precuneus (area 7; -13, -51, 56). All findings are expressed in Talairach coordinates as x, y, z.

In the explicit memory experiment, behavioral data confirmed that subjects remembered many more words in the High Recall condition (79%) than in the Low Recall condition [35%; $F(1, 7) = 205.74$, $P < 0.0001$]. However, the percentage of words recalled did not differ significantly across the three test blocks ($F < 1$). To examine associated blood flow changes, we compared the High and Low Recall conditions directly to one another, collapsing across the three test blocks. The logic of the comparison holds that brain regions that are specifically associated with the conscious recollection of a word should show significant blood flow increases in the High Recall minus Low Recall comparison, whereas regions that are specifically associated with the effort involved in trying to retrieve a

Table 1. Primary regions of interest exhibiting significant change in blood flow associated with the priming and explicit memory conditions [all additional findings ($Z \geq 3.00$) listed in text]

Contrast	Region	Z score (max pixel value)*	Max pixel coordinates[†]
Priming contrasts			
Priming minus Baseline	Right area 19	−3.10	33, −74, 0
	Left area 19	−3.23	−33, −79, 24
Explicit memory contrasts			
Low Recall minus High Recall	Left prefrontal (area 10/46)	3.81	−31, 43, 8
	Left anterior cingulate	3.25	−7, 15, 32
	Right precuneus (area 19)	3.70	5, −72, 32
High Recall minus Low Recall	Right hippocampal	2.82	25, −34, 0
High Recall minus Baseline	Left hippocampal	3.38	−19, −39, −4
	Right hippocampal	3.96	15, −37, 0
Low Recall minus Baseline	Right orbitofrontal (area 11)	3.25	5, 35, −12
	Right anterior cingulate	3.77	7, 34, 0
	Left prefrontal (area 10)	3.47	−35, 54, 8
	Right prefrontal (area 10)	3.12	30, 46, 8
	Right prefrontal (area 9)	4.04	12, 47, 28

*Values represent the maximum pixel value (Z score units) within the region of interest from the statistical parametric map.
[†]Coordinates in Talairach space (21), expressed as x, y, z; $x > 0$ is right of the midsagittal plane, $y > 0$ is anterior to the anterior commissure, and $z > 0$ is superior to the anterior commissure–posterior commissure plane.

recently studied word should show significant blood flow increases in the Low Recall minus High Recall comparison. Consistent with our hypothesis that prefrontal regions are related to retrieval effort, the Low Recall minus High Recall comparison revealed a significant blood flow increase in the left dorsolateral prefrontal cortex (Brodmann areas 10 and 46; Table 1). Previous studies have implicated this region in generating words and semantic associations (25–28). It is likely that attempts to generate candidate word responses occurred more frequently in the Low Recall condition than in the High Recall condition. This comparison also revealed significant increases in the left anterior cingulate and the right precuneus, which have been implicated previously in attentional processes, such as target selection (29, 30), that should have been

more relevant to the Low Recall than the High Recall condition.

Consistent with our hypothesis that the hippocampus is involved in some aspect of conscious recollection, the High Recall minus Low Recall comparison yielded only a single significant increase, in the right hippocampal formation (Table 1). The locus of this activation is nearly identical to the locus of the activation reported in the same-case condition of the earlier stem-cued recall study (5, 14).

To examine further the consistency of our results, we compared the High Recall and Low Recall conditions, with the Baseline condition in which subjects completed stems of nonstudied words with the first word that came to mind (Table 1/Fig. 1). The logic was similar to our reasoning in the previous

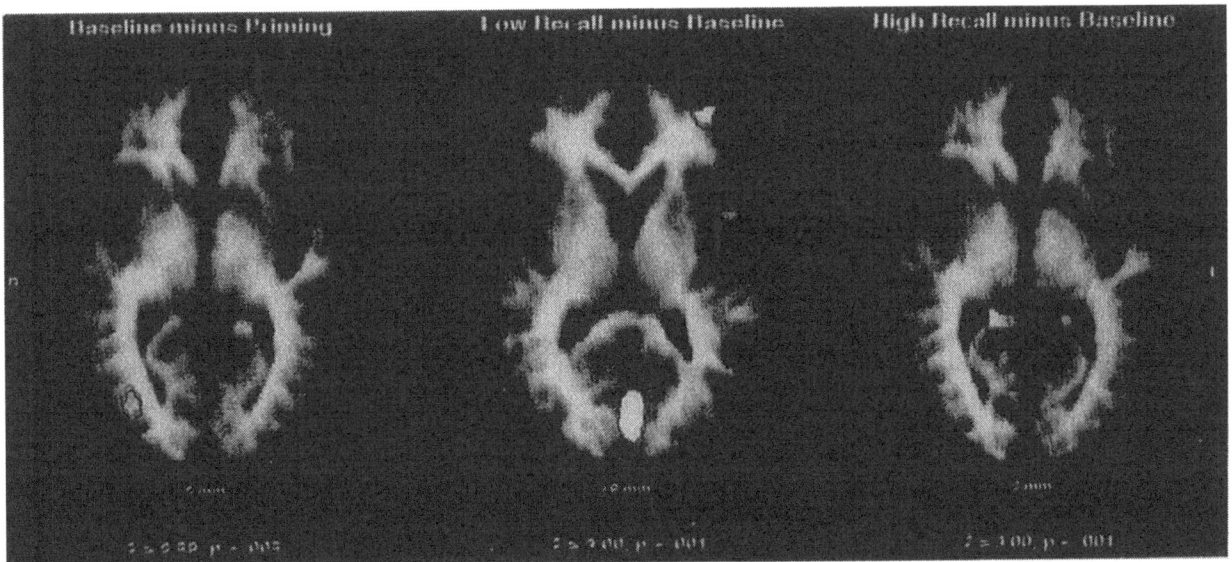

Fig. 1. PET statistical maps show territories of activation superimposed over averaged magnetic resonance images, transformed to Talairach space. Activations are thresholded to a Z score ≥ 2.58 for the Baseline minus Priming image and 3.00 for the Low Recall minus Baseline and High Recall minus Baseline images. Images are transverse sections, with z coordinates reflecting distance in millimeters from the anterior commissure–posterior commissure plane. The Baseline minus Priming image shows a region of significantly decreased blood flow (green) associated with priming in right visual association cortex (area 19). The Low Recall minus Baseline image shows regions of significantly increased blood flow (yellow) associated with high effort and low explicit recall (35% accuracy) in the left prefrontal cortex (area 10) and secondary visual cortex (area 18). The High Recall minus Baseline image shows regions of significantly increased blood flow (red) associated with high levels of explicit recall (79% accuracy) in bilateral hippocampal regions.

comparisons: brain regions associated with conscious recollection should show significant blood increases in the High Recall minus Baseline comparison, whereas regions associated with retrieval effort should show increases in the Low Recall minus Baseline comparison. In the High Recall minus Baseline comparison, there were extensive bilateral blood flow increases in the hippocampal formation, but no significant activations in the vicinity of the frontal lobes. The Low Recall minus Baseline comparison yielded extensive bilateral blood flow increases in the prefrontal cortex, especially in Brodmann area 10, but none in the vicinity of the hippocampal formation. Prefrontal cortex, particularly on the right side, has been activated in numerous previous PET studies of explicit retrieval (5, 10–14, 16), and the increases that we observed are close to previously reported ones. These results, together with the finding of significant left frontal activation in the Low Recall minus High Recall comparison, imply that the pervasive activation of frontal regions in previous memory studies reflects the effort involved in attempting to retrieve a past event. Both the High Recall and Low Recall minus Baseline comparisons yielded a number of other significant ($Z > 3.0$) blood flow increases that will be discussed in a separate report. High Recall minus Baseline: left cerebellum ($-26, -68, -12$), right cuneus (area 17; 3, $-71, 8$), bilateral supramarginal gyrus (area 40; $-47, -28, 20$; 47, $-22, 20$), and right visual association cortex (area 19; 26, $-82, 24$). Low Recall minus Baseline: left brain stem ($-10, -13, -12$), left cerebellum ($-24, -50, -12$), left secondary visual cortex (area 18; $-3, -76, 4$), right supramarginal gyrus (area 40; 41, $-3, 16$), left insular cortex ($-34, -16, 16$), and right cuneus (area 18; 7, $-82, 24$). All findings are expressed in Talairach coordinates as x, y, z.

DISCUSSION

Our major findings—that the hippocampal formation showed significant blood flow increases in the High Recall condition compared with the Low Recall and Baseline conditions, but no such increases during priming—provide new information about the role of the hippocampal formation in implicit and explicit memory. We first consider several puzzles that are clarified by our findings, and then we consider issues that remain to be clarified.

In view of our results, it now seems likely that previous findings of hippocampal activation during priming on the stem completion test were due to the influence of conscious recollection (5). Because frontal regions were not active during priming in the experiment of Squire *et al.*, this "contamination" from explicit memory probably reflects incidental or unintentional conscious recollection of words studied twice, under semantic encoding conditions, several minutes prior to the priming task. Our data support the idea that priming occurs independently of the hippocampal formation and depends instead on brain systems involved with the perceptual representation of words and objects (1–4).

Our data also help to clarify why the previous experiment by Buckner *et al.* (14) using the stem-cued recall test failed to detect significant blood flow increases in the vicinity of the hippocampus during explicit retrieval in both a different-case condition and a different modality condition. Our results suggest that hippocampal activation is more closely associated with the actual recollection of a past event than with the effort involved in attempting to remember the event. Simply instructing subjects to try to remember an event is probably not sufficient to produce significant blood flow increases in the hippocampal formation. These observations suggest that in the different-case and different-modality conditions of the experiments of Buckner *et al.*, the way in which subjects recollected studied items differed from the manner in which they recollected them in the same-case condition. Note that the absolute levels of recall in the different-case condition (73%) and

different-modality condition (62%) of Buckner *et al.* are closer to the levels of performance in our High Recall condition (79%) than in our Low Recall condition (35%). Although we must be cautious about between-experiment comparisons, these results suggest that the absolute level of recall may be less important in determining whether hippocampal activation is observed than the qualitative manner in which target events are remembered. Further research will be needed to specify exactly which features of recollection are most relevant to hippocampal activation.

This account is also consistent with the results of a study in which subjects studied and later tried to recognize structurally possible and structurally impossible novel visual objects (15). Right hippocampal activation was observed in association with explicit recognition of possible objects, but there was no corresponding activation in association with recognition of impossible objects. The possible objects were remembered more accurately than were the impossible objects. Our results thus suggest that differences in either the level or type of recollection associated with possible and impossible objects, respectively, account for the differential activation of the right hippocampal region during explicit recognition of the two types of objects.

Our hypotheses regarding conscious recollection and the hippocampus do not explain all relevant findings, however. We note first that factors other than conscious recollection, such as the novelty of a stimulus, can produce hippocampal activation (15, 31). The response of the hippocampal formation to a novel stimulus may be associated with its role in encoding and consolidation of new memories, whereas activations related to conscious recollection indicate a role for the hippocampus in memory retrieval.

However, in several studies that are quite similar to ours, where subjects presumably consciously recollected recently studied verbal materials, no hippocampal activations were observed (10–13). We make several observations. First, our study-test unit design used three separate replications for each subject of all critical comparisons to maximize power to detect hippocampal and other activations. Several of the experiments that failed to detect any evidence of hippocampal activation used only a single replication of critical comparisons (10–12), perhaps resulting in insufficient power to detect blood flow increases associated with hippocampal activity. Second, because the exact features of conscious recollection that are most relevant to hippocampal activation remain to be determined, it is possible that aspects of recollection that are most relevant to hippocampal activation played a more prominent role in our paradigm than in others. For instance, in one experiment that failed to observe hippocampal activation, some nonstudied items were presented with studied items during a single scan, possibly diluting the overall level of recollection (10). Other experiments used auditory presentation and test (12, 13, 16). Given the previously observed absence of hippocampal activation when modality and typographic case of stimuli differed at study and test (14), it is possible that reinstating visual information about a studied item, plus a high level of remembering, both contribute to blood flow increases in the hippocampal formation during explicit retrieval (see refs. 32 and 33 for data concerning visual information and recollective experience). Also, several experiments that failed to detect hippocampal activation used recognition tests (10, 11, 13, 16), whereas we used recall. Although the hippocampus was activated during recognition of novel visual objects in a study noted earlier (15), conscious recollection during recall and recognition may differ, such that it is more difficult to detect hippocampal blood flow increases in association with recognition than with recall. Additional studies will be needed to determine which of these factors, if any, are relevant to hippocampal activations in PET studies.

In contrast to inconsistent activation of the hippocampal formation in PET experiments, lesion studies with experimental animals and studies of human amnesic patients with hippocampal damage indicate a broader role for the hippocampus in explicit memory (for reviews, see refs. 3 and 34), which may reflect in part the hippocampal contribution to encoding and consolidation of memories alluded to earlier. By contrast, our results and the other PET evidence described in the preceding paragraph all bear on the role of the hippocampus in memory retrieval. The hippocampal formation may play a more limited role in retrieval than it does in encoding and consolidation. Alternatively, limitations on PET measurement techniques may account for some previous failures to detect hippocampal activity. While the exact role of the hippocampal formation in human memory retrieval remains to be specified, our study indicates that further exploration of specific aspects of conscious recollection is likely to be revealing.

Finally, our results also bear on the role of prefrontal cortex in explicit retrieval. Consistent with other recent PET data, they suggest that frontal regions play an important role in the retrieval effort associated with attempts to recall past events (11). The right anterior prefrontal cortex (area 10) in particular has been especially active during explicit retrieval (16). We observed activation of this area in the Low Recall minus Baseline comparison, but not in the Low Recall minus High Recall comparison, whereas left prefrontal cortex was active in both comparisons. One interpretation of this pattern is that right area 10 is especially relevant to shifting from semantic or lexical retrieval, which was required when subjects completed stems with the first word that came to mind in the baseline condition, to explicit or episodic retrieval, which was required when subjects tried to recall study list words. If so, it is curious that we did not see right frontal activity in the High Recall minus Baseline comparison, since the former involves episodic retrieval and the latter does not. This may be because words that have been studied four times in a semantic encoding condition, as in our High Recall condition, came to mind with little retrieval effort during the cued-recall test. An important problem for future research is to specify the conditions under which both right and left prefrontal regions play a greater or lesser role in efforts to retrieve recently experienced episodes.

We thank Kimberly Nelson for help with preparation of the manuscript and Brian Rafferty for experimental assistance. This research was supported by grants from the Charles A. Dana Foundation, Grant RO1 AG08441-06 from the National Institute on Aging, Grant T32 CA09362 from the National Cancer Institute, and Grants MH01215 and MH01230 from the National Institute of Mental Health.

1. Moscovitch, M. (1994) in *Memory Systems 1994*, eds. Schacter, D. L. & Tulving, E. (MIT Press, Cambridge, MA), pp. 269–310.
2. Schacter, D. L. (1994) in *Memory Systems 1994*, eds. Schacter, D. L. & Tulving, E. (MIT Press, Cambridge, MA), pp. 233–268.
3. Squire, L. R. (1992) *Psychol. Rev.* **99**, 195–231.
4. Tulving, E. & Schacter, D. L. (1990) *Science* **247**, 301–306.
5. Squire, L. R., Ojemann, J. G., Miezin, F. M., Petersen, S. E., Videen, T. O. & Raichle, M. E. (1992) *Proc. Natl. Acad. Sci. USA* **89**, 1837–1841.
6. Graf, P., Squire, L. R. & Mandler, G. (1984) *J. Exp. Psychol. Learn. Mem. Cognit.* **10**, 164–178.
7. Schacter, D. L. (1987) *J. Exp. Psychol. Learn. Mem. Cognit.* **13**, 501–518.
8. Shimamura, A. P. (1986) *Q. J. Exp. Psychol.* **38A**, 619–644.
9. Schacter, D. L., Bowers, J. & Booker, J. (1989) in *Implicit Memory: Theoretical Issues*, eds. Lewandowsky, S., Dunn, J. C. & Kirsner, K. (Erlbaum, Hillsdale, NJ), pp. 47–69.
10. Andreasen, N. C., O'Leary, D. S., Arndt, S., Cizadlo, T., Hurtig, R., Rezai, K., Watkins, G. L., Boles, Ponto, L. L. & Hichwa, R. D. (1995) *Proc. Natl. Acad. Sci. USA* **92**, 5111–5115.
11. Kapur, S., Craik, F. I. M., Jones, C., Brown, G. M., Houle, S. & Tulving, E. (1995) *NeuroReport* **6**, 1880–1884.
12. Shallice, T., Fletcher, P., Frith, C. D., Grasby, P., Frackowiak, R. S. J. & Dolan, R. J. (1994) *Nature (London)* **368**, 633–635.
13. Tulving, E., Kapur, S., Markowitsch, H. J., Craik, F. I. M., Habib, R. & Houle, S. (1994) *Proc. Natl. Acad. Sci. USA* **91**, 2012–2015.
14. Buckner, R. L., Petersen, S. E., Ojemann, J. G., Miezin, F. M., Squire, L. R. & Raichle, M. E. (1995) *J. Neurosci.* **15**, 12–29.
15. Schacter, D. L., Reiman, E., Uecker, A., Polster, M. R., Yun, L. S. & Cooper, L. A. (1995) *Nature (London)* **29**, 587–590.
16. Tulving, E., Kapur, S., Craik, F. I. M., Moscovitch, M. & Houle, S. (1994) *Proc. Natl. Acad. Sci. USA* **91**, 2016–2020.
17. Rauch, S. L., Savage, C. R., Alpert, N. M., Miguel, E. C., Baer, L., Breiter, H. C., Fischman, A. J., Manzo, P. A., Moretti, C. & Jenike, M. A. (1995) *Arch. Gen. Psychiatry* **52**, 20–28.
18. Kosslyn, S. M., Alpert, N. M., Thompson, W. L., Chabris, C. F., Rauch, S. L. & Anderson, A. K. (1994) *J. Cognit. Neurosci.* **5**, 263–287.
19. Kops, E. R., Herzog, H., Schmid, A., Holte, S. & Feinendegen, L. E. (1990) *J. Comput. Assist. Tomogr.* **14**, 437–445.
20. Woods, R. P., Cherry, S. & Mazziotta, J. (1992) *J. Comput. Assist. Tomogr.* **16**, 620–633.
21. Talairach, J. & Tournoux, P. (1988) *Co-Planar Stereotaxis Atlas of the Human Brain* (Thieme, New York).
22. Alpert, N. M., Belrdichevsky, D., Weise, S., Tang, J. & Rauch, S. L. (1993) *Quantification of Brain Function: Tracer Kinetics and Image Analysis in Brain PET* (Elsevier, Amsterdam), pp. 459–463.
23. Graf, P., Mandler, G. & Haden, P. (1982) *Science* **218**, 1243–1244.
24. Bowers, J. S. & Schacter, D. L. (1990) *J. Exp. Psychol. Learn. Mem. Cognit.* **16**, 404–416.
25. Petersen, S. E., Fox, P. T., Posner, M. I., Mintum, M. & Raichle, M. E. (1988) *Nature (London)* **331**, 585–589.
26. Frith, C. D., Friston, K., Liddle, P. F. & Frackowiack, R. S. J. (1991) *Neuropsychologia* **29**, 1137–1148.
27. Kapur, S., Craik, F. I. M., Tulving, E., Wilson, A. A., Houle, S. & Brown, G. M. (1994) *Proc. Natl. Acad. Sci. USA* **91**, 2008–2011.
28. Raichle, M. E., Fiez, J. A., Videen, T. O., MacLeod, A. M., Pardo, J. V., Fox, P. T. & Petersen, S. E. (1994) *Cereb. Cortex* **4**, 8–26.
29. Frith, C. D., Friston, K., Liddle, P. F. & Frackowiak, R. S. J. (1991) *Proc. R. Soc. London B* **244**, 241–246.
30. Posner, M. I. & Dehaene, S. (1994) *Trends Neurosci.* **17**, 75–79.
31. Tulving, E., Markowitsch, H. J., Kapur, S., Habib, R. & Houle, S. (1994) *NeuroReport* **5**, 2525–2528.
32. Brewer, W. F. (1988) in *Remembering Reconsidered: Ecolocial and Traditional Approaches to the Study of Memory*, eds. Neisser, U. & Winograd, E. (Cambridge Univ. Press, New York), pp. 21–90.
33. Dewhurst, S. A. & Conway, M. A. (1994) *J. Exp. Psychol. Learn. Mem. Cognit.* **20**, 1088–1098.
34. Milner, B. (1972) *Clin. Neurosurg.* **19**, 421–466.

16.
An Exploration of the Neural Bases of Memory Representations of Reward and Context

Raymond P. Kesner

An Exploration of the Neural Bases of Memory Representations of Reward and Context

Raymond P. Kesner

Department of Psychology
University of Utah

The question to be addressed in this conference is how memory representations can aid in the appreciation of reinforcement. There are likely to be multiple memory components associated with the representation of reinforcement and its consequences. Some of the more important components would include affect as influenced by reward value, spatial-temporal context, and sensory-perceptual and linguistic content of the information to be remembered. How are these components of memory represented in the brain and what neural circuits are involved? It is currently accepted by many theoreticians that there are multiple forms of memory, often based on a dual memory system with one system dependent upon the hippocampus and interconnected neural regions and the other system dependent upon the rest of the brain. For example, distinctions have been made between a hippocampus dependent declarative, relationship-based, flexible, explicit, working and data-based memory system versus a hippocampus independent nondeclarative, procedural, nonrelationship based, inflexible, implicit, reference, and knowledge-based memory system (Cohen and Eichenbaum, 1993; Kesner and DiMattia, 1987; Olton et al., 1979; Schacter, 1987; Squire, 1992, Squire, 1994; Tulving, 1983). The models differ in terms of the exact nature of the memory representation within the hippocampus with either an emphasis on only spatial information (O'Keefe and Nadel, 1978), spatial, temporal and linguistic information (Kesner, 1990), or all component information (Cohen and Eichenbaum, 1993; Olton et al., 1979; Squire,1994). The models also vary in terms of the process or processes that the hippocampus utilizes to represent mnemonic information. These processes include consolidation of new information (Squire, 1994), working, data-based or short-term memory representations (Kesner, 1990; Olton et al., 1979), or selective filtering of incoming information requiring the elimination of interfering stimuli (Shapiro and Olton, 1994), or all three processes (Cohen and Eichenbaum,1993; Kesner, see present chapter; Rolls,1989). Finally, models differ on the relationship between the two dual system components varying from dynamic interactions based on sequential processing or independent parallel processing of information. I have emphasized the idea that memories can be subdivided into multiple attributes including time, space, affect, response, sensory-perception and language and that memories are organized either based on new data, short-term or working memory information or on knowledge, long-term or reference memory information. Furthermore, different neural systems mediate different attributes and these systems can operate independent of each other (Kesner and DiMattia, 1987). For example, it can be shown that the hippocampus mediates time, space (external context) and language, the caudate-putamen mediates response, the amygdala mediates affect, and certain cortical and parahippocampal regions mediate sensory-perception information (Kesner et al., 1992; Kesner et al., 1993;

Kesner and Williams, 1995). For the purposes of the present chapter, I will concentrate on the hippocampus as a neural system that is important for providing memory for the spatial-temporal and linguistic context within which reward value or reinforcement is embedded and the amygdala as a neural system that is important for providing memory for reward value or reinforcement upon which affect might be based.

HIPPOCAMPUS

Structure

Does the hippocampus represent all attribute information in memory or is it restricted to spatial, temporal and linguistic attribute information? It can clearly be demonstrated that single cells within the hippocampus are activated by most sensory inputs, including vestibular, olfactory, visual, auditory, and somatosensory as well as by higher-order integration of sensory stimuli (Cohen and Eichenbaum, 1993). However, the question of importance is whether these sensory inputs have a memory representation within the hippocampus. Thusfar, it appears that short- term memory or data-based memory for odor or visual object information is not altered by lesions of the hippocampus (Aggleton et al., 1986; Jackson-Smith et al.,1993; Mumby et al., 1992; Otto and Eichenbaum, 1992), implying that sensory-perceptual information is not represented in memory within the hippocampus. In order to test this idea further, rats were trained on a recognition memory task for visual objects using a delayed non-matching-to-sample task procedure. In the visual object memory task, the animal was tested in a runway divided by a single door. In the study phase of a trial, the animal was given the opportunity to approach a unique visual object, push it aside, and receive a reinforcement. Immediately after consuming the reinforcement, the animal was tested on the other side of the runway and given a choice between the same unique object and a novel unique object (test phase). The animal had to choose the novel object in order to receive a reinforcement (win-shift or non-matching-to-sample rule). Each animal received 10 trials per day until it reached a criterion of performance of ≥80% on 30 consecutive trials. They then received large (dorsal and ventral) electrolytic hippocampal or cortical control lesions dorsal to the dorsal hippocampus. Following recovery from surgery they were retested in the same task. The results indicate that animals with hippocampal or control lesions readily relearn the task. The animals were then given trials with a 10- or 20-sec delay between the study and test phase. Again, there were no differences in performance between the two groups (Kesner et al. 1993). Thus, object information per se is not represented in short-term memory within the hippocampus. It is more likely that object information is used as a marker for delineating spatial location and temporal order information. In order to demonstrate this possibility, rats were trained in a memory for allocentric distance experiment (Long and Kesner, in press). In this experiment rats were trained to remember the distance between two identical objects using a matching-to-sample procedure. In the study phase the two stimuli were placed either 2 or 7 cm apart with food reward located underneath both objects. In the test phase the rat was either presented with an allocentric distance that matched the study phase (reward condition) or one that did not match the study phase (non-reward condition). The absolute spatial locations of the stimuli was varied in order to avoid the potential use of absolute spatial location information to solve the task. Latency to respond to the reward and nonreward condition was measured. A cut-off score of 10 sec was used. The rats were trained to a criterion of at least a 3 sec shorter latency on match in comparison with mismatch trials. After reaching criterion, rats received large (dorsal and ventral) electrolytic lesions or cortical control lesions dorsal to the dorsal hippocampus. Following recovery from surgery the rats were retested. The results are shown in Figures 1A and B and indicate that hippocampal lesioned rats are clearly impaired in memory for allocentric spatial distance as indicated by smaller latency differences between match and mismatch trials on post-surgery tests), whereas control rats continue to perform the task without any difficulty. New rats were trained to discriminate between the two distances used in the memory experiment. Following hippocampal or control lesions, there were no impairments in performing the discrimination. Thus, hippocampal lesioned rats have no difficulty in discriminating between two spatial distances, but they cannot remember a spatial distance. In this experiment, object information was necessary to define a spatial distance.

In a second experiment rats were trained to remember a single spatial location defined by a single object in one of four spatial locations within an enclosed Plexiglas box using a matching-to-sample procedure (Long and Kesner, 1994). In the study phase the rat was presented with a stimulus placed at one of the four possible spatial locations. In the test phase the rat was presented with the same stimulus either at the same spatial location (reward) or at a different spatial location (non-reward). Latency to respond to the reward and nonreward condition was measured. A cut-off score of 10 sec was used. The rats were trained to a criterion of at least a 3 sec shorter latency on match in comparison with mismatch trials. After reaching criterion, rats received large (dorsal and ventral) electrolytic lesions or cortical control lesions dorsal to the dorsal hippocampus. Following recovery from surgery the rats were retested. The results are shown in Figures 2A and B and indicate that hippocampal lesioned rats are clearly impaired in memory for spatial location information as indicated by smaller latency differences between match and mismatch trials on post-surgery tests, whereas control rats continue to perform the task without any difficulty. The same rats were then trained to discriminate between two spatial locations. There were no differences between the two groups. Clearly, hippocampal lesioned rats can discriminate between spatial locations even though they cannot remember spatial locations. Both experiments support the idea that spatial information can be based on objects as markers and involve the hippocampus in short-term memory only for spatial, but not object information.

In a different experiment short-term memory for the temporal attribute as duration was tested (Jackson-Smith et al, 1994). Rats were trained on a memory for duration task using a delayed conditional discrimination procedure. The rats had to learn that a black rectangle stimulus that was visible for 2 sec would result in a positive (go) reinforcement for one object (a ball) and no reinforcement (no go) for a different object (a bottle). However, if the black rectangle stimulus was visible for 8 sec then there would be no reinforcement for the ball (no go), but a reinforcement for the bottle (go). After rats learned to respond differentially in terms of latency to approach the object, they received large (dorsal and ventral) lesions of the hippocampus or cortex dorsal to the dorsal hippocampus. Following recovery from surgery they were retested. The results are shown in Figures 3A and B and indicate that there were in contrast to cortical control lesions major impairments following hippocampal lesions as indicated by smaller latency differences between positive and negative trials on post-surgery tests. In order to ensure that the deficits observed with hippocampal lesions were not due to a discrimination problem, new rats were trained in an object (black rectangle) duration discrimination task. In this situation the rats were reinforced for either a 2 or 10 sec exposure (duration) of the black rectangle. The stimulus was presented and remained visible for either

2 or 10 sec, following which the door was raised and latency to move the stimulus was measured. Half of the animals in each group received a piece of Froot Loop on trials with a short stimulus duration and the other half was reinforced on those trials with a long stimulus duration. After rats learned to respond differentially in terms of latency to approach the object, they received large (dorsal and ventral) lesions of the hippocampus or cortex dorsal to the dorsal hippocampus. Following recovery from surgery the rats were retested. The results indicate that after hippocampal lesions, there was an initial deficit followed by complete recovery. Thus, the hippocampus mediates only memory for duration, but does not mediate duration discrimination. The data are consistent with previous research which indicated that fimbria-fornix lesioned rats are impaired in remembering the duration of a stimulus across a short delay interval, even though there is only a small change in estimating the passage of time. (Olton, 1986; Meck, Church, Olton, 1984). In rats it appears that the hippocampus is actively involved in representing in short-term memory both spatial and temporal attribute information based on the use of in this case objects as markers for spatial location, distance, or the beginning and end of the presence (duration) of an object, but is not directly involved in representing in memory information concerning specific objects. A similar case can be made for the absence of a role for memory for odor information (Otto and Eichenbaum, 1992). Odor information involves the hippocampus only when is there is a need for flexible use of odor information (Bunsey and Eichenbaum, in press), often because of a high degree of ambiguity or excessive interference.

To what extent can one generalize from hippocampal function in rats to humans with respect to memory representation of spatial and temporal attribute information, but not sensory-perceptual attribute information? In order to answer this question, a number of experiments were conducted using humans that have been exposed to hypoxia due to a variety of causes, but primarily carbon monoxide poisoning (Hopkins and Kesner, 1994). These subjects have a profound anterograde amnesia and, based on MRI data, have bilateral damage to the hippocampus, but no detectable damage to entorhinal cortex, parahippocampal gyrus, or temporal cortex. They also show no signs of prefrontal cortex dysfunction based on normal performance on tests of fluency and Wisconsin Card sorting. The hypoxic subjects and age matched controls were tested for short-term memory for duration of a visual object. Subjects were presented with a single object (square, circle, etc) on a computer screen for a duration of 1 or 3 sec. They were instructed to remember the duration of presentation of the object. After a delay of 1, 4, 8, 12, 16, or 20 sec, the same object appeared for the same or different duration. The subjects were asked to indicate whether the duration was the same or different from the duration shown in the study phase. The results are shown in Figure 4A and indicate that the hypoxic subjects were impaired relative to control subjects in short-term memory for duration for all delays. In order to determine whether the deficits may have been due to impaired memory for the objects per se, a control task was administered to the same subjects. They were presented with a single object for a duration of 1 or 3 sec and were asked to remember the object. After a delay of 1, 4, 8, 12, 16,or 20 sec. either the identical or a different object appeared on the screen. The subjects were asked if it was the same or a different object. The results are shown in Figure 4B and indicate that there were no significant differences between the hypoxic and control subjects. The impairment could not be due to an inability to estimate time accurately, because in an additional experiment with objects the subjects were asked to estimate the time elapsed before each of the 1,4, 8, 12, 16, 20 sec delay intervals. Memory estimates were accurate up to 8 sec followed by some underestimation with longer delays, so that memory for the duration of 1 or 3 sec could not be due to difficulty in estimating time. Thus, the results suggest that like rodents, humans with hippocampal damage have difficulty in representing short-term memory for duration of an object, but not short-term memory for a single object.

In an additional experiment these same subjects were tested for memory for distance between two points. During the study phase, on a computer screen the subjects were presented with distances of .5, 1, 1.5, or 2.0 cm. between two points. After a delay of 1, 4, 8, 12, 16, or 20 sec, the same or different distance appeared. The subjects were asked to judge whether the distance was the same or different from the distance shown in the study phase. The results are shown in Figure 4C and indicate that the hypoxic subjects were impaired relative to control subjects in short-term memory for all the long delays, but with normal memory for the 1 sec delay. The results suggest that like rodents humans with hippocampal damage have difficulty in representing short-term memory for spatial distance information. Additional support for the idea that the hippocampus in humans represent spatial rather than sensory-perceptual attribute information comes from the observation of a) Smith and Milner (1981) who demonstrated that right temporal lobe resected patients which includes hippocampus, are impaired in short-term memory for the spatial location of an array of objects, but do not have any difficulty in free recall of the same objects, b) Pigott and Milner (1993) who have demonstrated that right temporal lobe resected patients are impaired in remembering that the location of an object was changed within a scene, but had no difficulty in remembering that an object was changed for a specific location within a scene, and c) Gabrieli et al (1995) and Keane et al (1995) who demonstrated that a patient with bilateral occipital lobe lesions was impaired in implicit short-term memory for sensory-perceptual information, but a patient with bilateral medial temporal lobe lesions had no difficulty. In summary, these data suggest that the hippocampus mediates short-term memory for spatial and temporal attribute, but not sensory-perceptual (object) attribute information.

Process

Based on ample evidence that almost all sensory information is processed by hippocampal neurons perhaps to provide for sensory markers for time and space and that the hippocampus mediates temporal and spatial information, it is likely that one of the main process functions of the hippocampus is to encode the temporal and spatial order of events. This would ensure that newly highly processed sensory information is organized within the hippocampus and enhances the possibility of remembering and temporarily storing one event as separate from another event in time (temporal order) and one place as separate from another place (spatial order). There are at least three processes associated with hippocampal function that are necessary to accomplish the goal of coding the temporal and spatial order of events. These processes include a) selective filtering of event information to reduce interference. This process is akin to the idea that the hippocampus is involved in representational differentiation (Myers et al., 1995) and indirectly in the utilization of relationships (Cohen and Eichenbaum, 1993), b) data-based, short-term or working memory representation of spatial and temporal information, and c) consolidation of new spatial-temporal information.

The key paradigms that require hippocampal function assess the role of selective filtering and reduced interference and can be labeled as temporal order and spatial order. In the spatial order task, rats are required to remember a spatial location dependent upon the distance between the study phase object and an object used as a foil. More specifically, during the study phase an object which covers a baited food well was randomly positioned in one of six possible spatial locations on a cheese board. Rats exited a start box and displaced the object in order to receive a food award and were then returned to the start box. On the ensuing test phase rats were allowed to choose between two objects which were identical to the study phase object. One object was baited and positioned in the previous study phase location (correct choice), the other (foil) was unbaited and placed in a different location (incorrect choice). Five distances (min =15 cm, max =105 cm) were randomly used to separate the foil from the correct object. Following the establishment of a criterion of 75 % correct averaged across all separation distances, rats were given either large (dorsal and ventral) hippocampal or cortical control lesions dorsal to the dorsal hippocampus. Following recovery from surgery the rats were retested. The results are shown in Figures 5A and B and indicate that whereas control rats matched their pre-surgery performance for all spatial distances, hippocampal lesioned rats displayed impairments for short (15 cm-37.5 cm) and medium (60 cm) spatial separations, but performed as well as controls when the spatial separation was long (82.5- 105 cm). It can be shown that the ability to remember the long distances was not based on an egocentric response strategy, because if the study phase was presented on one side of the cheese board and the test originated on the opposite side, the hippocampal lesioned rats still performed the long distances without difficulty. It is clear that in this task it is necessary to separate one spatial location from another spatial location. Hippocampal lesioned rats cannot separate these spatial locations very well, sothat they can perform the task only when the spatial locations are far apart.

Similar deficits have been observed in patients with hippocampal damage due to an hypoxic episode for new geographical information. In this experiment spatial information was presented in the context of various locations on a map of New Brunswick. During the study phase the subjects were shown a series of eight cities from North to South located randomly on a map of New Brunswick. The name and location of each city was presented for 5 seconds. After each sequence hypoxics or control subjects were asked to make decisions about which of two cities were closer to the North. In the test phase the subjects were given only the names of the two cities in the absence of the map. The distance between cities was systematically varied. Other sequences included cities from South to North, East to West and West to East. The results are shown in Figure 6A and clearly indicate that control subjects perform well for all distances, but hypoxic subjects are markedly impaired on all spatial distances, but with a slight improvement for the longest spatial distance. The pattern is remarkably similar to what was seen for rats with hippocampal lesions (see Figure 5B). One method that can be used to attenuate interference among spatial distances is to present subjects with highly familiar information. It would then be predicted that hypoxic subjects would show little if any impairments, because excessive interference should not affect memory performance in such a task. To accomplish this aim, the same procedures were used as described above, but in this case the map of the United States was used. The results are shown in Figure 6B and indicate that even though there was a ceiling effect for the control subjects, the hypoxic subjects were not impaired relative to controls. Thus, it appears that the hippocampus in both animals and humans plays a significant role in reducing interference produced by proximally close spatial locations in new situations.

In the temporal order task, rats are required to remember an event (e.g. spatial location, visual object) dependent upon the temporal order of occurrence of events. More specifically, on an eight arm maze during the study phase of each trial rats were allowed to visit each of eight arms once in an order that is randomly selected for that trial. The test phase required the rats to choose which of two arms occurred earlier in the sequence of arms visited during the study phase. The arms selected as test arms varied according to temporal lag or distance (0-6) or the number of arms that occurred between the two test arms in the study phase. After the rats reached a criterion of 75% or better performance on all the distances but zero, the rats received large (dorsal and ventral), small (dorsal) hippocampus or cortical control lesions dorsal to the dorsal hippocampus. Following recovery from surgery, the rats were retested. The results are shown in Figures 7A, B, C and indicate that for both pre- and post-surgery tests, the control rats performed at chance at a temporal distance of zero, but their performance was excellent for the remaining temporal distances. In contrast, on post-surgery tests dorsal hippocampal lesions disrupted performance for temporal distances of 0 and 2, but did not affect performance for the longest temporal distances of 4 and 6. Furthermore, on post-surgery tests large (dorsal plus ventral) hippocampal lesions produced a marked deficit for all temporal distances with a slight improvement for the longest temporal distance. In this task it is necessary to separate one event from another. Hippocampal lesioned rats cannot separate events across time, because of an inability to inhibit interference that is likely to accompany sequential occurring events. The resultant increase in temporal interference impairs the rat's ability to remember the order of specific events. It appears that the larger the damage to the hippocampus, the greater the temporal interference. It is possible to reduce the presence of temporal interference by presenting the rats with a constant sequence of eight spatial locations followed by temporal distance tests. In this case large hippocampal lesions following training did not result in any significant deficits (Chiba et al., 1994; Chiba et al., 1992).

The events do not have to be based on only spatial location information. Similar deficits have been observed with lists of visual objects in rats and lists of spatial locations and words in patients with hippocampal damage due to an hypoxic episode or temporal lobe resection. Patients with right or left temporal lobe resection, hypoxics with damage to the hippocampus, epileptic controls or age matched controls were presented with a list of eight word sentences or eight spatial locations (Xs) on a grid on a Macintosh computer and tested for memory for temporal distances (Chiba et al. ,1996; Hopkins et al.,1995).

In the spatial task during the study phase a series of random (novel) sequences of eight Xs appeared on the screen, for a period of 5 sec each. Subjects were instructed to pay attention to the locations of the Xs as well as to the order in which they occurred. In the test phase subjects were presented with two Xs and were asked to determine which one occurred earlier in the study phase. Unlimited time was allowed for the subjects to make their choices. Temporal distances of 0, 2, 4, and 6 were assessed with eight observations for each distance for each type of task. Temporal distance is determined by the number of items in the study phase that occur between the two test items. For each study phase, four tests were given, one for each temporal distance. The results are shown in Figures 8A and B and indicate that relative to controls, epileptic subjects, and subjects with left temporal lobe resections, the hypoxic subjects showed an impairment for all temporal distances, but with some slight improvement for the longest distance, whereas the right temporal lobe resected patients showed a deficit only for the shorter but not for the longer distances. It appears that the larger the damage to the hippocampus, the greater the temporal interference. These results are similar to what was found in rats and suggest that patients with right or bilateral hippocampus dysfunction have difficulty in encoding or remembering temporal contexts for spatial information due to increased interference in the temporal resolution of specific spatial locations.

To test this idea further the same subjects were given sequences where the Xs appeared in a meaningful geometric pattern. An example of a meaningful and familiar pattern would be the presentation of the locations in the pattern of a large X. The first X would be presented in the top left corner of the grid and then in each subsequent square down the diagonal ending in the bottom right corner. The next X would be presented in the bottom left corner of the grid and in each subsequent square up the diagonal ending in the top right corner. It was assumed that familiar structure would attenuate temporal interference and thus reduce the deficits observed in the hypoxic and right temporal lobe resected patients for the novel temporal sequences. The results indicated that the deficit was markedly attenuated with no deficit for the right temporal lobe resected patients and a very slight reduction in performance for the hypoxic patients.

In the language (verbal) task during the study phase eight-word meaningless (novel) sentences were presented on a computer screen. The novel sentences were neither syntactically nor semantically correct, but consisted of eight random words that formed a nonsense sentence (e.g., "But falls eat cord that tall educate nor."). The words were presented one at a time for 5 sec each. Subjects were asked to remember the sentences and the order that the words were presented. After each sentence was presented, the subjects were shown two words that occurred in the study phase and were asked to select the word that occurred earlier in the sequence. Unlimited time was allowed for the subjects to make their choices. Temporal distances (the number of words that occurred between the two test words) of 0, 2, 4, and 6 were assessed with eight observations for each temporal distance. For each study phase four tests were given, one for each temporal distance. The results are shown in Figures 9A and B and indicate that relative to controls, epileptic subjects, and subjects with right temporal lobe resections, the hypoxic subjects showed an impairment for all temporal distances, but with some slight improvement for the longest distance, whereas the left temporal lobe resected patients showed a deficit only for all but the longest distance. These results suggest that patients with left or bilateral hippocampal dysfunction have difficulty in encoding or remembering temporal contexts for linguistic information due to increased interference in the temporal resolution of specific words. It appears that the larger the damage to the hippocampus, the greater the temporal interference. To test this idea further the same subjects were given sequences of words that generated familiar sentences that were both syntactically and semantically correct (e.g., "The boy passed a very difficult math test.") and less-familiar sentences that were syntactically correct but not semantically correct (e.g., "The table was entertained by various empty vegetables."). It was assumed that familiar structure would attenuate temporal interference and thus reduce the deficits observed in the hypoxic and left temporal lobe resected patients for the novel temporal sequences. The results indicated that the deficit was markedly attenuated with no deficit for the left temporal lobe resected patients and a slight reduction in performance for the hypoxic group. These data are consistent with the observation of temporal order deficits in amnesic patients (Hubert and Piercy, 1976, Hirst and Volpe, 1982, Squire et al, 1981, Chiba et al., 1990). Only one study, Sagar et al (1990) found no deficit for patient H.M. who had a bilateral temporal lobectomy. However, Sagar used a different testing procedure for assessing temporal order memory. Reanalyzing Sagar's data using a temporal distance measure resulted in the same pattern of results. The data are also consistent with other studies that have reported that the right hippocampus mediates spatial attribute information, whereas the left hippocampus mediates linguistic attribute information (Milner, 1971; Chiba et al, 1990). These data support the idea that the hippocampus might function to reduce temporal interference between events. Thus, hypoxic and temporal lobe resected patients perform more poorly, especially at short distances, because of increased interference in the temporal resolution of events.

All the studies thusfar have emphasized the role of the hippocampus in separating time and space using short-term or data-based memory representations. The hippocampus also plays a role in consolidating new information to be stored permanently in probably neocortex. There is an extensive literature to support the idea that the hippocampus is important for consolidating new information. For example, there are many studies demonstrating that animals with hippocampal lesions are impaired in the acquisition of a variety of spatial tasks including the water maze (O'Keefe and Nadel, 1978; Morris et al., 1982). Furthermore, it has been shown that post-trial disruption of normal hippocampal function with, for example, electrical brain stimulation, results in time-dependent memory impairments (Kesner and Wilburn, 1974).

Does spatial and temporal interference play a role in the acquisition (consolidation) of a variety of hippocampal dependent tasks? A few examples will suffice. Because rats are started in different locations in the standard water maze task, there is a great potential for spatial interference. Thus, the observation that hippocampal lesioned rats are impaired in learning and subsequent consolidation of important spatial information in this task could be due to enhanced spatial interference. Support for this idea comes from the observation of Eichenbaum et al. (1990) who demonstrated that when fimbria-fornix lesioned rats are trained on the water maze task from only a single starting position (less spatial interference) there are hardly any learning deficits, whereas training from many different starting points resulted in learning difficulties. In a different study, McDonald and White (1995) used a place preference procedure in an eight arm maze. In this procedure food is placed at the end of one arm and no food is placed at the end of another arm. In a subsequent preference test normal rats prefer the arm that contained the food. In this study fornix lesioned rats acquired the place preference task as quickly as controls, if the arm locations were opposite each other, but the fornix lesioned rats were markedly impaired, if the locations were adjacent to each other. Clearly, it is likely that there would be greater spatial interference when the spatial locations are adjacent to each other rather than far apart. Thus, spatial interference can play a role in the acquisition of new spatial information. In a final study, it has been shown that hippocampal lesioned rats do not have a problem in learning (consolidating) a single object-pair discrimination, but have difficulty in learning eight-pair concurrent object discriminations (Shapiro and Olton, 1994). In contrast to the one-pair discrimination, there is a heightened temporal interference in learning eight pairs simultaneously. This increased temporal interference could account for the observed impairment in hippocampal lesioned rats.

If short-term memory and consolidation processes associated with mnemonic processing of spatial information are both subserved by the hippocampus is it possible to dissociate the two? The answer to the question is positive. It has been shown that phencyclidine (an NMDA antagonist) injections into the dentate gyrus of the hippocampus at dose levels to block electrical stimulation of the medial entorhinal cortex-induced LTP disrupts consolidation of new learning in a dry land version of the water maze, but the same dose of phencyclidine has only a mild effect on a short-term memory task for spatial location information in an 8 arm maze (Kesner and Dakis, 1995). In contrast, naloxone (an opiate antagonist) injections into the dentate gyrus of the hippocampus at dose levels to block electrical stimulation of the lateral entorhinal cortex-induced LTP disrupts completely performance within the short-term memory task, but has no effect on consolidation of new learning in the dry land version of the water maze spatial navigation task (Dakis et al., 1992). These results suggest that short-term memory and consolidation processes can operate independent of each other and that perhaps each process is mediated by a different form of LTP.

There are other interference paradigms that emphasize the scaling or discrimination of odors, color or size of objects. It is assumed, but still needs to be tested, that the hippocampus does not play an important role in these paradigms, rather one would expect the perirhinal and entorhinal cortex to be of importance. Thus, the role of the hippocampus is specific only to reducing spatial and temporal interference.

In summary, the hippocampus appears to be important in processing spatial and temporal information in terms of short-term memory representations and in terms of promoting consolidation of new information. This is accomplished in part by selective attention or filtering of interfering spatial and temporal information and thus accentuate the temporal and spatial resolution of events.

AMYGDALA

Structure

Does the amygdala represent all attribute information in memory or is it restricted to affect (reward value) attribute information? A case has been made elsewhere (Kesner, 1992) indicating that the amygdala does not contribute to short-term memory representation of spatial location, time (duration), response, or sensory-perceptual attribute information. A few explicit examples will suffice.

With respect to spatial location information a) Rafaelle and Olton (1988) reported that rats with amygdala lesions were not impaired in performance of delayed, left-right alternation task in a T maze, b) Kesner et al.(1990) reported that rats with amygdala lesions were not impaired in remembering a list of 5 spatial locations, c) Parkinson et al. (1988) showed that monkeys with amygdala lesions were not impaired in performance of a delayed non-matching to sample task for object-place associations.

With respect to time (duration) information Olton et al.(1987) have shown that rats with amygdala lesions are not impaired in remembering the duration of a stimulus across a short delay interval.

With respect to sensory-perceptual information Sutherland and McDonald (1990) have reported that rats with amygdala lesions are not impaired in memory for odor or visual information using a non-matching- to-sample task. Thus, the amygdala does not appear to mediate spatial location, time, and sensory-perceptual attribute information. The amygdala, however, does represent new affect attribute information. Support for this assumption comes from a variety of sources. First, single cells within the amygdala of monkeys primarily respond to positive and negative stimuli and these responses can be altered depending on the significance and affective salience of the stimulus (Nishijo, Ono, & Nishino, 1988; Ono, Nishijo, and Uwano, 1995). Second, pre-or post-training electrical stimulation, chemical stimulation or lesions of the amygdala produce profound memory deficits in a variety of tasks in which reinforcement contingencies of sufficiently high magnitude are used (Gold et al., 1975; Hitchcock and Davis, 1986; Kesner et al., 1975; Kesner and Andrus, 1982; McDonough and Kesner, 1971; Phillips and LeDoux, 1992; Todd and Kesner, 1978). Third, rats with amygdala lesions are impaired in a food reward dependent cue preference task (Everitt et al., 1991; McDonald & White, 1993). In most of the studies cited above the emphasis has been on the role of the amygdala on learning (consolidation) of new affect-laden information. A new study was designed to directly measure the role of the amygdala in representing affect attribute information within a short-term or data-based memory framework. To accomplish this goal a new task was developed (Kesner and Williams, 1995). In the study phase of the task, rats were given one of two cereals. One cereal contained 25% sugar; the other 50% sugar. One of the two cereals was always designated as the positive stimulus and the other as the negative stimulus. This study phase was followed by the test phase in which the rat was shown an object which covered a food well. If the rat was given the negative food stimulus during the study phase, no food was placed beneath the object. If the rat was given the positive food stimulus during the study phase, another food reward was placed beneath the object. Latency to approach the object was used as the dependent measure. Rats learn to approach the objects quickly when they expect a reward and they are slow to approach the object when they expect no reward. After they reached criterion of at least a 5 sec difference between the positive and negative trials, the rats were given amygdala or control lesions. The results are shown in Figures 10A and B and indicate that in contrast to controls, the amygdala lesioned rats displayed a deficit in performance as indicated by smaller latency differences between positive and negative trials on post-surgery tests. This deficit persisted at both short and long delays. In additional experiments, it was shown that the amygdala lesioned rats,like controls, had similar taste preferences and transferred readily to different cereals containing 25% or 50% sugar. It should be noted that rats with hippocampal lesions are not impaired on this task. A similar result was reported by Kesner et al. (1989) who showed that amygdala lesioned rats were impaired in short-term memory performance for 1 vs 7 pieces of food associated with different spatial locations on an 8 arm maze. Thus, the amygdala appears to mediate short-term or data-based memory for affect-laden information based on the reward value (magnitude) of reinforcement.

To what extent can one generalize from amygdala function in rats to humans with respect to affect attribute information. Previous research has shown that bilateral damage to the amygdala in humans impairs recognition of affect embedded within facial expressions (Adolphs et al., 1994). In order to elaborate further on the role of the amygdala in humans Chiba et al. (1996) developed a liking test based on the mere exposure effect described by Zajonc (1968). Based on this principle, a computerized liking task was designed to test the presence of the mere exposure effect. The liking task consisted of eight abstract pictures and eight obsolete words that were sequentially presented on the computer screen. Following the individual presentation of each of these 16 study stimuli, 16 liking trials were presented. In each liking trial, two stimuli, one study stimulus and a matched lure, were simultaneously presented on the computer screen. Subjects were then asked which of the two stimuli they liked better. Four groups of subjects were tested on this task, college students as control subjects, subjects with partial complex epilepsy of temporal lobe origin, subjects who had undergone unilateral temporal lobe resections, including the temporal cortex and the hippocampus, and subjects who had undergone unilateral temporal lobe resections including the temporal cortex, hippocampus, and amygdala. Results are shown in Figures 11A and B for mean percent preference for abstract pictures and obsolete words and indicate that all subject groups showed a stable liking or mere exposure effect for both sets of stimuli, with the exception of those who sustained amygdala damage. It appears that the integrity of the amygdala is critical to the existence of the liking effect.

Thus, it is likely that the amygdala of animals and humans is involved in a short-term memory representation of the affective quality and quantity (reward value) of stimuli. This idea is an extension of earlier theoretical notions that the amygdala is involved in the interpretation and integration of reinforcement (Weiskrantz, 1956), serves as a reinforcement register (Douglas and Pribram, 1966), mediates stimulus-reinforcement associations (Jones and Mishkin, 1972) and serves to associate stimuli with reward value (Gaffan, 1992).

Process

Based on ample evidence that all sensory including internally generated visceral sensory and hormonal information is processed by amygdala neurons perhaps to provide markers for experiencing reward value, it is likely that one of the main process functions of the amygdala is to encode the reward value of events. This would ensure that newly processed sensory information is organized within the amygdala and enhances the possibility of remembering and temporarily storing one event with its associated reward value as separate from another event with its associated reward value. There are at least four processes associated with amygdala function that are necessary to accomplish the goal of coding affect (reward value) associated with events. These processes include a) selective filtering of reward value information to reduce interference among different qualitative and quantitative values of reward, b) data-based, short-term or working memory representation of affect (reward value) information, c) consolidation of new reward value information, and d) promotion of consolidation of other types of information via the activation of attentional and arousal systems.

There is very little evidence thusfar to provide for a clear evaluation of the role of the amygdala in utilizing selective filtering in order to separate values associated with magnitude of reward. It is clear that amygdala lesions disrupt short-term memory for magnitude of reward differences (Kesner and Williams, 1995), but no psychophysical experiments in which the dimension of reward is varied across a large number of values has yet been carried out. The amygdala also plays a role in consolidating new information. Support for this idea comes from a variety of sources including the observation that pre-or post-training electrical stimulation, chemical stimulation or lesions of the amygdala produce profound memory deficits in a variety of learning tasks in which reinforcement contingencies of sufficiently high magnitude are used (Gold et al., 1975; Hitchcock and Davis, 1986; Kesner et al., 1975; Kesner and Andrus, 1982; McDonough and Kesner, 1971; Hitchcock and Davis, 1986; Phillips and LeDoux, 1992; Todd and Kesner, 1978), and rats with amygdala lesions are impaired in the acquisition (consolidation) of a food reward dependent cue preference task(Everitt et al., 1991; McDonald & White, 1993).

It has also been suggested that the amygdala modulates memory by promoting the consolidation of other memory attributes (McGaugh et al., 1992). This is likely to be accomplished by direct amygdala activation of neural circuits that mediate attention and arousal processes (Kapp et al., 1992). The best evidence in support of amygdala modulation of the consolidation of other forms of memory representations comes from a study by Packard et al., 1994. They showed that posttraining intrahippocampal injections of d-amphetamine facilitated retention of a spatial task, but had no facilitatory effect on a cued task. In contrast, posttraining intracaudate injections of d-amphetamine facilitated retention of the cued task, but had no facilitatory effect on the spatial task. Posttraining intraamygdala injections of d-amphetamine enhanced retention of both tasks, even though amygdala lesions did not affect performance in the spatial and cued tasks. These results suggest that the amygdala might indeed modulate consolidation of attribute information that is dependent upon mediation by other neural regions.

In summary, the amygdala appears to be important in processing affect (reward value) information in terms of short-term memory representations and in terms of consolidation of reward value information as well as modulation of consolidation of other attribute information. This is accomplished in part by selective attention or filtering of interfering values of reward and thus accentuate the specificity of an appropriate association of reward with new events.

DISSOCIATIONS BETWEEN THE HIPPOCAMPUS AND AMYGDALA

Based on the above mentioned analyses of hippocampal and amygdala function, it is clear that both the hippocampus and the amygdala contribute to the memory representation of reinforcing experiences. Each neural region uses similar processes (selective filtering, short-term memory and consolidation) to represent appropriate information to be remembered, but the amygdala promotes, in addition, a process that allows for modulation of mnemonic processes of other attributes of memory. Yet, each neural region appears to contribute differentially in that the hippocampus provides memory for the external context based on spatial, temporal and linguistic attribute information within which reward value, affect or reinforcement is embedded and the amygdala provides for memory for the reward value or reinforcement as constituents of affect attribute information. To what extent do the two neural regions (hippocampus and amygdala) interact with each other or operate independently of each other? Even though based on neuroanatomical connections there are likely to be many functional interactions between the hippocampus and amygdala, there is also a great deal of support for parallel and independent processing of attribute information. Research with rats reveals a double dissociation between the hippocampus and the amygdala. The hippocampus appears to be involved in tasks that require memory for spatial location and memory for duration of events, whereas the amygdala appears to be involved in tasks that require memory for taste aversion and magnitude of reinforcement. Rats with lesions or electrical stimulation of the hippocampus are impaired in performance on an 8-arm maze, memory for spatial locations, memory for duration of events, but not impaired in taste aversion learning or memory for magnitude of reinforcement. In contrast, rats with lesions or electrical stimulation of the amygdala are impaired in taste aversion learning and memory for magnitude of reinforcement, without altering 8-arm performance, efficiency of spatial location memory and memory for duration of an event (Best and Orr, 1973; Kesner & Berman, 1977; Kesner, et al., 1975; Kesner and Williams, 1995; McGowan, et al., 1972; Nachman & Ashe, 1974; Murray, 1990; Olton, 1983; Olton & Wolf, 1981; Olton et al., 1987; Swanson & Isaacson, 1969). Thus, for rats there appears to be a double dissociation between the amygdala and hippocampus implying that they can operate independent of each other.

Research with monkeys also reveals a double dissociation between the hippocampus and amygdala. The hippocampus appears to be involved in tasks that require memory for spatial locations of objects, whereas the amygdala appears to be involved in tasks that require memory for object-reward associations or object-object associations across modalities. Monkeys with lesions of the hippocampus, but not amygdala, have an impaired memory for object-place associations (Parkinson, Murray & Mishkin, 1988). In contrast, monkeys with lesions of the amygdala, but not hippocampus, have an impaired memory for object-reward associations (Spiegler & Mishkin, 1981) and cross modal object-object associations (Murray & Mishkin, 1985; Murray, 1990).

Research with humans also indicates that there is a double dissociation of function between the hippocampus and amygdala. The hippocampus appears to be involved in tasks that require memory for the temporal order of critical events, whereas the amygdala appears to be involved in tasks that require memory for an affective experience as measured by an autonomic response or memory based on the liking effect. For example, Bechara et al., 1995 showed that a patient with a bilateral lesion of the amygdala was impaired in learning a conditioned autonomic response to visual or auditory stimuli, but learned very readily which stimuli were paired with the unconditioned response. In contrast, a patient with bilateral damage to the hippocampus was impaired in learning which stimuli were paired with the unconditioned response, but learned very readily the conditioned autonomic response to the visual and auditory stimuli. In another study presented above, Chiba et al. (1996), and Chiba et al. (1996) demonstrated that humans with unilateral amygdala and hippocampal damage were impaired in displaying the liking effect and memory for temporal order of novel spatial location and linguistic information, whereas humans with only hippocampal damage were impaired in memory for temporal order of novel spatial location and linguistic information, but showed no impairment in displaying the liking effect. Thus, for rats, monkeys and humans there appears to be a double dissociation between the amygdala and hippocampus implying that they can operate independent of each other.

In conclusion, based on an analysis of the structural and dynamic process components of memory representations, data have been presented in animals and humans to support the idea that the hippocampus is important for providing the spatial-temporal and linguistic context that can be associated with reward value or reinforcement and that the amygdala is important for providing memory for reward value or reinforcement.

References

Adolphs, R., Tranel, D., Damasio, H., & Damasio, A. (1994). Impaired recognition of emotion in facial expressions following bilateral damage to the human amygdala. Nature, **372**, 669-672.

Aggleton, J.P., Hunt, P.R., & Rawlins, J.N.P. (1986). The effects of hippocampal lesions upon spatial and non-spatial tests of working memory. Behavioral Brain Research, **19**, 133-146.

Bechara, A., Tranel, D., Damasio, H., Adolphs, R., Rockland, C., & Damasio, A.R. (1995). Double dissociation of conditioning and declarative knowledge relative to the amygdala and hippocampus in humans. Science, **269**, 1115-1118.

Best, P.J., & Orr, J. (1973). Effect of hippocampal lesions on passive avoidance and taste aversion conditioning. Physiology & Behavior, **10**, 193-196.

Bunsey, M., & Eichenbaum, H. (in press). Conservation of hippocampal memory function in rats and humans. Nature.

Chiba, A.A., Johnson, D.L., & Kesner, R.P. (1992). The effects of lesions of the dorsal hippocampus or the ventral hippocampus on performance of a spatial location order recognition task. Society for Neuroscience Abstracts, **18**, 1422.

Chiba, A.A., Kesner, R.P., Matsuo, F., & Heilbrun, M.P. (1990). A dissociation between verbal and spatial memory following unilateral temporal lobectomy. Society for Neuroscience Abstracts, **16**, 286.

Chiba, A.A., Kesner, R.P., Matsuo, F., Heilbrun, M.P., & Plumb, S. (submitted). A dissociation between affect and recognition following unilateral temporal lobectomy including the amygdala.

Chiba, A.A., Kesner, R.P., Matsuo, F., Heilbrun, M.P., & Plumb, S. (submitted). A double dissociation between the right and left hippocampus in processing the temporal order of spatial and verbal information.

Chiba, A.A., Kesner, R.P., & Reynolds, A.M. (1994). Memory for spatial location as a function of temporal lag in rats: Role of hippocampus and medial prefrontal cortex. Behavioral and Neural Biology, **61**, 123-131.

Cohen, N.J., & Eichenbaum, H.B. (1993). Memory, amnesia, and hippocampal function. MIT Press: Cambridge.

Dakis, M., Martinez, J.S., Kesner, R.P., & Jackson-Smith, P. (1992). Effects of Phencyclidine and naloxone on learning of a spatial navigation task and performance of a spatial delayed non-matching to sample task. Society for Neuroscience Abstracts, **18**, 1220.

Douglas, R.J., & Pribram, K.H. (1966). Learning and limbic lesions. <u>Neuropsychology</u>, **4**, 197-220.

Eichenbaum, H., Stewart, C., & Morris, R.G.M. (1990). Hippocampal representation in spatial learning. <u>Journal of Neuroscience</u>, **10**, 331-339.

Everitt, B.J., Morris, K.A., O'Brien, A., & Robbins, T.W. (1991). The basolateral amygdala-ventral striatal system and conditioned place preference: Further evidence of limbic-striatal interactions underlying reward-related processes. <u>Neuroscience</u>, **42**, 1-18.

Gabrieli, J.D.E., Fleischman, D.A., Keane, M.M., Reminger, S.L., & Morrell, F. (1995). Double dissociation between memory systems underlying explicit and implicit memory in the human brain. <u>Psychological Science</u>, **6**, 76-82.

Gaffan, D. (1992). Amygdala and the memory of reward. In J.P. Aggleton (Ed.), The Amygdala: Neurobiological Aspects of Emotion, Memory, and Mental Dysfunction. Wiley-Liss: New York.

Gold, P.E., Hankins, L.L., Edwards, R., Chester, J., & McGaugh, J.L. (1975). Memory interference and facilitation with posttrial amygdala stimulation. Effect on memory varies with footshock level. <u>Brain Research</u>, **86**, 509-513.

Hirst, W., & Volpe, B. (1982). Temporal order judgments with amnesia. <u>Brain and Language</u>, **1**, 294-306.

Hitchcock, J. & Davis, M. (1986). Lesions of the amygdala, but not of the cerebellum or red nucleus, block conditioned fear as measured with the potentiated startle paradigm. <u>Behavioral Neuroscience</u>, **100**, 11-22.

Hopkins, R.O., & Kesner, R.P. (1994). Short-term memory for duration in hypoxic subjects. <u>Society for Neuroscience Abstracts</u>, **20**, 1075.

Hopkins, R. O., Kesner, R.P., & Goldstein, M. (1995). Memory for novel and familiar spatial and linguistic temporal distance information in hypoxic subjects. <u>Journal of the International Neuropsychological Society</u>, **1**, 454-468.

Hubert, F.A., & Piercy, M. (1976). Recognition memory in amnesic patients: Effect of temporal context and familiarity of material. <u>Cortex</u>, **12**, 3-20.

Jackson-Smith, P., Kesner, R.P., & Amann, K. (1994). Effects of hippocampal and medial prefrontal lesions on discrimination of duration in rats. <u>Society for Neuroscience Abstracts</u>, **20**, 1210.

Jackson-Smith, P., Kesner, R.P., & Chiba, A.A. (1993). Continuous recognition of spatial and nonspatial stimuli in hippocampal lesioned rats. <u>Behavioral and Neural Biology</u>, **59**, 107-119.

Jones, B., & Mishkin, M. (1972). Limbic lesions and the problem of stimulus-reinforcement association. Experimental Neurology, **36**, 362-377.

Kapp, B.S., Whalen, P.J., Supple, W.F., & Pascoe, J.P. (1992). Amygdaloid contributions to conditioned arousal and sensory information processing. In J.P. Aggleton (Ed.), The Amygdala: Neurobiological Aspects of Emotion, Memory, and Mental Dysfunction. Wiley-Liss: New York.

Keane, M.M., Gabrieli, J.D.E., Mapstone, H.C., Johnson, K.A., & Corkin, S. (1995). Double dissociation of memory capacities after bilateral occipital-lobe or medial temporal-lobe lesions. Brain, **118**, 1129-1148.

Kesner, R.P. (1990). Learning and memory in rats with an emphasis on the role of the hippocampal formation. In R.P. Kesner & D.S. Olton (Eds.), Neurobiology of Comparative Cognition. Erlbaum: Hillsdale, NJ.

Kesner, R.P. (1992). Learning and memory in rats with an emphasis on the role of the amygdala. In J.P. Aggleton (Ed.), The Amygdala: Neurobiological Aspects of Emotion, Memory and Mental Dysfunction. Wiley-Liss: New York.

Kesner, R.P., & Andrus, R.G. (1982). Amygdala stimulation disrupts magnitude of reinforcement contribution to long-term memory. Physiological Psychology, **10**, 55-59.

Kesner, R.P., Berman, R.F., Burton, B, & Hankins, W.G. (1975). Effects of electrical stimulation of amygdala upon neophobia and taste aversion. Behavioral Biology, **13**, 349-358.

Kesner, R.P., Bolland, B.L., & Dakis, M. (1993). Memory for spatial locations, motor responses, and objects: triple dissociation among the hippocampus, caudate nucleus, and extrastriate visual cortex. Experimental Brain Research, **93**, 462-470.

Kesner, R.P., Crutcher, K.A., & Omana, H. (1990). Memory deficits following nucleus basalis magnocellularis lesions may be mediated through limbic, but not neocortical, targets. Neuroscience, **38**, 93-102.

Kesner, R.P., & Dakis, M. (1995). Phencyclidine injections into the dorsal hippocampus disrupt long- but not short-term memory within a spatial learning task. Psychopharmacology, **120**, 203-208.

Kesner, R.P., & DiMattia, B.V. (1987). Neurobiology of an attribute model of memory. Progress in Psychobiology and Physiological Psychology, Academic Press: New York.

Kesner, R.P., Hopkins, R.O., & Chiba, A.A. (1992). Learning and memory in humans, with an emphasis on the role of the hippocampus. In L.R. Squire & N. Butters (Eds.), Neuropsychology of Memory. The Guilford Press, New York.

Kesner, R.P., Walser, R.D., & Winzenried, G. (1989). Central but not basolateral amygdala mediates memory for positive affective experiences. Behavioral Brain Research, **33**, 189-195.

Kesner, R.P., & Wilburn, M.W. (1974). A review of electrical stimulation of the brain in context of learning and retention. Behavioral Biology, **10**, 259-293.

Kesner, R.P., & Williams, J.M. (1995). Memory for magnitude of reinforcement: Dissociation between the amygdala and hippocampus. Neurobiology of Learning and Memory, **64**, 237-244.

Long, J.M., & Kesner, R.P. (1994). The effects of parietal cortex and hippocampal lesions on memory for allocentric distance, egocentric distance, and spatial location in rats. Society for Neuroscience Abstracts, **20**, 1210.

Long, J.M., & Kesner, R.P. (in press). The effects of dorsal vs. ventral hippocampal, total hippocampal, and parietal cortex lesions on memory for allocentric distance in rats. Behavioral Neuroscience.

McDonald, R.J., & White, N.M. (1993). A triple dissociation of systems: hippocampus, amygdala, and dorsal striatum. Behavioral Neuroscience, **107**, 3-22.

McDonald, R.J., & White, N.M. (1995). Hippocampal and nonhippocampal contributions to place learning in rats. Behavioral Neuroscience, **109**, 579-593.

McDonough, J.R., & Kesner, R.P. (1971). Amnesia produced by brief electrical stimulation of the amygdala or dorsal hippocampus in cats. Journal of Comparative Physiological Psychology, **77**, 171-178.

McGaugh, J., Intrioni-Collison, I., Cahill, L., Kim, M., & Liang, K. (1992). Involvement of the amygdala in neuromodulatory influences on memory storage. In J. Aggleton (Ed.), The Amygdala. Wiley-Liss: New York.

McGowan, B.K., Hankins, W.G., & Garcia, J. (1972). Limbic lesions and control of the internal and external environment. Behavioral Biology, **7**, 841-852.

Meck, W.H., Church, R.M., & Olton, D.S. (1984). Hippocampus, time and memory. Behavioral Neuroscience, **98**, 3-22.

Milner, B. (1971). Interhemispheric differences in the localization of psychological processes in man. British Medical Bulletin, **27**, 272-277.

Morris, R.G.M., Garrud, J.N.P., Rawlins, J.N.P., & O'Keefe, J. (1982). Place navigation impaired in rats with hippocampal lesions. Nature, **297**, 681-683.

Mumby, D.G., Wood, E.R., & Pinel, J.P.J. (1992). Object recognition memory is only mildly impaired in rats with lesions of the hippocampus and amygdala. Psychobiology, 20, 18-27.

Murray, E.A. (1990). Representational memory in nonhuman primates. In R.P. Kesner & D.S. Olton (Eds.), Neurobiology of Comparative Cognition. Lawrence Erlbaum: Hillsdale, NJ.

Murray, E.A., & Mishkin, M. (1985). Amygdalectomy impairs crossmodal association in monkeys. Science, 228, 604-606.

Myers, C.E., Gluck, M.A., & Granger, R. (1995). Dissociation of hippocampal and entorhinal function in associative learning: A computational approach. Psychobiology, 23, 116-138.

Nachman, M., & Ashe, J.H. (1974). Effects of basolateral amygdala lesions on neophobia, learned taste aversions, and sodium appetite in rats. Journal of Comparative Physiological Psychology, 87, 622-643.

Nishijo, H., Ono, T., & Nishino, H. (1988). Single neuron responses in amygdala of alert monkey during complex sensory stimulation with affective significance. Journal of Neuroscience, 8, 3570-3583.

O'Keefe, J., & Nadel, L. (1978). The Hippocampus as a Cognitive Map. Clarendon Press: Oxford.

Olton, D.S. (1983). Memory functions and the hippocampus. In W. Seifert (Ed.), Neurobiology of the Hippocampus. Academic Press: New York.

Olton, D.S. (1986). Hippocampal function and memory for temporal context. In R.L. Isaacson & K.H. Pribram (Eds.), The Hippocampus, Vol. 3. Plenum Press: New York.

Olton, D.S., Becker, J.T., & Handelmann, G.H. (1979). Hippocampus, space, and memory. Behavioral and Brain Sciences, 2, 313-365.

Olton, D.S., Meck, W.H., & Church, R.M. (1987). Separation of hippocampal and amygdaloid involvement in temporal memory dysfunctions. Brain Research, 404, 180-188.

Olton, D.S., & Wolf, W.A. (1981). Hippocampal seizures produce retrograde amnesia without a temporal gradient when they reset working memory. Behavioral Biology, 33, 437-452.

Ono, T., Nishijo, H., & Uwano, T. (1995). Amygdala role in conditioned associative learning. Progress in Neurobiology, 46, 401-422.

Otto, T., & Eichenbaum, H. (1992). Complementary roles of the orbital prefrontal cortex and the perirhinal-entorhinal cortices in an odor-guided delayed-nonmatching-to-sample task. Behavioral Neuroscience, **106**, 762-775.

Packard, M.G., Cahill, L., & McGaugh, J.L. (1994). Amygdala modulation of hippocampal-dependent and caudate nucleus-dependent memory processes. Proceeds of the National Academy of Science, **91**, 8477-8481.

Parkinson, J.K., Murray, E.A., & Mishkin, M. (1988). A selective mnemonic role for the hippocampus in monkeys: Memory for the location of objects. Journal of Neuroscience, **8**, 4159-4167.

Phillips, R.G., & LeDoux, J.E. (1992). Differential contribution of amygdala and hippocampus to cued and contextual fear conditioning. Behavioral Neuroscience, **106**, 274-285.

Pigott, S. & Milner, B. (1993). Memory for different aspects of complex visual scenes after unilateral temporal- or frontal-lobe resection. Neuropsychologia, **31**, 1-15.

Plunket, R.P., Faulds, B.D., & Albino, R.C. (1973). Place learning in hippocampalectomized rats. Bulletin of the Psychonomic Society, **2**, 79-80.

Raffaele, K.C. & Olton, D.S. (1988). Hippocampal and amygdaloid involvement in working memory for nonspatial stimuli. Behavioral Neuroscience, **102**, 349-355.

Rolls, E. (1989). Functions of neuronal networks in the hippocampus and neocortex in memory. In J.H. Byrne & W.O. Berry (Eds.), Neural Models of Plasticity: Theoretical and Empirical Approaches. Academic Press: New York.

Sagar, H.J., Gabrieli, J.D.E., Sullivan, E.V., & Corkin, S. (1992). Recency and frequency discrimination in the amnesic patient H.M.. Brain, **113**, 581-602.

Schacter, D.L. (1987). Implicit memory: History and current status. Journal of Experimental Psychology: Learning, Memory, and Cognition, **13**, 501-518.

Shapiro, M.L., & Olton, D.S. (1992). Hippocampal function and interference. In D.L. Schacter & E. Tulving (Eds.), Memory Systems 1994. MIT Press: Cambridge.

Smith, M.L., & Milner, B. (1981). The role of the right hippocampus in the recall of spatial location. Neuropsychologia, **19**, 781-793.

Spiegler, B.J., & Mishkin, M. (1981). Evidence for the sequential participation of inferior temporal cortex and amygdala in the acquisition of stimulus-reward learning. Behavioural Brain Research, **3**, 303-317.

Squire, L.R. (1992). Memory and the hippocampus: A synthesis from findings with rats, monkeys, and humans. Psychological Review, **99**, 195-231.

Squire, L.R. (1994).   Declarative and nondeclarative memory: Multiple brain systems supporting learning and memory.   In D.L. Schacter & E. Tulving (Eds.), Memory Systems 1994. MIT Press: Cambridge.

Squire, L.R., Nadel, L., & Slater, P.C. (1981). Anterograde amnesia and memory for temporal order. Neuropsychologia, **19**, 141-145.

Sutherland, R.J., & McDonald, R.J. (1990). Hippocampus, amygdala, and memory deficits in rats. Behavioral Brain Research, **37**, 57-79.

Swanson, A.M., & Isaacson, R.L. (1969). Hippocampal lesions and the frustration effect in rats. Journal of Comparative Physiological Psychology, **68**, 562-567.

Todd, J.W., & Kesner, R.P. (1978). Effects of posttraining injection of cholinergic agonists and antagonists into the amygdala on retention of passive avoidance training in rats. Journal of Comparative Physiological Psychology, **92**, 958-968.

Tulving, E. (1983). Elements of Episodic Memory. Clarendon Press: Oxford.

Weiskrantz, L. (1956). Behavioral changes with ablation of the amygdaloid complex in monkeys. Journal of Comparative Physiological Psychology, **49**, 381-391.

Zajonc, R.B. (1968). Attitudinal effects of mere exposure. Journal of Personality and Social Psychology, **9**, 1-27.

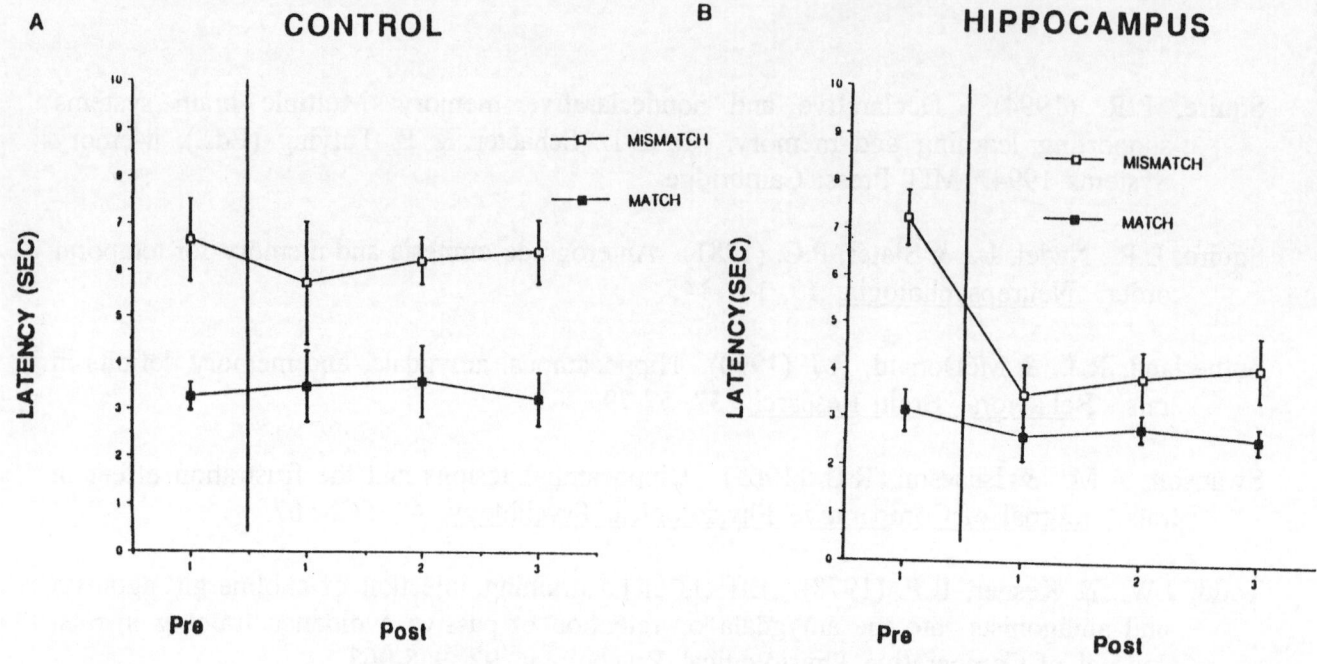

Figure 1.   Mean latency (±SE) to respond pre- and post-surgery on match and mismatch trials for control (A) and hippocampus (B) lesions within a short-term memory for allocentric space task.

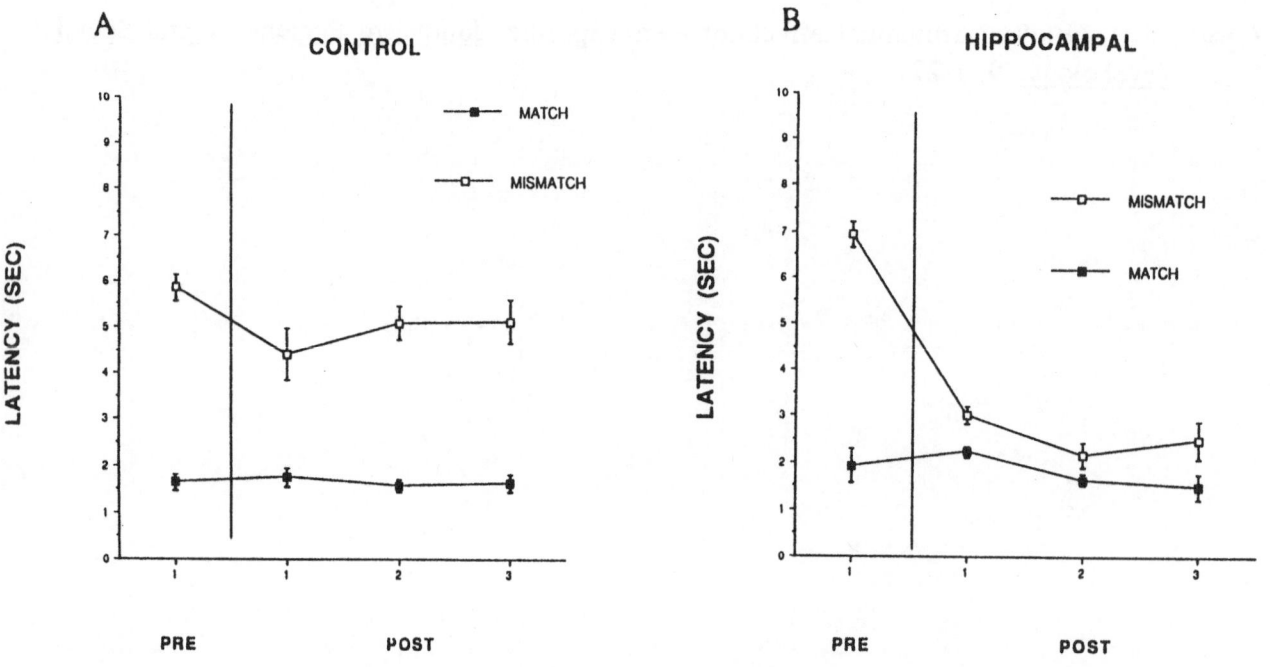

Figure 2.   Mean latency (±SE) to respond pre- and post-surgery on match and mismatch trials for control (A) and hippocampal (B) lesions within a short-term memory for spatial location task.

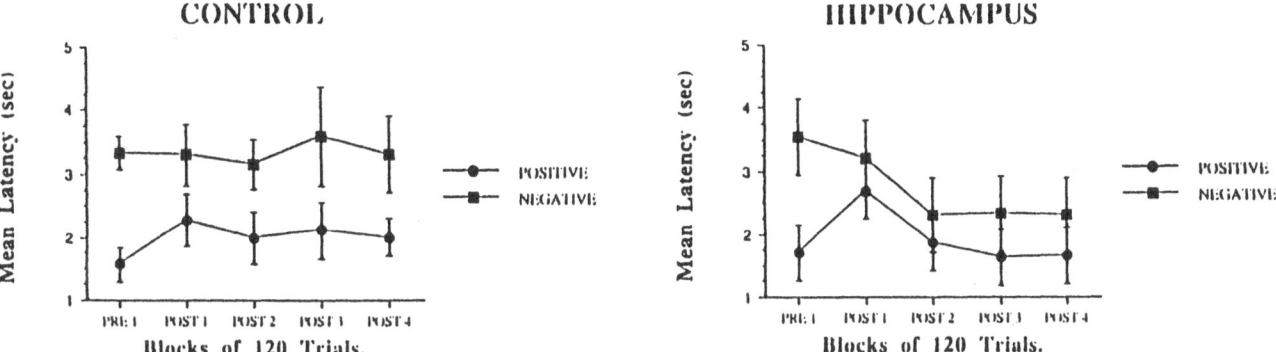

Figure 3. Mean latency (±SE) to respond pre- and post-surgery on positive and negative trials for control (A) and hippocampus (B) lesions within a short-term memory for duration task.

A   **Short Term Memory for Duration**

B   **Short Term Memory for a Single Object**

C   **Short Term Memory for Distance**

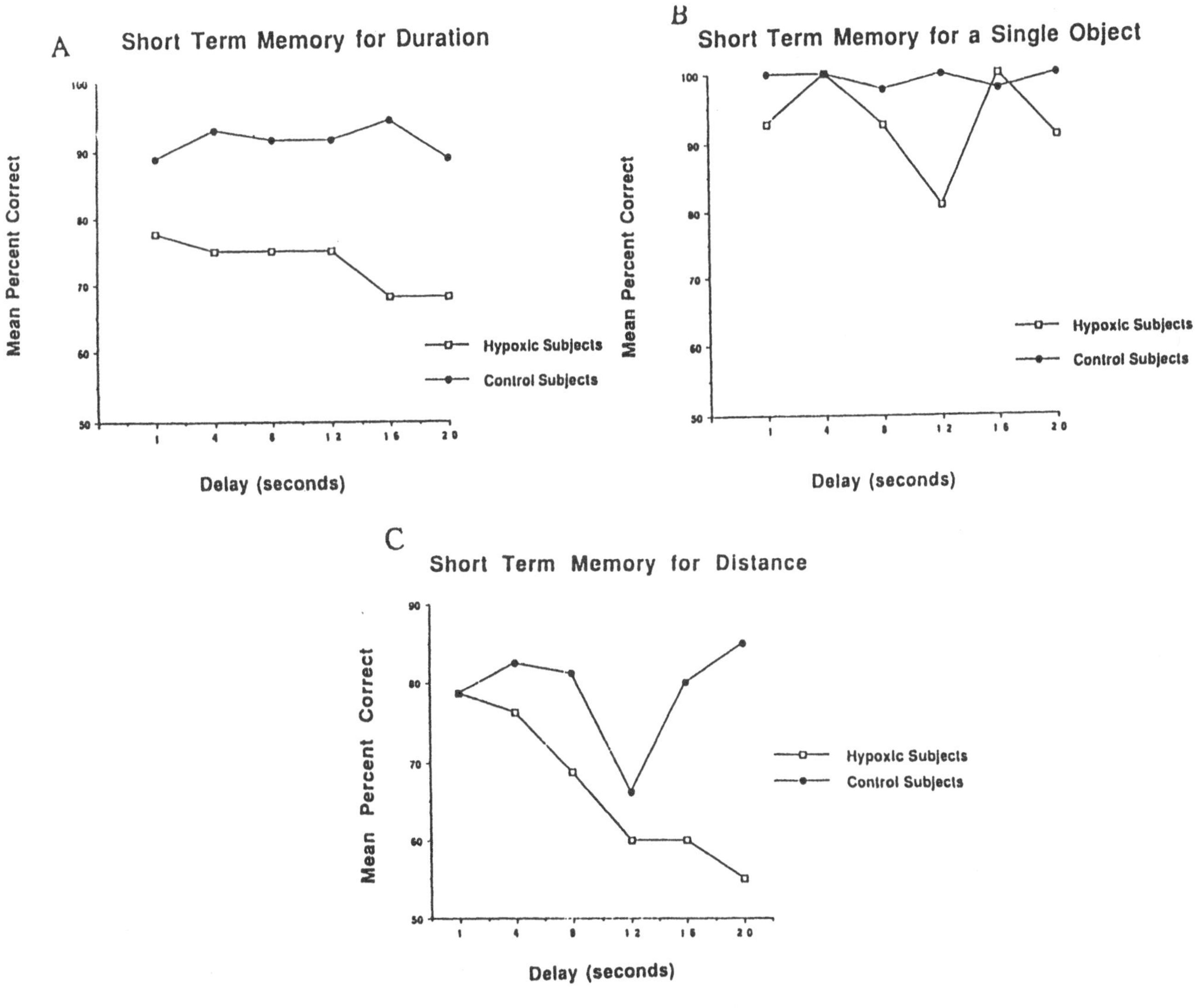

Figure 4. Mean percent correct performance as a function of delay for hypoxic and control subjects within a short-term memory for duration task (A), within a short-term memory for a single object task (B), and within a short-term memory for distance task (C).

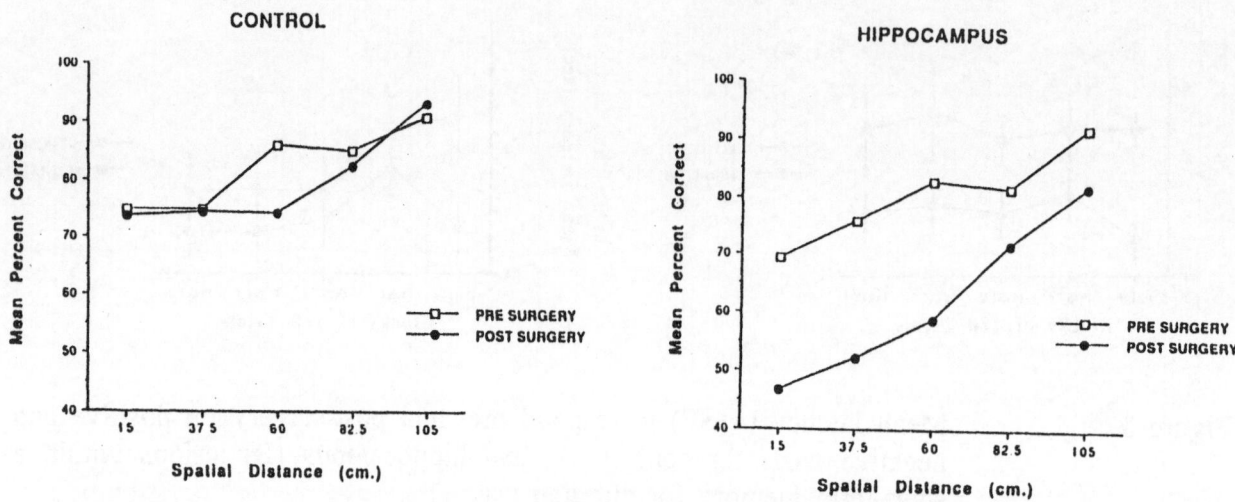

Figure 5.   Mean percent correct performance pre- and post-surgery as a function of spatial distance for control (A) and hippocampus (B) lesions within a short-term memory for spatial location task.

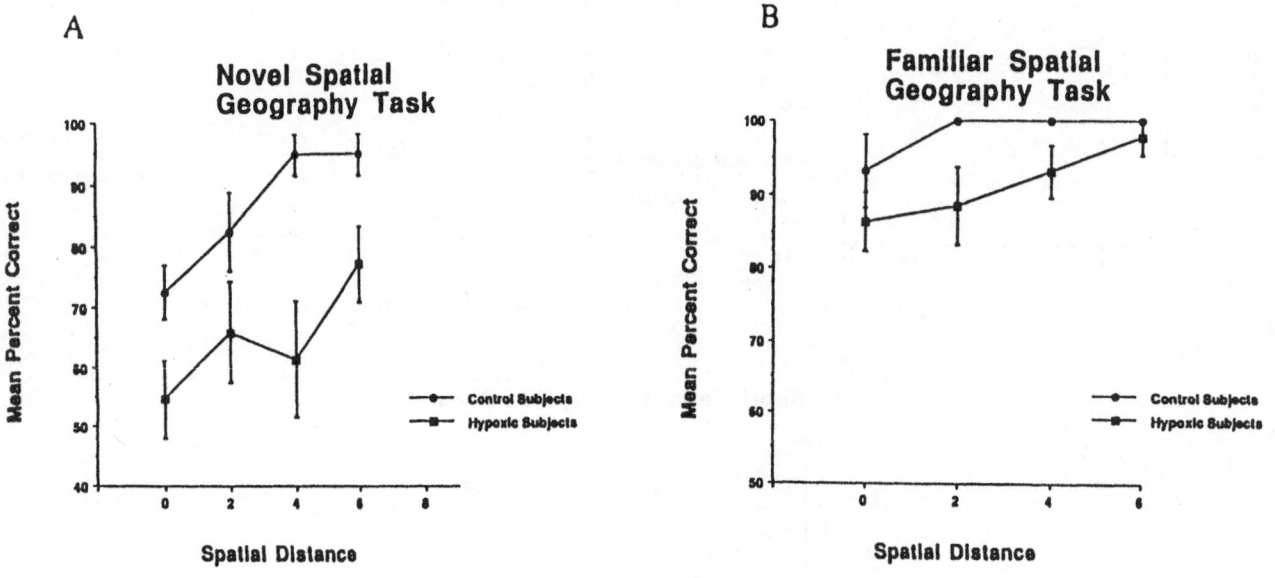

Figure 6.   Mean percent correct performance (±SE) as a function of spatial distance for hypoxic and control subjects within a novel spatial geography task (A) and a familiar spatial geography task (B).

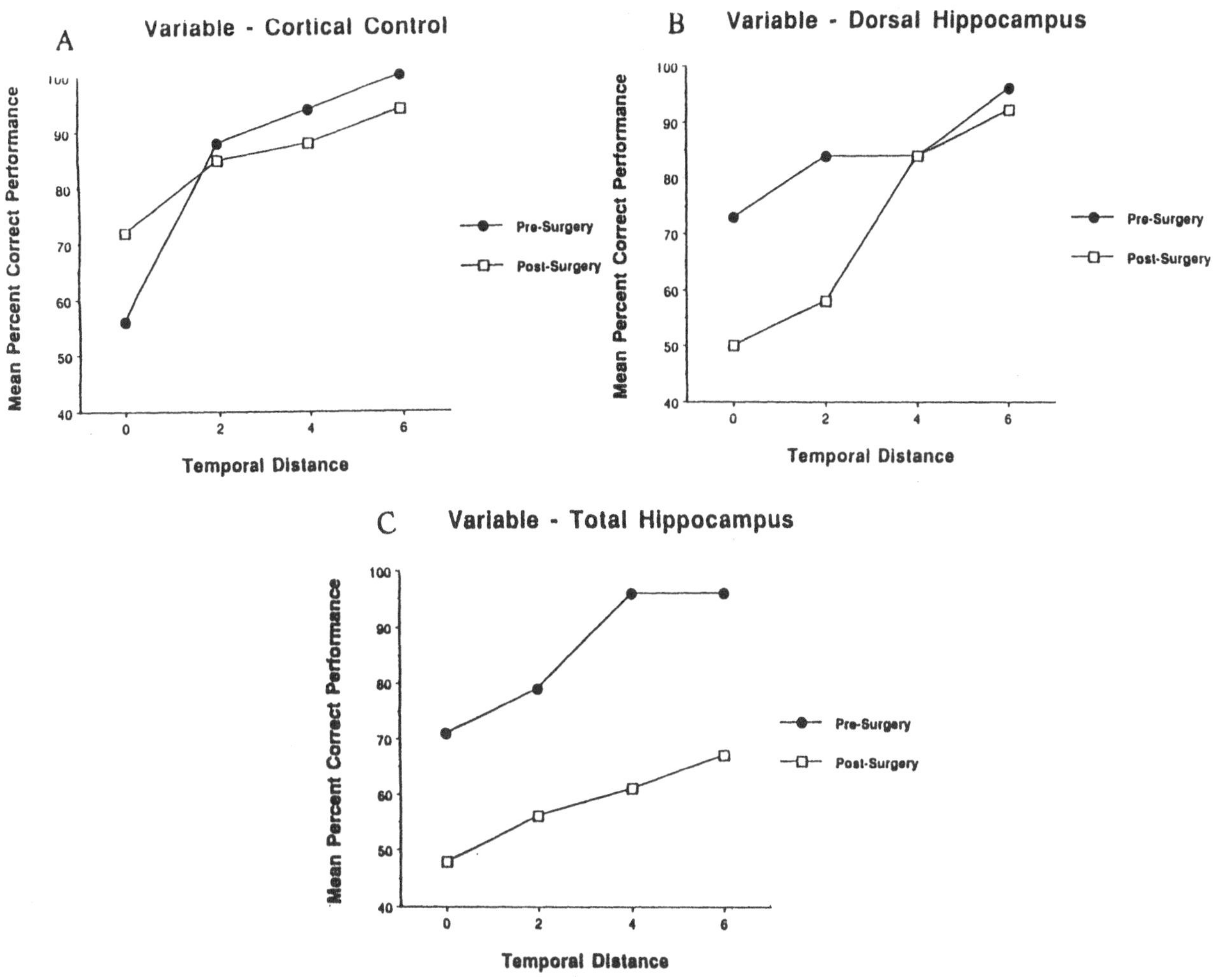

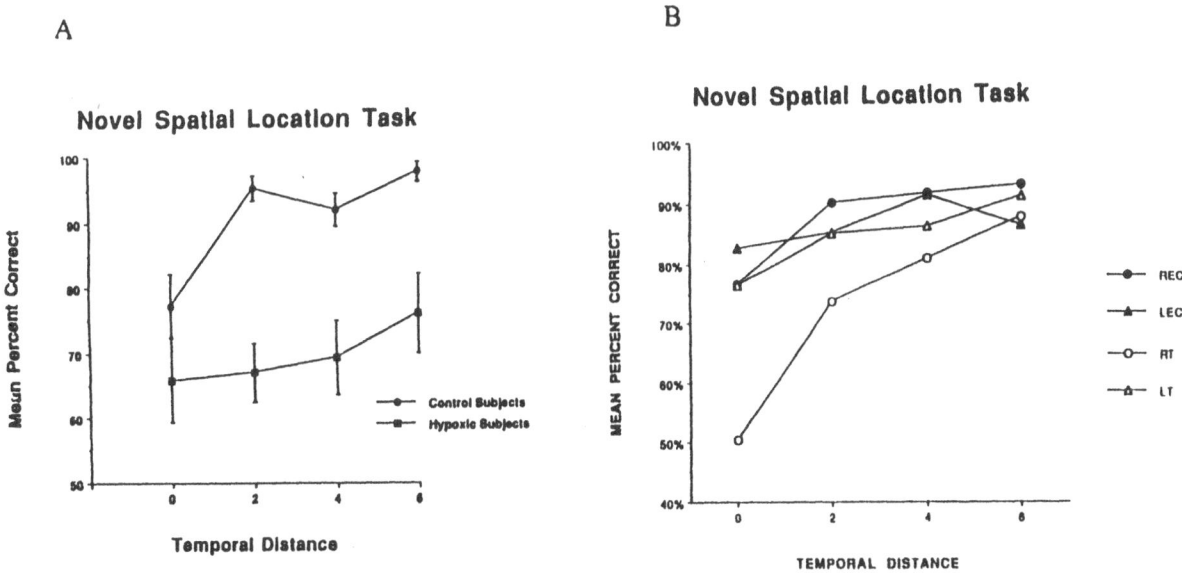

Figure 7.   Mean percent correct performance pre- and post-surgery as a function of temporal distance for variable cortical control (A), variable dorsal hippocampus (B) and variable total (dorsal and ventral) hippocampus (C) lesions within a short-term memory for temporal order of a spatial location information task.

Figure 8.   Mean percent correct performance (±SE) as a function of temporal distance for (A) hypoxic and control subjects and (B) right and left epileptic controls (REC,LEC) and right and left temporal lobe resected subjects (RT,LT) within a novel spatial location task.

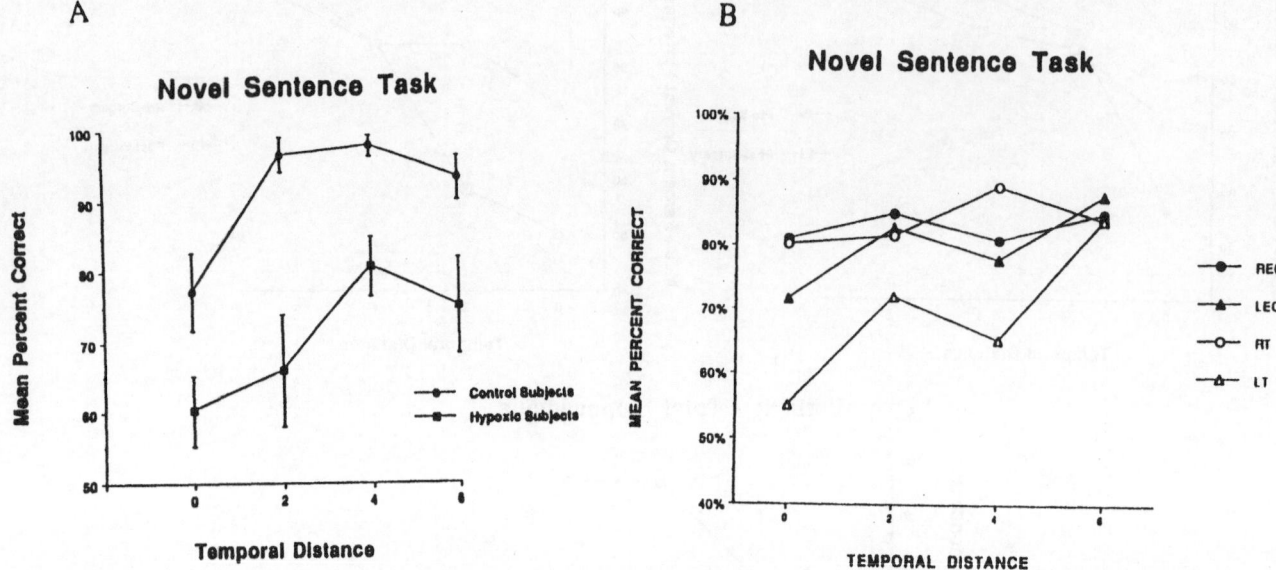

Figure 9.   Mean percent correct performance (±SE) as a function of temporal distance for (A) hypoxic and control subjects and (B) right and left epileptic controls (REC,LEC) and right and left temporal lobe resected subjects (RT,LT) within a novel sentence task.

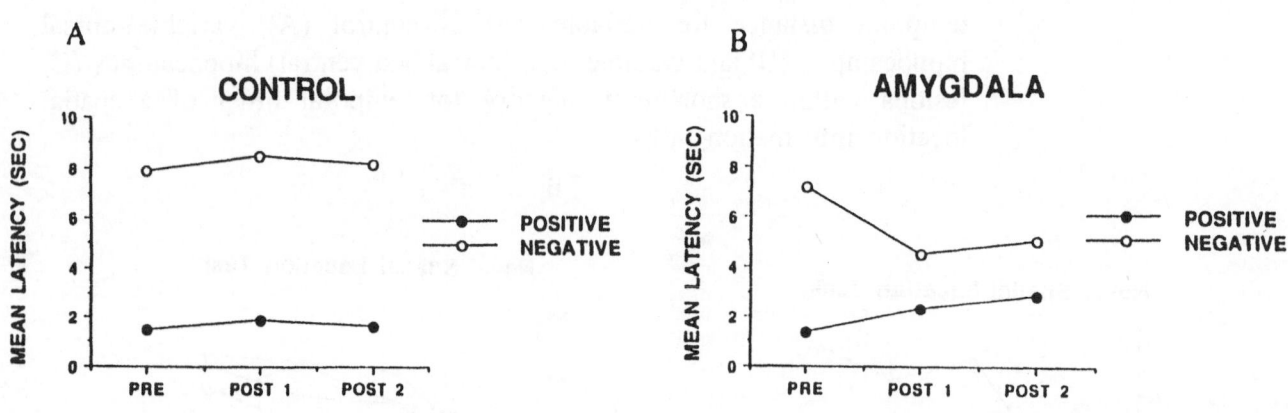

Figure 10.   Mean latency to respond pre- and post-surgery on positive and negative trials for control (A) and amygdala (B) lesions within a short-term memory for magnitude reward (affect) task.

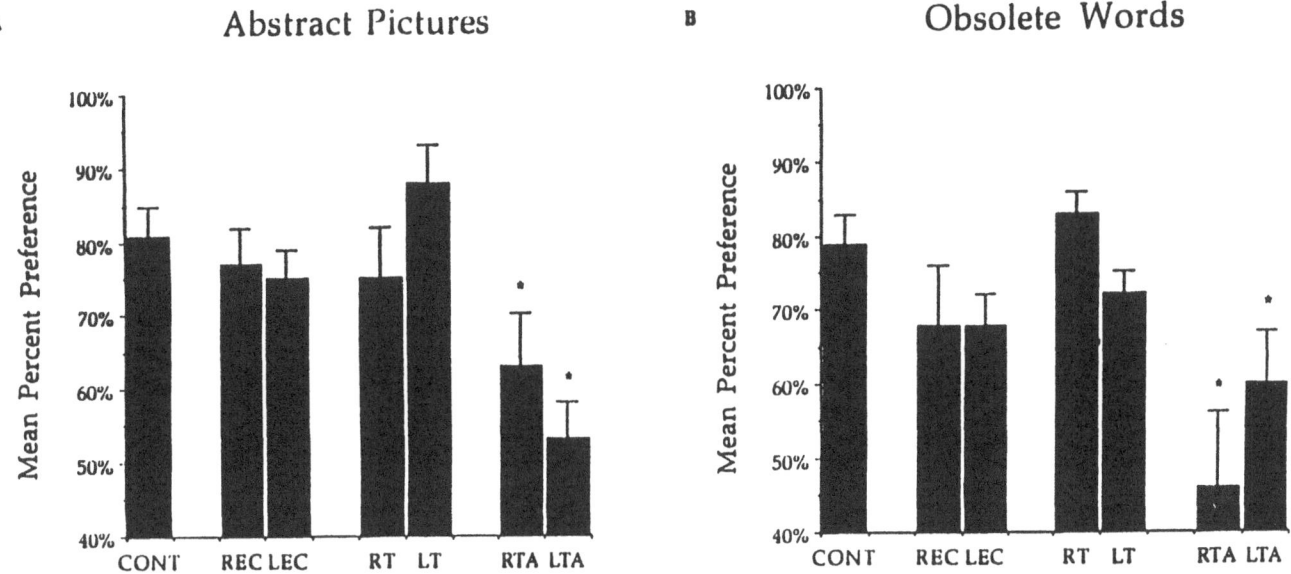

Figure 11.   Mean percent correct performance (±SE) for (A) abstract pictures and (B) obsolete words for controls (CONT), right and left epileptic controls (REC,LEC), right and left temporal lobe resected subjects with hippocampal damage (RT,LT), and right and left temporal lobe resected subjects with hippocampal and amygdala damage (RTA,LTA)

17.
Emotion and the Self-Organization of Semantic Memory

Daniel D. de Grandpre and Don M. Tucker

Emotion and the Self-Organization of Semantic Memory

Daniel D. de Grandpre1
and
Don M. Tucker1,2

1Department of Psychology, University of Oregon
2Electrical Geodesics, Inc.

Address for correspondence:
Daniel de Grandpre
Psychology Department
University of Oregon
Eugene, OR 97403 USA.
Supported by NIMH research grants MH42128 and MH42669 to the University of Oregon and
small business innovation research grants MH50409 and MH51069 to Electrical Geodesics, Inc.

The development of emotion as a research discipline has undergone a remarkable resurgence in the last decade. This 4th Appalachian Conference on Behavioral Neurodynamics seeks to explore how the motivation for learning represents self-organization rather than self-gratification. In this spirit, we will consider how learning and memory are integrated with the brain's motivational processes. We believe that an understanding of the semantic organization of the human mind cannot be achieved without a basic understanding of the individualized, emotional content of semantics at both cognitive and neural levels. Such an understanding may be an important step in characterizing human personality.

We begin by attempting to integrate two constructs, one of emotional valence and one of emotional activation, in terms of Freud's (1940) concept of "cathexis." We propose that cathexis entails a vector of motivation. Because the cortico-limbic mechanisms of motivation also determine the process of memory storage, every representation in memory must contain a cathected element, as illustrated by both behavioral evidence and the structure of limbic cortex. Ultimately, differences in cathexis across individuals will influence cognition and behavior in a systematic fashion, forging the elementary structure of personality. In making these arguments, we explore the concept of cathexis at both theoretical and neuroanatomical levels, drawing on neurophysiological concepts of cortico-limbic interaction, links between anterior cingulate and medial frontal lobe function, and synaptic plasticity.

Cognition and Emotional Priming

The human memory system may be organized around vectors of emotional valence. A functional separation may exist in the semantic network, dividing memories for emotionally positive versus negative representations (del Grandpre, 1996). This notion of valenced semantic priming provides the foundation for our theory of semantic development.

Emotion's Role in Memory

The notion of uniqueness in the processing of emotion within cognition is not novel. Zajonc (1980) has argued for a functional separation of affect and cognition. Wickens (1972) and Posner and Snyder (1975) have proposed that emotion has a special salience in memory constructs.

Psychometric evidence has born out such distinctions. Experimental studies have demonstrated a special role for emotion in memory. For example, Johnson, Petzel, Hartney, and Morgan (1983) found that depressed subjects recall more failure experiences than non-depressed subjects, and that success experiences were better recalled by non-depressed subjects than depressed. Tucker, Chung, West, Potts, and Liotti (in press) demonstrated that the emotional direction of induced moods are predictive of expectancy in story completion; subjects in bad moods expected a story to end badly, and subjects in good moods expected a story to end well. Blaney (1986) and Eich (1995) in particular have attempted to characterize such "mood-congruency" or "mood-dependent" effects and describe their limitations, particularly in terms of the idiosyncrasy of their robustness.

The principal explanatory account of these empirical findings of mood-congruency is Bower's (1981) associative network theory of mood and memory. In his model, Bower proposes that information is represented as nodes, and associative links exist between these nodes in a fashion similar to Collins and Loftus' (1975) model of spreading activation. When information is encoded, links are created between its memory representation and the emotion that was being experienced at the time. These emotions are also nodes. Consequentially, when particular emotion nodes are active, representations linked to them are also activated, creating the response and recall biases described in the mood-congruency literature. Bower named eight primary emotion nodes, based on a factor analysis by Plutchik (1980): joy, acceptance, fear, surprise, sadness, disgust, anger, and anticipation.

Bower's (1981) model, however, suffers from several inherent difficulties (see Eich, 1995; Teasdale, 1993). Accordingly, Teasdale (1993; Barnard & Teasdale, 1991; Teasdale & Barnard, 1993) has developed a theory to account for mood congruent effects based on Barnard's (1985) Interacting Cognitive Subsystems (ICS) framework. Effects of mood, according to Teasdale (1993), occur at a more generic level of representation. That is, two levels of meaning exist for mental codes: Implicational and Propositional. Propositional code is specific in meaning, e.g., "Roger has brown hair." Implicational code is generic and holistic in meaning. Implicational codes do not map directly onto language but are instead abstract concepts made up of lower-order patterns of specific meanings (Teasdale, 1993). Access to a higher-order, implicational code can activate lower-level representations, and vice versa. Teasdale's model differs from Bower's (1981) in that rather than operating at the level of specific emotions and their associates, mood works at a higher level of abstraction as captured in the notion of ICS's schematic models of experience. Teasdale (1993) argues that his theory therefore captures more of the variance in mood-congruency effects than does Bower's (1981).

Affective Valence Priming

It may not be necessary for memory to be organized in such nodal or specific/abstract patterns. de Grandpre (1996) has described a theoretical emotional structure to memory that differs from Bower's (1981) and Teasdale's (1993) models but still explains the effects of mood-congruency. Termed Affective Valence Priming, this model supposes that each representation in memory, or node, is encoded with an emotional "tag" indicating the hedonic or antagonistic value of that representation; i.e., these tags mark each representation as positive if emotionally "good" or negative if emotionally "bad"; the Affective Valence Priming system has a corresponding positive affect component and negative affect component. The critical aspect of this mechanism is that Affective Valence Priming serves as an emotional bypass to the standard semantic network and primes all representations attached to the system in parallel, providing a more global priming than that of a purely "semantic" network which operates at a more local level. This priming is specifically positive or negative. So, for example, a happy stimulus like a field of daisies may create a priming bias of a cheerful time that a parent has spent with a newborn daughter.

Such a system of separated affective valence explains the mood-congruency effect admirably. Moods activate corresponding Affective Valence Priming components; a good mood energizes the positive affect component, and a bad mood energizes the negative affect component, leading to priming of attached nodes and therefore the response biases consistent with the mood-congruency effect. The very nature of such broad, parallel priming consolidates Teasdale's (1993) distinction between lower- and higher-order representations by having one priming mechanism for both.

Moreover, mood is not essential for Affective Valence Priming. de Grandpre (1996) asserts that this emotional activation is based on the primitive, elementary structure of memory. Consequentially, access through the positive or negative affect components occur at a most basic level and is not necessarily directed by which component is currently more active.

Also, this system does not require conscious processing. While Eich (1995) has suggested that mood-congruency effects are most robust in explicitly emotional tasks rather than implicit ones, others (e.g., Tobias, Kihlstrom & Schacter, 1992) have demonstrated that emotional experience has robust effects on implicit memory as well.

Finally, priming may occur independent of the semantic associative network. For example, the words "cancer" and "tumor" are both emotionally negative words. They are also related in associative meaning: a malignant tumor is created by cancerous growth. However, "cancer" and "stingy" are both emotionally negative words but are not related in associative meaning, or very weakly so. But cancer and stingy both have emotionally negative tags and are attached to the negative affect system. Therefore, they could prime each other.

How can a dissociation between valence and associative priming be demonstrated empirically? While the body of literature for the mood-congruency effect provides ample evidence, further work has suggested that Affective Valence Priming exists independent of mood measurements or manipulations (Hill & Kemp-Wheeler, 1989; Kemp-Wheeler & Hill, 1992). A popular test in the literature for semantic activation has been priming and lexical decisions. It is well established that words close in meaning prime each other (Meyer & Schvaneveldt, 1971; Meyer, Schvaneveldt & Ruddy, 1975; Neely, 1977). For example, when a person sees the word "nurse" preceded by "doctor," that person responds more quickly to nurse in a lexical decision (word/non-word) or naming task than when preceded with an unrelated word, like "wall." Previous effects of mood on lexical decisions have been well established (Challis & Krane, 1988; Clark, Teasdale, Broadbent & Martin, 1983). But more specifically, Hill and Kemp-Wheeler (1989; Kemp-Wheeler & Hill, 1992) demonstrated priming effects for not only word pairs like doctor and nurse but also cancer and stingy as well, indicating that cancer primes stingy. Because cancer and stingy are not reasonably associated words, only the similarity in emotional valence can account for the priming effect, providing evidence for the separation of Affective Valence Priming and traditional associative priming. Hill and Kemp-Wheeler's two studies suffer from a confound, however: Because they chose to block rather than mix conditions, Hill and Kemp-Wheeler may have actually manipulated mood by giving multiple, continuous presentations of negative words. So mood-congruency, rather than a system of affective valence separated from associative memory, might explain their effects (de Grandpre, 1996). In fact, in a mixed design with different stimuli, de Grandpre (1996) was unable to replicate Hill and Kemp-Wheeler's (1989) finding. We are currently trying to characterize these effects in our laboratory.

Tonic and Phasic Systems

The Affective Valence Priming model can be compared with the tonic activation and phasic arousal systems proposed by Tucker and associates (Tucker, 1992; Tucker, Vannatta & Rothlind, 1990; Tucker & Williamson, 1984). Drawing on the work of Pribram and McGuinness (1975), Tucker and his colleagues proposed that two types of regulatory influences direct attention and working memory: tonic activation and phasic arousal. These tonic and phasic systems are proposed to have inherent affective valences. Tonic activation engages motor readiness in fight-flight conditions; it simultaneously produces the affective qualities of hostility and/or anxiety. Phasic arousal engages perceptual responsivity and orienting response processes; at high levels it simultaneously produces the affective state of elation.

In addition to affective features, there are inherent attentional influences of these "arousal" systems. Tonic activation is a sustained, steady stream of activity. The tonic activation system supports preparedness by maintaining a steady activating influence. Tonic activation enhances redundancy in the processing of information. This serves as a bias against change. Changes that do occur would then require substantial determination. Other aspects of alerting, as in an orienting response, may be supported by a phasic arousal system. Phasic arousal is an energetic and rapid arousal level. This system provides a habituation bias that results in a reactivity to novelty (Tucker, 1992).

Using this distinction, the most critical influence of the tonic activation system may be its engagement under threatening conditions (Tucker et al., 1990). Tucker et al. assert that as the brain encodes memories, it tags them with the potential to elicit an affective/attentional state appropriate to threat avoidance. On later occasions, the engagement of an anxiety/hostility state via tonic activation is sufficient to prime all threat-related representations, in parallel and through a separate mechanism from the semantic associations to the current perceptual experience. Tonic activation therefore primes the negative affect system. The network of negative, or "bad," tags prevalent in Affective Valence Priming provides a network of association engaged under periods of sustained stress.

Phasic arousal serves a different purpose by monitoring the hedonic value of stimuli in ongoing perception. Thus, a stimulus is encoded in memory not only for its sensory qualities but also for the degree of elation it engenders. The influence on the memory system is compatible with the Affective Valence Priming model, in that the effect is a broad, parallel priming of positive (hedonic) domains of memory. There is interaction between phasic arousal and cognitive retrieval; becoming elated at a point in time facilitates access to the hedonic object, and accessing the object in memory augments elation.

Although we emphasize that there is a dominant affective valence associated with each specific arousal mechanism, we recognize the complexity of the quality of affect that is not captured by a simple valence dimension. For example, the emotion of fear engages both the phasic arousal (orienting response) and tonic activation (defensive response) systems. Also, the anticipation of a pleasurable experience often creates anxiety, and tonic activation. But by proposing that there is an inherent, valenced, subjective quality to each of the activation and arousal mechanisms, we take issue with the traditional argument that "arousal" is unidimensional and unvalenced, and that the valence of all emotional experience is determined in the cognitive domain (Schachter & Singer, 1962).

Cathexis

Motivational processes must determine the extent to which tonic activation or phasic arousal are elicited by internal states and by ongoing behavior in the environment. These processes must dictate whether a representation ascribes to what an organism desires, and is therefore "good," or is counterproductive to the organism's intentions, and is "bad," or sometimes neither. We propose that cathexis is the emotional charge that activates semantic networks in a motivated fashion.

Freud (1940) developed the concept of cathexis to describe how emotional significance regulates memory. In his "Project for a Scientific Psychology," he theorized that internal homeostatic processes apply influences on mental representation, to direct cognition in service of biological needs.

Within our framework of Affective Valence Priming, we can describe cathexis as the valence and arousal, or priming, level of a representation. But cathexis is also the process that must be initiated to implement the systems of arousal and affect. Various mental processes determine what an organism's goals are and how they are best achieved. These processes in turn activate the Affective Valence Priming system by sending a signal comprising the cathexis. This signal contains two components of information that direct activation processes. Each component indicates how the system will be energized.

The first component of cathexis is its valence, or direction. A motivational charge must either be towards an object if that object facilitates the organism in reaching some goal, or away from an object if it impedes progress towards goal fulfillment. Thus, cathexes prime affective valence differentially depending on the motivational significance of a stimulus.

The second component of cathexis can be described as its arousal, or strength. Is the cathexis initiating a sustained positive or negative state, such as depression or satiation? Or does it activate rapid, energetic arousal systems, like those in anxiety, fear, and elation? This strength of arousal, sustained or elated, is the strength of the cathexis.

Cathexis can thus be seen as a vector of motivation. Not only does it determine the direction of motivational activation as positive or negative, but it also determines the strength of arousal of such activation. Cathexis is therefore not only integrated with every representation as its emotional "tag" but also acts as an independent agent in memory that can serve to determine and later activate pathways based on its cathected dimensions. Of course, a lack of emotion can be attached to the representation; the strength and direction of the cathexis as a tag are determined by previous cognitive determination of the emotional significance of the representation (Schachter & Singer, 1962).

Tucker and Williamson's (Tucker & Williamson, 1984) distinction between levels of arousal are characterized by cathected processes. Tonic activation is cathexis of a steady and sustained nature that is negative in valence; phasic arousal is a sharp and short-lived cathexis that is positive in valence.

Cathexes serve to change the pathways of hedonic or antagonistic value of memory representations in a constructive or deconstructive manner. When a cathexis activates a representation, it travels the link between that representation and the corresponding component of the Affective Valence Priming system. This enervation, in turn, strengthens that link. A lack of use of a link leads to decay of that link. For example, the initial tagging of a stimulus as positive may not be reliable when other experiences with that stimulus yield negative consequences. So, cathexes reroute the nodal connection of that stimulus, linking it to the negative affect component. The link to the positive affect component gradually dissipates, for the strength of these links is only maintained when traveled by cathexes.

Biases in the Memory System

In our cognitive model of human memory and emotion, the ability of cathexis to shape memory pathways is the essence of how biases in these pathways are formed.

The maintenance of memory is a recursive process in the brain. One's memory is geared to one's motivational processes, which are in turn developed based on previous memories. That is, when a cathexis is formed in response to a stimulus, it is created not only from that experience with the stimulus but from an amalgam of previous experiences with that stimulus or similar stimuli. For example, if experiences with that stimulus had been good, the cathexis takes on a positive direction, and if that experience was elating, the cathexis activates phasic arousal. This activates the memory system in an adaptive fashion so that other joyous stimuli are active in memory and the organism is better prepared to recognize and respond to them.

Continuing the example, the cathexis strengthens the link of the stimulus to the positive affective valence component and phasic arousal system. This makes the stimulus more likely to activate the systems and potentiates the motivational system's ability to signal the significance of the stimulus in the future. Thus, the motivational system is biased towards the stimulus.

How is this accomplished? During encoding, a new representation is tagged with its motivational significance: a direction and a strength (the cathexis). These tags lead to links to positive or negative affective valence components and perhaps tonic or phasic systems of arousal. So during a state with a particular affective valence and arousal, the system is better able to access the encoded representation if those variables are congruent. This access in turn strengthens the links and together with other associations (e.g., semantic priming), increases the significance of the representation. If this representation is not accessed, the consolidation process inherent to memory storage requires the link to decay as newer and stronger links are formed. The more degraded the pathway, the less likely the representation is activated during recall, creating a use-it-or-lose-it system.

Elements of Personality

While the basic system of cathexis may indeed be present across all individuals, differences in its engagement could be indicative of differences in how people access memory and consequentially of how they perceive their environment. The biases described above could create the individual differences that make up personality.

Temperament in particular may be a function of what pathways have been biased by cathected processes. Lazarus (1991) noted that temperament refers to stable individual differences in hedonic tone and arousal (cathexis, in our framework) with regard to discrete emotions like anger or fear. Thus, when a set of pathways for a set of associated representations are biased towards a particular cathected direction, future experiences with those representation are correspondingly primed. Cathexis can then in extreme cases lead to neuroses and psychoticism. For example, a person with anxious, negative experiences in novel social interactions during childhood might become shy.

This shy person would avoid activating those networks involving speaking to strangers, which are correspondingly linked to negative emotions like fear and self-hate. New, positive experiences with strangers may be emotionally "washed out" by memory's propensity to consolidate experience within the set of negative biases already present, making it exceedingly difficult to overcome the person's agoraphobic tendencies and "relearn" coping with novel social interactions.

Neural Mechanisms of Cathexis

We propose that the neural mechanisms underlying this separation of the tonic and phasic systems differ from person to person, and these differences in organization lead to differences in personality. A distributed circuit of processing is described, detailing how cathexis is formed on a global scale in the brain. But it is the plasticity of the neurons in this circuit that is critical for individualized pathways, and consequentially systematic individual biases in processing.

Cortico-Limbic Resonance

One so far unaddressed question is how and where motivational processes are created. The synergy between cortex and limbic structures is most critical in forming memory and determining motivation and cathexes.

A general principle of brain connections illustrates the unique opportunity for synergy across the three levels of the human brain (MacLean, 1973). Called "top-down fan-in" (Derryberry & Tucker, 1991), this principle notes that the limbic system is situated midway between the cortex and the control systems of the brainstem. In terms of control, the brain's self-regulation is achieved as limbic mechanisms alter the functioning of the brainstem's neuromodulator systems, which in turn project to wide areas of cortex through a bottom-up fan-out pattern of connection. Thus, a cathected process created through frontal and limbic decisions permeates cortex through this top-down fan-in system, providing system-wide response orientation.

Derryberry and Tucker (1991) proposed that representations formed at the paralimbic level may characterize the emotional significance of events, indicating a possible neuronal center for the cathected process. As the paralimbic networks tune the activity of brainstem neuromodulator systems, those networks engage modes of self-control like the nonadrenergic phasic arousal or dopaminergic tonic activation systems (Tucker & Williamson, 1984). Together, cathected information is created by arousal and Affective Valence Priming and is likely implemented through these nonadrenergic and dopaminergic systems.

The dense activity and connectivity between amygdala (LeDoux, 1987) and hippocampal (Squire, 1992) structures obviates the role of emotion in memory. The parallel and distributed connectivity between emotional and memory structures, specifically amygdala and hippocampus, suggests that semantic processing must be expressed in a similarly parallel and distributed fashion. Affective Valence Priming provides, in a cognitive model, the functional capacity that the neuronal organization of the limbic system implicates.

The Role of Frontal Lobes

The role of the frontal lobes in determining motivational and therefore cathected processes is one of established importance. In particular, working memory is seen as a biasing mechanism for motivational significance, where the strength and direction of a cathexis is maintained with task demands when interpreting a complex environmental situation.

The motivational direction of the self is realized in limbic and brainstem projections to the frontal lobes. The densely interconnected paralimbic cortices serve to maintain a global motivational context within which specific actions are articulated and sequenced within frontal neocortical networks (Tucker, Luu & Pribram, 1995).

The network architecture of the frontal lobe reflects the dual limbic origins of frontal cortex, in the dorsal and ventral paleocortical structures. These two cortico-limbic pathways apply different motivational biases to direct the frontal lobe representation of working memory, a functionally established component of the frontal lobe (Goldman-Rakic, 1988). Given the essential role of cortico-limbic interaction in memory consolidation, we can assume that frontal connections with limbic networks will be necessary to consolidate the cognitive operations that support extended motor planning (Tucker et al., 1995).

Working Memory and Executive Systems

While working memory may be partly situated in the frontal lobes, a cognitive model of working memory shows its neural realization to be more complex and, in turn, an ideal functional and neuroanatomical point for motivational and emotional processes to take shape.

Baddeley (1992) has described working memory as the temporary storage of information in connection with performing other, more complex tasks. He has divided working memory into three functionally separate components. The *central executive* is an attentional control system that is functionally aided by either visual material (the *visuospatial sketchpad*) or verbal material (the *phonological loop*) (Baddeley, 1992; Baddeley & Hitch, 1994).

Baddeley's component architecture maps onto brain areas that are functionally significant to cathected processes. The limbic system, particularly the amygdala and hippocampal structures (Davis, 1992; LeDoux, 1989; Squire, 1992), provide motivational and emotional responses to anterior cingulate and other midline and lateral frontal areas, which together may perform central executive functions (Posner & DiGirolamo, in press). This in turn biases neocortical structures towards the strength and direction of the limbic response, including object working memory (part of the phonological loop) in left prefrontal cortex (Cohen et al., 1994; Seeck et al., 1995; Smith et al., 1995) and spatial working memory (the visuospatial sketchpad) in right prefrontal cortex (Jonides et al., 1993; Smith et al., 1995). These areas of working memory maintain the prepotent bias determined by limbic processes, thereby influencing the interpretation of incoming sensory information. The frontal cortex also parcels out the now-biased sensory information and corresponding motor response to SMA and other brain areas requiring such information.

Such an ability to redistribute and process sensory information is critical for the human response system to selectively act upon and interpret competing stimuli. The function of emotion in attention is to identify critical aspects of the environment and activate associated thoughts, plans, and memories (Bower, 1992). The amygdala and hippocampal structures are thought to be able to determine which stimuli, from a dangling spider to a funny joke, must be responded to by interpreting their emotional significance (LeDoux, 1987; McGaugh, 1992; Mesulam, 1990; Rolls, 1995). The amygdala also signals reflexive responses to potentiate any physical avoidance via hypothalamus (LeDoux, 1994). Anterior cingulate and other midline and lateral frontal areas are then given this information of emotional significance and activate in awareness those stimuli above some emotional threshold. This determines what stimuli will be active in a cognitive response, while frontal systems maintain the priorities and signal motor responses (Tucker et al., 1995).

Taken together, these three brain areas (limbic, frontal, and cingulate) form a circuit for motivational biasing to emerge. However, an examination of the pathways at a more reductionist level may begin to explain how these circuits can, over the long term, form functional differences in motivational processes that determine individual differences in emotional reactivity.

At the Neuronal Level

How might our distributed network of motivational biasing occur at the cellular level? Cathexis may direct the anatomical self-organization of synaptic patterns that instantiate memories. It is generally believed that the critical neural changes underlying memory take place at the synaptic level (Hebb, 1949; Kandel & Spencer, 1968; Lynch & Baudry, 1984; Thompson, 1986). Cathexis may be realized as the genesis of individualized emotional construction of representations because of its importance in regulating neural plasticity. As consolidation of memories occurs in the growing infant, subtle differences develop in the quality and perhaps quantity of the neural pathways in the developing brain. Such consolidation forms the framework of personality.

The differentiated anatomy of cortical networks is apparently realized through overabundant production of synaptic connections followed by a competitive reduction that pares connections that fail to be effective (Singer, 1987). Ineffective synapses and neural processes are eliminated (Innocenti, 1983). This dynamic process of selective pattern formations in cortical networks has been termed Neural Darwinism, indicating that the survival of patterns in neural tissue depends on effective competition for information (Changeux & Dehaene, 1989).

One of us (Tucker, 1992) has argued that these dynamic processes of Neural Darwinism are not only physiological but are formed through experience, or more metaphorically "commerce with the environment." The cognitive processes of the global network have the inherent capacity to instantiate their preferred pathways within neural tissue. This is the fundamental assumption of a neuronal genesis for cathected processes. As we have suggested, cathexis causes links to be strengthened or expire based on the emotional and motivational salience of the experience. When cathexis activates a representation, it travels the link between that representation and the Affective Valence Priming system. Such travel strengthens that link. A lack of use of a link leads to decay of that link. These links are theoretical representations of actual neural connections, and these physical pathways suffer from the same gains and losses that the representational pathway endures.

This system of Neural Darwinism is of particular importance in the growing infant. Much of its experience with emotional circumstances are formed by the formation of neural patterns based on the cognitive experiences with its environment. This learning is similar to the process of kindling, where repeated low-level electrical stimulation may result in exaggerated reactivity of limbic structures (Doane & Livingston, 1986). A prepotent reactivity is formed in neural connections by cathected processes. This in turn dramatically influences perceptions later in life because the neural pathways are well-trained to accept new representations in a manner that befits established conformity.

While the neuronal circuits sculpted by cathected processes may appear dedicated or "burned in," they retain a degree of plasticity. However, it requires a great deal of cathected momentum to shift cathected pathways, to alter the functional boundaries inherent in the network constructed in the critical stages of human development. Therefore, while it is especially difficult to stray from the neuronal framework that determines behavioral tendency, it is possible to do so with pathway energization that is continuous and directed. For example, if the infant has developed a strong positive link to the representation "mother," a great deal of negative cathexes must be spawned to shift the fundamental neural connections of "mother" to a negative emotional response.

Another concern in this framework is the continuing emergence of research of synaptic plasticity in the relevant neural substrates. The general feature of synaptic plasticity is still being demonstrated in the frontal lobes, amygdala, and cingulate structures. Long-term potentiation (LTP), a form of synaptic plasticity that has been associated with learning and memory, has been most extensively studied in the hippocampus (see Squire, 1992). Less is known of LTP in neural structures specific to our hypotheses, although Clugnet and LeDoux (1990) and Chapman and Brown (1988) have demonstrated LTP in the amygdala. As research in neural learning moves from occipital and hippocampal structures to other areas, LTP or some other form of plasticity may be traced in such a fashion as to complement our theory.

Conclusions

In this chapter we have theorized that every representation in memory is charged by some degree of emotional significance. This significance may be hedonic or aversive, but the charge is unlikely to be completely neutral. Functional and neural pathways are formed based on this emotional significance, and these pathways create the bias for accessing motivated representations.

What is the function of this organizational scheme? In response to emotional task demands, most notably fear, human behavior seems geared to minimize response time (de Grandpre, 1996; Lang, Bradley & Cuthbert, 1990; LeDoux, 1994; Öhman, 1993; Tucker et al., 1990). An individual would naturally be benefitted by recognizing threats or hedonic opportunities as quickly as possible. The system we have described allows speeded access to emotionally salient events. Cathected events are usually those that have the greatest consequences for the individual.

Concepts of motivation have not been integrated well with recent theories of semantic memory. In contrast, classical models of cognitive psychology considered the emotional and motivational dimensions to be fundamental (Osgood, 1957). We propose that dimensions of emotional arousal and motivational activation are integral to the control of neural activity, and therefore to the control of activity-dependent synaptic plasticity. At the psychological level, these mechanisms form the basis for the self-organization of personality.

References

Baddeley, A. D. (1992). Working memory. Science, 255, 556-559.

Baddeley, A. D., & Hitch, G. J. (1994). Developments in the concept of working memory. Neuropsychology, 8(4), 485-493.

Barnard, P. (1985). Interacting cognitive subsystems: A psycholinguistic approach to short-term memory. In A. Ellis (Ed.), Progress in the psychology of language, (Vol. 2, pp. 197-258). London: Erlbaum.

Barnard, P. J., & Teasdale, J. D. (1991). Interacting cognitive subsystems: A systemic approach to cognitive-affective interaction and change. Cognition and Emotion, 5, 1-39.

Blaney, P. H. (1986). Affect and memory: A review. Psychological Bulletin, 99, 229–246.

Bower, G. H. (1981). Mood and memory. American Psychologist, 36, 129-148.

Bower, G. H. (1992). How might emotions affect learning? In S.-A. Christianson (Ed.), The handbook of emotion and memory: Research and theory, (pp. 3-31). Hillsdale, NJ: Lawrence Erlbaum Associates.

Challis, B. H., & Krane, R. V. (1988). Mood induction and the priming of semantic memory in a lexical decision task: Asymmetric effects of elation and depression. Bulletin of the Psychonomic Society, 26(4), 309-312.

Changeux, J., & Dehaene, S. (1989). Neuronal models of cognitive functions. Cognition, 33, 63-109.

Chapman, P. F., & Brown, T. H. (1988). Long-term potentiation in amygdala brain slices. Society for Neuroscience Abstracts, 14, 566.

Clark, D. M., Teasdale, J. D., Broadbent, D. E., & Martin, M. (1983). Effects of mood on lexical decision. Bulletin of the Psychonomic Society, 21, 175-178.

Clugnet, M. C., & LeDoux, J. E. (1990). Synaptic plasticity in fear conditioning circuits: Induction of LTP in the lateral nucleus of the amygdala by stimulation of the medial geniculate body. The Journal of Neuroscience, 10, 2818-2824.

Cohen, J. D., Forman, S. D., Braver, T. S., Casey, B. J., Servan-Schreiber, D., & Noll, D. C. (1994). Activation of the prefrontal cortex in a nonspatial working memory task with functional MRI. Human Brain Mapping, 1, 293-304.

Collins, A. M., & Loftus, E. F. (1975). A spreading-activation theory of semantic processing. Psychological Review, 82, 407-428.

Davis, M. (1992). The role of the amygdala in fear and anxiety. Annual Review of Neuroscience, 15, 353-375.

de Grandpre, D. D. (1996). Valence priming and affective context: Effects of emotional similarity in a lexical decision task. (Technical Report TR-96-06). Eugene, OR: Institute of Cognitive and Decision Sciences, University of Oregon.

Derryberry, D., & Tucker, D. M. (1991). The adaptive base of the neural hierarchy: Elementary motivational controls on network function. In R. Dienstbier (Ed.), Nebraska Symposium on Motivation, (pp. 289-342). Lincoln: University of Nebraska Press.

Doane, B. K., & Livingston, K. E. (1986). The limbic system: Functional organization and clinical disorders. New York: Raven Press.

Eich, E. (1995). Searching for mood dependent memory. Psychological Science, 6(2), 67-75.

Freud, S. (1940). An outline of psychoanalysis. (Vol. 23). London: Hogarth Press, 1953 (First German Edition, 1940).

Goldman-Rakic, P. W. (1988). Topography of cognition: Parallel distributed networks in primate association cortex. Annual Review of Neuroscience, 11, 137-156.

Hebb, D. O. (1949). The organization of behavior. New York: Wiley.

Hill, A. B., & Kemp-Wheeler, S. M. (1989). The influence of context on lexical decision times for emotionally aversive words. Current Psychology: Research & Reviews, 8(3), 219-227.

Innocenti, G. M. (1983). Exuberant callosal projections between the developing hemispheres. In R. Villani, I. Papo, M. Giovanelli, S. M. Gaini, & G. Tomei (Eds.), Advances in Neurotraumatology. Amsterdam: Excerpta Medica.

Johnson, J. E., Petzel, T. P., Hartney, L. M., & Morgan, L. M. (1983). Recall and importance ratings of completed and uncompleted tasks as a function of depression. Cognitive Therapy and Research, 7, 51-56.

Jonides, J., Smith, E. E., Koeppe, R. A., Awh, E., Minoshima, S., & Mintun, M. A. (1993). Spatial working memory in humans as revealed by PET. Nature, 363, 623-625.

Kandel, E. R., & Spencer, W. A. (1968). Cellular neurophysiological approaches to the study of learning. Physiology Review, 48, 65-134.

Kemp-Wheeler, S. M., & Hill, A. B. (1992). Semantic and emotional priming below objective detection threshold. Cognition & Emotion, 6(2), 129-148.

Lang, P. J., Bradley, M. M., & Cuthbert, B. N. (1990). Emotion, attention, and the startle reflex. Psychological Review, 97, 377-395.

Lazarus, R. S. (1991). Emotion and adaptation. New York: Oxford University Press.

LeDoux, J. E. (1987). Emotion. In F. Plum (Ed.), Handbook of physiology. Section 1: The nervous system. Volume V. Higher functions of the brain. Part 1, (pp. 49-459). Bethesda, MD: American Physiological Society.

LeDoux, J. E. (1989). Cognitive–emotional interactions in the brain. Cognition and Emotion, 3, 267-289.

LeDoux, J. E. (1994). Emotion, memory, and the brain. Scientific American, 270(6), 50-59.

Lynch, G., & Baudry, M. (1984). The biochemistry of memory: A new and specific hypothesis. Science, 224, 1057-1063.

MacLean, P. D. (1973). A triune concept of the brain and behavior. Toronto: University of Toronto Press.

McGaugh, J. L. (1992). Affect, neuromodulatory systems, and memory storage. In S.-A. Christianson (Ed.), The handbook of emotion and memory, (pp. 245-268). Hillsdale, NJ: Lawrence Erlbaum Associates.

Mesulam, M. (1990). Large-scale neurocognitive networks and distributed processing for attention, language, and memory. Annals of Neurology, 28(5), 597-613.

Meyer, D. E., & Schvaneveldt, R. W. (1971). Facilitation in recognition pairs of words: Evidence of a dependence between retrieval operations. Journal of Experimental Psychology, 90, 227-234.

Meyer, D. E., Schvaneveldt, R. W., & Ruddy, M. G. (1975). Loci of contextual effects on visual word recognition. In P. M. A. Rabbityt & S. Dornic (Eds.), Attention and performance, (Vol. V). London: Academic Press.

Neely, J. H. (1977). Semantic priming and retrieval from lexical memory: Roles of inhibitionless spreading activation and limited-capacity attention. Journal of Experimental Psychology: General, 106(3), 226-254.

Öhman, A. (1993). Fear and anxiety as emotional phenomena: Clinical phenomenology, evolutionary perspectives, and information-processing mechanisms. In M. L. Lewis & J. M. Haviland (Eds.), Handbook of emotions, (pp. 511-536). New York: The Guilford Press.

Osgood, C. E., Suci, G. J., & Tannenbaum, P. H. (1957). The measurement of meaning. Urbana, Ill: University of Illinois Press.

Plutchik, R. (1980). A language for the emotions. Psychology Today, 68-78.

Posner, M. I., & DiGirolamo, G. J. (in press). Conflict, target detection and cognitive control. In R. Parasuraman (Ed.), The Attentive Brain. Cambridge, MA: MIT Press.

Posner, M. I., & Snyder, C. R. R. (1975). Attention and cognitive control. In R. L. Solso (Ed.), Information processing and cognition. The Loyola Symposium, (pp. 55-85). New York: Wiley-Liss, Inc.

Pribram, K. H., & McGuinness, D. (1975). Arousal, activation, and effort in the control of attention. Psychological Review, 82, 116-149.

Rolls, E. T. (1995). A theory of emotion and consciousness, and its application to understanding the neural basis of emotion. In M. S. Gazzaniga (Ed.), The cognitive neurosciences, (pp. 1091-1106). Cambridge, MA: MIT Press.

Schachter, S., & Singer, J. E. (1962). Cognitive, social and physiological determinants of emotional state. Psychological Review, 69, 379-399.

Seeck, M., Schomer, D., Mainwaring, N., Ives, J., Dubuisson, D., Blume, H., Cosgrove, R., Ransil, B. J., & Mesulam, M.-M. (1995). Selectively distributed processing of visual object recognition in the temporal and frontal lobes of the human brain. Annals of Neurology, 37(4), 538-545.

Singer, W. (1987). Activity-dependent self-organization of synaptic connections as a substrate of learning. In J. P. Changeux & M. Konishi (Eds.), The neural and molecular basis of learning. New York: Wiley-Liss, Inc.

Smith, E. E., Jonides, J., Koeppe, R. A., Awh, E., Schumacher, E. H., & Minoshima, S. (1995). Spatial versus object working memory: PET investigations. Journal of Cognitive Neuroscience, 7(3), 337-356.

Squire, L. R. (1992). Memory and the hippocampus: A synthesis from findings with rats, monkeys, and humans. Psychological Review, 99(2), 195-231.

Teasdale, J. D. (1993). Selective effects of emotion on information-processing. In A. Baddeley & L. Weiskrantz (Eds.), Attention: Selection, awareness, and control. New York: Oxford University Press.

Teasdale, J. D., & Barnard, P. J. (1993). Affect, cognition and change: Remodeling depressive thought. Hove: Erlbaum.

Thompson, R. F. (1986). The neurobiology of learning and memory. Science, 233, 941-947.

Tobias, B. A., Kihlstrom, J. F., & Schacter, D. L. (1992). Emotion and implicit memory. In S.-A. Christianson (Ed.), The handbook of emotion and memory: Research and theory, (pp. 67-92). Hillsdale, NJ: Lawrence Erlbaum Associates.

Tucker, D. M. (1992). Developing emotions and cortical networks. In M. Gunnar & C. Nelson (Eds.), Minnesota Symposium on Child Development: Developmental Neuroscience. New York: Oxford.

Tucker, D. M., Chung, G., West, P., Potts, G. F., & Liotti, M. (in press). Emotional expectancy: Psychometric and brain electrical studies of mood-congruent cognitive bias. In M. Spitzer & B. Mahler (Eds.), Studies in experimental psychopathology, . Cambridge: Cambridge University Press.

Tucker, D. M., Luu, P., & Pribram, K. (1995). Social and emotional self-regulation. In J. Grafman, K. J. Holyoak, & F. Boller (Eds.), Annals of the New York Academy of Sciences, (Vol. 769, pp. 213-239). New York: New York Academy of Sciences.

Tucker, D. M., Vannatta, K., & Rothlind, J. (1990). Arousal and activation systems and primitive adaptive controls on cognitive priming. In N. Stein, B. Leventhal, & T. Trabasso (Eds.), Psychological and biological approaches to emotion, (pp. 145-166). Hillsdale, NJ: Lawrence Erlbaum Associates.

Tucker, D. M., & Williamson, P. A. (1984). Asymmetric neural control systems in human self-regulation. Psychological Review, 91(2), 185-215.

Wickens, D. D. (1972). Characteristics of word encoding. In A. W. Melton & E. Martin (Eds.), Coding processes in human memory, (pp. 191-215). Washington, D.C.: V.H. Winston & Sons.

Zajonc, R. B. (1980). Feeling and thinking: Preferences need no inferences. American Psychologist, 35, 151-175.

The Social Level: The Organization of Self in Society

18.
Learning and Unlearning in the Formation of Social Bonds

Walter J. Freeman

Learning and Unlearning in the Formation of Social Bonds[*]

Walter J. Freeman
Department of Molecular and Cell Biology
University of California at Berkeley

Abstract

Learning is the set of processes by which intentional structures stretch forth and change themselves through self-organizing, chaotic dynamics. It leads to solipsistic isolation of individual brains progressively with increasing complexity of learned tasks. In order for understanding between brains to develop as the basis for trust and cooperation, neurohumoral mechanisms are postulated to exist in mammals, which bring about unlearning through a meltdown of intentional beliefs, without loss of procedural and declarative memories. Evidence for unlearning exists in the physiology of conversion ("brain washing") and in the secretion of oxytocin and vasopressin in human and other mammalian brains during behaviors directed toward reproduction and the care of altricial offspring.

Introduction

Measurements of the electrical activity of brains show that dynamical states of neuroactivity emerge like vortices in a weather system, triggered by physical energies impinging onto sensory receptors, drifting in time, and changing with the context of the subjects (Freeman 1992). These dynamical states determine the structures of intentional actions and the patterns constructed upon destabilization of the perceptual systems, as shaped by corollary discharge when expected sensory input arrives. These neuroactivity patterns may provide the physical basis for thoughts and meanings. In order for this to be so, neuroactivity in brains must act back onto the intentional structure and change it in accordance with three properties of intentionality: to stretch forth and modify the self in conformance with the world; to seek wholeness in growth to maturity; and to maintain the unity of the self.

Learning and self-organization

In the educationist view a person learns by gaining knowledge, skills, and understanding. In the behaviorist view a system learns by changing its response upon repeated presentations of the same stimulus given to form a reflex and elicit a response. In spite of its failure as a system for understanding brain function, the behaviorist approach has given a powerful set of rules for describing, predicting and, to some extent, controlling behavior, without ever having to understand the aims, desires, and other aspects of minds. This simplicity makes it possible for cell biologists to apply the same rules not to a whole animal but to a neuron or part of a neuron as a system. The focus for studies of learning by cells is the modifiable synapse between two neurons. As Freud (1895) wrote, the logical place to look for cellular changes with learning is at the "contact barrier". The criterion for learning is that the conditions of a change in the synaptic strength must conform to the behaviorists' rules. The method is to find and analyze a synapse in which repeated stimulation of presynaptic neurons leads to *use-dependent* change in amplitude of responses of postsynaptic neurons, such as occurs with PTP, LTP, LTD, etc.

---

[*] Excerpted and adapted from Chapter 6 in Freeman WJ (1995) "Societies of Brains. A Study in the Neuroscience of Love and Hate". © Lawrence Erlbaum Associates, Inc., Hillsdale NJ

The reductionists have a Sisyphean task in attempting to construct an "alphabet" (Alkon 1992) of cellular and molecular events with which to write the neural words of learning. An increase in efficacy or sensitivity at a synapse may (or may not) require an increase in amounts or rates of release of transmitter, or in number of synaptic vesicles, or in membrane locations or conformations of postsynaptic receptor proteins. It is certain to require increases in synaptic membrane surface fluxes, rates of turnover in lipids and proteins, replication of the mitochondria to meet increased energy demands, sprouting of new spines on dendrites and new axon terminals, and growth of supporting glia and capillaries to increase the blood supply. All these changes require activation of the genome, which increases RNA production and protein synthesis with learning.

Learning also requires that discrimination be attended by habituation, which may include atrophy of axonal and dendritic branches and the programmed death of whole neurons (*apoptosis*). Too many synaptic connections are as bad as too few. During development the neurons sprout their branches profusely and make connections indiscriminately. Intentionality then prunes and shapes them. Other changes with learning, that are not yet known, must occur in respect to maintenance of long-term stability of cortex. Cumulative changes in excitatory or inhibitory synapses must be regulated in some way to avoid growing imbalances, and some kind of homeostatic feedback mechanisms must serve this need. We have no idea what the mechanisms might be or how to recognize them, but we surely will encounter them, if we have not already done so in LTP and related phenomena such as *kindling*. Among these mechanisms, which are "essential" for learning? It appears that all of them play essential roles, even if merely janitorial. New theory is needed to organize and interpret the data.

Artificial neural networks and digital computers

That need for theory appeared to have been met with emergence of neural networks as an alternative to conventional AI. Artificial neural networks have an architecture that more closely resembles that of cortex than computers. It consists of a large number of simple components that are connected in parallel. A component occupies a *node*, a point of convergence and divergence of connections to other components. Each node resembles a neuron in receiving input from many other nodes simultaneously in parallel, multiplying each input by a number assigned to each input line (a connection weight), adding the products from all lines, putting upper and lower bounds on the sum by use of bilateral saturation (the sigmoid curve), and distributing its output to many other nodes.

The interesting properties of artificial neural networks lie in their patterns of connection between nodes. In a feedforward net the nodes are arranged in layers with the flow of numbers going from an input layer through one or more *hidden layers* to an output layer, so that each node in a layer receives input only from a preceding layer and sends its output to a next layer. These nets are also known as *parallel distributed processors* and as *multi-layered perceptrons* (MLPs, Anderson and Rosenfeld 1988). The operations they perform are equivalent to those done by fuzzy logic (Kosko 1993) and in statistics by nonlinear regression for gradient descent. In a feedback net each node has input lines from a layer of sensors. It transmits output to other nodes in the same layer, and to an output layer that conveys the state of the middle, computational layer (Amari 1977; Hopfield 1982). The feedforward nets operate as filters on their inputs, whereas the feedback nets operate as dynamical systems, maintaining a set of basins each with an attractor for a class of input, and constructing a pattern of output that serves to represent the class to which each input belongs. The feedback net based in cooperativity is more closely related to brains.

Artificial neural networks learn in one of three ways. In learning with a *teacher* a desired output is stored as a representation in a matrix of numbers. A "training" pattern is put into the net, and the difference is calculated between the observed and desired outputs. Then each of the connection weights is changed by a small amount, and the input is repeated. If the difference decreases, the weights are changed in the same direction, and if it increases they are changed oppositely. By repeated steps the difference between observed and desired outputs is minimized. Then test inputs are given to find out how well the network generalizes. The algorithm is known in statistics as error minimization by gradient descent. The neural network eases the task of finding solutions by automating the weight adjustments.

In learning without a teacher the desired output is not set in advance as a pattern to be simulated, but instead the network is made to seek out clusters of similar patterns in a data set. This algorithm is known in statistics as cluster analysis. Typically the clustering is done blindly, because it isn't known in advance how many clusters there are. An arbitrary cutoff is made between one extreme of a single cluster that contains all the items to be classified and the other extreme of as many clusters as there are items. In the third type, learning by reinforcement, a network is not given patterns to match but, instead, examples of patterns to be classified, as behaviorists do in training animals. The model forms its own criteria by which to extract features of the examples, and if it works, it generalizes from training patterns to test inputs. A widely used rule governing use-dependent change is that the connection strength between two nodes increases, up to a limit, in proportion to the correlation of activity of the two nodes, a Hebbian synapse leading to a "cell-assembly" when repeated firing of one cell accompanies firing of other cells. The Hebbian synapse is the element by which theorists in artificial neural networks and experimentalists in long-term potentiation (LTP) can understand and interact with each other.

The theory of neural networks is inadequate for the simulation of biological learning, mainly because the nets don't interface with the world effectively. The greater part of the work of building and using artificial neural networks resides in preprocessing input data from sensors by hand to get them into a form a network can use. This is precisely the role of sensory cortex: to serve as the interface between the infinite environment and the interior of unitary brains. Moreover, the mathematics to handle large systems of nonlinear differential delay equations, which are dimensionally infinite, does not exist. Study of intentional brains and behaviors still provides the best hope for new theory.

Arousal, motivation, and reinforcement

In neurodynamics *learning* is defined as a directed change in an intentional structure that accompanies neuroactivity. There is no other way to distinguish it from the hypertrophy of a muscle under weight training. Study of an isolated ganglion or slab of cortex is a good way *not* to find it, because intentional change is nonlocal. This biological premise is reflected in the existence of global neurochemical systems in brains. In all vertebrates, from the simplest to the most advanced, the brain stem has collections of neurochemically specialized neurons that send their widely branched axons throughout the forebrain. The nuclei form a double chain resembling the architecture of invertebrate brains, but they are within the wall of the neural tube. Typically their axons form no synapses. The chemicals that they release diffuse widely through the neural tissue and bathe the neural populations in cortex and striatum.

The actions of brain stem chemicals are to modify the synaptic efficacy of conventional transmission. They are *neuromodulators* as distinct from *neurotransmitters*. In a double row of chemical factories each nucleus makes its own neuromodulator, which interacts with other neuromodulators in complex ways. The constancy of the architecture of this system throughout the phylogenetic tree and its global nature in each brain indicate that it appeared early in evolution to provide the essential neurochemical basis for intentionality. Evolutionary processes acted on the whole of the system, not on its parts, with modulation of behavior as the test of success. Therefore, as researchers unravel the components and test the modulators one at a time, the results of experiments are intriguing but baffling. It is unlikely that we will understand the neural basis for learning until a solid theory of intentionality has been derived, by which to assemble the components into a system and to orchestrate application of neuromodulators to isolated slabs.

The neuromodulators are grouped by their chemical structures into two main classes: the neuroamines (Kandel, Schwartz and Jessup 1993) and the neuropeptides (Gorman 1989). The lists are long and still growing, and the lesser classes such as amino acids and diffusible gases (for example, nitric oxide and carbon monoxide) clamor for attention. Each interacts in complex patterns with other modulators and transmitters in different contexts.

In associative learning by classical conditioning the unconditioned stimulus activates a nucleus in the brain stem, which releases norepinephrine throughout the forebrain. Blocking the action of norepinephrine prevents the olfactory bulb from forming new EEG patterns and prevents the animal from learning a CR to an odor CS (Gray, Freeman and Skinner 1986). Acetylcholine is thought to be involved in memory formation and retrieval, because the memory loss in Alzheimer's disease is associated with disappearance of the cholinergic system in the forebrain. It is suspected that dopamine and some endorphins are also involved in associative learning, but it is not known in what ways. Clearly each learning experience involves changes in many parts of the forebrain, and, owing to the broadcast of the modulators, it is likely that synapses in all parts are involved at all times. The global actions of the neuromodulators give good reason to propose that the neuropil of each hemisphere is the material basis of the unity of its intentional structure, so that our EEG recordings may manifest limited aspects of the global actualization of that structure in transient thoughts.

Isolation, unlearning, and pair bonding

The synaptic changes incurred during a single trial in a learning experience are widely dispersed and weak in comparison to the whole, but they constitute a finite step along a trajectory of the whole, particularly for eidetic images, which are stock in trade for scholarly research. From its beginning in and from the womb the path of an intentional structure is unique, growing from the genetically determined groundwork by the grasping for available sensory input from within and outside its own body. The same steps are not replicated in any other brain, and the structure that evolves is self-organized, with its own frames of reference and its unique patterns. We see this reflected in the EEG patterns that lack any recognizable geometry or relations between animals and are meaningless to us, except in the context of observing the behavior of the individual. It is in this sense that each brain is epistemologically solipsistic. Its knowledge comes from its cumulative constructions induced by the sensory milieu, which is also unique for each individual in the common environment, due to individual differences in physical and emotional perspective and expectancy.

The question arises: How can any two intentional structures change and converge so as to support cooperative behavior? The requirement is for a mechanism to bring about an extensive but selective and properly directed modification in both of the participating structures. The most pressing need arises during the process of reproduction by those species in which the altricial infants require prolonged care by their parents, particularly the mother. The requirement is for a mechanism in mammals (birds have different hormones) that depends on the release into brains of neuromodulators, which can selectively dissolve an intentional structure and open the way for new construction to meet environmental and developmental crises. The dissolution of an intentional structure is *unlearning*, the opposite of imprinting, which is the irreversible laying down of structure in the young. Dissolution opens the way for new learning, which imprinting does not allow.

The most promising candidate for this mechanism is the peptide oxytocin, which is released in mammals during parturition and lactation. The peripheral role of oxytocin (known clinically as *pitocin*) in promoting uterine contractions and the production of milk is well known. The behavioral effects have only recently come under intensive study (Pedersen et al. 1992). At the onset of lactation oxytocin floods through the hypothalamus in rats and modulates in complex ways the neural actions of a broad array of neurotransmitters and other neuromodulators (Arletti et al. 1992; Caldwell 1992). It is also released in the olfactory bulbs of sheep, particularly after the first litter (Kendrick et al. 1992), suggesting that the imprinting in the dam onto each new litter requires that prior imprints be dissolved, and not merely be written over, as the basis for reliable perceptual recognition, and that intentional changes take place in cortex, striatum and hypothalamus, such that the mother can recognize and nurse her young rather than attack and eat them. Changes in the functional connectivity of neurons in the breast area of the somatosensory cortex have been found with the onset and termination of lactation (Xerri et al. 1994), indicating that the perceptual systems of neocortex are also modified, perhaps most easily observed in cortical areas receiving sensory inputs from the soft tissues that surround the nipples, but probably taking place also in areas serving hearing and vision and, by hypothesis, throughout the forebrain.

Since the infant at birth has an unformed and nonfunctioning cortex , the bonding of mother and child can begin unilaterally. Formation of a bond between parents is more complex. Owing to their shared genetic and experiential determinants, the intentional structures of littermates are the most similar to each other, but the genetic advantages of outbreeding require that young adults mate with individuals other than siblings. Thereby they incur the task of adapting the structures of their behaviors (minds), so as to succeed in attempts at cooperation with those with whom they choose to mate. Experimental evidence is accruing that the same neurohumoral mechanism for unlearning may have been adapted for this role through evolution. Elevated levels of oxytocin have been found in the hypothalamus of male rats (Argiolas 1992) and in the circulation of human males at the time of ejaculation during sexual intercourse, presumably related to orgasm.

I speculate that release of oxytocin is part of an orchestration of neuromodulators upon sexual arousal prior to and during intercourse, and that it may be responsible for the profound changes that take place in behavior and belief structures, when children after puberty become parents and take responsibility for the next generation (Moore 1992; Insel 1992). Their systems of values and priorities must change abruptly. New behaviors are not instilled by neuropeptides. When the old behaviors are momentarily dissolved, the new behaviors must be learned by cooperative interaction expressed through complementary, goal-directed movements of persons having a shared environment and a common aim. Dancing is a good metaphor. A pair bond cannot form without the opportunity for extended interplay, because without re-education after dissolution, an intentional structure lapses by default to the old way. The male or female who solipsistically departs the morning after remains untouched by the night before. The deepest meaning of sexual experience lies not in pleasure, or even in reproduction, but in the opportunity it affords to surmount the solipsistic gulf, opening the door, so to speak, whether or not one undertakes the work to go through. It is the afterplay, not the foreplay, that counts in building trust.

Conversion and social bonding

Mammalian pair bonds between human parents are powerful but short-term, typically lasting only long enough to get the human infant firmly onto its feet, about 2 to 4 years according to the divorce statistics in Finland and the United States compiled by Helen Fisher (1992). Goldbart and Wallin (1994) describe the intentional structure as "an internal map" in each person. Prolonged learning is required for the "deep mapping" between mental structures formed in infancy and childhood, by which early attraction ripens to lasting attachment. "When we are puzzled about a current interaction in love and where to take it, we unconsciously turn inward, where our own idiosyncratic map provides an interpretation and a sense of direction" [p. 4]. The white heat of oxytocin may precede a warm glow of endorphins, and with practice, as the internal maps converge, the solipsistic gulf may seem to disappear under habitual routine.

But this is only half the story. The formation of a new unity in a pair bond draws a boundary around the couple, which serves to classify outsiders as potential threats. The appearance of aggressive behavior toward sexual rivals is accompanied by the release of another hypothalamic neuropeptide, vasopressin. Insel (1992) and his co-workers (Winslow et al. 1993) find that prairie voles release vasopressin when forming bonds between fathers and pups. The researchers find anatomical patterns of location of oxytocin in brains differing diametrically between an aggressive, monogamous species of voles and a polygamous species showing neither attachment nor conspecific aggression. The bonding of a pair and the attendant aggressiveness toward outsiders appear to be elaborated by differing neurochemicals.

Moreover, the boundaries of each paired self never disappear but become fractal in their indistinctness and fluid in a tug of war, like interdigitations of land and water in tidal estuaries. When the time comes to "fall out of love" or when love turns to hate, the aggression is focused not on a rival but instead onto the partner in a different kind of dance. Virtually nothing is scientifically known about chemistries of jealousy, shame, humiliation, hatred, and despair, because the experiences are too painful and too unique in each lifetime to be subject to experimental repetitions and controls. The present research climate does not even allow research on orgasm through brain imaging by PET or functional MRI, because subjects would have to lie motionless, while a research assistant masturbates them.

Social bonding, conversion, and brain washing

Pair bonds obviously cannot account for the formation of groups larger than the nuclear family, particularly same-sex groupings, except for harems maintained by alpha males at the expense of useless drones, rogues and outcasts. Groupings among humans and pack animals such as wolves and dogs may be explained by another form of dissolution of intentional structure, which takes place during various forms of socialization and religious or political conversion. Scientific study of the process began with an accident in the laboratory of Ivan Pavlov (1955), in which some of his trained dogs nearly drowned. The River Neva in St. Petersburg was flooded by an ice jam in a spring thaw, and the dogs were rescued as they swam with their noses at the tops of their cages in the cellar. Afterward, Pavlov's workers found that the dogs had completely forgotten their prior training and had to be re-trained.

Pavlov's systematic exploration of the phenomenon revealed the biological conditions sufficing to induce this unlearning. There were three stages, starting with heightened excitation by strenuous physical activity, sensory overload by continuous light or sound stimulation, and sleep deprivation. In the second stage the continuing assault on the senses produced *paradoxical inhibition*, which physiologists have observed when muscles are pushed beyond the peak of an inverted "U" input-output curve, such that stronger stimuli give weaker responses and vice versa. The third stage of *transmarginal inhibition* is marked by collapse, demoralization, and sometimes apparent coma. The process is facilitated by isolation from normal environments, chemical stresses by starvation, emetics, and purgatives, and assaults by intense emotions such as rage, shame, and fear. Pavlov was led to his approach and findings by the context of Russian neurophysiology, which had roots in 19th-century German science, but grew in isolation from the West after World War I in a distinctive direction. This may account for failure of Western physiologists to discover the phenomenon.

The breakdown by overload was mere prelude to re-education, by which new patterns of behavior were instilled. The most important step was release from the ordeal, by a change in venue, or by some small act of kindness offering surcease and hope, and a feeling of gratitude.

The usefulness of this technology was easily recognized by Pavlov's grant managers for their program of perfecting model Soviet citizens. The proof of its efficacy came in astonishing reversals of belief structure by individuals in the Moscow show trials of the 1930's. The procedures and their results have been vividly described by Arthur Koestler(1950) and George Orwell (1948, "Nineteen Eighty-four"). They have been stigmatized as "brainwashing".

Since declarative and procedural memories are part of an intentional structure, it is problematic how they are protected or recovered. In more extreme forms of transmarginal inhibition subjects lose motor coordination, clarity of speech, and recollection of events. Some suffer complete collapse and apparent loss of consciousness. Others exhibit catatonic trances, epileptiform motor discharges, glossolalia, and incoherent shrieking. However, there is a remarkable resiliency of intentional structure, in that, if the subjects return to their normal environments, within a few hours or days their pre-existing behaviors and recollections recur. Perhaps the process is an exaggerated form of sleep with dreaming, which has its own mechanisms enabling an intentional structure each night to re-assert its unity, as it sorts through and digests new material it has incorporated during the preceding day (Hobson 1994). If so, release of a master neuropeptide, oxytocin, may accompany somatostatin in sleep. Instead of being known as the hormone of orgasm, it should be called the hormone of "satisfaction" (Pedersen at al. 1992) or "gratitude". Its central role may give new substance to the concept of gratitude in philosophy.

Acknowledgements

This work was supported by grants from the National Institute of Mental Health MH 06686 and from the Office of Naval Research N00014-93-1-0938.

References

Alkon D (1992) Memory's Voice: Deciphering the Mind-Brain Code. New York: Harper.

Amari S (1977) Neural theory of association and concept formation. Biological Cybernetics 26: 175-185.

Anderson JA & Rosenfeld E (1988) Neurocomputing: Foundations of Research. Cambridge MA: MIT Press.

Argiolas A (1992) Oxytocin stimulation of penile erection. In CA Pedersen, JD Caldwell, GF Jirikowski & TR Insel (Eds.), Oxytocin in Maternal, Sexual, and Social Behaviors. Annals of the New York Academy of Sciences, Vol. 652: 194-211.

Arletti R, Benelli AA & Bertolini A (1992) Oxytocin involvement in male and female sexual behavior. In CA Pedersen, JD Caldwell, GF Jirikowski & TR Insel (Eds.), Oxytocin in Maternal, Sexual, and Social Behaviors. Annals of the New York Academy of Sciences, Vol. 652: 180-193.

Caldwell JD (1992) Central oxytocin and female sexual behavior. In CA Pedersen, JD Caldwell, GF Jirikowski & TR Insel (Eds.), Oxytocin in Maternal, Sexual, and Social Behaviors. Annals of the New York Academy of Sciences, Vol. 652: 166-179.

Fisher HE (1992) Anatomy of Love: The Natural History of Monogamy, Adultery, and Divorce. New York, Norton.

Freeman WJ (1992) Tutorial in Neurobiology. International Journal of Bifurcation and Chaos 2: 451-482.

Freeman WJ (1995) Societies of Brains. A Study in the Neuroscience of Love and Hate. Hillsdale NJ, Lawrence Erlbaum Associates.

Freud S (1895/1954) The project of a scientific psychology. In M Bonaparte, A Freud & E Kris (Eds.), The Origins of Psycho-Analysis (E Mosbacher & J Strachey, Trans.). New York: Basic Books.

Goldbart S & Wallin D (1994) Mapping the Terrain of the Heart: The Six Capacities That Guide the Journey of Love. New York: Addison-Wesley.

Gorman J (1989) The Man With No Endorphins, and Other Reflections on Science. New York: Penguin.

Gray CM, Freeman WJ & Skinner JE (1986) Chemical dependencies of learning in the rabbit olfactory bulb. Behavioral Neuroscience 100: 585-596.

Hobson JA (1994) The Chemistry of Conscious States: How the Brain Changes Its Mind. New York: Little, Brown.

Hopfield JJ (1982) Neuronal networks and physical systems with emergent collective computational abilities. Proceedings of the National Academy of Sciences 81: 3058-3092.

Insel TR (1992) Oxytocin: A neuropeptide for affiliation. Evidence from behavioral, receptor autoradiographic, and comparative studies. Psychoneuroendocrinology 17: 3-35.

Kandel ER, Schwartz JH & Jessel TM (1993) Principles of Neural Science (3rd ed.). New York: Elsevier.

Kendrick KM, Levy F & Keverne EB (1992) Changes in the sensory processing of olfactory signals induced by birth in sheep. Science 256: 833-836.

Kosko B (1993) Fuzzy Thinking. New York: Hyperion.

Moore FL (1992) Evolutionary precedents for behavioral actions of oxytocin and vasopressin. In CA Pedersen, JD Caldwell, GF Jirikowski & TR Insel (Eds.), Oxytocin in Maternal, Sexual, and Social Behaviors. Annals, New York Academy of Sciences, Vol. 652: 156-165.

Orwell G (1949) Nineteen Eighty-Four. New York: Harcourt Brace.

Pavlov IP (1955) Selected Works. (J Gibbons, Ed.; S Belsky, Trans.). Moscow: Foreign Languages Publishing House.

Pedersen CA, Caldwell JD, Jirikowski GF & Insel TR (Eds.) (1992) Oxytocin in Maternal, Sexual, and Social Behaviors. Annals of the New York Academy of Sciences, Vol. 652.

Winslow JT, Hastings N, Carter CS, Harbaugh CR & Insel TR (1993) A role for central vasopressin in pair bonding in monogamous prairie voles. Nature 365: 545-548.

Xerri C, Stern JM & Merzenich MM (1994) Alterations of the cortical representation of the rat ventrum induced by nursing behavior. Journal of Neuroscience 14: 1710-1721.

19.
Language as an Instrument for Self Reorganization

Alan Gregory

Language as an Instrument for Self-Reorganization:
A New and Unique System

Alan Gregory, President
Dynamic Careers, Inc.
P. O. Box 560040
Montverde, FL 34756-0040

With the following
PREFACE

by Karl H. Pribram

Sigmund Freud initiated the practice of using systematic verbal trains of associations to effect a change in a client's perceptions and behavior. Freud studied with Brentano and through this contact, he had probably become aware of the findings of the Würtzburg school of experimental psychology working under the direction of Külpe: though the answer to a question can be foreseen (provided the question is thoroughly understood), the process of thinking varies from individual to individual and is, therefore, unpredictable. Freud reconceptualized this finding into what he called the technique of "free association." The patient was to verbally associate with the aim of answering the question of what ails him/her. Thus, the flow of associations was not totally "free" but constrained by the problem to be solved. Freud was convinced that by tracking the course of the associations, the person's cognitive and deeper personality structure would become apparent: the process of thinking, being based on memory, would lay bare the motivations that guided behavior.

Carl Jung took this "flow of associations" procedure a step further. Jung presented words to his subjects and asked them for an association. Utilizing Ebbinghaus' technique of measuring reaction time, Jung found that when the associations were straight-forward, responses occurred quickly. When, however, a great number of connections were involved, especially if these were emotion-laden, reaction times were prolonged. Jung called the associations thus uncovered "complexes."

Americans are not as patient as Europeans and are less interested in the psyche of their clients. Rather, Americans prefer to devise techniques that more quickly result in changes of awareness and behavior. But Americans have also learned from their European forbearers not to be prescriptive: Prescription doesn't work in psychotherapy. If the technique of association is to be invoked, it should be a "guided association," but the guide must be the client himself. To be a guide, one needs to be well acquainted with the territory through which one travels.

Alan Gregory has combined these requirements in a set of techniques that allow a client to guide himself to what he/she considers to be a better way of life. He has done this in typically American fashion: we pride ourselves on obeying the rule of law rather than the rule of authoritarian power. Gregory asked: What are the rules of law for verbal association? The Dictionary and the Thesaurus. Gregory therefore uses their associative structures to help form the paths through which you, the client, will guide yourself to a better life. Gregory presents the outlines of his system in the following essay:

INTRODUCTION

When circumstances make life less than bearable, and finances and circumstances allow, various forms of intervention can be very useful. However, a suitable intervener and/or financing may not be available.

Many attempts have been made to provide personal growth systems to aid people in transcending their current state, or to provide an ongoing method of personal growth and self controlled intervention. Some methods that have been used include psychoanalysis, psychotherapy, EST, Transactional Analysis (TA), Transcendental Meditation (TM), and meditations of all kinds.

Our belief system, combined with what we do as a vocation (or avocation), tend to make up who we think we are (identity). Yet our search for identity never seems to end. In the same way, we search for why we are here in the first place. When certainty is added to the belief system, then learning (for all practical purposes) ceases. Add skepticism (destructive doubt), opinions (especially the type without reason), and complacency to the mix makes opening the door to the Self Reorganization process very difficult, even with professional intervention. The article which follows shows how the use of the certain concepts <u>will</u> open the door to the Self Reorganization process.

The objective is to lead you to consider some alternate methods of gaining control over yourself and to show you how to be your own intervener. The methods outlined here are based on the premise that in order to know who you are and who you are not, you must first know what everyone in the world is made up of and how everyone else in the world ticks. This can be accomplished through the system I call "The Tool Boxes of Learning". Learning includes but is not limited to Self Reorganization and Personal Development using the American version of the English language.

We all know all of the answers. The problem is that we do not know what questions to ask to get at those answers. I hope that what you are about to read will generate a myriad of questions for you to answer.

We all have the ability to see and understand how things work. This article outlines for you my system of understanding *how people function internally, and how things work for them within the environment where they choose to live.*

Patents: in process of being applied for

My Mentor in the work presented here (Dr. Karl Pribram) encouraged me for ten years to write about my work. In examining the reasons for my reluctance to do so, I discovered that there was a basis in reality. The system that I will to explain and explicate *must be personally experienced,* not just read. People *must experience* the language system or they will not understand it.. Two problems frequently appear: First, it is difficult for people to listen. Although they hear my words, they are listening to the words that my words are producing in their heads. Second, and most important, they are trying to validate, invalidate or apply the system from the body of knowledge that they bring to the experience. What I present to you is original (in it's use) and practical.

REQUISITES FOR LEARNING AND SELF-REORGANIZATION

The following steps present a new and unique system of maps for using language for Self Development and Self Reorganization. Under-standing of this system can only occur when you actually experience a change of attitude resulting in some form of self development reorganization for yourself. The system is simple and anyone can learn it. It is based on the American English language. There is no need to coin a new word or system of words --it uses words already coined.

STEP ONE: You must set aside (momentarily) any temptation you may have to validate or invalidate the new, even strange concepts being presented. This will be necessary as there is a strong possibility that you may not have many frames of reference to be able to qualify or disqualify this system. If you are a skeptic, or if you should become skeptical about anything you are about to read, use your skill of skepticism against your own skeptical view. This is one of the best tools you can use to stay open minded. If not, you will negate the possibility of understanding this presentation.

STEP TWO: There must be an acceptance that you are who you are because of what you do not know. You become different than who you are by learning what you do not now know.

STEP THREE: Adjustment of the belief system. All the knowledge you have accumulated is true and untrue at the same time. You must accept that the truth you know is true unto itself, but untrue in the sense of its incompleteness. A major path to the process of learning and Self Reorganization is *disillusionment.*

STEP FOUR: Disillusionment. When questioned as to their feelings about being disillusioned, people commonly answer that they prefer to avoid being disillusioned. On the other hand, the same people when asked if they would prefer knowing the truth (even if painful) to not knowing the truth, will usually answer without hesitation in the affirmative. Would you? Yet, I could not disillusion you or anyone else if there had not been an illusion in the first place. If the illusion is a not reality or is a non truth, then the person seeking learning and Self Reorganization must learn to welcome disillusionment as their greatest tool for Self Reorganization.

STEP FIVE: Listening. Most people when asked if everyone listens 100% of the time, will answer (in various forms of incredulousness), that the idea is absurd. Yet when someone says, "You aren't listening", the answer is usually "Yes, I am." The question should be posed, Listening to whom? To the speaker? Or to yourself? But you are listening, either to yourself or to the other person. A clear mind (absent of thoughts that are associatively generated as we hear or read the material(s) being presented) must be achieved. Learn the discipline of not allowing thoughts to distract, validate/invalidate, or disagree with anything new being presented.

STEP SIX: Use the tool boxes of our language as every day learning tools, not just on rare occasions. To overcome the old habit of being who you are, and becoming who you are not, this new habit is essential.

STEP SEVEN: You must adjust your (very common) belief that when something is depicted as "Probably True", it is "almost true" but not as desirable as "true" itself. What is really true is that when something is "probably true", what is known is exactly true, but probably incomplete. Therefore, PROBABLY has the potential (when used in front of every concept of definite connotation in both thought and speech) to become the strongest concept key to self reorganization and self development. Every belief known to be not completely true (yet not in error on its own) is a pointer or link to what is incomplete and therefore not known; otherwise, the deception of certainty closes the mind to the reality that is potentially available in the data base of human knowledge and experience.

LEARNING THEORY QUESTIONS ABOUT THIS PRESCRIBED ORDER TO LEARNING ABOUT THIS SELF REORGANIZATION PROCESS

QUESTIONS: Is it possible that the reason you stop learning is because at some point in your learning process you begin to do what you have learned? Is it possible that (as contented) doing is performed in the exact opposite Sequence to that of the learning sequence, so that formal learning ceases?

ANSWER 1. THE LEARNING SEQUENCE: a) You recognize that something exists. b) You recognize how much you do not know about it. c) You learn what it is made up of (it's SUBSTANCE). d) You learn what it is not made up of. e) You determine it's essential nature (INTRINSIC) as it relates to you, and what it is worth.

ANSWER 2. THE DOING SEQUENCE: e) You desire it so it is worth relating to intrinsically. d) You can avoid those things that would interfere. c) You know what it is made up of and can use that to make it happen. b) You have learned what you did not know about it. a) You (choose to or can) make it exist for you.

QUESTIONS: Is it possible that there is a scientific order to language and that this order holds the answers to almost any human problem, while also being able to explain the conditions, state, and/or any relationships presented to it? Is it possible that language could be used in it's reverse order as an instrument for Self Reorganization and Personal Development? Is it possible that the logistical result of the Freudian experimental model could be tapped into directly in each moment in time, thereby reducing the psychological, analytical, and research processes to a minimum? Is it possible that wisdom, sanity, reality, and effective mind power are all available in simple form for personal guidance? Is it possible that there is an order of the senses, the intellect, behavior, and

456

emotions, that can be used to show the internal biological mapping that could be used for Self Reorganization and attitudinal corrections?

THE ANSWER TO ALL OF THESE QUESTIONS IS YES. The answers can be found in one or combinations of the Tool Boxes.

THE THREE MAJOR "UNCOVERIES"

An "UNCOVERY" was found within WEBSTER'S THIRD NEW INTERNATIONAL DICTIONARY - *G & C Merriam Co.* This tool is very useful in determining intuitively what learned (and missing or misunderstood) concepts cause people to say what they say and to think what they think. It is also valuable in learning about charged words and the method of discharging them; levels of understandings about the effect of the hidden implications, subtleties, causation, and nuances (to name a few); how you can intuitively discover the programmed differences between you, the people you ordinarily relate with, and the world at large; and how you can adjust and exchange ideas (Self Reorganization) by accessing the simpler terms in any hierarchial sequence of concepts.

Another "UNCOVERY" was made within the contents of Crowell's Fourth INTERNATIONAL ROGET'S THESAURUS, edited by Professor Robert L. Chapman, of Drew University. The rights to publish this thesaurus have been purchased by Harper & Row. This tool is a representation of the order of how things work in our society in the exact order it occurs (or where applicable) or does not occur. This paradigm of the collective interpersonal knowledge of man is very useful for comparing individual abstracted reality with our societies reality. This tool becomes especially powerful when concepts and attitudes achieved using Tool Box 1 are located in this Tool Box exponentially manifesting the self realization and self actualization Self Reorganizing process.

NOTE. Dr. Chapman has completed and had published the Fifth edition. Time has not allowed the study and research necessary to integrate the Fifth edition into the system. It appears that Dr. Chapman has changed Roget's original catholic approach to language to his own view as centered around the spoken word. The implications are not yet clear except for the possibility that there can be a separate central theme for major natural phenomenon. Although it appears that the system for use of the Fourth Edition being presented should be transferable to the Fifth, that conclusion is still to be reached. Purchases of Fourth editions must therefore be made at used book stores.

The third "UNCOVERY" was the wisdom found in the FUNK AND WAGNALL'S HAND BOOK OF SYNONYMS/ANTONYMS & PREPOSITIONS - *Harper & Row* - Fernald.

This tool fits into the system in a number of ways and is almost instantaneously useful for bringing balance, calmness, and mental continuity to a troubled person. It is also useful for finding redirection (and distraction) from what we want to reduce or eliminate or want more of, for connecting the first two tool boxes to each other, and for fine tuning the developments that are achieved in using those tool boxes. The system consists of a combination and integration of three basic maps (uncoveries or tool boxes):

TOOL BOX I and how it works.

WEBSTER'S THIRD NEW INTERNATIONAL DICTIONARY, ITS USE AND ITS APPLICATION.

The first system (or tool box) is based on the application of WEBSTER'S THIRD NEW INTERNATIONAL DICTIONARY - G & C Merriam Company. It is the ultimate multiple choice question and answer document. Unfortunately it is used in this society in a very limited way, i.e. to define words. Beyond that, there exists a system that can be intuitively accessed to determine what makes people say what they say, think what they think, and act the way they act. This component is the primary Tool Box that can be used for the requisite process of discharging all "charged" words. It is common for people to be "charged" on many concepts in the language. "A "charged" word is one for which you consciously and consistently use only one of all its definitions, to the exclusion of its other definitions. The most significant limiting language problem that I have encountered is the "charged word". A word is "charged" when a person's internal system is not aware of all or at least many of the definitions of a particular word. This causes a resistance when that word is used and can seriously interrupt communications. The degree to which a person's system remains charged is the degree to which a human being in this society tends to function out of order. It is the cause of most people difficulties, it interferes with experiencing a superior quality of life, and it interferes with personal development and Self Reorganization.

CHARGED CONCEPT (WORD) THEORY:

Hypothesis - sample charged concepts - missing links of the development process. *Definitions and applications of sample charged words.* What would you answer if you were asked if you liked or desired being: **Manipulated, Controlled, Stupid, Asked to give up the right to disagree, Ordinary (the same as everyone else rather than different than everyone else), Used, Romantic, 100% wrong or at fault, having liabilities**. It is conceivable that you would find these accusations make you uncomfortable when in actuality the worthwhileness hidden behind the "charge" you may have about the concept could be extremely useful to personal growth and development. "Uncharging" **each of the above possibly charged ideas, and the process of "uncharging" other charged concepts, is the way you begin**. You deal with the "uncharging" process by accessing WEBSTER'S THIRD and defining these words for yourself. The connotation you have attached to these concepts probably will be changed by the many other definitions you will find. WEBSTER'S THIRD INTERNATIONAL is especially useful as it (more than any other dictionary) puts definitions in the most commonly order. This will give you the chance to compare the definition you are "charged" with to that sequence of definitions.

SOME "UNCHARGING EXAMPLES: **manipulated**. The connotation is a poor one, yet if "manipulation" were to be eliminated, the world would come to a standstill. On the other hand, the kind of manipulation that would be undesirable is rarely experienced. This entire document is designed as a manipulation. Allow the system to manipulate you into a different attitude, and the Self Reorganization processes. **controlled**. This word means to many people the loss of a certain amount of liberty. In essence, it is a privilege to find someone worthy and to whom you can turn over control. The loss of never finding a superior person to whom control can be given is immeasurable. Also what is generally not understood is that a person must have control over themselves before they can give up control to another. Anyone capable of having enough control to give themselves over to someone has the same amount of control available to take that control back at any time. Harvard University offers a course in Mentoring, including the skills of being a mentee. Their theme is that "No One Can Make It Without A Mentor". **Told that you are being, or acting, stupid**. Contrary to the connotation, "stupid" is honorable because it is a relative term. Surround yourself with inferior people and they will appear stupid to you. Be the one who surrounds yourself with superior people -- the more superior they are the more stupid you will become. The dictionary does not define stupid as being dumb, but rather as being a slow learner. **Asked to give up your right to disagree**. This is one of the more controversial concepts. It is generally felt that without the right to disagree, how could one prove one's point to others? Yet disagreement is one of the more hostile concepts in the language. The ROGET'S THESAURUS has "inappropriate" as one of disagreement's complexes. When you disagree you are saying that the person you are hearing does not know what they are talking about. What was just said is not what you would have said. When you give up the right to disagree (in spirit, mind, and body) you can gain more in Self Reorganization terms than any other way. The next time you are tempted to disagree, instead ask the person what is making them make that statement (this is especially effective if your question includes the exact words they used). Become interested in how your fellow man processes information and why. You will amazed what you will learn. **Asked to become ordinary** (the same as everyone else) rather than being special and different usually brings protests. Yet the ultimate result of being special and different is to be unique or one of kind -- a very lonely place to be. The salesman who has a thousand faces has learned to be the same as everyone else. The person trying to be different is trying to be minus different. The person who learns the process of being ordinary (in order) like every one else is the person who is (or is becoming) plus different. Accumulate the processes of how other people think, thereby developing more choices and therefore becoming more liberated. It is better to be the observed than the observer. This component is also the essential tool for developing the String process that can be found in the section describing *Strings*. used is one of the better concepts in the language. If you will look at the dictionary definitions, you will not find a single definition that you will feel is inappropriate. You might be confusing it (the charged word) with misused, or abused. **Romantic** is a word that is supposed to make the world go around. In our society it is usually associated with love and entanglements. What is not generally seen is that there are two kinds of love, the romantic (Roman) and the good will (Greek). Most relationships formed on a romantic base will dissolve if not propped up by other interests. However, love in the good will sense (friendship) that is supported by other interests will usually stand the test of time. It is the basis for redeeming relationships. **100% wrong or at fault** usually connotates something wrong or undesirable. This misconception can make improvement improbable since a fault is something that is missing and needs to be filled in rather than something wrong that needs to be fixed or replaced. **Pointing out your liabilities** connotates something negative within the system. Yet according to the ROGET'S THESAURUS (section 175) a person's liabilities are their greatest asset. No asset can be achieved without it first having been a liability. All children are liable -- liable to turn their lack of knowledge into an ongoing learning process. In addition ROGET'S THESAURUS shows that liabilities are aptitudes and are the basis for all getting, gaining, and acquiring.

HIDDEN WITHIN THE DICTIONARY

The secret of consequential thinking, speech, and actions.

The process of consequences (con-sequence) can anticipate what people will say or do. once you know the causal sequence of thoughts, speech, or actions, the consequence of that sequence can be found. This can determine the options for what you (or people) will think, speak or how you (or people) will act. The skilled person in the use of the consequence process can develop and anticipate consequential happenings. This is accomplished by using the dictionary in reverse. Using the word you have defined, you can now search for the word that this word is a definition of. Those words are the consequences of any word you have defined. I use the complete unabridged OXFORD and MICROSOFT'S (DOS version) of BOOKSHELF (both on CD-ROM) to browse the language in all of it's directions. In theory, the use of the dictionary under the rules of the system can depict the logical order of how things might be arranged in an individual's memory banks (much like programming). This system has the capability to compare different people's cognitive processes permitting Self Reorganization via a process of logical change and exchange of sequences of words I call "Strings". This process can be used as a supplement to all present forms of intervention.

How *String* theory phase I works.

The WEBSTER'S UNABRIDGED DICTIONARY is a hierarchical reflection of how concepts (in their electrical value state) are arranged in our brain computer. The primary internal uses include our programming (the logistical memory created by our learning and experiences), verbal and written communications, mental facility for ideation, associative thought (thinking), deductive and inductive reasoning, and identifying the mental pictures and images we can mentally visualize with the "Mind's Eye". Although the dictionary is arranged in alphabetical order, there exists another order which I call the order of Strings. The connective levels are as follows: a) The concept with which we start is usually selected because we want to know what a person has in mind as close to the core beginnings as we can practically reach. Consider that each level below (b) becomes further removed as we progress down the intuitive String. This process may be likened to the way the subconscious is regarded in formal therapy. The difference here is that one is dealing with the result (effect\affect) of those experiences. In theory one is taping into the precise experiential caused complexes.

The first word that you pick is a complex of the level(s) below, the word that is intuitively picked out of the multiple choice of the definitions is it's simplex, b) explanation/explication, and continues further down the string to c) the implication, d) the connotation/denotation e) the subtlety, f) it's causal nature g) probability (possible/improbable) aspect, h) the figment, and i) the nuance, j) the finite k) the genesis (genius), l) and unsounded levels. Many more intermediary levels can be found as differences in degree.

The letters used above are important in understanding the relationships and the creation of the String order. For example, select your first word to define. Consider this first word to be the President of a company as in a company organization chart. Read the definitions of that word slowly and carefully. Try not to think about it, just sense which word makes the word you are defining true for you. Consider each of the important definitions to be Vice Presidents (the b level) laid out horizontally under the President. Pick one or more of these Vice Presidents as your first step in creating a *String*. Then start the process over again. Take the definition you have picked (or select one if you have picked more than one), and define that word in the same way you did the (a) first definitions and call that (the c level) an Assistant Vice President. Note: When selecting the c) level you must keep in mind a), as c) must intuitively associate with a) as well as with b). Keeping the levels above in mind while you develop a string is essential to the correct development of the string.

Restated, b) is the simplex of the complex a). c) is the simplex of the complex b) while b) remains the simplex of the complex a).

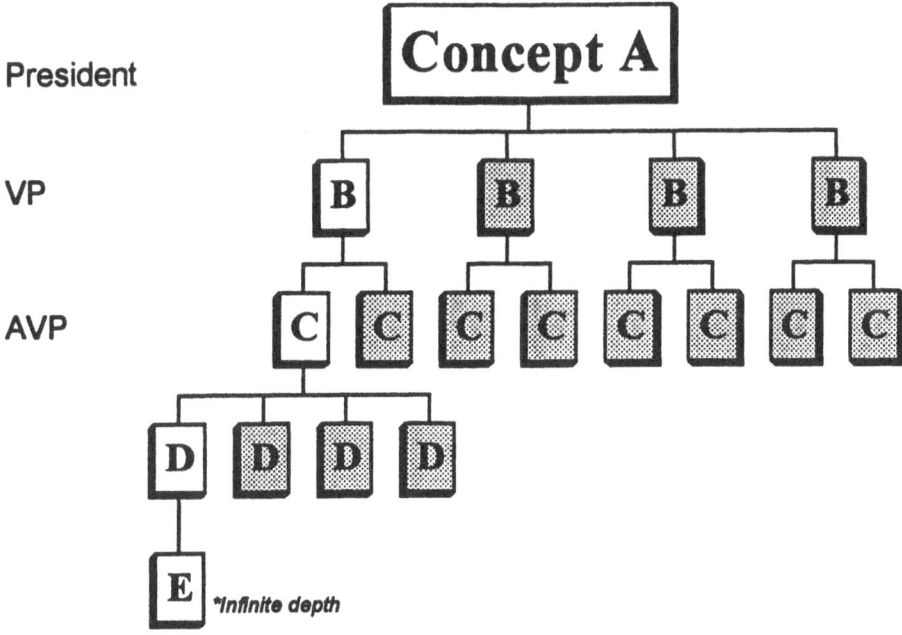

Figure 1. An organization chart type rendition of *String* development.

You can continue the process defining the AVP <u>ad infinitum</u>. I can visualize no end to the process except perhaps that the word picked as President would eventually become a definition of another word at the consequential level of a).

In simple terms, it is the definitions (or combination of definitions) of a word that make it true. Therefore if b) makes a) true, and c) makes b) true, and d) makes c) true, then d) is in indirectly making b) and a) true as well as directly making c) true.

461

Applying this sequence to the list above, b) is the explanation/explication of a). c) is the implication of a). d) is the connotation/denotation of a). e) is the subtlety of a). f) is causal to a). g) is the probability (possible/improbable) of a). h) is a figment of a). i) is a nuance of a). j) is a finite of a). k) is the geneses (potential genius level), l) is the unsounded level of a), the place where vibrations are forming. This same process can be repeated when you start with c) which is the simplex of b) and therefore the explanation/explication of b) where c) before was the implication of a).

How String theory phase II works.

The use of intuition in the creation of comparative *Strings*. The key to all three tool boxes is our innate ability to classify, i.e., to look at any situation and select one word that describes that situation. Using the same word would probably not describe the condition you have in mind to someone else, but for the purposes of using Language as a Tool For Self Reorganization it does very well. Although I have researched many dictionaries, I have found WEBSTER'S THIRD INTERNATIONAL published by G. & C. Merriam to be the most revealing. The only drawback in using the string process is its slowness, but compared to the amount of time other forms of intervention consume, this process can outdistance those as if at the speed of light. Caution: Because of the varying states of mind that people encounter, Strings developed on a particular day and in a particular mood may vary, but no matter how they vary each of the *Strings* can be very useful and productive.

The ability to perfect the use of *String* process varies from person to person primarily because of a) being busy in mind (occupied with thought); or b) having a mind process that brings out of memory terms and experiences associated with the words being explored. In the arena of Self-Reorganization and Self-Development there is no substitute for a quiet mind (thoughtless and empty headed). There are many mind disciplines (Silva Mind Control, Zen, The School of Practical Philosophy, etc.) for helping one to achieve quiet (quiescent) mind. Caution: Spend some quiet time selecting your first word to String out. It is important that you pick a quiet place with the least possibility that you will be either distracted or disturbed. Prepare yourself with adequate pen and paper. A computer will not do as it becomes distractive. Some suggestions are a word that describes a trait that you would like to have more of; that you have too much of; is an opinion about someone you either like or dislike; is a word that you like or do not like having done to you; or a word you think describes you to a "T". It is best to select these words while you are in the contemplative mode. Since you are just starting the process you may need to discard one track and start another. Open your dictionary and look up your first word. Write this word at the top of a ruled sheet of paper. It is the beginning of your *String*. Read very, very, slowly, one word at a time, carefully looking for the word that makes the word you have chosen true for you. Selecting more than one is OK, and can be used for a second track if the track you are pursuing runs out. If you have selected more than one, review the word that seems to make your a word truest. Place this word under the first word, leaving a line or two of space for notes. You now have the second word of your string. Define your second word in the same way you did your first word. Continue the process until you have fifteen words in the string, although ten can also be a good place to start. Caution: If you seem to run into a dead end, just go back to the place where you may have chosen more than one word and start over from that point down. The theory is that the word on the bottom of the *String* is the simplex or simplest of the terms. You arrived at that word deductively, but the direction in actual internal use is inductive (which will be from the last word in the string becoming the first and traveling inductively to the end, which was your beginning).

Now you begin the study and contemplative part of the process. Try to see the effect that your last word (not the first) has upon the second, the effect the second word has upon the first, etc. After you have studied the sequences for a period of time, put the *String* away. You can start on another or come back to the first one at a later time. Please do not expect results of any kind. Do not look for realizations, or judge your progress in this process for that is just what it is, a process. In order for something to be seen as a process, anticipating a result will surely block the procedure. Practice doing *Strings* without any concern. Let the purpose of your practice be "for its own sake". In other words, practice the process with only one purpose in mind: To get better at practicing the process.

How (comparative) *String* theory phase III works.

In order to use the comparative *String* process, you will need another person with whom to work. Although the *String* process can be effectively utilized when working alone, most of us need to make comparisons in order to see relationships that can be used to internalize adjustments. Get someone who is interested in Personal Development (preferably someone who might also be interested in Self Reorganization, or at least in experiencing the process). First get them to practice what you have been doing alone for at least enough time for them to become comfortable with the process. Then, using the same starting word a), each of you should (independently) read the definitions of a), and carefully (intuitively) select what will become b) for each of you. (NOTE: I personally work better when someone reads the definitions aloud.) Compare the word(s) your associate selected with the word(s) you selected. If you cannot see the differences, take the next step by defining both words again. The definitions should make it obvious as to why one is a more useful concept. If the definitions do not seem to make a difference, then you may each have to go back and find an alternate choice for you b), (this can occur at any of the levels as well as the b) level). Take the next step(s) by both you and your associate picking your c) word which you will use to define b) etc.. down the *String*. Note: When defining words down the string, try to keep the a) word in mind (as well as all of the concepts that you have developed to that point). The more of them that you keep in mind, the more accurate your intuitive selections will be as you begin to understand more about how and why you manifest your a) word. Try in the beginning to avoid picking definitions that you might like; instead, search for intuitive truth, the true word that makes you say or think a). There will be plenty of time for you to come back and make adjustments with the words that you like or see would have greater impact on a). This process can be expanded to as many people working on the same concept as you can manage to get interested in the process. The more times that you compare different people's ways of defining, the more you will fill in the spaces in your own cognitive map. The more you fill in the spaces in your own cognitive map, the more accurate and the more quickly you will be able to both think and communicate effectively, and the more you will exponentially gain the level and skill of Self Reorganization and Personal development achievable.

TOOL BOX II and how it works.

ROGET'S INTERNATIONAL THESAURUS

The second system or Tool Box is based on the order to be found in the ROGET'S INTERNATIONAL THESAURUS Fourth Edition - *Crowell* - edited by Robert L. Chapman. Do not confuse dictionary type editions and abstracts of this version. Those are generally used for the limited purpose of finding associated words and even antonyms and synonyms, but all of the important concepts have been deleted because no one has found a use for them, until now. There is a Fifth International edition but the order has changed so dramatically that updating this system to encompass that order is a task for the future. Although the information in the Fourth Edition is some 25 years old, things having to do with how people process information have not changed that much.

Hypothesis - the reality process at any given moment. The following classes abstracted from the ROGET'S THESAURUS are listed below for your consideration of the order and it's potential use. NOTE: Each concept in the language will appear in more than one class. As a term repeats itself within these different classes, so do definitions affect the term. Here is an example using the word PRESSURE, showing how it plays out through the system:

Class	As found in the ROGET'S	As I see it's contents
ONE	ABSTRACT RELATIONS	Anatomy and attributes of mind.
TWO	SPACE	Spatial relations & occurrences found in the environment.
THREE	PHYSICS	All of the natural sciences.
FOUR	MATTER	All of the physical things in life especially our functions as human beings.
FIVE	SENSATIONS	The Combined Senses.
SIX	INTELLECT	What mind does and how it works.
SEVEN	VOLITION	All human behavior.
EIGHT	AFFECTIONS	Feelings, Emotions, Morals, Moral Obligations, Moral Sentiments, Moral Conditions, Moral Practice, Moral Observance (the Law), & Religion.

Examples of Connections to Learning and Reorganization using aspects of PRESSURE by Classes as found in the ROGET'S THESAURUS:

000.00 =========	CLASS ONE	ABSTRACT RELATIONS
172.01 INFLUENCE		
179.00 =========	CLASS TWO	SPACE
198.02 SQUEEZING		
228.31 AVIATION		
283.02 THRUST		
325.00 =========	CLASS THREE	PHYSICS

464

352.07 BURDEN		
375.00 ==========	CLASS FOUR	MATTER
422.00 ==========	CLASS FIVE	SENSATION
425.02 TOUCHING		
466.00 ==========	CLASS SIX	INTELLECT
621.00 ==========	CLASS SEVEN	VOLITION
639.04 NECESSITY		
648.05 COMPEL		
648.05 URGING		
648.06 DRIVE		
648.14 URGE		
672.04 URGENCY		
729.01 ADVERSITY		
730.04 AUTHORITY		
755.03 INSISTENCE		
756.03 COERCION		
774.03 ENTREATY		
774.12 IMPORTUNE		
855.00 ==========	CLASS EIGHT	AFFECTIONS
859.03 TENSION		

CONTENTION: If you find these and the following items in the ROGET'S THESAURUS and WEBSTER'S DICTIONARY in order, then might it not follow that all other concepts from one end to the other. and back again, are also in order? Therefore they can be used as a guide for the purpose of personal Self Reorganization.

 The Fourth edition of ROGET'S THESAURUS depicts the order and the way things appear and occur in this society in the exact order that they work and the exact order that you integrate with that world. The order is displayed in hierarchical form and as it logistically appears and is used by mind and body to affect perception, thinking, behavior (both involuntary and voluntary), feelings and emotions. The order can also depict the probability of how all things work in this society and how you can choose to fit into them. The ROGET'S THESAURUS is also used as a mapping tool for actualizing Self Reorganization. Some examples of order abstracted from ROGET'S FOURTH INTERNATIONAL THESAURUS. Warning: If in examining this system you attempt to apply what you are reading (induction) instead of just understanding the process (deduction) your internal thinking mechanism (mind's eye) will be blinded by the confusion that ensues. If you catch yourself picturing using anything that you have read or will be reading, do your best to pause, clear your mind, and start again. Note: One of the formulas I have found interesting is what I call, THE CIRCLING FORMULA. Start with a simplex (and going forward in the book) the words that follow will in some way be a consequence, a complex, effect or affect, or an event of that simplex. Conversely, the complex, consequence, effect, or event, will work backwards to it's simplex as shown in the following example.

Study the sequential list of words under each of the headings on the next page (from a-h). It is important that you read and reread the listings until you can recognize the logical order. They appear here in the exact order that appear in the ROGET'S THESAURUS. You may

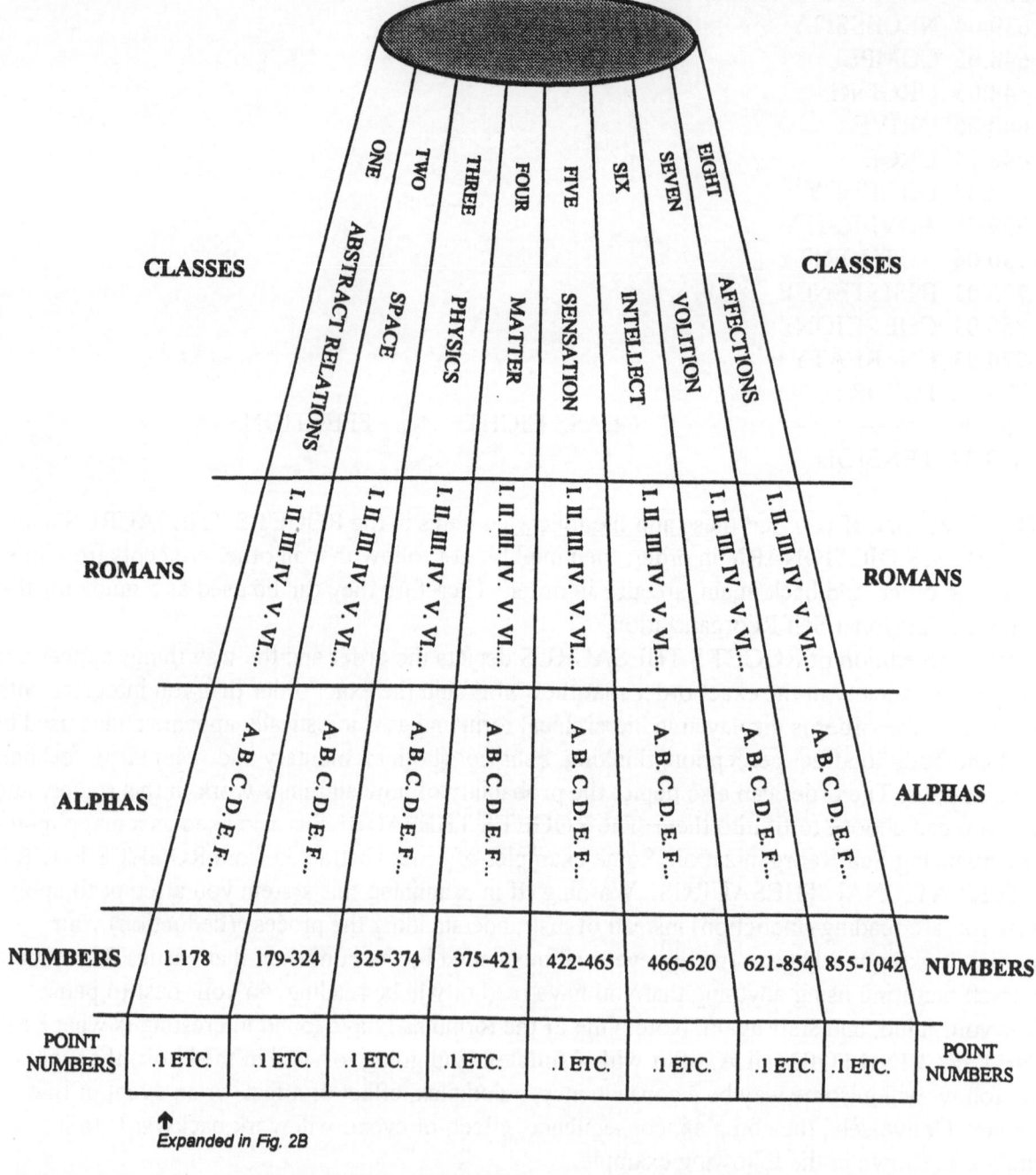

Figure 2a. A graphic representation of the ROGET' THESAURUS hierarchical tree showing levels of language for Eight Classes, divided by Roman numbers, further divided by Alpha letters, and then divided into the 1042 numbers depicting the common usage level. There is another level of divisions of point (.) numbers that establish the working language levels.

466

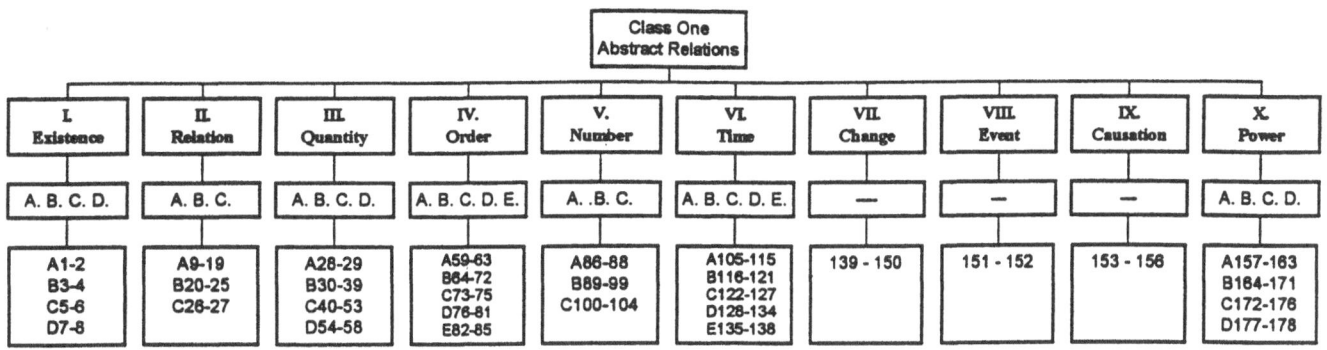

Figure 2b. A graphical representation of how Class ONE divides into Class, Roman, Alpha, Number and Point numbers. All other classes follow the same hierarchical tree structure. NOTE: A major importance is that words directly control the divisions that follow.

find it useful to set this section aside and come back to it over an over again until you begin to understand the order that this Tool Box offers you. If you can accept the order to be found in these selected examples., then you might see that all things in the ROGET'S THESAURUS (and in the other Tool Boxes as well) are in a similar useful order, from one end of our reality to the other and back again. Just the study of these processes can go a long way towards understanding the discipline of Self Reorganization. If a particular sequence does not make sense to you, go on to the next one and come back to it at another time. It may take practice for you to give up attachment to the order that you live with every day. You can learn to see past the veil of your personal reality within which you have become comfortable. NOTE: Although the concepts are in order, they do not necessarily mean "more'. They could be an option for less, or even to stop, or even to go in a different direction. The important thing to remember is that each concept that follows another concept is an option. If a concept is not useable at a given moment it can be skipped over. What is important and even essential is that all concepts are options (either in whole, in part, or degree). NOTE: Although ROGET'S THESAURUS (and Thesauruses in general) are usually associated with the antonym/synonym process, the organization is designed to bring associated words together in a simplex to complex generating process.

ORIENTATION TO THE ORGANIZATION OF ORDER

The ROGET"S THESAURUS begins at the top of the hierarchy with the eight classes outlined above. The following is a display of some of the sequences of Class ONE -ABSTRACT RELATIONS. Class ONE contains X Roman divisions, the X Roman divisions contain 26 Alpha letter divisions, and the Alpha letters have 178 number divisions. I consider Class ONE - ABSTRACT RELATIONS to be the Anatomy of the INTELLECT found in Class SIX. Here are laid out all of the attributes and capabilities in their abstracted form. The abstracted form is the form the attribute takes as a quality not yet put into use where these qualities will be manifested and quantified. NOTE: I consider that the reason 1) Existence is the first concept is that no other work or concept could exist without this as it's simplex. 2) Nonexistence follows Existence (where it might seem that it should come before existence) because we would not be aware of a nonexistent concept without some understanding of its existence. etc.

a) The instruction set (the sequence needed to enter any computer). It appears as the first nine major concepts in the ROGET'S THESAURUS as **the learning modules**: 1) Existence, 2) Nonexistence, 3) Substantiality, 4) Unsubstantiality, 5) Intrinsicality, 6) Extrinsicality, 7) State, 8) Circumstance, are all requisites for forming or missing 9) Relations. Being relative (relational) thinking animals, this is the originating process for forming relationships. The cognitive system is based on relative degrees of relationships. After one has established relationships one shifts from learning to using those relationships. More information is provided later in this article concerning the shift from the learning mode to the doing mode, and how this gradually inhibits, interferes with, and eventually renders formal learning ineffectual, leaving the student to learn by the questionable "experience" method.

b) Intermediary supportive concepts that build relations. 9) Relations, 10 Unrelatedness, 11) Relationships by Blood, 12) Relationships by Marriage, 13) Correlation, 14) Identity, 15) Contrariety, 16) Difference, 17) Uniformity, 18) NonUniformity, 19) Multiformity, 20) Similarity, 21) Dissimilarity, 22) Imitation, 23) Nonimitation (Original), 24) Copy, 25) Model. Although 9) appears in a) and also appears in b) occurs because in a) it is the end (complex) of a sequence, while in b) it is the beginning (simplex) of a sequence. Here in b) all the types of relationships develop until they either come into agreement with the environment or into disagreement and/or vacillate back and forth. In this case 26) agreement will actually be the complex while in c) it will be the simplex.

c) Progressive Relations (Doing). 26) Agreement, 27) Disagreement, 28) Quantity, 29) Degree, 30) Equality, 31) Inequality, 32) Mean, 33) Compensation, 34) Greatness, 35) Smallness, 36) Superiority, 37) Inferiority, 38) Increase, 39) Decrease, 40) Addition, 41) Adjunct, 42) Subtraction, 43) Remainder, 44) Mixture, 45) Simplicity, 46) Complexity, 47) Joining, 48) Analysis, 49) Separation, 50) Cohesion, 51) Noncohision, 52) Combination, 53) Disintegration 54) Whole, 55) Part, 56) Completeness, 57) Incompleteness, 58) Composition, 59) Order, 60) Arrangement, 61) Classification, 62) Disorder, 63) Disarrangement, 64) Precedence, 65) Sequence, 66) Precursor, 67) Sequel, 68) Beginning, 69) Middle, 70) End, 71) Continuity, 72 Discontinuity. The combinations in c) are prerequisite to the INTELLECTS (Class SIX) functioning as the SENSATIONS (Class FIVE), filter environmental input/output data. Continuity is essential for mental processing. Most mentally attributable problems can be attributed to c) out putting 72) Discontinuity.

d) Intellect (Class SIX). 466) Intellect, 467) Intelligence/Wisdom, 468) Wise Man, 470) Foolishness, 471) Fool, 472) Sanity, 473) Insanity/Mania, 474) Eccentricity, 475) Knowledge, 476) Intellectual, 477) Ignorance, 478) Thought, 479) Idea, 480) Absence of Thought, 481) Intuition, 482) Reasoning, 483) Sophistry, 484) Topic, 485) Inquiry, 486) Answer, 487) Solution, 488) Discovery, 489) Experiment, 490) Measurement, 491) Comparison, 492) Discrimination, 493) Indiscrimination, 494) Judgement, 495) Prejudgment, 496) Misjudgment, 497) Overestimation, 498) Underestimation, 499) Theory/supposition,

500) Philosophy, 501) Belief, 502) Credulity, 503) Unbelief, 504) Incredulity, 505) Evidence/Proof, 506) Disproof, 507) Qualification, 508) No Qualification, 509) Possibility, 510) Impossibility, 511) Probability, 512) Improbability, 513) Certainty, 514) Uncertainty, 515) Gamble, 516) Truth. 517 Maxim. This combination is especially significant as it clearly shows the pitfalls that the Intellect must circumnavigate to get to Truth. Truth in this case is an individual's Truth, not necessarily Reality as seen in Class SEVEN sequences. What is clearly seen in ROGET"S THESAURUS sequences is that we begin with the Intellect and Intelligence at the Wisdom level indicating that all men start off with the attributes for being a Wise Man, and as convolution (busy mind) occurs, goes down hill from there. NOTE: The last three major concepts of Class SIX (INTELLECT) are 618) Deception, 619) Deceiver, and 620) Dupe. There are many other such stories in between just like the following. Read them and see if you can begin to make sense out of the order for yourself.

e) Mind (Class SIX). 518, Error, 519) Illusion, 520, Disillusionment, 521) Assent, 522) Dissent, 523) Affirmation, 524) Negation/Denial, 525) Mental Attitude, 526) Broad-Mindedness, 527) Narrow Mindedness, 528) Curiosity, 529) Incuriosity, 530) Attention, 531) Inattention, 532) Distraction-confusion, 533) Carefulness, 534) Neglect, 535) Imagination, 536) Unimaginativeness, 537) Memory, 538) Forgetfulness, 539) Expectation, 540) Inexpectation, 541) Disappointment, 542) Foresight, 543) Presentiment, 544) Meaning, 545) Latent Meaning, 546) Meaninglessness, 547) Intelligibility.

f) The Working Process (Class SEVEN). 785) Aid, 786) Cooperation, 787) Associate, 788) Association, 789) Labor Union, 790) Opposition, 791) Opponent, 792) Resistance, 793) Defiance, 794) Accord, 795) Disacord, 796) Contention, 797) Warfare, 798) Attack, 799) Defense, 800) Combatant, 801) Arms, 802) Arena, 803) Peace, 804) Pacification, 805) Mediation, 806) Neutrality, 807) Compromise, 808) Possession, 809) Possessor, 810) Property, 811) Acquisition, 812) Loss.

g) Affections (Class EIGHT). 855) Feelings, 856) Lack of Feelings, 857) Excitement, 858) Inexitability, 859) Nervousness, 860) Unnervousness, 861) Patience, 862) Impatience, 863) Pleasantness, 864) Unpleasantness, 865) Pleasure, 866) Unpleasure, 867) Dislike, 868) Contentment, 869) Discontent, 870) Cheerfulness, 871) Solemnity, 872) Sadness, 873) Regret, 874) Unregretfulness, 875) Lamentation, 876) Rejoice, 877) Celebration, 878) Amusement, 879) Dancing, 880) Humorousness, 881) Wit/humor, 882) Banter, 883) Dullness, 884) Tedium, 885) Aggravation, 886) Relief, 887) Comfort, 888) Hope, 889) Hopelessness, 890) Anxiety, 891) Fear/Frighteningness, 892) Cowardice, 893) Courage.

h) Moral observance (Class EIGHT). 998) Legality, 999) Illegality, 1000) Jurisdiction, 1001) Tribunal, 1002) Judge/jury, 1003) Lawyer, 1004) Legal Action, 1006) Accusation, 1006) Justification, 1007) Acquittal, 1008) Condemnation, 1009) Penalty, 1010) Punishment, 1011) Instruments of Punishment, 1012) Atonement.

FUNK & WAGNALL'S STANDARD HANDBOOK OF SYNONYMS ANTONYMS & PREPOSITIONS

Hypothesis - the redirection process: FUNK & WAGNALL'S STANDARD HANDBOOK OF SYNONYMS ANTONYMS & PREPOSITIONS is the third system (or tool box) and is based on the application of the antonyms and synonyms, found in *Harper & Row's* FUNK & WAGNALL'S BOOK OF ANTONYMS & SYNONYMS by Fernald.
Unfortunately Fernald died and took his secrets with him. The book has not been substantively updated since 1947, but it I find it still the wisest book I have ever read. This tool is the one that interconnects ROGET'S THESAURUS to the WEBSTER'S DICTIONARY and back again. This book clearly depicts the wisdom of implication as applied to what people want and what they do not want or do not want more of. The redirection element of Self Reorganization pinpoints what you want less of and redirects you by pointing opposite (usually what you do not want). This sets the stage for the beginning of a personal development search or clinical research. Another interesting use of this tool box is that it can provide almost immediate help in creating mental health. Usually when a person is mentally troubled, it is because mind/brain has lost sight of what was throwing it out of balance. A person is generally able to describe a symptom. By locating the symptom using this tool, one can find then find its opposite (usually causing the problem because it is missing). This permits the system to regain its missing balance.

The use of this map is to tap into the wisdom as applied to what people want more or less of in their lives. It is also useful in achieving mental health and balance when applied to out of one's mind experiences. It is common to find that people who are bothered by a state of mind, an environmental condition, or a difficulty in sorting their relationships, will spend time and effort studying the syndrome that they wish to fix or change for the better. The more effort they put into learning about the deficiency, the more the system is reinforced in the opposite direction. **THEORY: If something is not working, then its opposite probably will**. The Funk & Wagnall's is used to determine an accurate and precise opposite, thereby providing you with what you want more of and directing you away from reinforcing the symptom, problem, or situation of which you want less. Now you can define this opposite concept using the dictionary and even do a String using the dictionary (as previously instructed). (The ROGET'S THESAURUS can be very useful at this point, permitting a view of the process steps that can be learned and that now become options for Self Reorganization as it relates the outside world.)

An example of what you might want less of through redirection to what you want more of: If you are experiencing "hate" and try to study the cause of your hate (a common therapeutic process), there is the tendency to become even more caught up in the cause when you to discover it. Instead, look up "hate" in the FUNK AND WAGNALL'S. It will be found on page 27, with it's antonyms on page 26. Exploring the ideas behind adore, applaud, approve, delight in, enjoy, esteem, extol, honor, love, respect, revere, reverence, venerate, and wonder, will cause continuity to replace the discontinuity in the system being disrupted by the hate.

Other examples to be found In the FUNK & WAGNALL'S: On Page 193 is an example of wisdom. *Faith* (a primary heading) is a union of *belief* and *trust*. It is a belief that is so strong that it becomes part of one's nature. *Belief* (on the other hand) as an intellectual process, is the acceptance of something as true on grounds other than personal observation and experience. How many of us have enough *faith* in ourselves to believe in what we have not seen, experienced, or are hearing for the first time (like much of the material in this article)? Only those who are personally developed to this point, can enjoy the benefits of transcending personal belief to a higher learning plane. On Page 117 in his another example of wisdom. To *change* is in some respect to distinctively make a thing other than what it has been. To *exchange* is to take or put something else in it's place. Personal Development can be retarded because one does not trust changing (because they might not be clear on what they will change into). On the other hand, "exchange" might be simpler to achieve because it implies choice. On Page 81 is another example of wisdom. "Bad" is the opposite of "good" in any one of it's many senses, and almost any negative adjective in the language may be in some connection a synonym of bad. (Departing at this point from the Funk & Wagnall's and connecting into the ROGET'S THESAURUS, one can fit a very important aspect of Human Nature and Behavior into a very positive mental health providing system. At this point the healthy system will see many human occurrences as "BAD" in one or more of it's many senses, misperceiving what in reality was merely some form of human frailty. To the degree that the system perceives something as "bad", the system will develop the corresponding degree of anger. On the other hand, if the occurrence is perceived as FRAIL, the system will develop the corresponding degree of COMPASSION.) On Page 45 is another example of wisdom: "Astonishment" especially affects the emotions, "amazement" the intellect.

THE HUMAN EQUATION

(As observed and abstracted from the tool box). WE BEGIN WITH WISDOM, BUT END UP IN SELF DECEPTION. The intellect has potential profoundness and wisdom (as from the mouth of babes). The clutter of learning and experiences crowds the belief system with abstractions and becomes an Ego driven state of self deception. It is possible with a program of Self Reorganization to learn how to stay the mind at the Wisdom level and discriminate instead of pinioning and being judgmental.

Ego. People commonly defend their belief system because their beliefs have transcended to certainty. This certainty (in the face of any new concept, differing or threatening idea) makes belief (and it's certainty aspect) primary and new data secondary. The person's primary data (with no frame of reference) is then used to validate or invalidate the new data. This causes destructive doubt and skepticism. This out of order process is the bane of learning. **Comfort** is the primary resistance to learning and self reorganization, and is the primary cause of continued self deception. Habit/Tradition is its name. **Probability**. The most powerful word in the vocabulary as it opens the mind to Self Reorganization. "Probably" in front of any concept (either in thought or in speech) is an important guarantee of open mindedness. It is a stronger concept than "sure" or "certain", although most people would choose "sure" or "certain" as stronger. "Probably true" is a stronger concept than "absolutely true" because "true" appears complete while "probably" means "true" is completely true while at the same time using the incompleteness of probably to keep the mind open, a requirement for Self Reorganization and Personal Development.

Illusionment/disillusionment. When asked the question, "Do you like being disillusioned?" would you have said "No", or "Of course not"? When asked the question, "Would you prefer knowing the truth rather than being treated kindly?", would you have said, "Of course"? Look at the contradiction in attitudes. On the one hand you say that you do not like being disillusioned and on the other hand you say that you prefer the truth. Yet disillusionment is the essence of Self Reorganization. *Belief*. Personal growth depends on the acceptance that everything that you believe in is a lie. *Belief* is spelled that way: "BE LIE". A lie is merely an angle, like the lie of the land. Consider that no one could be disillusioned if they had not been in an illusion in the first place. The System is primarily a source of disillusionment and Self Reorganization through lifting the veil of illusion. *Liability*. What you do not know that you need to learn are your liabilities. Another word for "liabilities" is potential (as is in electrical potential). Liabilities are the source of aptitudes. Developing liabilities to get, gain and acquire, makes our liabilities our greatest asset. *Timeliness*. The only way to be on time is to be early and to let the time arrive. *Mental Power*. Mental power depends upon one's ability to combine forces. Whoever can combine the greatest forces is the winner. You can combine forces with the tool boxes of the System and win. *Strong Will*. Strong will is unwillingness itself, the most difficult of Self Development obstacles. *Moderation*. Moderation may well be the most important God Force that we possess. Moderation is the position were the extremes are balanced and equidistant from each other. It is the extremeness of a concept that slows mind and body from their effective roles. Moderation is always the shortest distance between points of any set of concepts.

Who we are and who we are not?

Common Belief: We are a product of our experiences, what we have learned and what we do, combined in various ways with what we have inherited from our parents and our family tree. Correction Theory for Learning and Self Reorganization: What we really are is a product of what we have not experienced, what we have not learned, combined with what we have inherited from parents and our family tree. What you learn and experience from the day you are born to this moment in time appears to always represent who you are. Yet you become who you are as you learn. At what point in a person's life do they change from becoming who they are from what they are in the process of learning, to being who they are from what they have learned?

Theories about who you think you are, who you are not, who you really are, and, what makes self reorganization possible.

A common belief is that you are a product of what you have learned through experience, trauma, disappointment; and of your inherited traits. A common belief is that you can solve your problems or improve your Conditions by analyzing your upbringing and experiences.
Question: what if all of the above was merely extrinsic to the learning process, merely an ego illusion at the gross level of man? Question: what if instead, what is really true is that you are a product of what you do not know and what you have not yet learned or become conscious of? Consider that a baby at the moment of birth has no other references but mother and father, and is a product of what it does not know. It can also be observed that at age 3, 5, 10, 13, 16, 18, 21 etc. that the child, adolescent, and adult is and continues to be a product of what is not yet known. When then does the point occur where Ego begins to believe (in part or in whole) that it exists because of what it knows? Contention: This is where and when personal development slows and eventually stops, on a module by module basis.

The system you are being asked to become familiar with can keep a student (and perhaps even a disciple) for life, a mentee of the master language system, and in touch with all that you do not know about your personal explicate/implicate order.

Implicate/explicate order theory.

Question: Is the Worldly Process of the Implicate/explicate Order Applicable in Human Development? Consider that in the year 1900 all things known to the human race (even if known to only one person) represent the explicate order of the knowledge of mankind. Then consider that all things unknown to the human race in 1900 but are now known, comprised the implicate order of man at that time. Then consider that all things known to the human race at this time are the explicate order of mankind, and that all things that will be revealed to or by people any time in the future is the present state of the implicate order of man.

Answer: Each person at time of birth is a product of what he/she does not know or has not experienced (very little of the implicate order revealed). As the child grows through the ages of his/her life, that entity continues to have an implicate order. But if the person can find a way to interfere with the process of building and defending Ego, then the person (knowing that he/she is a product of what they do not know) can be open to the universe of their own implicate order. Therefore this language system of Tool Boxes can become the most perfect of masters for anyone in pursuit of Self Development and Self Reorganization, because this master system can provide all implicate developmental information you will need..

SOME USES FOR THE TOOL BOXES

There are many different ways to get to the same place. Some routes are more direct and efficient than others. Using the tool boxes will provide maps for you to effectively find your way through the maze of the Senses, Intellect, Volition, Feelings/Emotions, and the world in which you live. *Use the Language Tool Box for Personal Development.* You can test your attitudes against Reality Attitudes. You can tap into the Knowledge Data Base using the Language Tools. You can see that reality is the process of how things work at any given moment in time, and is the order in which things work. The big questions is, can you set aside (momentarily) any temptation you may have to validate or invalidate new concepts? This is necessary as there is a strong possibility that you do not have the frames of reference to qualify or disqualify this system. Understanding this system can only be achieved when you actually experience a change of attitudes resulting in some form or Self Development Reorganization for yourself.

OTHER ASPECTS OF THE LANGUAGE MODULES

WISDOM is basic to the intellect and to all beings (as from the mouth of babes). The system provides this basic wisdom at all levels. FUNDAMENTALS become available as the principles of how things work are explicated within the system. BASICS: The principles that are basic to human endeavor are all available. QUESTIONS/ANSWERS/SOLUTIONS to almost any human internal and interactive process are available. DIFFERENCES in the form of relationships can be easily seen, measured, and used. CAUSE AND EFFECT - The effect of any event is more attributable to its causation, than the cause itself. ATTRIBUTION of cause becomes the most significant key to correcting and solving human equations. Until you attribute (assign) the correct cause of a desired effect, there is no way to solve or change an undesirable effect, or gain the one that is more desirable. PERSUASION in European terms is "convincing" but in American terms is rarely long lasting. In American terms persuasion must be reasoned. The System clearly provides the reason for any position. The reasoning process provides the persuasion even if you are the one that ends up being persuaded. EDUCATION AS AN AID TO PSYCHOLOGICAL INTERVENTIONS: If the concept that we are a product of what we do not know and have not experienced can be accepted, then the System is the perfect solution for providing the tools for aiding all interventions. PRESENTIMENT: This is the protector, resister, bane, and most important singular cause of mental resistance. It is where suggestions, new ideas, apprehensions, misgivings, warnings, threats (real or imagined), are mired until meaning, intentions, purpose, value, tone, and possible effects, are clearly understood (or an ambiguity that seems meaningful is accepted). In the sales process, the communications process, or any other occurrence where one person (or group) is trying to influence another person (without offering the resistances of disagreement, rebuttal, or negative reaction) silence is essential while the person works through their own maze, in their own way, and in their own time. The Probabilities Are Endless. Name the subject and the system will reveal more than any seer or fortune teller.

CLAIMS

1. On a percentage basis, all processes, occurrences, and existing knowledge appear in these books (Tool Boxes) in the order in which they occur. This order includes your environment (and people as they function in it), the way you process the abstracted information within your version of the whole, and the way these interact with each other moment to moment.
2. The ability to use these tool boxes is built into every human being because people created these tool boxes. 3. This system does not create any words of its own. All concepts used in the system are basic English words already in common use in the United States. 4. The tool boxes are capable of aiding in the intervention process. They can provide an intuitive process of tracking the logistic (both mental and physical) positions of any person with results similar to those tracked by experiential examination. Analysis can take place in days or months, rather than months or years.
5. The system works from anyone's ability to classify using a single word or concept to describe a thing that is the key. From that key a sequence of causal and consequential concepts is developed. When these concepts are shown to the person doing the classifying, it will appear that the listener (practitioner) is mind reading.
6. All that the system asks is that you use your common sense to validate the order. You will find that it does not usually correspond to the order you have come to believe is true. 7. Each of the modules (tool boxes) of the system can be introduced to any open minded person in about two hours. Understanding the system generally takes no more than an hour.

BENEFITS

Through the process of using the language for Self Reorganizing you can gain:
1. The ability to clearly define your identity and expedite the search for who you are, by learning

who you are not, so you can choose who and what you would like to become. You can gain the options in life you have always wanted. 2. The ability to transcend the busy, logical, result-oriented perceptions to the higher and more powerful mental level of process. You can see what things actually look like (the process of reality) not what they seem to be from the logical perspective. Imagine the benefit of transforming anything obscure into the obvious. 3. The ability to Self Reorganize segment by segment, the senses, the intellect, volition, and even feelings and emotions, until all of your intelligences are going in the same direction at the same time. Imagine, no more confusion (the fusion of "concepts"). 4. The ability to develop stress prevention and stress reduction (not stress management) in any life situation or when internal disquiet or disruption exists. 5. The ability to know what causes a particular sequence of thought, speech, behavior, or feeling, and what can be done to interfere with, reduce or increase that particular occurrence. 6. The ability to communicate (verbally or in writing) in simple terms rather than the more common complex terms. 7. The ability to understand communications (written or spoken) in simple (rather than the more common complex) terms. 8. This can be used for Personal Growth in the Self Reorganizing process by a lay person, or by any professional. 9. Professional therapists and practitioners will be able to accomplish in hours what otherwise commonly takes months or years of experiential analysis. 10. There is a minimum regression process. What a person learns through language they will always remember (and the internal system will use). What a person is told about (or to do) will rarely be remembered in the way that it was heard. 11. The ability to tap into the tool boxes is instantaneously available on a day to day basis and thereby explain how any situation works in any personal or interpersonal dimension.

SUMMARY

All of the theory, hypotheses, contentions, notes, aspects, examples, observations, Tool Box suggestions, and order concepts have been tried and tested in real life situations over the last 20 years. A piece at a time, it all has evolved into the major system presented in this treatise. The biggest benefiter has been me. Everything you have read has worked for hundreds of clients and especially for myself. What is important is that you put persistent effort into the system. It takes repetition after repetition, reading and rereading, practice upon practice. You have years of beliefs and conditioning. Most importantly, do not expect results. The more you do the more discouraged you will become. Stick with the process. Let the results come when they come.

REFERENCES

1) Roget's International Thesaurus, Fourth Edition, Revised by Robert L. Chapman, New York, Thomas Y. Crowell, 1977.

2) Fernald, James C., Funk & Wagnall's Standard Handbook of Synonyms, Antonyms, and Prepositions, New York, Harper & Row, 1947.

3) Webster's Third New International Dictionary, Philip Gove editor, New York, G & C Merriam Publishing Company, 1972.

20.
Self-Organization and the Social Collective

Raymond Trevor Bradley and Karl H. Pribram

Self-Organization and the Social Collective

Raymond Trevor Bradley[1]

Institute for Whole Social Science, Carmel, CA

and

Center for Brains Research and Informational Sciences
Radford University

Karl H. Pribram

Center for Brains Research and Informational Sciences
Radford University

ABSTRACT   A theory of social communication is developed to explain the processes of self-organization by which stability is achieved in the social collective: to explain how energy expenditure interacts with control operations to form a self-organizing information processing system that results in a stable collective. Drawing on concepts and principles from thermodynamics and information control engineering, the theory shows how two orders of social relations, *flux* and *control*, act on the biosocial energy of the collective's members to create quantum-like, elementary units of information. Each unit of information enfolds a description of the collective's endogenous organization. The interpenetration between the two orders operates as a self-organizing communication system that *in*-forms (gives shape to) the expenditure of energy to produce stable collective organization. Results from a longitudinal study of 46 social collectives offer empirical support for the theory: only those interactions between flux and control that produced a path of least action--one which entailed the smallest amount of turbulence-- resulted in a stable social collective. By contrast, measures of the collectives' normative and structural organization and of the members' social characteristics were found to be unrelated to the stability of the collectives.

PURPOSE AND APPROACH

The purpose of this study was to develop and test a theory of social communication to explain the processes of self-organization (Nicolis and Prigogine, 1977; Prigogine and Stengers, 1984; Varela et al., 1974) by which the social collective achieves and sustains a stable pattern of organization. By *social communication* we mean a process by which information about the collective's internal organization is gathered, processed, and distributed throughout the collective as a *whole*; this concept is similar to the notion of communication that underlies the connectionist computational models of brain-style processing (e.g., Rumelhart, 1992). A social collective is defined as a durable arrangement of individuals distinguished by shared membership and interaction in relation to a common purpose or goal. An understanding of stability--the means by which structural integrity and functional viability are sustained--is of primary importance for social science in that a stable platform of organization is a necessary prerequisite for effective collective action.

---

[1]Address correspondence to: Raymond Bradley, Institute for Whole Social Science, 25400 Telarana Way, Carmel, CA 93923; Tel., (408) 626-8057; FAX, (408) 622-9423.

Most of the current work on social collectives has continued to follow earlier approaches-- either to pursue endogenous perspectives that emphasize the deterministic constraints of structure (Blau and Scott, 1962; Merton, 1940) or the intentional order created by purposeful individuals (Child, 1972; March and Olsen, 1976; Silverman, 1970), or to pursue exogenous perspectives that emphasize the ecological processes of natural selection (Aldrich, 1979; Hannan and Freeman, 1977) or the negotiated order of networks of collectives (Cook, 1977; Pfeffer and Salancik, 1978). But despite an enormous body of work, there is at the moment no empirically based model that predicts the stability of social collectives.[1] And while there is growing recent interest among some social scientists (e.g., Garud and Kotha, 1994; Hutchins, 1991; Loye and Eisler, 1987; MacKenzie, 1991; Morgan, 1986; Sandelands and Stablein, 1987; Weick and Roberts, 1993;) in discoveries regarding stability (at equilibrium, or far-from-equilibrium as in chaos theory) in systems behavior from physics, biology, and computer science, virtually all of this work is metaphorical rather than explanatory (some exceptions are Bradley, 1987; Bougon and Komocar, 1990; Dendrinos and Sonis, 1990; Glazer, 1986). As Weick and Roberts (1993) acknowledge, metaphorical reasoning is a "shaky basis" for a theory of collective organization.

In our approach we draw on the principles of self-organization and the concepts of *energy*, *least action*, and *information* from the natural sciences to inform our understanding of communication in the social collective. We do this because these concepts capture features of collective organization that appear to be fundamental in accounting for system organization and behavior. We use formalisms from these fields to develop a rigorous, testable theory of the behavior of bounded social collectives. The formalisms have proven a powerful means for describing physical and biological collective organization.

Two steps are usually required to obtain insights when borrowing concepts from other fields of inquiry. The first step is to discern commonalities in the behavior of collectives operating at different levels and to describe these in a common terminology. The second step seeks an understanding of the intimate relations that connect two adjacent levels of inquiry. Ideally, the operation of these relations, formally described as transfer functions (transformations), must account for the results obtained in the first step (see Nicolis and Prigogine, 1977; Pribram, 1991). In this study we take only the first of these steps and describe social phenomena in the terms of concepts and formalisms from physics and control engineering to help understand the communicative structure of small social collectives.

Previous analyses of the groups in this study (Bradley, 1987; Bradley and Roberts, 1989a, 1989b; Carlton-Ford, 1993; Zablocki, 1980) have shown that two patterns of social relations form the communicative structure (see Figure 1). One patterns is a dense web of reciprocated affective relations interconnecting virtually all members. This web is organized as a *field*, a distributed, massively parallel order of symmetrical ties in which individuals are essentially interchangeable. The second pattern is a densely interlocking order of power relations which also extends to connect virtually all individuals. This is a *hierarchy*, a highly stratified system of asymmetrical, transitively-ordered relations which defines, for each individual, a position that is spatially and temporally localized and, therefore, is unique. Following up on these earlier findings, the aim here is to understand how the interaction between field and hierarchy operates as a self-organizing system to generate and distribute information about the collective's endogenous organization.

Figure 1 (next two pages). **Sociometric Structure of "power" and "loving" Relations Selected Stable and Unstable Communes**

---

[1]Carley's (1991) work is no exception in that it is based upon computer simulations of the distribution of "information" by "individuals" in small social "groups."

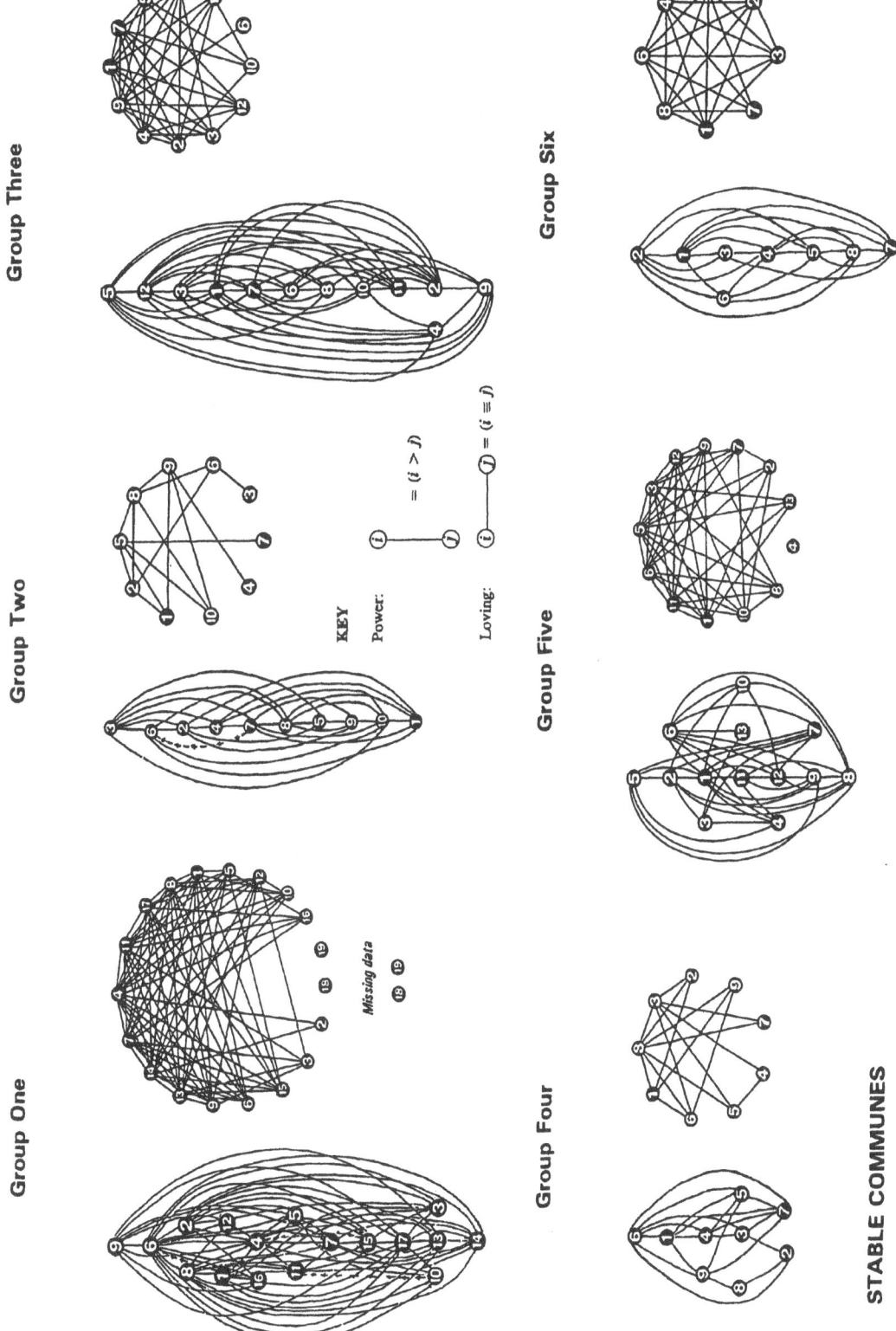

Group Three

Group Two

KEY

Power: ①——② = (i > j)

Loving: ①——② = (i ≡ j)

Group One

Missing data

Group Six

Group Five

Group Four

STABLE COMMUNES

481

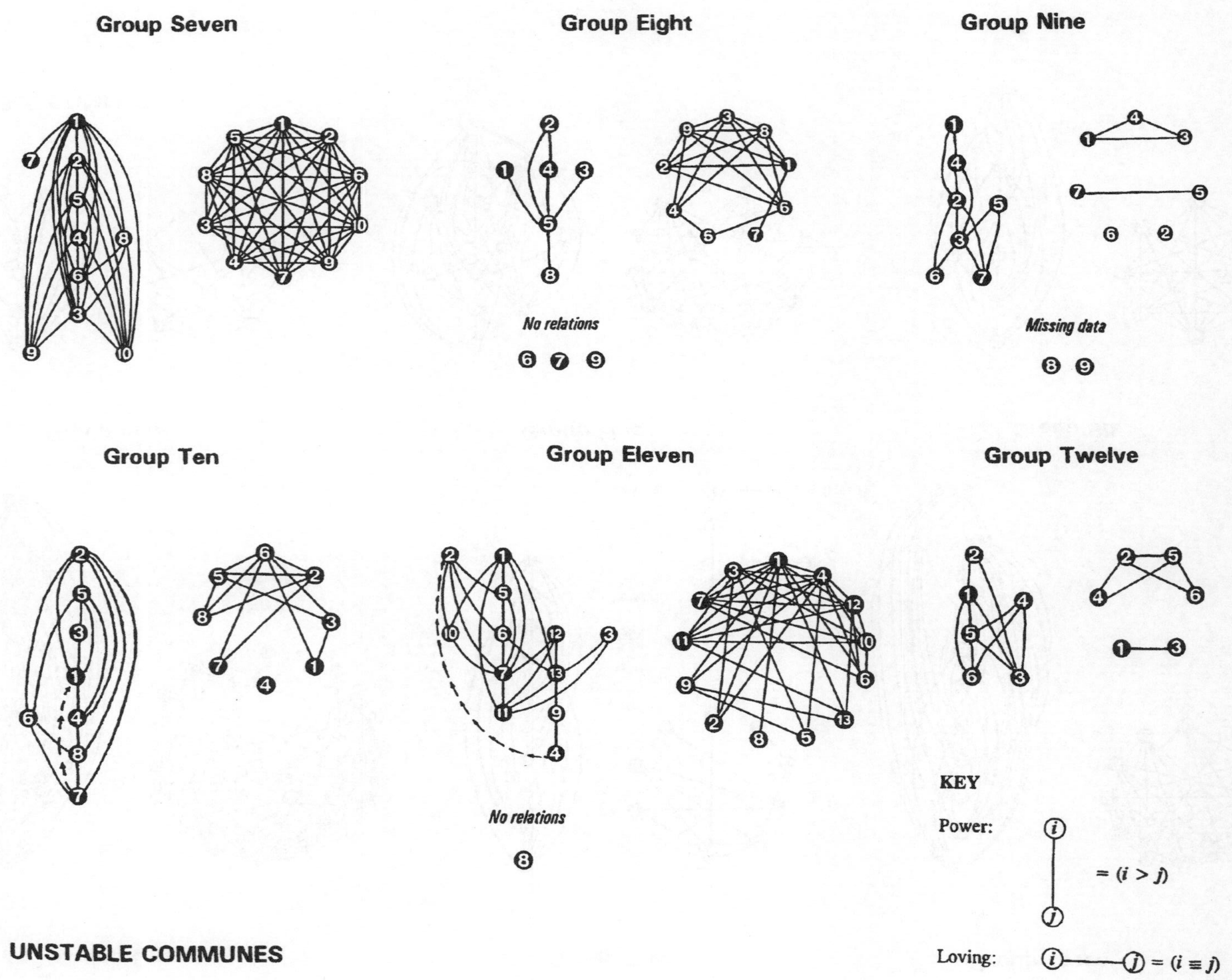

Group Nine

Group Eight

Group Seven

Group Twelve

Group Eleven

Group Ten

Missing data

No relations

No relations

KEY

Power: $i \quad j = (i > j)$

Loving: $i \quad j = (i \equiv j)$

UNSTABLE COMMUNES

THEORY[1]

We make two simplifying assumptions. First, our account does not consider any effects that the characteristics (gender, age, personality etc.) of the collective's members, as individuals, may have on the collective. Second, we restrict the theory's scope to communication involving events that are proximate in social space by limiting it to endogenous operations within bounded collectives. This means that we exclude partially bounded or open social systems like cliques or social networks.

Energy and Least Action

Ontologically, a rigorous concept of energy, or its equivalent, is fundamental to an understanding of collective organization. Energy is a measure of the means--the fuel--for maintaining order in the face of challenge (novelty) or changing an order in the face of inertia. As individual biological organisms, a collective's members possess the potential for work, measured as energy. To exist as an entity, a social collective *must* mobilize and appropriate the members' potential for work, their biological capacity for physical behavior and activity, and direct it toward collective ends. "Energy," so defined, also is the medium for information processing, the medium for encoding and relaying communications as signals back and forth among the elements of a system.

In applying the concepts of *energy* and *least action* we assume that the members of the social collective are biologically capable of work, and that this capability is measurable as potential energy. When activated by the collective, the members' potential energy becomes engaged in social interaction. To actualize the potential energy, to realize it in collective action, entails work; work is measured as kinetic energy. The tendency to energy conservation leads the social collective to strive towards an efficient use of energy. This requires effort to explore alternative *paths* towards order, patterns of actualization that allow collective work to proceed efficiently, that is with the least amount of dissipation (Pribram and McGuinness, 1975).

Within this framework, two processes can be identified which act to generate descriptions of the collective's internal organization. The first is *flux*, the constant transformation of energy throughout the collective. It occurs in the distributed, massively parallel field of *equi-valent* relations which connects all individuals to everyone else. The field operates to activate and unify affective attachments among individuals, thereby mobilizing their potential energy. The energy transforms continuously *throughout* the field as the collective adjusts and readjusts continuously to internal and external changes.

In the absence of other factors, initial conditions such as negative emotions like fear, hatred, or jealousy, will block the efficient conversion of potential energy to kinetic energy; in non-linear dynamics such systems are characterized by negative Liapunov exponents leading to stasis, ossification (complete [physical] equilibrium), or to regular fluctuations described by relaxation oscillators (Abraham, 1991). On the other hand, as elaborated below, initial conditions such as admiration, awe, or love create a kind of harmonic resonance (due to a positive Liapunov exponent) in the relations among members, which will enhance the conversion of potential to kinetic energy--a phenomenon Zablocki (1971 and 1980) observed in his studies of communes and called the "cathexis effect." The danger here, if this enhanced kinetic energy is unconstrained, is that undue dissipation of energy will ensue: in the language of non-linear dynamics, chaos will result (for examples, see Zablocki, 1980, Figure 4-5: 165).

The second process is *control*, the construction of a landscape of social constraints which

---

[1]See Bradley and Pribram (1996) for a full presentation of the theory.

efficiently directs the transformation of energy into collective action. This operation is achieved by the hierarchical order which is a densely interlocking stratified system of asymmetrical relations connecting all individuals. By differentially constraining the paths by which individuals expend their energy, both with respect to specific locations in space and with respect to particular moments in time, the controls render an *in*-formed pattern of collective organization.[1]

Information and Communication

To show how the interpenetration between flux and control acts as a communication system, we draw on Gabor's concept of information (Gabor, 1946): information as the *minimum* uncertainty with which a signal can be encoded as a *pattern of energy oscillations across a waveband of frequencies,* as in the encoding and transmission of vocal utterances for telephonic communication. Although virtually unknown in the social and psychological sciences, Gabor's concept is radically different than, though related to, the more commonly used measure of information developed by Claude Shannon (1949): information as *a reduction of uncertainty through choice among alternatives.*[2]

In his classic "Theory of Communication," Gabor (1946) shows that there is a restriction to the efficiency with which a set of telephone signals can be processed and communicated. The restriction is due to the limit on the precision to which concurrent measurements of spectral components (frequency, amplitude, and phase) and the (space)time epoch of the signal can be made. So that although accurate measurement of the signal can be made in time **or** in frequency, *it **cannot** be simultaneously made in both beyond a certain limit* (Gabor, 1946: 431-432). Using Heisenberg's uncertainty principle, Gabor gave this limit formal expression and showed that the signal that occupies this minimum area "is the modulation product of a harmonic oscillation of any frequency with a pulse in the form of a probability function" (Gabor, 1946: 435). Gabor called his unit a *logon*, or a *quantum of information.* This elementary unit of information both minimizes uncertainty and provides the maximally efficient compression of communication (the minimum space or time of transmission occupied by the signal which still maintains the fidelity of [telephonic] communication).[3]

The Gabor elementary function, as it is often referred to, which has been found to characterize perceptual processing in the cerebral cortex (see Pribram, 1991, Lectures 1-5, for a review of the evidence)[4] is, therefore, an alternative unit for biological information processing to Shannon's (1949) BIT of information. Moreover, two previous findings from the social collectives examined in this study, document an order of social communication that does not seem describable within the terms of Shannon's concept but appears more readily understood within Gabor's terms. The first finding is of a non-localized order in which information about

---

[1]This conception is similar to Bohm and Hileys' notion of "active information" (see Bohm and Hiley, 1993: 35-42, 59-71).

[2]See Cherry (1966) for an excellent review of these ideas, and Kaiser (1994) for a readable introduction to the physics of signal processing.

[3]Gabor's elementary unit of information, the **logon**, is a sinusoid variably constrained by space-time coordinates; it differs from Shannon's elementary unit of information, the **BI**nary digi**T**, which is the Boolean choice between alternatives (Pribram, 1991: 28).

[4]For example, in a series of recent studies on the barrel cortex of the rat (involving the stimulation of the rat's whiskers in terms of the spectral and spatial components of neural response activity), Pribram and his collaborators (King et al., 1994; Santa Maria et al., 1995) have shown that the response activity of receptive fields could be described in terms of spectral and spatial manifolds, and that each of these manifolds could be derived from Gabor-like functions.

the collective's global organization appears to be enfolded and distributed to all individuals; the second is that this holographic-like order was found to be constrained by a system of hierarchical relations (see Bradley, 1987, Chapters 8 and 9, respectively).[1]

Drawing on Gabor's concept of information, it is expected, therefore, that the operation of hierarchical controls on the distribution of flux generates a moment-by-moment--*quantized*--description of the collective in terms of both structure (spatial-temporal position) and flux (distribution of energy). By providing a succession of descriptions within space-time and spectral coordinates, quantum-like elementary units of information are constructed and communicated, via a holographic-like process (Pribram, 1991,) throughout the collective. Because each quantum of information overlaps with the unit that succeeds it--contains an "overlap (with) the future" (Gabor, 1946: 437)--each unit contains probabilistic information about the potential order of the collective. In this way, the communicative system *"anticipates"* the next moment of the collective's order (Bradley, 1996).

However, whenever there is an imbalance between the amount of distribution of flux and the amount of control, quantization breaks down resulting in a loss of information transmission. The reduction in information transmission impairs the efficient operation of the collective which, in turn, increases the likelihood of instability. This impairment is due to what Ashby (1956) characterizes as the necessity for "requisite variety" in cybernetic (information and control) systems. The logic of the theory is schematically represented in Figure 2.

Hypotheses

For the purposes of this analysis, five hypotheses were derived from the theory for empirical testing with data from 46 social collectives (described momentarily); the hypotheses assume that social connections linking all individuals have been established, and that the interpenetrations between flux and control are governed by the principle of least action (energy conservation).

---

[1]Musical notation is an example of a logon-like communication system that operates to inform the action of a musical [social] collective such as an orchestra, a band, or a choir. An individual "note" can be viewed as a direct analogue of a logon. It is composed of data "plotted" in a (written) musical score on the same two orthogonally-related dimensions as a logon: one dimension is *frequency*, varying oscillations of sound waves (energy vibrations) produced by the operation of a musical instrument; the second dimension is *time*, how long the note is to be played--its duration. The pattern of energy expenditure by which the music is actualized is prescribed on a musical score as a moment-by-moment sequence of operations on the musical instrument, for each musician, specified **both** in frequency and in time. Moreover, the score for all musicians contains a *spatial* component as well: it also specifies which subset of musicians, in relation to the whole orchestra, is to play at each moment. Thus a composer's written musical score represents a description of how the potential energy of a collective of musicians is translated into expenditures of energy, differentiated for each individual on the dimensions of frequency and time-space, to actualize a given composition as "music."

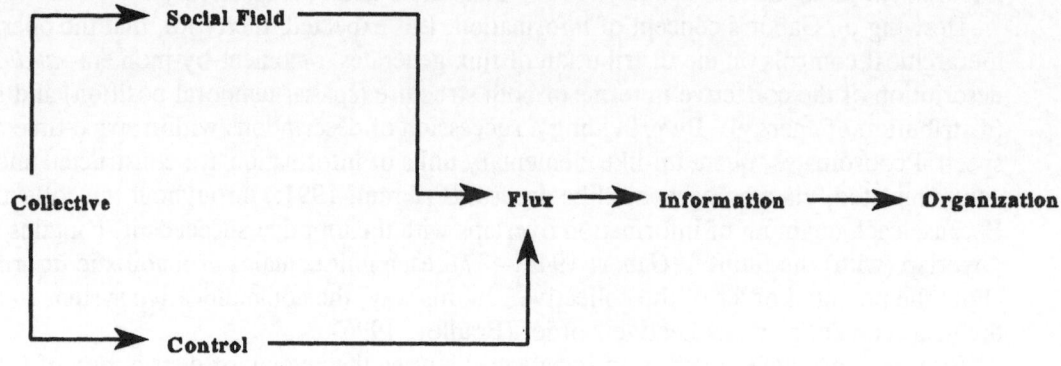

Figure 2. **Logic of Theoretical Model**

Because the movement of energy tends to become distributed, and because space-time constraints operate to inform this distribution and so make possible the transformation of potential energy into collective work, it was expected that:

>   *Hypothesis 1. The communicative structure for collective organization requires a certain amount of flux in interaction with a certain amount of control.*

There is an increased risk of turbulence as kinetic energy increases. Thus, when the conversion of potential to kinetic energy is enhanced at higher levels of flux, it was expected that:

>   *Hypothesis 2. The tendency toward least action results in an increasingly closer correspondence in the amounts of flux and of control at higher magnitudes of flux.*

Further, the total amount of information per unit of space or time must be both sufficient to inform the transformation of potential energy into effective collective action and also not exceed the information processing capacity of the communication system. Thus, it was expected that:

>   *Hypothesis 3. There would be limits on the total amount (a minimum and a maximum) of information that could be processed in a given unit of space or time, which, if exceeded, would threaten the stability of the collective.*

The fourth hypothesis follows from Gabor's energetic concept of information in which efficiency requires constraints on the relation between the frequency and (space)time components of communication. So that:

Hypothesis 4. *There would be both a lower and an upper limit on the amount flux and on the amount of control processed by an effective collective: combinations of flux and control that fall outside the limits (viz, high flux with low control, and low flux with high control) reduce the efficiency of information processing which, in turn, increases the likelihood of collective dysfunction.*

This is because dysfunction entails situations in which certainty (reliability) on one component of communication has been achieved at the cost of uncertainty on the second component.

For collectives with communicative structures operating within Gabor's limits, a fifth proposition was investigated. This is the expectation that:

Hypothesis 5. *The likely disposition of the collective at a future moment will be enfolded in the information processed by the communicative structure in the present.*

This follows from the overlap among the elementary units of information by means of which the present order is probabilistically *in*-formed (given shape to) by the order implicit in the series of succeeding moments. Thus, combinations of flux and control within Gabor's limits at a given moment are expected to yield an increased potential for stability in succeeding moments.

METHOD AND DATA

The data were gathered over a decade ago as part of a nation-wide longitudinal field study of sixty urban communes (Zablocki, 1980), stratified on a number of basic social characteristics, and sampled in equal numbers from six Standard Metropolitan Statistical Areas (Atlanta, Boston, Houston, Los Angeles, Minneapolis-Saint Paul, and New York). Formal and informal methods were used to collect two panels of data, twelve months apart, during the summers of 1974 and 1975. Data on commune survival status were also gathered for an additional two years. Data from 57 communes are used in this report (see Table 1; for further details, see Zablocki, 1980, and Bradley, 1987).

A sociometric instrument, the "Relationships Questionnaire" (see Bradley, 1987, Appendix B), was administered to map the structure of social relationships in each commune. Each adult member was asked about the content of his or her relationship to each other member, thus providing an exhaustive mapping of the N(N-1) possible pair-wise (dyadic) relations in the group (where N = the number of permanent adult members). However, eleven had unacceptable levels of missing sociometric data and, as in the original study (Bradley, 1987), were excluded from the structural analyses reported here.

Operationalization

To test the theory, sociometric procedures (following Bradley and Roberts, 1989b) were used to operationalize the primary concepts of flux and control. **Flux**, the activation of potential energy, was indicated by a reciprocated positive response (an answer of "yes") by both individuals to either the "loving," "improving," or "exciting" questions (Question 5g, of the "Relationships Questionnaire"). **Control**, the operation of constraints on the activation of potential energy, was measured by the "power" question (Question 5e); only those responses that indicated the asymmetric ordering of the relationship--i.e., which of the two individuals (the respondent, i, or the other member, j) held the "greater amount of power," were used.

The subsets of relations that met these two operational definitions were then translated into a symmetric and an asymmetric sociomatrix of relations of flux and control, respectively, to encode the disposition of these dyadic relations among all members in each group. A binary coding was used in which, for flux, a value of 1 (one) indicated the presence of a reciprocated relation, and for control, a value of 1 (one) indicated the presence of an ordered relationship (i.e., $i \rightarrow j = 1$, control flows from i to j; $j \rightarrow i = 1$, control flows from j to i); any other condition, for either flux or control, was indicated by a value of 0 (zero).

The final step entailed the use of triadic analysis (Holland and Leinhardt, 1976)--a technique for analyzing the structural organization of social networks--as the means to build structural indices of flux and control. This technique first subdivides the sociomatrix into triads, and then, through a census of all possible triadic configurations, classifies the array of triads for the group into 16 isomorphic triad types (Figure 3). Of the sixteen triad types, three are symmetric in form in that they are composed exclusively of reciprocated positive dyads-- the 102, 201, and 300 triad types. Aggregated across the "loving," "improving," and "exciting" relations, the mean sum of these three triads as a proportion of all possible triads in a commune was used to measure the amount of flux (mean sum = .629, standard deviation (SD) = .196). Seven other triad types are composed exclusively of asymmetric dyads, and three of these (the 021C, the 021D, and the 030T, summed and expressed as the proportion of all possible triads) were used to measure the amount of control. Aggregated for "power" relations, the three triad types constituted just over half (.509), on average, of all possible triads of control in the communes (mean sum = .510, SD = .218).[1]

Stability, the degree to which the collective is able to maintain structural integrity and functional viability as a self-sustaining entity, was measured by a commune's survival status at a specific moment in time. Classified into one of two categories, *survivor* or *nonsurvivor*, each commune's stability was determined at each of the four successive twelve month intervals from Time 1 through Time 5; Time 0 is the point in time when a commune was founded and Time 1 is moment of the first wave of data collection (August, 1974). Twenty-two (48%) of the 46 communes survived through Time 5; a pattern of declining instability over time was observed, from 24% at Time 2 to 8% at Time 5 (see the Appendix for the summary statistics for the measures of flux, control, and stability).[2]

---

[1]See Bradley (1987) for the details of these operational procedures.

[2]Although the communes ranged in group age from three months to nine years at Time 1, there is little evidence that "period effects" (differences in group age at the time data collection commenced) explain the variability in survival status. Dividing the sample into "young" (two or less years; N = 23) and "old" (more than two years; N = 23) categories of group age at Time 1, and cross-tabulating these classifications by survival status grouped in three categories (dissolved by Time 2 or Time 3: N = 17; dissolved in Time 4 or Time 5: N = 7; survived beyond Time 5: N = 22) shows non-existent (0%) to modest (12%) non-statistically significant differences between the "young" and "old" categories of communes (*chi*-square coefficient with two degrees of freedom = 1.260, pr. = .533).

Table 1
Urban Communes Sample
Social Characteristics of Adult Population and Communes

Characteristics of Adult Population (15 years and older; N = 545)

Median age	25 years
Percentage male	54%
Percentage single, never married	72%
Percentage with college diploma	50%
Percentage with white collar or professional occupation	63%
Percentage with FT or PT job	67%

Characteristics of Communes (N = 57)

Mean size (adult members)	9.9
Percentage existed two or more years	42%
Percentage with "many" rules	21%
Percentage assign or rotate chores	51%
Percentage have communal business or jobs	16%
Percentage requiring novitiate or trial membership	33%
Mean percentage members holding formal positions or office	41%[1]
Percentage ideology "important" to group	79%
Percentage without leaders	30%

Ideological Type:	%
Religious	40
Political or counter-cultural	26
Personal growth, household, or family	34
	100%

---

[1]N = 273, respondents to the "Long Form" interview.

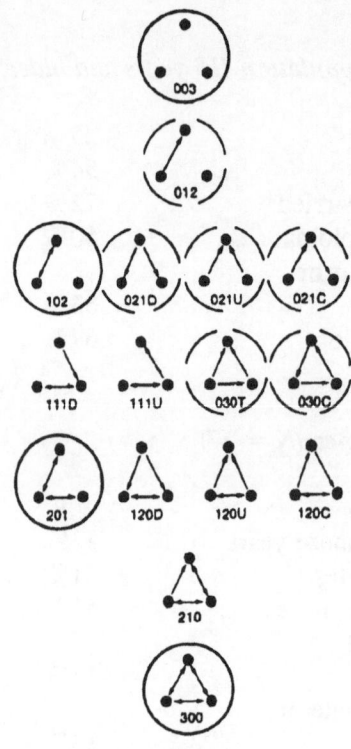

Holland and Leinhardts' sixteen isomorphic triad types: the 16 isomorphism classes for digraphs with $g = 3$ (that is, the triad types). Triad labeling convention: the first digit is the number of mutual dyads; the second digit is the number of asymmetric dyads; the third digit is the number of null dyads; trailing letters further differentiate among the triad types. Four symmetric triad types (unbroken circle) were used in the structural analysis of FLUX relations and seven asymmetric triad types (broken circle) were used in the analysis of CONTROL relations; the "vacuous" 003 triad was used in both analyses. Redrawn from Holland and Leinhardt (1976: 6, Figure 2).

Figure 3. **Isomorphic Triad Types Showing Symmetric and Asymmetric Triads**

EMPIRICAL ANALYSES

Verification

Testing the hypotheses entailed an analysis to determine the degree to which the observed patterns of commune behavior, with respect to our measures of flux, control, and stability, were consistent with the patterns expected by the hypotheses.

Hypotheses 1 and 2

The scatterplot in Figure 4 presents data on the relationship between flux (horizontal ordinate) and control (vertical ordinate). It can be seen that the null hypothesis--an equal or random distribution of groups--does not hold, and that the low non-significant correlation (Pearson's $r = .12$; pr. $> r$, $.43$) actually masks a non-linear association. Moreover, with the exception of five outlying cases (hollow dots in Figure 4), the observed pattern--a triangular distribution with a wide base and its apex in the high flux/high control region (upper-right quadrant)--is consistent with the pattern expected by Hypotheses 1 and 2.

Hypothesis 3

A measure of the total amount of information processed by a collective at a given moment in time was obtained by summing the measure of flux and the measure of control at Time 1, for each commune, and averaging the product (the mean for all communes $= .569$; median $= .552$, and SD $= .155$). The values for all communes were grouped into .10 intervals and, holding these values on this measure constant at Time 1, the sample was partitioned by survival status and the distribution of survivors and nonsurvivors was plotted on a time series of bar charts at twelve month intervals from Time 2 through Time 5 (see Figure 5).

The pattern of results show that the bell-shaped distribution at Time 1 gradually devolves into *two contrasting patterns* that are virtually the *inverse* of each other by Time 5: a single-peaked distribution for the twenty-two survivors with its mode in the .500 - .599 interval; a bi-modal distribution for the twenty-four nonsurvivors with its trough in the .500 - .599 interval and its twin peaks in the two adjacent intervals of .400 - .499 and .600 - .699. This difference in survival rates between the groups in .500-.599 interval and the other groups outside this range is statistically significant (*chi*-square $= 6.695$, pr. $= .010$).

Taken together, these two patterns appear to mark the bounds of a region where the probability of stability is maximized, that is in the .500 - .599 interval. When computed for the communes in the two sets of adjoining intervals at Time 1, the rate of instability by Time 5 for each grouping of communes is 63%, which is significantly different than the 18% for the eleven communes in the .500-.599 interval (*chi*-square $= 6.966$, pr. $= .035$). Thus, in line with Hypothesis 3, it would appear that the total amount of information in the intervals *above* .599 was excessive in terms of information processing capacity, whereas the amount of information in the intervals *below* .500 was insufficient to sustain a viable collective.

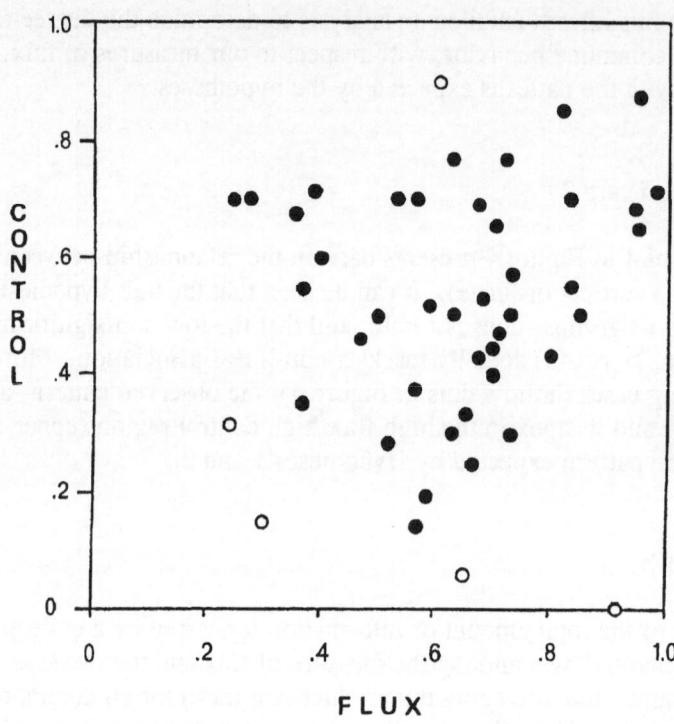

Figure 4. Scatterplot of All Communes (N = 46) by Flux and Control at Time 1
(outlying cases are shown as hollow dots)

Figure 5 (next page). Barcharts Showing Distribution of Communes for Total Amount of
Information at Time 1 by Survival Status at Time 2 Through Time 5

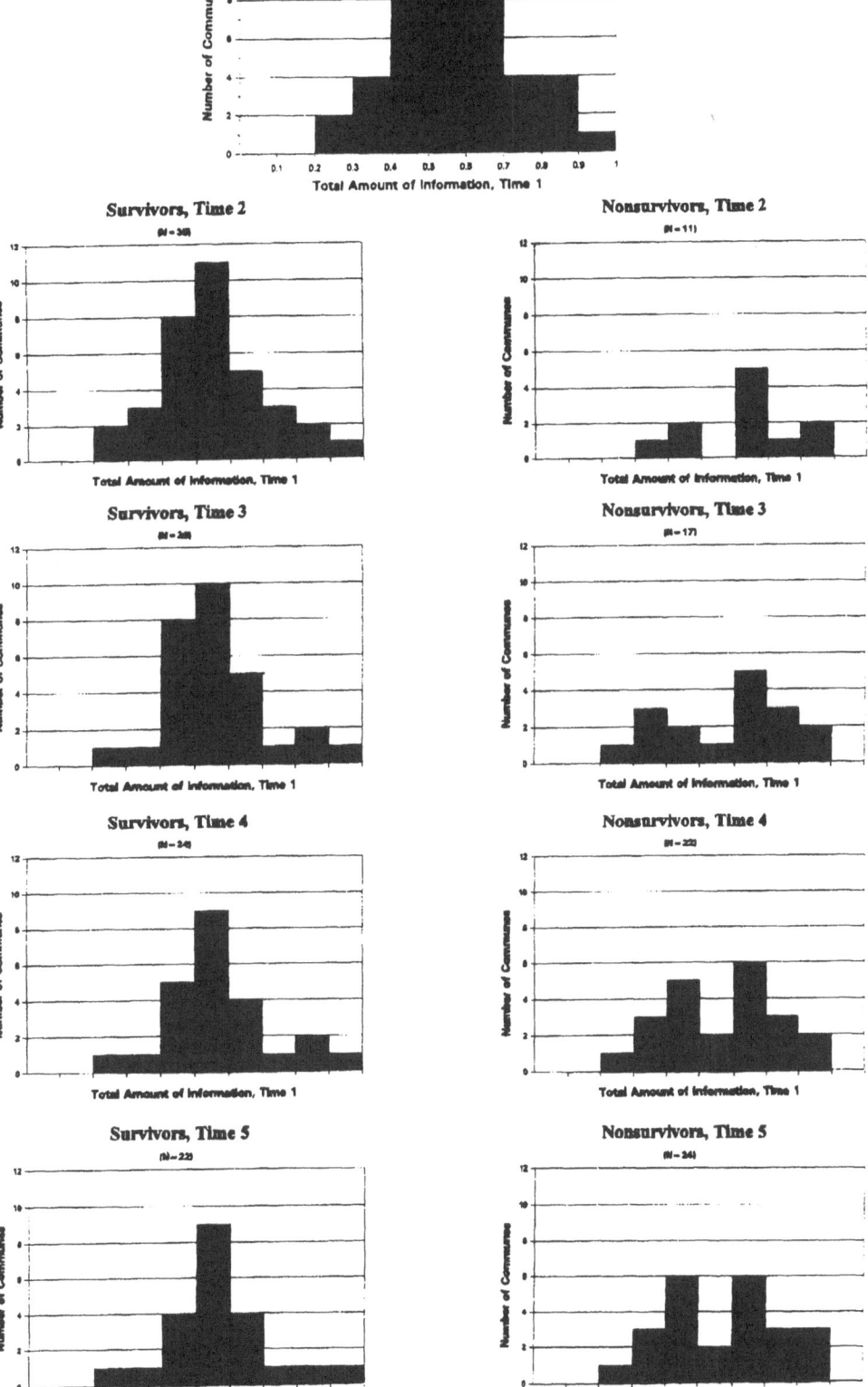

493

Hypotheses 4 and 5

Figure 6 presents a time-series of scatterplots showing the relationship between flux and control at Time 1 to stability at Time 2 and at Time 3--the relationship between the composition (in terms of flux and control) of the information provided by a collective's communication system at a given point in time, and the stability of the collective at two successive moments in the future. The scatterplot on the far left-hand side is for all communes plotted by their values for flux and control at Time 1. Holding the values for each commune on flux and control constant at Time 1, the scatterplots for Time 2 and Time 3 are divided into a plot for survivors (top row of scatterplots in Figure 9) and a plot for nonsurvivors (bottom row).

Starting with the baseline pattern at Time 1 for all communes, three patterns become increasingly evident as survival status is plotted at Time 2 and Time 3. First, as expected by Hypothesis 4, the probability of instability is highest for groups in the peripheral regions of the field--that is, for groups with the greatest imbalance between flux and control. Second, survivors tend to form a triangular pattern--in line with Hypothesis 3--with most groups clustered together in the mid-region. And third, consistent with Hypothesis 5, location in this mid-region at Time 1 is strongly related to survival at Time 3, twenty-four months into the future. What is most striking about the results for the mid-region is that the pattern for survivors is virtually the **complement** of that for nonsurvivors: *there is a complete **absence** of nonsurvivors in the mid-region where the greatest **concentration** of survivors is observed.*

Two bands of stability and two bands instability (orthogonal to the main axis) are apparent in the patterns for survivors and nonsurvivors at Time 3 (Figure 6). Immediately below the cluster of the three stable communes in the high-flux/high-control region is an upper-band of instability that separates the former from a set of stable communes in the mid-region. And beneath this stable region is a lower-band of unstable communes. These different bands of communes seem to distinguish functional from dysfunctional combinations of flux and control.

To test this interpretation, we divided the full sample of communes into stable and unstable sets such that the probability of survival was maximized for the former while being minimized for the latter. Operationally, this entailed establishing partitions that would mark the upper and lower bounds to the regions where stability is optimized.

The results are shown in the scatterplot for all communes in Figure 7. This scatterplot is identical to the scatterplot at Time 1 in Figure 4 with the following additions: first, the three lines separating the bands of stable and unstable regions, as just established, are indicated; and second, the survival status for each commune is shown at Time 3 (nonsurvivors are shown as hollow dots in Figure 7). It is clearly evident that the three partitions separate two areas of stability--one in the mid-region, consistent with Hypotheses 3, and one in the apex of the high-flux/high-control region, consistent with Hypothesis 4--from two adjoining regions characterized by a high probability of collective instability; the differences in the rates of instability, by Time 3, between the communes in the four regions is statistically significant (*chi*-square = 16.928, pr. = .0007). Moreover, in addition to its extraordinarily high stability *over* the twenty-four month period from the point of initial measurement--in line with Hypothesis 5--the mid-region also is distinguished by the lack of dispersion of communes along the low control-high flux/high control-low flux axis. Instead, there is a strong tendency to locate *between* these extremes of rapid (high) flux and rigid (high) control in the area expected to define efficient information processing--a finding consistent with Hypothesis 4.

Figure 6 (next page). **Scatterplots of Communes on Flux and Control at Time 1 by Stability (Survival Status) at Time 2 and Time 3**

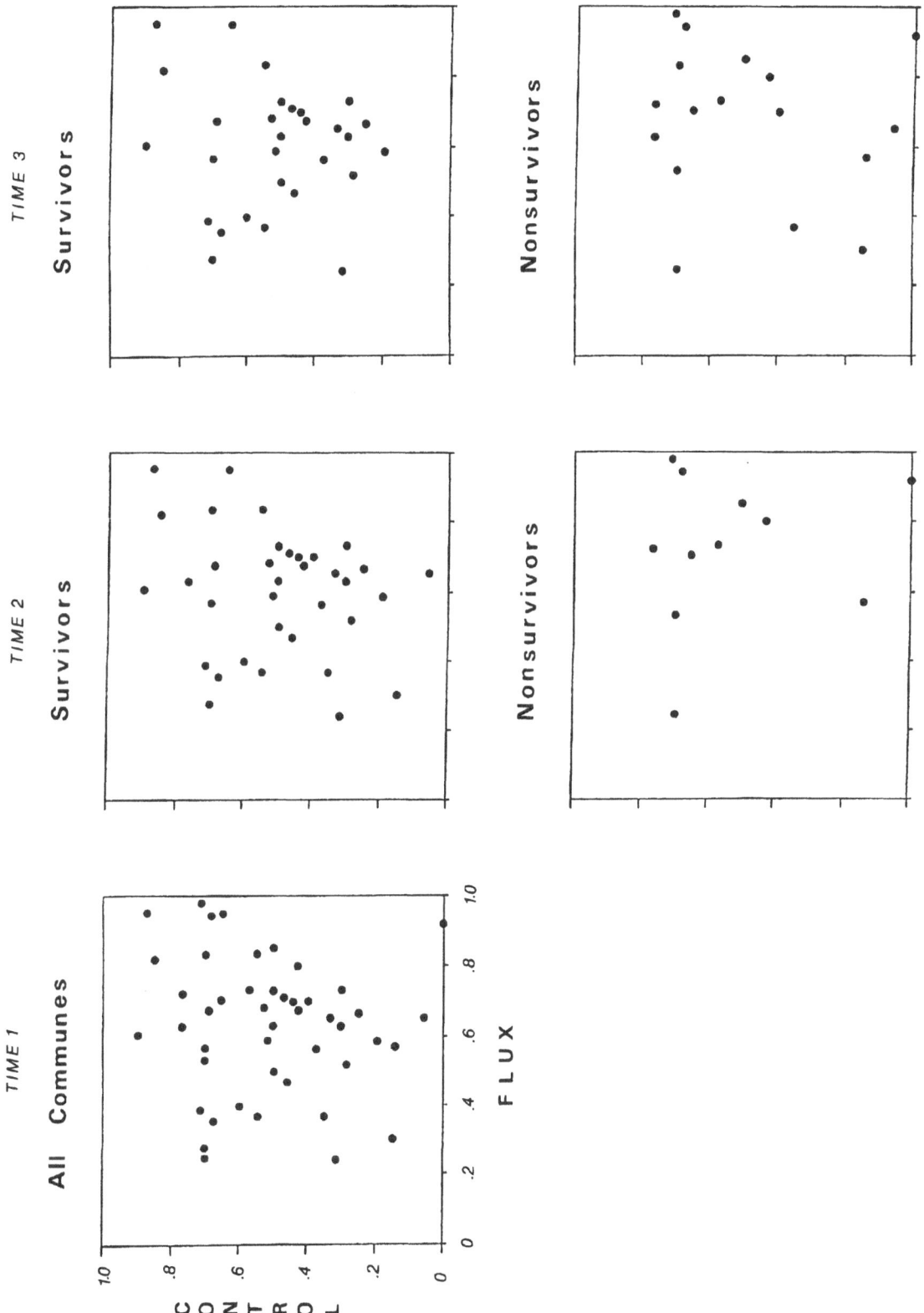

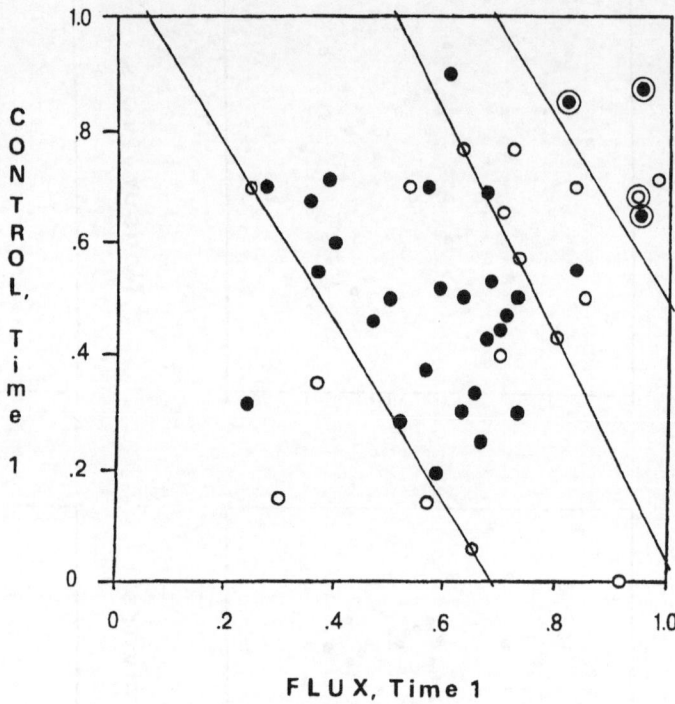

FLUX, Time 1

KEY:

● Survived through Time 3 (N=29)

○ Dissolved by Time 3 (N=17)

◉ Charismatic leader in residence (N=4)

Figure 7. Scatterplot of Communes on Flux and Control at Time 1 by Stability (Survival Status) at Time 3, and Showing Transformational Communes (Charismatic Leader in Residence)

Finally, also shown in Figure 7 are four communes, out of the whole sample, which had a charismatic leader living in residence with the group (circled in Figure 7). Of all communes in the sample, these were the collectives most intent on achieving a radical restructuring of social order. All four of these transformation-oriented (charismatic) communes--three of which were still in existence by Time 3--are concentrated *exclusively* in the apex of the high flux/high control region; the fifth group (a nonsurvivor) is a noncharismatic commune whose members expressed a strong desire for charismatic leadership as the means to facilitate their efforts at social change. As established elsewhere (Bradley, 1987: 167-193; 264-268), charismatic leadership is not only correlated with enormous increases in flux and control, but when these two are conjoined as a balanced-coupling charismatic leadership also is associated with an increase in stability.

A Multivariate Model of Optimality

We subjected these bi-variate results to a stronger statistical test with *discriminant function analysis*. This procedure aims to construct a multivariate linear (discriminant) function that maximizes the separation between two or more mutually exclusive groupings of data and offers a test of predictive power by comparing the *a priori* group classifications against those made by the discriminant function/s. As a measure of *statistical optimality*, it thus provides a rigorous means of testing the finding that, in relation to other factors, our measures of flux and control provide the *best* means of predicting optimal collective stability.

To perform the discriminant analysis, the communes were classified into one of the four categories of stability at Time 3 shown in Figure 7: namely, location in the upper band of stability, location in the mid-band of instability, location in the mid-band of stability, or location in the lower band of instability--referred to forthwith as *stable-transformational, unstable-turbulent, stable-optimal,* and *unstable-insufficient*, respectively. Using this classification as the dependent variable, two stepwise discriminant analyses[1] were conducted: one using nine sociological variables as independent variables; the second was conducted with the addition of flux and control. The univariate statistics (mean, SD, Wilks' Lambda, and univariate F-ratio) are given in Table 2a.

The first analysis, conducted on the nine sociological variables alone, was not successful. Only one variable, the mean proportion of members "who would reject an offer of $10,000 to leave the commune" (Prpn. Reject $10K), met the selection criteria for the step-wise analysis, and it was insufficient for the analytic task at hand; a minimum of two discriminant functions are required to discriminate among more than two groupings of data.

Adding the two measures of flux and control to the nine variables examined in the first analysis, a second discriminant analysis was conducted (see Table 2b). The *only* two variables selected in the stepwise procedure were the two measures of flux and control: flux was entered at the first step (min. D-squared = 1.378, pr. = .0033; Wilks' Lambda = .445, pr. = .0000); control at the second step (min. D-squared = 6.281, pr. = .0003; Wilks' Lambda = .142; pr. =.0000). The F-test of the differences between each pair of groupings after Step 2 (which range from F = 10.213, pr. = .0003, to F = 91.603, pr. = .0000) shows that there are differences between each pair which cannot be explained by chance.

---

[1]Maximizing the minimum Mahalanobis distance (min. D-squared--a measure of separation) between the four groupings of communes, was the selection rule used for the stepwise multivariate analysis; the statistical significance of the F-statistic was used as the criterion to enter (pr. $\leq$ .050) and remove (pr. $\geq$ .100) the independent variables.

Table 2a

Discriminant Function Analysis of Stability Classification By Selected Characteristics: Univariate Statistics

	Stability Groupings											
Variable	Unstable-Insufficient		Stable-Optimal		Unstable-Turbulent		Stable-Transformational		Total		Wilks' Lambda[2]	Univariate F-ratio Pr.[3]
	Mean	SD[1]	Mean	SD	Mean	SD	Mean	SD	Mean	SD		
(N)	(6)		(24)[4]		(10)		(5)		(45)			
Mmbrshp. Criteria	2.00	.89	1.92	.93	1.80	1.03	2.00	1.00	1.91	.92	.994	.972
Part of Larger Orgn.	.33	.52	.46	.51	.30	.48	.40	.55	.40	.50	.981	.848
Degree of Authority	1.33	.52	1.50	.51	1.30	.48	1.80	.45	1.47	.50	.914	.290
Control	.286	.231	.462	.180	.654	.143	.755	.101	.514	.218	.589	.000
Flux	.396	.174	.578	.152	.739	.089	.928	.064	.628	.198	.445	.000
Formal Rules	1.50	.55	1.42	.50	1.30	.48	1.60	.55	1.42	.50	.969	.724
Group Age, yrs.	2.67	1.51	3.38	1.95	2.20	1.14	2.20	1.10	2.89	1.71	.898	.216
Group Size, # adults	8.67	2.73	9.08	4.50	8.20	2.30	8.20	1.79	8.73	3.60	.987	.912
Ideolological Agrmnt.	1.00	0	1.42	.50	1.30	.48	1.80	.45	1.38	.49	.826	.047
Prpn. Old Mmbrs.	.41	.35	.47	.30	.37	.34	.70	.22	.47	.31	.912	.280
Prpn. Reject $10K	.29	.19	.58	.32	.55	.24	.98	.05	.58	.32	.704	.002

[1] Standard Deviation.

[2] U-statistic.

[3] Statistical significance, with 2 and 43 degrees of freedom.

[4] Excludes 1 case with a "missing value;" the mean value was assigned to this case for the classification analysis.

Table 2b
Discriminant Function Analysis of Stability Classification By
Selected Characteristics: Stepwise Results and Canonical Analyses

Summary of Stepwise Analysis*

Variable	Step	Wilks' Lambda	Pr.	Minimum D-squared	Pr.	Equivalent F-statistic	Pr.
Flux	1	.445	.0000	1.378	.0033	9.726	.0033
Control	2	.142	.0000	6.281	.0003	10.213	.0003

*Maximum significance of F-statistic to enter = .050; minimum significance of F-statistic to remove = .100.

Test of Differences Between Pairs of Groupings After Step 2

Pairs of Stability Groupings	F-statistic	Significance*
Optimal/Insufficient	20.331	.0000
Optimal/Turbulent	28.245	.0000
Optimal/Transformational	57.818	.0000
Transformational/Turbulent	10.213	.0003
Transformational/Insufficient	91.603	.0000
Insufficient/Turbulent	61.731	.0000

*With 2 and 40 degrees of freedom.

Canonical Discriminant Functions

	Function 1	Function 2
Canonical Correlation	.926	.085
Squared Canonical Correlation	.857	.007
Percent of Variance	99.88%	.12%
Eigenvalue	5.988	.007

Unstandardized Canonical Discriminant Function Coefficients

	Function 1	Function 2
Control	7.014	3.590
Flux	9.406	-3.402
(Constant)	-9.513	.293

The results of the classification analysis (Table 2c) show that the two discriminant functions were able to correctly predict the stability grouping for each commune in 45 of 46 cases, an overall success rate of 98%: 5 (83%) correct of 6 in the unstable-insufficient category; 25 (100%) correct of 25 in the stable-optimal category; 10 (100%) correct of 10 in the unstable-turbulent category; and 5 (100%) correct of 5 communes in the stable- transformational category (obviously, these prediction rates are substantially higher than the prior probabilities: 0.13, 0.54, .22, and 0.11, respectively).

A split-sample analysis (not shown here due to space restrictions), to test the reliability of these results, was conducted by randomly dividing the 46 communes into two samples of 23 cases each, replicating the stepwise multivariate discriminant analysis on the nine sociological variables plus flux and control, and using the discriminant functions constructed on the first-half sample to predict the classification of cases in the second half-sample into the four *a priori* stability groupings of communes.

Again, the only variables selected in the stepwise analysis were flux and control; all of the other variables failed the selection criteria. As before, flux, had the strongest discriminating power and was entered at the first step (min. D-squared = 0.882, pr. = .1242; Wilks' Lambda = .405, pr. = .0005); control entered at the second step (min. D-squared = 6.012, pr. = .0303; Wilks' Lambda = .097; pr. = .0000). The reduction in Wilks' Lambda suggests that most of the association observed between the four groupings of communes had been removed, and the F-test of the differences between each pair of groupings after Step 2 (which ranged from F = 4.271, pr. = .0303, to F = 48.0633, pr. = .0000) indicates that the differences between each pair were statistically significant. Finally, the two canonical discriminant functions constructed were able to correctly classify 19 (83%) of the 23 communes into their four *a priori* groupings[1]. In short, as a statistical means for testing our findings on an independent sample of collectives, these results offer strong corroboration.

Overall then, the results of the discriminant function analysis confirms our conclusion based on more simple statistical procedures: namely, that flux and control are predictive of collective stability.

DISCUSSION

Self-Organization of Endogenous Communication

Drawing on the thesis and empirical results presented above, a model of the social collective as a self-organizing communication system was constructed (see Figure 8). In the terms of this model, the communicative structure is formed by the interpenetration of networks of endogenous relations organized along two dimensions in which the values allocated in each dimension define points within a social field. The values ascribed to the horizontal dimension represent flux, the amount of activation of potential energy in a social collective. The values ascribed to the vertical dimension represent the amount of control (the degree to which individuals are interconnected by a transitively ordered network of relations) exercised at that location. The

---

[1] The breakdown of cases correctly classified in each category is: all three (100%) cases belonging to the stable-transformational grouping, three (75%) of four cases in the unstable-turbulent grouping, thirteen (93%) of fourteen cases in the stable-optimal grouping, and none (0%) of the two cases in the unstable-insufficient grouping.

**Discriminant Function Analysis of Stability Classification By
Selected Characteristics: Classification Results**

Predicted Group

Actual Group	Unstable-Insufficient		Stable-Optimal		Unstable-Turbulent		Stable-Transfor-mational		Total	Prior Pr.
	N	%	N	%	N	%	N	%	N	
Unstable-Insufficient	5	83.3	1	16.7	0	0	0	0	6	.13
Stable-Optimal	0	0	25	100.0	0	0	0	0	25	.54
Unstable-Turbulent	0	0	0	0	10	100.0	0	0	10	.22
Stable-Transfrmtnl.	0	0	0	0	0	0	5	100.0	5	.11
Total	5		26		10		5		46	1.00

---

coordinates representing the dimensions bound a phase space within which each value represents an amount of information (in Gabor's terms) characteristic of the communicative structure of the collective.

Two regions of stability, separated by a turbulent gap, can be distinguished within the phase space. These are regions associated with viable patterns of social communication. They are located within a larger region in which the minimum values for efficient social communication are not met so that various forms of collective dysfunction result (areas labeled as *insufficiency*, *volatility*, and *ossification* in Figure 8).

All regions are separated from each other, marked, in the terms of non-linear dynamics, by a phase transition from psycho-social instabilities to [far-from-(physical)-equilibrium] psycho-social stabilities in collective organization (Jantsch, 1980; Prigogine and Stengers, 1984). The region of optimal function represents, therefore, a qualitative change in psycho-social organization. The phase transition from dysfunctional to stable collective forms (which includes the area between the two stable regions) is described by fluctuations in potential and control which end in a point (the bi-furcation point) where the patterns of energy activation and

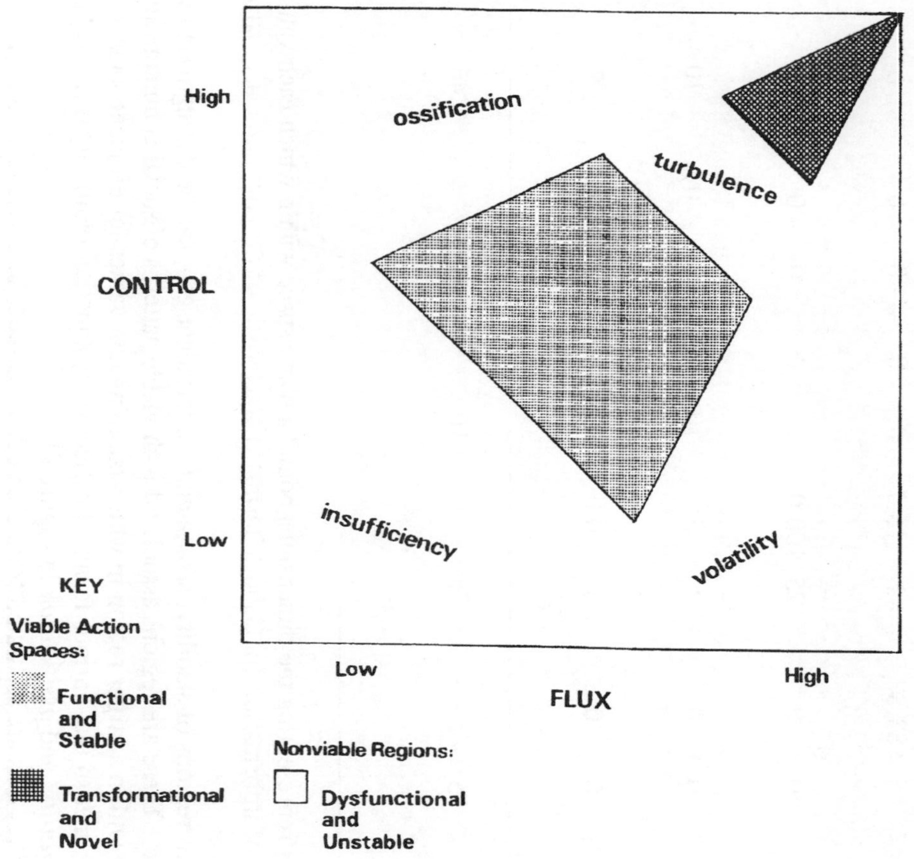

Figure 8. Model of Dynamics of Communication and Collective Organization

expenditure no longer dissipate into the environment--no longer average out to equal the energy levels of the surrounding context--but coalesce to crystallize as an emergent stable collective order.

The region of dysfunction surrounds the region of optimal collective function which is centered along the main diagonal of the phase space, and in which there is a progressive narrowing of optional structures for stable collective organization based on the increasingly close articulation between flux and control. Thus the shape of the space of optimal collective function is triangular. To defy the tendency toward entropy (disorder), and sustain a stable collective order, requires minimizing the fluctuations by linking the activation of potential to the control operations so that the energy expenditure of *all* members is *in*-formed in relation to the collective's organization. Thus, in terms of the data presented in Figure 1, stable organization requires a minimum of flux *and* a minimum of control. This region of optimal function is consistent with thermo-dynamically inspired connectionist models of neural networks (e.g., Hopfield, 1982, and Hinton and Sejnowski, 1986). In such models *efficient pattern matching is found to occur in a region between total randomness and total organization*: in our terms, between rapid flux and rigid control.

Separated by a turbulent gap, at the apex of the optimal region, is a small subregion (labeled *transformational* in Figure 8) defined by an almost one-to-one relationship between flux and control. To assure stability a tight coupling between the two must be maintained, taking much effort. Often, when such an effortful course is in operation, a sudden organizational spasm occurs. The spasm has two possible outcomes: one is structural transformation in the pattern of information processing, resulting in a totally reorganized, qualitatively different collective; the other is structural devolution, the complete breakdown and collapse of the collective as a viable organization.

CONCLUSION

Our model concerns the internal structure of the collective. This internal structure is conceived to be based on the biological potential of the individuals composing the collective to engage in physical work, measured as energy. When activated by the collective, this biological energy is made available for social interaction as a field of potential biosocial energy. We have labeled this dimension of the endogenous order, flux.

In the other dimension, individuals are connected hierarchically. We have labeled this dimension as control because it appears to direct and regulate the activation of the biosocial energy of the collective. Controls over the activation and distribution of flux result in social communication by way of quantum-like units of information (logons)--moment-by-moment descriptions, in terms of space-time and spectral coordinates, of the collective's endogenous organization. Because these elementary units of information overlap as a series, the collective's order at a given moment is informed by the order probabilistically implicit (as "expectations") in the units of succeeding moments.

The efficiency of the internal dynamics, and its relationship to the collective's action, was found to display an optimal (energy conserving) combination of flux and control which is associated with stable collective action. Our results thus show that for the group to survive as an effective working unit, an efficient communicative structure was required. Only those configurations that produce a path of least action--one which entailed the smallest amount of turbulence--resulted in a stable, efficient collective. This finding, and our finding of the absence of a relationship between stability and measures of the collective's normative and structural organization, suggest a new and promising approach to the study of social collectives based on the principles of self-organization.

APPENDIX

Summary Statistics for the Operational Procedures Used to Measure Flux, Control, and Stability
(N = 46 communes)

	Mean Dyadic Density[1]	Triadic Structure (Mean Proportions) Symmetric Triad Types				
		003	102	201	300	Total
Flux						
Loving (L)	.44	.260	.341	.208	.192	1.001
Improving (I)	.46	.232	.348	.224	.196	1.000
Exciting (E)	.17	.622	.285	.067	.027	1.001
Mean (L, I, E / 3)	.36	.371	.325	.166	.138	1.000

		Asymmetric Triad Types						
		003	012	021D	021U	021C	030T	030C	Total
Control									
Power	.30	.097	.261	.137	.113	.129	.243	.020	1.000

Stability
Survival status, Time 1 - Time 5 (12 month intervals)

	T1 1974		T2 1975		T3 1976		T4 1977		T5 1978		Total	
	N	%	N	%	N	%	N	%	N	%	N	%
Survived	46	100	35	76	29	83	24	83	22	92	22	48
Disintegrated	0	0	11	24	6	17	5	17	2	8	24	52
Total	46	100	46	100	35	100	29	100	24	100	46	100

[1]Number of relations of a selected dyad type / all possible relations. For the three indicators of flux (*loving*, *improving*, and *exciting*) the numerator was the number of relations formed as a dyad of reciprocated positive relations (i.e., both *i* and *j* answered "yes"); for the indicator of control (*power*) the numerator was the number of dyads for which an asymmetric ordering was evident in the relationship between *i* and *j* (i.e., either *i* had greater power in the relationship than *j*, or *j* had greater power than *i*).

REFERENCES

Abraham, F. D. (1991). *A Visual Introduction to Dynamical Systems Theory for Psychology*. Santa Cruz, CA: Aerial Press.

Aldrich, H. (1979). *Organizations and Environments*. Englewood Cliffs, NJ: Prentice-Hall.

Ashby, W. R. (1956). *An Introduction to Cybernetics*. London: Chapman & Hall.

Blau, P. M. & R. Scott (1962). *Formal Organizations*. San Francisco: Chandler.

Bohm, D., & B. J. Hiley, (1993). *The Undivided Universe*. London: Routledge.

Bougon, M., & J. Komocar (1990). Directing strategic change: A dynamic wholistic approach In A. Huff (ed.), *Mapping Strategic Thought*, pp. 135-163. New York: Wiley.

Bradley, R. T. (1987). *Charisma and Social Structure: A Study of Love and Power, Wholeness and Transformation*. New York: Paragon House.

Bradley, R. T. (1996; in press). The anticipation of order in biosocial collectives. *World Futures: The Journal of General Evolution*.

Bradley, R. T., & K. H. Pribram (1996; in press). Communication and optimality in biosocial collectives. In D. S. Levine & W. R. Elsberry (eds.), *Optimality in Biological and Artificial Networks*. Hillsdale, NJ: Lawrence Erlbaum Associates.

Bradley, R. T., & N. C. Roberts (1989a). Relational dynamics of charismatic organization: The complementarity of love and power. *World Futures: The Journal of General Evolution*, **27**: 87-123.

Bradley, R. T., & N. C. Roberts (1989b). Network structure from relational data: Measurement and inference in four operational models. *Social Networks*, **11**: 89-134.

Carlton-Ford, S. (1993). *The Effects of Ritual and Charisma*. New York: Garland Publishing.

Carley, K. (1991). A theory of group stability. *American Sociological Review*, **56**: 331-354.

Cherry, C. (1978). *On Human Communication: A Review, a Survey, and a Criticism*. Cambridge, MA: The MIT Press.

Child, J. (1972). Organization structure, environment and performance: The role of strategic choice. *Sociology*, **6**: 1-22.

Cook, K. S. (1977). Exchange and power in networks of interorganizational relations. In J. K. Benson (ed.), *Organizational Analysis: Critique and Innovation* (pp. 64-84). Beverly Hills, CA: Sage.

Dendrinos, D., & M. Sonis (1990). *Chaos and Socio-Spatial Dynamics*. New York: Springer-Verlag.

Gabor, D. (1946). Theory of communication. *Journal of the Institute of Electrical Engineers*, **93**: 429-457.

Garud, R., & S. Kotha (1994). Using the brain as a metaphor to model flexible production systems. *Academy of Management Review*, **19**: 671-698.

Glazer, R. (1986). A holographic theory of decision making. Unpublished paper, Graduate School of Business, Columbia University, New York.

Hannan, M., & J. Freeman (1977). The population ecology of organizations. *American Journal of Sociology*, **82**: 929-64.

Hinton, G. E., & T. J. Sejnowski (1986). Learning and relearning in Boltzmann machines. In D. E. Rumelhart & J. L. McClelland (eds.), *Parallel Distributed Processing: Explorations in the Microstructure of Cognition* (Vol. 1, pp. 282-317). Cambridge, MA: MIT Press.

Holland, P. W., & S. Leinhardt (1976). Local structure in social networks. In D. R. Heise (ed.), *Sociological Methodology 1976* (pp. 1-45). San Francisco: Jossey-Bass.

Hopfield, J. J. (1982). Neural networks and physical systems with emergent collective computational abilities. *Proceedings of the National Academy of Sciences*, **79**, 2554-2558.

Hutchins, E. (1991). The social organization of distributed cognition. In L. B. Resnick, J. M. Levine, and S. D. Teasley (eds.), *Perspectives on Socially Shared Cognition*. Washington, DC: American Psychological Association.

Jantsch, E. (1980). *The Self-Organizing Universe: Scientific and Human Implications of the Emerging Paradigm of Evolution*. Oxford: Pergamon Press.

Kaiser, G. (1994). *A Friendly Guide to Wavelets*. Boston: Birkhauser.

King, J. S., M. Xie, B. Zheng, & K. H. Pribram (1994). Spectral density maps of receptive fields in the rat's somatosensory cortex. In *Proceedings of the Second Appalachian Conference on Behavioral Neurodynamics*, pp. 557-571. Hillsdale, NJ: Lawrence Erlbaum Associates.

Loye, D., & R. Eisler (1987). Chaos and transformation: Implications of disequilibrium theory for social science and society. *Behavioral Science,* **32**: 53-65.

MacKenzie, K. D. (1991). *The Organizational Hologram: The Effective Management of Organizational Change*. Norwell, MA: Kluwer Academic Publishers.

March, J. G. & J. P. Olsen (1976). *Ambiguity and Choice in Organizations*. Bergen, Norway: Universitetsforlaget.

Merton, R. K. (1940). Bureaucratic structure and personality. *Social Forces,* **18**: 560-568.

Morgan, G. (1986). *Images of Organization*. Beverly Hills, CA: Sage Publications, Inc.

Nicolis, G., & I. Prigogine (1977). *Self-Organization in Nonequilibrium Systems: From Dissipative Structures to Order Through Fluctuation*. New York: Wiley-Interscience.

Pfeffer, J., & G. R. Salancik (1978). *The External Control of Organizations: A Resource Dependence Perspective*. New York: Harper and Row.

Pribram, K. H. (1991). *Brain and Perception: Holonomy and Structure in Figural Processing*. Hillsdale, NJ: Lawrence Erlbaum Associates.

Pribram, K. H., & D. McGuinness (1975). Arousal, activation and effort in the control of attention. *Psychological Review*, **82**, 116-149.

Prigogine, I., & I. Stengers (1984). *Order out of Chaos: Man's New Dialogue with Nature*. New York: Bantam Books.

Rumelhart, D. E. (1992). Towards a microstructural account of human reasoning. In S. Davis (ed.), *Connectionism: Theory and Practice*: 69-83. New York: Oxford University Press.

Sandelands, L. E., & R. E. Stablein (1987). The concept of organization mind. In S. Bacharach and N. DiTomaso (eds.), *Research in the Sociology of Organizations*, **5**: 135-161. Greenwich, CT: JAI Press.

Santa Maria, M., J. King, M. Xie, B. Zheng, K. H. Pribram, & D. Doherty (1995). Responses of somatosensory cortical neurons to spatial frequency and orientation: A progress report. In J. King & K. H. Pribram (eds.), *Scale in Conscious Experience: Is the Brain too Important to Left to Specialists to Study?* Pp. 157-168. Mahwah, NJ: Lawrence Erlbaum Associates.

Shannon, C. E. (1949). The mathematical theory of communication. In C. E. Shannon & W. Weaver, *The Mathematical Theory of Communication* (pp. 3-91). Urbana, IL: The University of Illinois Press.

Silverman, D. (1970). *The Theory of Organizations*. Exeter, NH: Heinemann.

Varela, F., H. R. Maturana, & R. Uribe (1974). Autopoiesis: the organization of living systems, its characterization and a model. *Biosystems*, **5**: 187-196.

Weick, K. E., & K. H. Roberts (1993). Collective mind in organizations: heedful interrelating on flight decks. *Administrative Science Quarterly*, **38**: 357-381.

Zablocki, B. D. (1971). *The Joyful Community*. New York: Penguin Books.

Zablocki, B. D. (1980). *Alienation and Charisma: A Study of Contemporary Communes*. New York: The Free Press.

The Transcendental Level: Self Organization on a Grand Scale

21.
Reflections in Clouded Mirrors: Selfhood in Animals and Machines

Subhash C. Kak

Reflections In Clouded Mirrors: Selfhood In Animals and Machines

Subhash C. Kak
Department of Electrical and Computer Engineering
Louisiana State University
Baton Rouge, LA 70803-5901

Technical Report 92/12 ECE-LSU December 1, 1992
Revised October 24, 1994

Abstract

This essay presents a summary of the Vedic theory of consciousness in juxtaposition with the parallel ideas of quantum mechanics and neuroscience. The ancient Vedic tradition of philosophy of consciousness that goes back to at least 2000 BC posits that analytical approaches to defining awareness or personhood end up in paradox. In this tradition one views awareness in terms of the reflection that the hardware of the brain provides to an underlying illuminating or awareness principle called the *self*. This tradition allows one to separate questions of the tools of awareness, such as vision and hearing and the mind, from the *person* who obtains this awareness. This tradition is reviewed regarding the question of the definition of personhood in animals and machines.

1   Introduction

To place contemporary research in consciousness, artificial intelligence (AI), and self-awareness in context we review the early Vedic theory of consciousness (Aurobindo 1949, Sinha 1958) and show how its insights parallel those of quantum mechanics and neuroscience. In this theory, which dates back to at least 2000 BC (Kak 1994a), one views awareness in terms of the reflection that the hardware of the brain provides to an underlying illuminating or awareness principle called the *self*. This approach allows one to separate questions of the tools of awareness, such as vision, hearing and the mind,

511

from the person who obtains this awareness. The person is the conscious self, who is taken to be a reservoir of infinite potential. But the actual capability of the animal are determined by the neural hardware of its brain. This hardware may be compared to a mirror. The hardware of the human brain represents the clearest structure to focus the self, which is why humans are able to perform in ways that other animals cannot. Within the framework of this theory humans and other animals are persons and their apparent behavioral distinctions arise from the increased cloudedness of the neural hardware of the lower animals. Self-awareness is an emergent phenomenon which is grounded on the self *and* the associations stored in the brain.

We begin with a review of life from a modern scientific viewpoint. Living systems are dynamic structures, that are defined in terms of their interaction with their environment, and their neural structures incorporate self definition. Their behavior reflects their past history in terms of instincts, which is why a bird is afraid of a cat. Living systems can also be defined recursively in terms of living sub-systems. Thus for ants one may consider their society, an ant colony, as a living superorganism and in turn the ant's sub-systems are also living. Such a recursive definition appears basic to life. Machines, on the other hand, are based on networking of elements so as to instrument a well-defined computing procedure. Since they lack a recursive self definition, it appears that machines cannot be conscious.

There is considerable evidence that supports the notion of a mediating consciousness in human behavior. This is evident not only from the fact that responses are different in sleepwalking and awake states but from the considerable experimentation with split-brain patients (Trevarthen 1990). Then there are the experiments of Kornhuber (1974) that indicate that it takes about eight-tenths of a second for the readiness potential to build up in the brain before voluntary action begins. Furthermore, the experiments of Libet (1973) show that the mind extrapolates back in time by about half a second or so the occurrence of certain events. Since consciousness is not an epiphenomenon and since it possesses a unity, it should be described by a quantum mechanical wavefunction. But no representation in terms of networking of classical objects such as threshold neurons can model a wavefunction.

But the concept of a quantum mechanical wavefunction comes with its own problems of interpretation. The collapse of the wavefunction, upon measurement, leads to logical difficulties in any unified scenario, as is exemplified by the Schrödinger cat paradox. The complementarity interpretation of quantum mechanics avoids such paradoxes at the cost of an abandonment of a single unified picture.

Eugene Wigner (1967) argued that the laws of quantum mechanics may not apply to consciousness. In a variant of the setting of the Schrödinger cat experiment, he visualized two conscious agents, one inside the box and another outside. If the inside agent makes an observation that leads to the collapse of the wavefunction, then how is the linear superposition of the states for the outside observer to be viewed? Wigner argued that in such a case, with a conscious observer as part of the system, linear superposition must not apply. This result, now called the Wigner's friend paradox, and others have led many quantum theorists to argue that basic advances in physics would eventually require one to include consciousness in the scientific framework. A recent, influential, exposition of this view has been given in Roger Penrose's *The Emperor's New Mind.* But quantum mechanics has not yet been challenged and it appears that quantum mechanical models of consciousness cannot resolve the paradoxes mentioned above. It is thus likely that consciousness cannot be fitted in a traditional kind of a scientific theory.

The Vedic system, which was an earlier attempt to unify knowledge, was confronted by similar paradoxes. It is well known that Schrödinger's development of quantum mechanics was inspired, in part, by Vedanta, the full-blossomed Vedic system (Schrödinger 1961). His debt to the Vedic views is expressed in an essay he wrote in 1925 *before* he created his quantum theory:

> This life of yours which you are living is not merely a piece of this entire existence, but is in a certain sense the "whole"; only this whole is not so constituted that it can be surveyed in one single glance. This, as we know, is what the Brahmins express in that sacred, mystic formula which is yet really so simple and so clear: *tat tvam asi,* this is you. Or, again, in such words as "I am in the east and the west. I am above and below, *I am this entire world.* (See Schrödinger 1961; Moore 1989, pages 170-3)

Schrödinger used Vedic ideas also in his immensely influential book *What is Life?* (Schrödinger 1965) that played a significant role in the development of modern biology. According to his biographer Walter Moore, there is a clear continuity between Schrödinger's understanding of Vedanta and his research:

> The unity and continuity of Vedanta are reflected in the unity and continuity of wave mechanics. In 1925, the world view of

physics was a model of a great machine composed of separable interacting material particles. During the next few years, Schrödinger and Heisenberg and their followers created a universe based on superimposed inseparable waves of probability amplitudes. This new view would be entirely consistent with the Vedantic concept of All in One. (Moore 1989, page 173)

In view of this connection between the Vedic system and quantum mechanics and the fact that quantum mechanical models of consciousness are being attempted, it is important to see how the Vedic philosophers developed their classificatory model of consciousness. A summary of this classificatory model is the main focus of the paper. We begin with further issues related to the self, describe the Vedic model, and discuss implications for animal and machine intelligence.

2 Paradox and Unification of Knowledge

Models of the mind (man's inner space) show several paradoxical characteristics. Some of these are unique and others have close parallels with the paradoxes associated with the outer or physical reality. This is not surprising, since the knowledge of nature is obtained by the mind and therefore it is to be expected that this knowledge will be expressed in categories that relate to the nature of the mind.

The links between our understanding of the physical world and the cognitive categories of the mind are illumined by a variety of sources. This includes human and animal behavior and linguistics and psychology, and the properties of logical structures of mathematics and physics. Investigations of the neural structure of the brain may be related to a hypothesized nature of the mind. If brain structure is neuronal then cognitive capabilities should be found for networks of neurons. But the limitations of neural models have been highlighted by Sacks (1990) and others who point out that these models do not take into account the notion of self.

Self, biology, psychology

Neural network models, studied extensively in cognitive science, have almost exclusively used the threshold circuit model of the neuron (McGeer et al 1987). A threshold neuron is a device that functions like an electronic circuit and a network of such devices can only execute algorithms. There is evidence that the threshold model does not represent biological reality closely enough (Libet 1986).

The neural system of the biological organism develops in three stages: (i) proliferation, specification and migration, (ii) axon and dendritic growth, synapse formation, and (iii) nerve cell death, adult connectivity changes. The organization of the neural structure is a genetically determined consequence of its interaction with other bodily systems and the environment. The behavior of living organisms expresses, in many ways, the experience and response of the species in its past history. As example, living organisms have internal clocks or rhythms that are matched to different astronomical phenomena. Even isolated neurons generate circadian rhythm (Michel et al 1993).

The phenomenon of phantom limbs reveals a lot about the limitations of neural network models of the brain and self. A phantom limb is commonly experienced after the loss of an arm or a leg. Phantoms of body parts feel real. What is fascinating is that the phantoms of body parts that are deformed at birth can be normal Clearly all animals have a notion of self which is why a snake does not have lunch on its own tail!

The limitations of current theories of psychology were recently well summarized by the distinguished Canadian psychologist Melzack:

> The field of psychology is in a state of crisis. We are no closer now to understanding the most fundamental problems of psychology than we were when psychology became a science a hundred years ago. Each of us is aware of being a unique "self", different from other people and the world around us. But the nature of the "self", which is central to all psychology, has no physiological basis in any contemporary theory and continues to elude us. The concept of "mind" is as perplexing as ever... There is a profusion of little theories–theories of vision, pain, behaviour-modification, and so forth–but no broad unifying concepts... Cognitive psychology has recently been proclaimed as the revolutionary concept which will lead us away from the sterility of behaviourism. The freedom to talk about major psychological topics such as awareness and perceptual illusions does, indeed, represent a great advance over behaviourism. But on closer examination, cognitive psychology turns out to be little more than the psychology of William James published in 1890; some neuroscience and computer technology have been stirred in with the old psychological

ingredients, but there have been no important conceptual advances... We are adrift, without the anchor of neuropsychological theory, in a sea of facts–and practically drowning in them. We desperately need new concepts, new approaches. (Melzack, 1989, pp. 1-2)

Melzack himself proposes an anatomical substrate of the body-self to explain the phenomenon of phantom limbs. But his model does not explain other aspects of the self.

Cognitive abilities arise from a continuing reflection on the perceived world and this question of reflection is central to the brain-mind problem, the measurement problem of physics, and the problem of determinism and free-will (see, for example, Kak 1986, Penrose 1989). A dualist hypothesis (for example Eccles 1986) to explain brain-mind interaction or the process of reflection meets with the criticism that this violates the conservation laws of physics. On the other hand a brain-mind identity hypothesis, with a mechanistic or electronic representation of the brain processes, does not explain how self-awareness could arise. At the level of ordinary perception there exists a duality and complementarity between an autonomous (and reflexive) brain and a mind with intentionality. The notion of self seems to hinge on an indivisibility akin to that found in quantum mechanics.

Complementarity

The wave-particle duality encountered in quantum phenomena led Neils Bohr in 1927 to introduce the notion of complementarity as an approach to the study of the individuality of quantum and other indivisible phenomena. Complementarity is the principle that description of reality in any of the mutually contradictory pictures is incomplete; but between them such pictures form a complete, complementary description. This principle also presupposes that experiments can be unambiguously described only in classical terms. Considering the question of logical foundations of biology Bohr concluded that life (and also cognitive) processes are likewise subject to complementarity. The complementarity exhibited by life may be expressed most fundamentally between structure and behavior.

The recognition of the limitation of mechanical concepts in atomic physics would rather seem suited to conciliate the apparently contrasting viewpoints of physiology and psychology. Indeed, the necessity of considering the interaction between the measuring instruments and the object under investigation in atomic mechanics exhibits a close analogy to the peculiar difficulties in psychological analysis arising from the fact that the mental content is invariably altered when the attention is concentrated on any special feature of it (Bohr 1961).

Bohr suggested an interesting analogy between neural (thought) and quantum processes. The instantaneous state of a thought may be compared with the position of a particle, whereas the direction of change of that thought may be compared with the particle's momentum. This is described by Bohm as follows:

> Part of the significance of each element of a thought process appears to originate in its indivisible and incompletely controllable connections with other elements. Similarly, some of the characteristic properties of a quantum system (for instance, wave or particle nature) depend on indivisible and incompletely controllable quantum connections with surrounding objects. Thus, thought processes and quantum systems are analogous in that they cannot be analyzed too much in terms of distinct elements, because the *intrinsic* nature of each element is not a property existing separately from and independently of other elements but is, instead, a property that arises partially from its relation with other elements (Bohm 1951).

There is also a similarity between the thought process and the classical limit of the quantum theory. The logical process corresponds to the most general type of thought process as the classical limit corresponds to the most general quantum process. In the logical process, we deal with classifications. These classifications are conceived as being completely separate but related by the rules of logic, which may be regarded as the analogue of the causal laws of classical physics. In any thought process, the component ideas are not separate but flow steadily and indivisibly. An attempt to analyze them into separate parts destroys or changes their meanings. Yet there are certain types of concepts, among which are those involving the classification of objects, in which we can, without producing any essential changes, neglect the indivisible and incompletely controllable connection with other ideas.

Complementarity is required at different levels of description. But just as one might use a probabilistic interpretation instead of complementarity for atomic descriptions (for example, Bohm 1951), a probabilistic description may also be used for cognitive behavior. However, such a probabilistic behavior is inadequate to describe the behavior of individual agents, just as notions of probability break down for individual objects.

As an epistemological principle complementarity has been criticized for not providing a unifying picture. But from an operational point of view complementarity, by considering all kinds of responses, becomes a very useful approach. When analyzed in terms of local interactions the framework of quantum mechanics suffers from other paradoxical characteristics. This shows up in non-local correlations that appear in the manner of action at a distance (Bell 1987). But quantum mechanics remains a very successful theory in its predictive power.

The Vedic System of Knowledge

Now we consider a brief historical background to the study of the self. This history goes back at least as far as the Rigveda, conservatively dated to the late third or early second millennia BC (Kak 1994a, 1994b). The Rigveda and the other Vedic books do not present a logical resolution of the paradox but assert that knowledge is of two types: it is superficially dual but at a deeper level it has a unity. The Vedic theory implies a complementarity by insisting that the material and the conscious are aspects of the same transcendent reality. The modern scientific tradition is like the Vedic tradition since it it acknowledges contradictory or dual descriptions but seeks unifying explanations.

The Vedic approach to knowledge was based on the assumption that there exist equivalences of diverse kinds between the outer and the inner worlds. This prompted a deep examination of the human mind. In the description of physical reality the Vedic philosophers noted several paradoxes (Kak 1986). If matter is divisible, each atom must be point-like because otherwise it would be further divisible. But how do point-like atoms lead to gross matter with size? Space is neither continuous nor discontinuous, for if it were continuous its points would be non-enumerable, but if it is discontinuous then how do objects move across the discontinuity? A popular way to express these difficulties was to talk about the riddle of being and becoming. The basic question here is how does an entity change its form and become another?

The philosophical systems that arose in India early on were meant to help one to find clues to the nature of consciousness. It was recognized that a complementarity existed between different approaches to reality, presenting contradictory perspectives. That is why philosophies of logic (nyāya) and physics (vaiśeṣika), cosmology and self (sāṃkhya) and psychology (yoga), and language (mimāṃsa) and reality (vedānta) were grouped together in pairs. The system of Sāṃkhya considered a representation of matter and mind in different enumerative categories. The actual analysis of the physical world was continued outside of the cognitive tradition of Sāṃkhya in the sister system of Vaiśeṣika, that deals with further characteristics of the gross elements. The atomic doctrine of Vaiśeṣika can be seen to be an extension of the method of counting in terms of categories and relationships. The reality in itself was taken to be complex, continuous and beyond logical explanation. However, its representation in terms of the gross elements like space, mass (earth), energy (fire) and so on that are cognitively apprehendable, can be analyzed in discrete categories leading to atomicity. The cosmology of Sāṃkhya is really a reflection of the development of the mind, represented in cognitive categories.

The Greek philosophers also spoke of paradoxes inherent in descriptions. For example, we have Zeno's famed paradoxes on motion. But the Greek tradition does not appear to have dealt with the problem of consciousness.

3 The Vedic Model of the Mind

A review of the different aspects of the Vedic model of the mind may be seen in the book by Sinha (1958). This model is expressed by the famous metaphor of the chariot in Kaṭha Upaniṣad and the Bhagavad Gītā. A person is compared to a chariot that is pulled in different directions by the horses yoked to it; the horses represent the senses. The mind is the driver who holds the reins to these horses; but next to the mind sits the true observer, the self, who represents a universal unity. Without this self no coherent behaviour is possible.

3.1 The Five Levels

The individual is represented in terms of five different sheaths or levels that enclose the individual's self. These levels, shown in an ascending order, are:

- The physical body (annamaya kośa)

- Energy sheath (prāṇamaya kośa)

- Mental sheath (manomaya kośa)

- Intellect sheath (vijñānamaya kośa)

- Emotion sheath (ānandamaya kośa)

These sheaths are defined at increasingly finer levels. At the highest level, above the emotion sheath, is the self. It is significant that emotion is placed higher than the intellect. This is a recognition of the fact that eventually meaning is communicated by associations which are influenced by the emotional state.

The energy that underlies physical and mental processes is called prāṇa. One may look at an individual in three different levels. At the lowest level is the physical body, at the next higher level is the energy systems at work, and at the next higher level are the thoughts. Since the three levels are interrelated, the energy situation may be changed by inputs either at the physical level or at the mental level. When the energy state is agitated and restless, it is characterized by rajas; when it is dull and lethargic, it is characterized by tamas. The state of equilibrium and balance is termed sattva.

Prāṇa, or energy, is described as the currency, or the medium of exchange, of the psychophysiological system. The levels 3, 4, and 5 are often lumped together and called the mind.

The key notion is that each higher level represents characteristics that are emergent on the ground of the previous level. In this theory mind is an emergent entity, but this emergence requires the presence of the self.

3.2 The Structure of the Mind

Now we consider the structural characteristics of the mind. The mind is viewed as consisting of five components: manas, ahaṃkāra, citta, buddhi, and ātman.

Manas is the lower mind which collects sense impressions. Its perceptions shift from moment to moment. This sensory-motor mind obtains its inputs from the senses of hearing, touch, sight, taste, and smell. Each of these senses may be taken to be governed by a separate agent.

Ahaṃkāra is the sense of I-ness that associates some perceptions to a subjective and personal experience.

Once sensory impressions have been related to I-ness by ahaṃkāra, their evaluation and resulting decisions are arrived at by buddhi, the intellect.

Manas, ahaṃkāra, and buddhi are collectively called the internal instruments of the mind.

Next we come to citta, which is the memory bank of the mind. These memories constitute the foundation on which the rest of the mind operates. But citta is not merely a passive instrument. The organization of the new impressions throws up instinctual or primitive urges that creates different emotional states.

This mental complex surrounds the innermost aspect of consciousness which is called ātman. It is also called the self, brahman, or jīva. Ātman is considered to be beyond a finite enumeration of categories.

3.3 Hierarchical Levels Within the Brain

Since the state of mind is mediated by the pranic energy, it becomes useful to determine how this is related to the focus on the various parts of the body. In the Vedic system seven points of primary focus which are called chakras (Sanskrit cakra) are described. Their positions appear to be areas in the brain some of which map to different points on the spinal cord. The lowest one is located at the bottom of the vertebral column (mūlādhāra chakra). The next chakra is a few inches higher at the reproductive organs (svādhiṣṭhāna chakra). The third chakra (maṇipūra chakra) is at the solar plexus. The heart region is the anāhata chakra. The throat has the fifth chakra called the viśuddhi chakra. Between the eyebrows is the ājñā chakra. At the top of the head is the sahasrāra chakra.

It is believed that the stimulation of these chakras in a proper way leads to the development of certain neural structures that allow the I-ness to experience the self. In other words, the chakras are points of basic focus inside the brain that lead to the explication of the cognitive process.

3.4 Further Universal Categories

If the categories of the mind are taken to arise from pattern recognition of shadow mental images, then how are these categories associated with a single 'agent', and how does the mind bootstrap these shadow categories to find the nature of reality?

These questions were considered by the Shaivite philosophers of India who further developed the earlier Vedic ideas (Abhinavagupta 1989, Dyczkowski 1987). For these philosophers Shiva (Sanskrit Śiva) is the name for the absolute or transcendental consciousness. Ordinary consciousness is bound by cognitive categories related to conditioned behavior. By exploring the true springwells of ordinary consciousness one comes to recognize its universal (Shiva). This brings the further recognition that one is not a slave (paśu) of creation but its master (pati).

According to the ancient doctrine of Sāṃkhya, reality may be represented in terms of twenty five categories. These categories form the substratum of the classification in Shaivism. These categories are:

- (i) five elements of materiality, represented by earth, water, fire, air, ether;

- (ii) five subtle elements , represented by smell, taste, form, touch, sound;

- (iii) five organs of action, represented by reproduction, excretion, locomotion, grasping, speech;

- (iv) five organs of cognition, related to smell, taste, vision, touch, hearing;

- (v) three internal organs, being mind, ego, and intellect; and inherent nature (prakṛti), and

- consciousness (puruṣa).

These categories define the structure of the physical world and of agents and their minds. But this classification is not rich enough to describe the processes of consciousness as it is mentioned as a single category.

Shaivism enumerates further characteristics of consciousness:

- (vii) sheaths or limitations of consciousness, being time (kāla), space (niyati), selectivity (rāga), awareness (vidyā), creativity (kalā), self-forgetting (māyā), and

- (viii) five principles of the universal experience, which are correlation in the universal experience (sadvidyā, śuddhavidyā), identification of the universal (īśvara), the principle of being (sādākhya), the principle of negation and potentialization (śakti), and pure awareness by itself (śiva)

The first twenty five categories relate to an everyday classification of reality where the initial five characteristics relate to the physical inanimate world, and the next eighteen define the characteristics of the conscious organism. The inherent nature of the individual is called prakṛti while puruṣa represents self.

The next eleven categories characterize different aspects of consciousness which is to be understood in a sense different to that of mental capacities (categories 21,22,23). One of these mental capacities is akin to artificial intelligence of current computer science, which is geared to finding patterns and deciding between hypotheses. On the other hand categories 26 through 36 deal with interrelationships in space and time between these patterns and deeper levels of comprehension and awareness.

Any focus of consciousness must first be circumscribed by coordinates of time and space. Next, it is essential to select a process (out of the many defined) for attention (category 28). The aspect of consciousness that makes one have a feeling of inclusiveness with this process, followed later by a sense of alienation is called māyā (category 31). Thus māyā permits one, by a process of identification and detachment, to obtain limited knowledge (category 29) and to be creative (category 30).

3.4.1 Universal Experience

How does consciousness ebb and flow between an identity of self and an identity with the processes of the universe? According to Shaivism, a higher category (number 32) permits comprehension of oneness and separation with equal clarity. On the other hand category 33 allows a visualization of the ideal universe. Category 33 allows one to move beyond mere comprehension into a will to act. The final two categories deal with the potential energy that leads to continuing transformation (35) and pure consciousness by itself (36). Pure awareness is not to be understood as similar to everyday awareness of humans but rather as the underlying schema that the laws of nature express. The laws themselves define the śakti tattva.

The cognitive categories of Shaivism are of relevance in artificial intelligence. At present only a subset of these categories can be dealt with by the most versatile computing machines. Current research is focused on the lower categories such as endowing machines with action capacities (as in robotics) and powers of sense perception (as in vision). At the higher levels, while machines can be endowed with some capacity for judgment that typically involves computation of suitably framed cost functions, or finding patterns, of choosing between hypotheses, the capacities of concretization and especially self-awareness seem to be completely out of the realm of present day computing science.

As regards sheaths of consciousness, constraints due to time, space, and selectivity are incorporated in many computers by pre-selecting the data to be processed of by tuning in the sensors to specified time windows. In general even the most elementary constraints require active intervention by human agents.

3.5   A Theory of Speech and Cognition

Rigveda 1.164.45 describes that speech and its concomitant cognition is of four kinds. The names of these kinds of speech are described by Bhartṛhari (circa 450 CE) in his Vākyapadīya to be vaikharī, madhyamā, paśyantī, and parā (Abhyankar and Limaye 1965, Coward 1976). Vaikharī represents gross sound; madhyamā is the level of mental images; paśyantī represents that gestalt or undifferentiated whole that sounds emerge from in the process of speaking and into which they merge in the process of hearing; parā is the unmanifest sound that resides in one's self or universal consciousness.

Bhartṛhari argues that reality (sampratisattā) when seen through the window of language reduces to a formal reality (aupacārikī sattā). Language can only deal upto the level of paśyantī, the gestalts underlying mental constructs, and it remains limited because parā speech lies beyond it.

Bhartṛhari calls the word or sentence considered an an indivisible meaning-unit as the sphoṭa. He bases this concept on the Vedic theory that speech (vāk) is a manifestation of the primordial reality. The word-sphoṭa is thus contrasted from word-sound. Meaning is obtained at a deep level based on the sequence of sounds.

The discovery of a very large number of phonetic symmetries in the first hymn of the Rigveda that cannot be conceived to have been deliberately introduced gives support to the thesis that language captures only some of the symmetries that nature's intelligence can express. Raster (1992) summarizes this discovery thus:

> In our search for phonetic symmetries in the first sūkta of the Rigveda, we examined the occurrence frequencies of more than 50 sound classes. Of these more than 40 sound classes were found with occurrence frequencies which are integral multiples of 8 and more than 20 sound classes with occurrence frequencies which are even integral multiples of 24... Moreover, in many cases, the occurrence frequencies of phonetically related sound

classes form simple integral ratios, for example, the ratio of 2:1 between the frequencies of voiced and voiceless consonants and the ratio of 1:2 between the frequencies of long and short vowels. Thus, fundamental oppositions of the phonological system are reflected in the quantitative distribution of sounds in the text... The order which has been found underlying the phonological structure of the text is a hidden order. It cannot be perceived consciously while reading or listening to the text... Although the order found in the distribution of the sounds in the text is unfolded sequentially in time, it is in itself not a linear, but a global phenomenon...[and] it is multidimensional. [Page 38]

The parallels between Bhartṛhari's theory and the quantum framework are clear. Language is to be compared to the observables of the quantum theory.

3.6 Vedic Evolution

The Vedic theory of consciousness speaks of a process of evolution. In this evolution the higher animals have a greater capacity to grasp the nature of the universe. In such a process the primary mechanism is the development of the neural hardware to deal with more knowledge.

The actual neurophysiological structures were seen to evolve in a recursive manner. The nature of these structures was also taken to encode knowledge. Biological clocks, in synchrony with various astronomical processes, appear to have inspired the search for correspondences between the outer and the inner (Kak 1994b). Parenthetically, it may be seen how such a notion would lead to the development of astrology since the motions of the planets would be mirrored by processes inside the mind and the body. The processes in the two sides of the brain were considered to parallel, in a metaphorical sense, the motions of the sun and the moon. These motions are not synchronized, which is why the lunar and the solar years are different in duration. Therefore it was taken that the natural state of the two hemispheres was to be in asynchrony and the idea of meditation was to synchronize the brain and the mind for higher cognition (Springer and Deutsch 1985).

Vedic evolution is not at variance with Darwinian evolution but it has a different focus. The urge to evolve into higher forms is taken to be inherent in nature. A system of an evolution from inanimate to progressively higher life is clearly spelt out in the system of Sāṃkhya. At the mythological level this is represented by an ascent of Vishnu through the forms of fish, tortoise, boar, man-lion, the dwarf into man.

4   Implicate Order

The ancient resolution of the paradoxes of time, space, and the self was through the postulation of a unity that lies behind observed phenomena. Yet this resolution also asserts that this unity cannot be comprehended in logical terms. What this suggests is that there is a divide between the categories of cognition and that of reality. In other words, there exists a hidden order that underlies the observed phenomena.

Explanations are eventually sketched in terms of metaphors. Therefore it is not surprising that the metaphor to explain the contradictions embraced by complementarity could be similar to this ancient argument. In quantum mechanics this metaphor goes back to the pilot-wave theory of de Broglie, which attempted to present a causal interpretation of the field. This was formulated more systematically by David Bohm (see Hiley and Peat 1987) in a theory that has been called the hidden-variable theory. But the measurement problem remains unresolved since measurement defines a break in the flow of time and fixes a definite direction for it. Measurements can be comprehended only in terms of connections between the past, the present, and the unfolding future. Classical, deterministic physics also cannot do this since its rules are time-symmetric. Recognizing this difficulty Bohm suggests an infinitely nested framework of an implicate order that expresses characteristics of the universe.

To the extent that the notion of implicate order implies a theory of the universe where the properties of the parts cannot be separated from the properties of the whole, we notice that neural processes have a similar structure. Eccles (1986) has summarized evidence that shows how global mental states (states of the mind) affect the behavior of individual neurons. However, this does not solve the mind-body problem, owing to the infinite regress in implicate orders.

5   Animal and Artificial Intelligence

Consider first the question of artificial intelligence which has often been vitiated by semantic problems. There are many algorithmic tasks that are associated with intelligence, and clearly a machine should be able to do them.

But the fundamental difference between humans and machines is that, unlike humans, machines are programmed for a static universe. Consciousness refers to an aware state where one is able to deal with the local and the global on a quickly changing canvas. If it is correct to represent consciousness by a quantum mechanical type of function then machines could never be conscious. It is not surprising then that machines have been unable to handle problems such as natural language processing, where it is not enough to analyze sentences, as one must also track the constantly changing context.

Investigations of subhuman animal intelligence have presented new riddles since animals appear to do as well as humans at many cognitive tasks (Herrnstein 1985) that seem to be beyond the pail of any machine. For example, pigeons understand the abstract notion of a tree; in other words, pigeons can identify pictures that contain trees, irrespective of their shape and size. No machines can do this.

It had long been thought that the cognitive capacities of the humans were to be credited in part to the mediating role of the inner linguistic discourse. Research has shown that animals do think but cannot master language. Terrace (1985) summarizes the evidence thus:

> Ample evidence is available of the existence of human thought. There is considerable evidence that apes, considered by many psychologists to be man's most intelligent relatives, are unable to master the basic features of human language. That state of affairs raises a fascinating question that can be asked of animals in general: ' what is the non-linguistic medium of animal thought?'

One may postulate that just as the wavefunction in quantum mechanics represents a sum of potentialities of object behavior, consciousness in an animal represents a sum of potentialities of thought behavior. These thought potentialities refer to different types of cognitive function and awareness. The actual capacity of the animal would be determined by the structure of its neural hardware. This implies that while apes may not have the same linguistic ability as humans but nevertheless one must take them to be persons. In other words, the mirror that reflects the self is in animals more clouded than that of humans. It also suggests that apes and other animals could be trained to extend their capability by the use of technology.

Machines would never be conscious however, since they do not come with a set of potentialities that consciousness provides. Machines are therefore like the neural hardware that provides extension. A machine that is so designed so that it has infinite set of potentialities would be alive.

5.1 The Phantom Limb Phenomenon and an Action Field

We now consider the notion of self from the perspective of the phantom limb phenomenon where a missing body part continues to have a vivid presence in the mind of the subject. The phantom limb phenomenon persists for decades and it is clearly produced by brain processes that defined self as distinct from non-self. In the words of Melzack (1989):

> Just as extraordinary as the persistent experience of a limb after it has been amputated is the converse—the denial that a part of one's body belongs to one's self. Typically, the person, after a lesion of the right parietal lobe or any of several other brain areas...denies that a side of the body is part of himself and even ignores the space on that side. There are several descriptions of patients who topple out of their hospital bed because they believe that a strange leg is in their bed, which they try to throw out of bed with the consequence, of course, that the rest of the body follows the leg. Generally, the leg (or arm, or whole side of the body) is treated negatively—as undesirable. On one occasion, however, a patient thought that the leg belonged to an attractive woman and was happy to have it share his bed.

Neural networks in the brain generate the experience of the self although this experience is modulated by the inputs from the body; furthermore, this experience has a unity. Other experiments have led to the significant conclusion that this experience is primarily determined genetically. In an experiment "newborn monkeys, within hours after birth, underwent complete sensory-root deafferentation of both forelimbs and were blinded by suturing the eyelids. Astonishingly, by the age of three months, these monkeys had spontaneously developed the ability to walk and clasp objects. They were also trained to make precise hand-to-mouth movements and to discretely extend the arm toward the front after a tap on the upper lip. The only reasonable explanation of such spatially coordinated behaviour in the absence of vision and sensory feedback from limbs after birth is that the monkeys possess built-in brain mechanisms for a phantom body capable of meaningful actions in three-dimensional space." (Melzack 1989).

One might generalize from the phantom limb phenomenon and postulate that in addition to the self there also exists an action field. This action field becomes operative in terms of the context of the life history of the individual. Just as the individual knows if his body is complete and if it is not then the brain supplies the phantom, so does the individual know if his actions define a full life and if it is not then the brain compensates for the deficit in terms of inner extra-rational experiences. Language is one of the dimensions of this action field.

6 Concluding Remarks

Psychology shows that all animals have a notion of self. Arguments regarding the origin of this characteristic turn out to be circular. We cannot consider any organism to be entirely reflexive because then it will not be able to distinguish itself from other organisms. For a living organism one must visualize hierarchically distinct networks that use nested maps and such a nesting will have to go on endlessly. The iterative system of these maps amounts to the definition of a self of infinite regress!

A systematic analysis of neuron behavior presents paradoxes, and an approach such as that of implicate order does not address the fundamental difficulty related to the mind-body problem. Any critical reasoning of consciousness leads to dualistic phenomena. But this dualism cannot be explained away by mechanisms of self-reference, of which the split-brain dichotomy is one. Approaches such as the one in which life is postulated as a fundamental notion, solve one set of problems but create others. A solution to the mystery of the mind in logical terms does not appear to be possible. But a continuing study of the neurophysiological basis of cognitive processes is bound to reveal new insights.

If we consider the mind to be a unity, we are compelled to a quantum mechanical type of a model. This opens the door to examine the characteristics of quantum neural computers (Kak 1995).

Aliens and apes could be taken to be persons if they can articulate, through linguistic expression or through other behavior, their reflection of the self. One cannot predicate the notion of personhood only on linguistic competence because that would declare a deaf/mute not to be a person. On the other hand, silicon machines, whether they use neural networks or not, cannot be taken to be alive.

The classificatory system developed by Vedic philosophers does not address the paradoxes of consciousness. Rather it defines categories, such as that of universal experience, that can be seen to be explain the "complementary" nature of human experience. These categories clearly assign central role to selectivity, or context, and change. The Vedic system takes the mind to be emergent on the ground of the neural hardware of the brain, but this emergence is contingent on the principle of the self.

Acknowledgement

This paper was presented at the Symposium on *Aliens, Apes, and Artificial Intelligence: Who is a person in the postmodern world?*, Southern Humanities Council Annual Conference, University of Alabama in Huntsville, February 13, 1993.

References

Abhyankar, K.V. and Limaye, V.P., 1965. *Vākyapadīya of Bhartṛhari.* Poona: University of Poona.

Abhinavagupta, R., 1989. *A Trident of Wisdom.* Albany: State University of New York Press.

Aurobindo, S., 1949. *The Life Divine.* New York: The Greystone Press.

Bell, J.S., 1987. *Speakable and Unspeakable in Quantum Mechanics.* Cambridge: Cambridge University Press.

Bohm, D., 1951. *Quantum Theory.* New York: Prentice-Hall.

Bohr, N., 1961. *Atomic Physics and Human Knowledge.* New York: Science Editions.

Coward, H.G., 1976. *Bhartṛhari.* Boston: Twayne Publishers.

Dyczkowski, M.S.G., 1987. *The Doctrine of Vibration.* Albany: SUNY Press.

Eccles, J.C., 1986. Do mental events cause neural events analogously to the probability fields of quantum mechanics? *Proc. Royal Society London,* B 227, 411-428.

Herrnstein, R.J., 1985. Riddles of natural categorization. *Phil. Trans. R. Soc. London* B 308, 129-144.

Hiley, B.J., and Peat, F. David, 1987. *Quantum Implications - essays in honour of David Bohm.* London: Routledge & Kegan Paul.

Kak, S.C., 1986. *The Nature of Physical Reality.* New York: Peter Lang.

Kak, S.C., 1994a. *The Astronomical Code of the Ṛgveda.* New Delhi: Aditya.

Kak, S.C., 1994b. From Vedic science to Vedānta. *Fifth International Congress of Vedānta,* Miami University, Oxford, Ohio, August 1994.

Kak, S.C., 1995. On quantum neural computing. *Information Sciences,* 82, 1995. (in press)

Kornhuber, H.H., 1974. Cerebral cortex, cerebellum, and basal ganglia: An introduction to their motor function. In W. Schmitt (ed.), *The Neurosciences: Third Study Program.* Cambridge: MIT Press.

Libet, B., 1973. Electrical stimulation of cortex in human subjects, and conscious sensory aspects. In A. Iggo (ed.), *Handbook of sensory physiology: Vol. II. Somatosensory System.* New York: Springer-Verlag.

Libet, B., 1986. Nonclassical synaptic functions of transmitters. *Federation Proceedings, American Societies for Experimental Biology,* 45, 2678-2686.

Michel, S., Geusz, M.E., Zaritsky, J.J., and Block, G.D., 1993. Circadian rhythm in membrane conductance expressed in isolated neurons. *Science,* 259, 239-241.

McGeer, P.L., Eccles, J.C., and McGeer, E.G., 1987. *Molecular Neurobiology Of The Mammalian Brain.* New York: Plenum.

Melzack, R., 1989. Phantom limbs, the self and the brain. *Canadian Psychology,* 30, 1-16.

Moore, W., 1989. *Schrödinger: Life And Thought.* Cambridge University Press.

Penrose, R., 1989. *The Emperor's New Mind: Concerning Computers, Minds, And The Laws Of Physics.* Oxford University Press.

Raster, P., 1992. *Phonetic Symmetries in the First Hymn of the Rigveda.* Innsbruck.

Sacks, O., 1990. *Awakenings, A Leg To Stand On, The Man Who Mistook His Wife For A Hat.* New York: Book of the Month Club.

Schrödinger, E., 1961. *Meine Weltansicht*. Wien: Paul Zsolnay.

Schrödinger, E., 1965. *What Is Life?*. New York: Macmillan.

Sinha, J., 1958. *Indian Psychology*. Calcutta.

Springer, S.P., and Deutsch, G., 1985. *Left Brain, Right Brain*. New York: W.H. Freeman and Company.

Terrace, H.S., 1985. Animal cognition: thinking without language. *Phil. Trans. R. Soc. Lond.* B 308, 113-128.

Trevarthen, C. (Ed.), 1990. *Brain Circuits and Functions of the Mind: Essays in Honor of Roger W. Sperry*. Cambridge: Cambridge University Press.

Wigner, E., 1967. *Symmetries and Reflections*. Bloomington: Indiana University Press.

Symposium on Aliens, Apes, and Artificial Intelligence, The University of Alabama in Huntsville, February 13, 1993.

22.

Chance, Choice, and Consciousness: A Causal Quantum Theory of the Mind/Brain

Henry P. Stapp

Chance, Choice, and Consciousness
A Causal Quantum Theory of the Mind/Brain *

Henry P. Stapp

Lawrence Berkeley Laboratory

University of California

Berkeley, California 94720

Abstract

An analysis of the measurement problem of quantum theory points to the need to introduce consciousness, per se, into physics. Conversely, an analysis of the mind-brain problem of cogsci/neuroscience/psychology and philosophy-of-mind points to the need to introduce quantum theory into our treatment of the human mind/brain: an adequate theory of the mind/brain should make a person's conscious thoughts, per se, efficacious, and, in particular, allow his descriptions of the quality and content of his conscious thoughts to flow dynamically from the quality and content of these thoughts themselves. These demands cannot be met within the ontologically and dynamically monistic conceptualization of the world provided by classical-mechanics, but are achievable in a natural way within a dualistic quantum-mechanical conceptualization of nature. A causal quantum mechanical theory of the mind/brain that meets these demands is described.

Invited paper for the conference:

Toward a Scientific Theory of Consciousness 1996

University of Arizona, Tucson, April, 1996.

---

*This work was supported by the Director, Office of Energy Research, Office of High Energy and Nuclear Physics, Division of High Energy Physics of the U.S. Department of Energy under Contract DE-AC03-76SF00098.

Disclaimer

This document was prepared as an account of work sponsored by the United States Government. While this document is believed to contain correct information, neither the United States Government nor any agency thereof, nor The Regents of the University of California, nor any of their employees, makes any warranty, express or implied, or assumes any legal liability or responsibility for the accuracy, completeness, or usefulness of any information, apparatus, product, or process disclosed, or represents that its use would not infringe privately owned rights. Reference herein to any specific commercial products process, or service by its trade name, trademark, manufacturer, or otherwise, does not necessarily constitute or imply its endorsement, recommendation, or favoring by the United States Government or any agency thereof, or The Regents of the University of California. The views and opinions of authors expressed herein do not necessarily state or reflect those of the United States Government or any agency thereof or The Regents of the University of California and shall not be used for advertising or product endorsement purposes.

Lawrence Berkeley Laboratory is an equal opportunity employer.

1. Introduction.

Two topics of intense current research activity are the problem of measurement in quantum theory and the problem of the connection between mind and brain in cogsci/neuroscience/psychology and the philosophy of mind. I begin by arguing that each of these problems leads to the other, and thence to the need for a quantum dynamical description of the mind/brain in which our conscious thoughts, per se, influence our actions, and are, in particular, direct causes of our normal truthful reports about the content and qualities of our experience. A causal quantum dynamical model of the mind/brain that meets these demands will then be described.

1.1 Why the problem of measurement in quantum theory points to the need to introduce conscious thoughts, per se, into mind/brain dynamics.

The orthodox 'Copenhagen' of the interpretations of quantum theory already brings the experience of 'the observer' into physical theory, in a fundamental, though non-dynamical, way. Niels Bohr says that:

"In our description of nature the purpose is not to disclose the real essence of phenomena but only to track down as far as possible relations between the multifold aspects of our experience."[1]

This focus upon "our experience" was the foundation of Bohr's quantum philosophy, and was a central point of contention in the famous Bohr-Einstein debate. Einstein's contrary opinion was that:

"Physics is an attempt conceptually to grasp reality as it is thought independently of its being observed."[2]

Heisenberg also brings into quantum theory the subjective aspect of nature in the form of 'our knowledge': the wave function, which is the basic quantity in quantum theory, collapses when 'our knowledge' changes, and it is therefore essentially subjective; it lives in the world of our thoughts. Yet this collapse is related also to certain objective aspects the external real world. In particular, Heisenberg speaks of the 'possible' the 'actual', and says that:

"the transition from 'possible' to 'actual' takes place as soon as the interaction between the object and the measuring device, and thereby the rest of the world, has come into play..."[3]

A deficiency in Heisenberg's account of this objective process of actualization is that the concept of 'the measuring device' is left undefined. Wigner, giving credit to von Neumann, emphasizes that there is nothing in the physical world alone that specifies where the line of demarcation between the quantum system (the object) and the 'measuring device' should be drawn, and he suggests that the only natural place to draw the line between the observed system and the observing system is where consciousness enters. Pauli and Schroedinger also argue that the quantum aspect of nature indicates that consciousness must be brought into our dynamical description of the world.

The awkwardness of this quantum splitting of nature into two parts, the quantum system and the observing system, with some transition from 'possible' to 'actual' occurring at the interface, leads naturally to the notion, spelled out in some detail by Everett[4], that there is no such transition: that the Schroedinger equation never fails. However, this suggestion focuses attention even more strongly than ever on the problem of our consciousness experience. For in this interpretation there is a huge disparity between the objective world, which is represented by the evolving state of the universe, and our subjective experiences of it. The basic problem with this interpretation is that the needed psychological, i.e., experiential, properties of brains do not follow from the Schroedinger equation: the latter generates independently evolving 'branches' of the wave function of the brain, with different branches corresponding to different conscious thoughts, but these branches are *conjunctively* present. However, in order to obtain the statistical predictions of quantum theory, which pertain to our experiences, the experiential branches that correspond to these different physical branches must be *disjunctive*. That is, the objective physical state will contains branch A *and* B, etc., whereas to get statistical statements about our subjective experiences one needs the logical structure of experience A *or* experience B, etc.. This means that 'mind' needs an ontology and dynamics that does logically follow from the Schroedinger equation that controls the 'brain'. This need for a second level of reality and dynamics that does not follow logically from the first appears to nullify the advantage that the Everett interpretation seemed at first to provide. In any case, a spotlight becomes focused more strongly than ever on the problem of the connection between the subjective-experiential and objective-material aspect of the mind/brain.

Looking at the evaluations by physicists who are pursuing environmental decoherence effects, and other essentially 'Everett' ways of approaching the problem of quantum measurement we find Zurek[5] saying, of these approaches, that they do not allow us to understand how we as 'observers' fit in, and hence they appear to him to be merely "a hint about how to proceed rather than the means to settle the issue quickly." Joos[6] says " Of course the central problem remains unsolved: Why are there local observer's?" Gell-mann and Hartle[7] emphasize that: "If history dependence can be properly introduced into the explicit treatment of quantum mechanics, then we may be able to handle individuality [of observers] with the care that it deserves". Omnes[8], who gives perhaps to most comprehensive description of these Everett-type theories says, about the Everett proposal, that he feels "it impossible to accept as a satisfactory answer to the problem of actuality." So almost forty years after the Everett paper appeared it is acknowledged by these workers, and I think by all others who have examined the matter with sufficient care, that the problem of 'the observer' and the nature of his actual experience has not been resolved by that approach. This issue, namely the problem of how our individual experiences fit into nature, remains the central unsolved problem, and the Everett approach makes this fact even more clear than ever.

1.2 Why an adequate theory of consciousness requires a quantum description of the mind/brain.

The inadequacy of classical mechanics for the study of the mind-brain problem is made manifest by the 'Zombie Dilemma'.

A 'Zombie' is defined to be a creature that is just like a conscious human being but lacks consciousness: it is identical to a conscious human being physically, but has no qualia; no experience; no inner light of consciousness.

The 'Zombie dilemma' has two horns. The first arises from the following fact: if the principles of classical mechanics were to give a completely adequate description of the human mind/brain then our conscious thoughts would be nonefficacious. For, according to the principles of classical mechanics (CM) the behaviour of any system that CM describes correctly is completely determined by the principles of CM and the magnitudes of the physical variables of CM, namely the positions and velocities of the particles, and the values of the local fields. The principles of CM do not themselves specify whether or not qualia are present, and the calculations of physicists will be the same, and lead to the same conclusions about behaviour, independently of any belief that the physicists might hold, one way or the other, on the issue of whether or not qualia are present. Therefore, within the framework of classical mechanics 'it does not matter' whether or not the qualia are present. Thus if classical mechanics were to give a completely adequate description of the behaviour of the mind/brain system then qualia, per se, could have no effect on the behaviour of the classically describable system: the qualia would lack causal efficacy; they would be epiphenomenal.

This argument shows that if the human mind/brain were adequately described by classical mechanics then human consciousness would be epiphenomenal, and Zombies would, therefore, be logically possible. This is the first horn of the dilemma.

The second horn is this. Suppose we ask the Zombie 'Are you conscious?' He will answer 'Yes!' But this answer cannot flow dynamically from the psychological fact that it purports to report, because for Zombies there are no psychological facts. Since a Zombie is physically identical to his conscious twin, the truthful answer 'Yes!' of the conscious twin likewise cannot flow dynamically from the relevant actual fact of the matter, namely that he is conscious. Thus there would need to be, in conscious human beings, some sort of weird mechanical linkage that would cause truthful answers to questions about a person's qualia to be spoken without there being any appropriate dynamical flow from the relevant facts.

The idea of nonefficacious consciousness, and consequently of Zombies, is therefore not reasonable: it is completely unreasonable to suppose that there is some strange mechanism that makes a conscious human being speak words that correctly describe the contents and quality of his consciousness without those words being caused by the aspects of consciousness that they purport to describe. In a reasonable theory the actions of human beings should flow dynamically from their efficacious thoughts.

The rational conclusion to be drawn from the 'Zombie dilemma' is therefore that the principles of CM cannot provide an adequate framework for the study of the mind/brain: one must turn to a theoretical framework that allows consciousness to be a bona fide cause of behaviour.

Of course, the notion that classical mechanics could give an adequate account of the human mind/brain was unreasonable to begin with, for CM does not describe the properties of the materials from which human brains are made. Quantum theory is needed for that.

2. The Program of John Bell.

The late John Bell[9] made a plea for conceptual clarity in physics. He identified the source of the muddled and confused state of the foundations of physics with the 'FAPP arguments' that abound within it: at a certain point in the usual arguments it is claimed that 'for all practical purposes' concept A is indistinguishable from concept B, and hence that the two need not be distinguished. But this sort of reasoning conflates matters of logical principle with practical matters associated with experimental difficulties. Bell argued that, even though scientific theories must at certain points make contact with empirical evidence, logical clarity should nonetheless be maintained, and that FAPP arguments should therefore be avoided. He held that professional physicists ought to be able to do better than to build their theories on FAPP arguments.

A key question then becomes: WHEN should we face up to the central problem of constructing a rational theory of the connection of brains to minds? In ten years? A hundred years? A thousand?

The answer, I say, is 'Now'! This issue is currently 'in the wind': it is an important converging research interest in many fields, namely cogsci, neuroscience, psychology, philosophy of mind, and physics. The time has finally come for us to face the brain-mind issue squarely.

3. The Physics Problem

Einstein[10] illustrated the central logical problem in contemporary quantum theory with a simple example. It involves a radioactive source, a detector of some product of the decay, a pen that draws a line on a moving strip of paper, and that makes a blip when the decay is detected, and a human observer of the blip. If one uses the Schroedinger then one finds that the system evolves into a continuous superposition of states corresponding to all possible positions of the blip on the strip of paper. But when the human observer looks, he sees the blip in one well defined place. Thus the Schroedinger equation is not telling the whole story. If one wants to have an account of what is actually happening, then something else needs to be added, namely Heisenberg's 'transition from possible to actual' (or some substitute for it) that allows the many possibilities generated by the Schoedinger dynamics to be reduced to the single actually experienced reality.

But if there really is in nature some 'second process', involving transitions from possible to actual, then the Scroedinger equation is telling only half the story: contemporary physics is limping along on one leg. What are the effects of this 'second process' on cosmological processes? What are its effects on biological processes?

Once one begins to develop a detailed theory for the second process the question arises as to whether the irreducible element of chance that enters into the pragmatic Copenhagen account must persist when one goes over to a more detailed description of what is actually happening: should one not entertain the idea that the underlying dynamics, though presumably highly non-classical and non-local, is nevertheless causal in some sufficiently comprehensive conceptualization. It is hard to believe that, in the final analysis, a definite result simply pops out of thin air: i.e., that there is absolutely no cause, or sufficient reason, beyond an irreducible element of pure chance, for the experienced reality to be what it turns out to be.

4. Objectives of Causal Quantum Mind/Brain Dynamics.

William James[11] says that his study of "the particulars of the distribution of consciousness" ... "will show that consciousness is at all times primarily a *selecting agency*". Notice that the basis for this conclusion is not our perhaps fallible or misleading introspection, but rather an objective study of certain particulars. This identification of consciousness as "primarily a selecting agency" meshes well with the notion that it is closely connected with the quantum selection of the 'actual' from among the many 'possible'.

The objectives of the theory of the mind/brain to be described here are to:

1. Make consciousness a bona fide selecting agency.

2. Make consciousness efficacious (resolve the Zombie Dilemma).

3. Explain how consciousness does what it apparently does, namely select and initiate courses of body/brain action.

4. Explain why consciousness feels like it feels.

5. Explain the connection of consciousness to memory.

6. Explain the Survival Advantages Conferred by Consciousness.

7. Eludidate Free-Will

The position adopted here is that the Heisenberg process of actualization is a real process in nature, which occur both inside and outside brains, but that the process inside human brains is the best place to start the study, because if the actualizations that occur there are closely linked to human conscious experience, which seems likely, then we have, in the case of human brains, two extra kinds of pertinent empirical data, namely our objective reports of our experiences, and also our personal subjective experiences themselves. Also, because of its importance in medicine and mental health, and to our idea of what we human beings are, the brain and its connection to mind is now, and is going to continue to be, even in times of shrinking research budgets, a field of intensive empirical activity, and hence a topic supported by an immense and rapidly growing amount of scientifically acquired data.

5. Chance, Observation, and Experience

Chance enters into orthodox quantum theory in connection with our observations. If quantum theory is applied to the universe, then in the absence of observations the evolution of the universe would be governed by local laws that are natural generalizations of the laws of classical mechanics: the universe is conceived to be an aggregate of localized properties, and the rate of change in each such property is governed exclusively by nearby properties. Observations, however, are associated with a "second process" that is logically required to be highly nonlocal [12], and therefore fundamentally different from the first process.

As regards the role of chance in this second process Bohr has this to say:

"The circumstance that, in general, one and the same experimental arrangement may yield different recordings is sometimes picturequely described as a 'choice of nature' between such possibilities. Needless to say, such a phrase implies no allusion to a personification of nature, but simply points to the impossibility of ascertaining on accustomed lines directives for the course of a closed indivisible phenomena. Here, logical approach cannot go beyond the deduction of the relative probabilities for the appearance of the individual phenomena under given conditions.[13]"

543

Bohr carefully avoids affirming that there actually is in nature herself an irreducible element of chance. He says, rather, that the entry of chance (into the orthodox quantum theory that he describes) is due to difficulties that arise from trying to apply customary (local-reductionistic) thinking to closed indivisible phenomena. Since he also says that new ideas will be needed to encompass biological systems it seems well within his philosophy to accept that when theoretical physics reaches the point of dealing squarely with the details of the nonlocal and indivisible aspects of the 'second process' then the notion of an irreducible element of chance will no longer be needed. That is the attitude adopted here. Indeed, the central idea of this work is to replace, as the immediate cause of the quantum actualizations in human brains, the abstract notion of pure chance by an empirically known reality, namely our conscious thoughts. In this formulation of quantum brain dynamics our (mathematically represented) conscious experiences exercise genuine control over brain activity. Analogous elements should occur in all biological systems, due to the enormous survival advantage they can confer. But in lower life forms, and also in the inanimate part of nature, these elements will, because of the absence in them of the specific structures created by human brains, be very different from human conscious experiences.

6. Quantum Searching and Survival.

Survival, at least in the animal kingdom, depends on rapidly finding and executing appropriate behaviors. Options are generally available, and the organism must reject those not appropriate in the specific situation in which it finds itself, and pursue one that is appropriate. The process of searching for an appropriate behavior can be likened to a search for the way out of a maze. The classical search procedure is essentially to try, at some mental level, each of the possibilities until a blockage is encountered, and then to back off and try another. This can be very time consuming, and an organism that uses it is likely to be devoured by one that employs a faster process. Massive parallel (and interconnected) processing may offer advantages, but it introduces the compensating problem of keeping the whole system operating in a coordinated way.

For rapid searching the exploitation of the quantum character of brains can confer a huge advantage. Quantum dynamics is essentially hydrodynamics [14].

The contrast between classical and quantum search procedures can be likened to the contrast between the particle and hydrodynamical solutions to the problem of getting out of a maze: in the particle solution the particle bounces randomly around the maze in the hope of finding the small opening; in the hydrodynamical solution the maze is filled with water, which then rushes out through the opening. The essential point is that in classical-particle dynamics what the particle does is completely unaffected by what it would have done if it had been on a nearby trajectory, whereas the flow of water is affected by what is happening nearby: if water rushes out at one place, leaving a void, then nearby water rushes in to fill the void, sucking in water from further away.

This point can be illustrated by considering a circular trough that has also a circular cross section in each radial plane. Suppose this trough is filled with a statistical ensemble representing alternative possibilities for the position and velocity of one particle. Each element of the ensemble oscillates in a radial plane, with no angular motion. Suppose we open a small angular section of the trough so that the particles in that section flow out. The remaining particles, which represent the alternative possibilities, will continue to oscillate forever. But if one fills the trough with water and opens the section then all the water runs out. The quantum probability function for one particle behaves like water, not like the statistical ensemble of independent particles.

A physicist who wants to see this in the equations can consider a wave function for a particle confined to a circle. The time-dependent Schroedinger has on the left the operator i times the derivative with respect to t, and on the right the kinetic energy term. To represent the opening in the maze (i.e., the solution that is not blocked by negative feed-back) add on the right the term b times minus i times a Dirac delta function of the (cyclic) argument x(mod 1). Then the rate of loss of probability in the ring is 2b times the square of the magnitude of the wave function at x=0. This is non-negative, and more detailed calculations show that the probability is rapidly sucked to the point x=0, where it disappears.

(A more realistic model would have in place of the Dirac delta function a function with a flat central plateau bounded on each side by a sharp gaussian fall-off. The rate of flow of probability from the surrounding region into the region of probability loss is controlled by the sharpness of the gaussian walls.)

This way of searching for an appropriate response should be particularly rapid and effective in a brain organized in the way described in [15], because in that system the unblocked flow out of the maze (of alternatives, most of which are blocked by negative feed-back) creates a template for action, which then automatically evolves into the corresponding action itself. There is no need to convert the solution represented by the unblocked flow into a plan of action, and then to create the corresponding sets of instructions to muscles etc.: the unimpeded flow produces a template for action that, if not blocked, automatically evolves into the appropriate action itself. So the basic problem of rapidly producing an appropriate action is precisely that of rapidly getting all of the probability into an unblocked channel, i.e., of keeping the search process from getting hung up exploring the blocked channels. The hydrodynamical character of the quantum law of evolution (together with quantum tunneling) provides an efficient way to solve this problem.

Notice also that the quantum mechanism does not involve a sudden 'all or nothing' leap in phylogenetic development: even a little bit of sucking of the probabilities into unblocked channels will aid survival, and the organism can gradually evolve in a way that tends to enhance the process.

7. Decoherence

It has often been observed that the coupling of a system to its environment has a tendency to make interference phenomena that are present in principle within quantum systems difficult to observe in practice. Phase relationships, which are essential to interference phenomena, get diffused into the environment, and are difficult to retrieve. The net effect of this is to make a large part of the observable phenomena in a quantum universe similar to what would be observed in a world in which certain collective (i.e., macroscopic) variables are governed by classical mechanics. This greatly diminishes the realm of phenomena that require for their understanding the explicit use of quantum theory.

These decoherence effects will have a tendency to reduce, in a system such as the brain, the distances over which the idea of a simple single quantum system holds. This will reduce the distances over which the simple hydrodynamical considerations described above will hold. However, the following points must be considered.

a) A calcium ion entering a bouton through a microchannel of diameter x must, by Heisenberg's indeterminacy principle, have a momentum spread of $\hbar/x$, and hence a velocity spread of $(\hbar/x)/m$, and hence a spatial spread in time t, if the particle were freely moving, of $t(\hbar/x)/m$. Taking t to be 200 microseconds, the typical time for the ion to diffuse from the microchannel opening to a triggering site for the release of a vesicle of neurotransmitter, and taking x to be one nanometer, one finds the diameter of the wave function to be about 0.04 centimeters, which is huge compared to the 1/100000000 centimeter size of the calcium ion. There is, therefore, in brain dynamics a powerful counterforce to the mechanisms that tend to diminish quantum coherence effects. More generally, one cannot, without collapses, keep the wave function of a calcium ion confined to a region that is not huge compared to the size of the ion. This entails that classical ideas cannot be adequate: the brain must, if no collapes occur, evolve into an amorphous superposition of states corresponding to a continuum of different possible macroscopic behaviours.

b) The normal process that induces decoherence arises from the fact that a collision of a state represented by a broad wave function with a state represented by a narrow wave packet effectively reduces the coherence length in the first state to a distance proportional to the width of the second state. But in an aqueous medium in which all the states of the individual systems have broad packets this mechanism is no longer effective: coherence lengths can remain long.

c) Even if the coherence length were only a factor of ten times the diameter of the atom or ion involved in some process, the cross section involved would be a hundred times larger. The search processes under consideration here involves huge numbers of atoms and ions acting together, and the cross-section factors multiply. Thus even a small effect at the level of the individual atoms and ions could give, by virtue of the hydrodynamical effect, a large quantum enhancement of the efficiency of an essentially aqueous macroscopic search process.

8. General Description of Brain/Mind Dynamics

Before going into the specific mathematical details of the model, I give a general overview of the main elements, and how they fit together.

8.1 Body-World Schema.
It is accepted here (or postulated) that there is in a person's brain a high-level represention of his body and its environment:

i.e., that a person's body and its environment are represented in his brain by patterns of neurological and other brain activity. This representation in the person's brain of his body and its environment is called the 'body-world schema'. It is expanded to include representations of beliefs, and hence is sometimes called the body-world-belief schema, but I shall stick to the shorter name. This pattern of activity has sub-patterns that represent information coming from the various senses, and from various processing activities in the brain, and this input information is dynamically intertwined with associated outputs that are effectively directing the collection of pertinent information, and controlling the activity of the brain and body. These various sub-patterns, or components, become dynamically tied together into modules in the course of a person's life, due to reinforcements and facilitations, and the architecture of the brain, and these modules become tied together in a momentary configuration, the body-world schema, under the constraints and influences of the immediate situation in which the organism finds itself. In the succession of such momentary configurations there is a large carry-over of structure from one overall pattern to the next, and there is built into each momentary pattern a 'temporal' ordering that orders the components with respect earlier and later times of entry into the sequence, so that, for example, the sequence of recently heard notes in a short tune are 'ordered' in the momentary body-world schema, as are the sequence of movements that the schema calls forth. All of this is described in much more detail in ref. 15.

8.2 Facilitation, Associative Recall, and Control. Built into the conception of brain activity described above is the idea that the persistence of certain patterns of brain activity 'etches' this pattern into the physical structure of the brain, in the sense that this pattern is 'facilitated' (made easier to activate), and that a later activation of part of the pattern tends to spread to the whole. This facilitation and spreading effect provides the basic mechanism for an explanation of associative recall, and of the control aspect of the body-world schema.

8.3 The Effect of Quantum Theory

The conception of brain action sketched above is classical. The effect of quantum theory is essentially the same as it was in the Einstein example described earlier: the evolution controlled by the Schroedinger equation will produce, instead of one single classically describable situation, rather a continuum, consisting of a superposition of all of the classically describable possibilities, with no one possibility singled out as the one that is actually experienced. Thus, for example, for every possibility in which a 'synaptic event' —the release of a vesicle of neuro-transmitter— occurs there will be other superposed possibilities in which this event does not occur; and for every situation in which an action potential spike exists at one place along an axon there will be other superposed possibilities in which the spike is a little earlier, or a little later, and still others in which it is much earlier, or much later. To extract from this amorphous conglomerate of superposed possibilities the actually experienced reality one needs, according to the Heisenberg ontology accepted here, a transition from 'possible' to 'actual'. This transition is called an actualization event: it selects and actualizes one of the alternative possibilities generated by the Schroedinger-equation-controlled evolution.

There is a widespread tendency for people, and even scientists, to suppose that these actualization events must occur at a microscopic level: that these events must actualize realities that are very tiny on the scale of a human brain. However, there is no good theoretical reason for this to be so, and absolutely no empirical evidence that this is the case. Any such evidence would, in fact, be evidence *against* the correctness or completeness of contemporary quantum theory, and no such evidence has ever been found. Von Neuman's analysis of the process of measurement (fortified, if one wishes, by recent decoherence arguments) shows that there is no reason at all why an actualization event cannot actualize a collection of properties that extends over large parts of the brain, and that corresponds to an individual conscious thought.

The core idea of the present work is that each conscious event is associated with a quantum event that actualizes a state of the brain that corresponds, in a certain specific way, to this conscious event. Specifically, each human conscious event is taken to be the *cause* of the quantum actualization event that actualizes the body-world schema that defines the picture of the body-world that is experienced in that conscious event: the conscious experience effectively bootstraps itself into existence in a manner made possible by the dual nature of Heisenberg's quantum ontology.

8.4 The Dynamics of the Selection of the Actualization Event

Each of the actualization events under consideration here actualizes a body-world schema. It actualizes a pattern of brain activity that represents a picture of the body in its environment, or some aspect or generalization of this picture. This picture constitutes an up-dating with 'projective' aspects: the picture is, in part, a picture of what is intended, and in this regard the body-world schema is a 'template for action': the actualization of the schema actualizes a pattern of brain activities that will tend to bring the projected picture into being. Thus when I am about to serve a tennis ball the actualized picture is a sequence of pictures of what I am intending to do, and these pictures drive my actions. The basic question is this: how does one formulate, mathematically, a physical theory that allows the experienced thought itself to drive the physical action, instead of being a mere witness to it?

There is no problem with the assertion that the experienced picture 'corresponds' to the schema, for the schema is supposed to contain all the information that is present in the picture: there is a mapping from the space of the possible schemas to the space of the possible pictures. Also, there is no problem with the fact that the (experienced) pictures actually exist: experiences are, in fact, the only things that we really know exist. But how can an experience *cause* an actualization? Indeed, an experience actually exists only insofar as the associated schema is actualized. We do not wish, in this naturalistic approach, to have thoughts floating around unattached to brains, or to be not specified in terms of the corresponding patterns of brain activity. But then how, rationally, can an experienced picture *cause* the actualization of the very schema that defines this picture itself?

Heisenberg's quantum ontology makes this possible. For, according to this ontology there exists, prior to the actualization, an objectively real 'potentia', consisting of all of the 'possible' realities existing together in a virtual or un-actualized state. This means that the 'possible' experienced pictures, which are mathematically imbedded in the various virtually existing schemas, likewise exist virtually, before the actual one emerges from the collection of possibilities. These 'virtual experiences' are part of what is mathematically represented by the quantum state. So instead of letting pure chance do the selecting, we allow this job to be done by a dynamical process involving these virtual pictures.

The Schroedinger equation itself act on properties that exist in this realm of virtual realities. However, the Schroedinger-controlled process does not complete the job. So postulating the existence of a 'second process' that acts in this same virtual realm, but at the level of more complex entities, is neither irrational, nor unphysical: the basic known dynamical law of physics already operates in this 'virtual' realm, and this dynamics needs to be completed in order to a become a rational theory that is compatible with our actual experience.

A second dynamical process can be introduced that will allow the experience that actually occurs to extract itself out of the milieu of virtual possibilities on the basis of an 'optimal' property , which is defined in relation to the overall aims and physical possibilities of the organism. An important point here is that this second dynamical physical process is formulated directly in terms of mathematically specified entities that are representions of whole indivisible experiencable pictures: the dynamics is not expressible, in any direct or simple way, in terms of localizable entities, or in a semi-classical framework.

9. Mathematical Formulation of the Model

My aim here to provide a mathematical model of causal quantum brain dynamics in which the quantum selection process is governed by our conscious thoughts, rather than by pure chance; i.e., where the notorious stochastic selection process of quantum mechanics, called the "irrational" element by Pauli, is replaced by a causal process in which our conscious thoughts, acting as whole entities not reducible to aggregates of local properties, become the bona fide selecting agents.

Quantum electrodynamics (extended to cover the magnetic properties of nuclei) is the theory that controls, as far as we know, the properties of the tissues and the aqueous (ionic) solutions that constitute our brains. This theory is our paradigm basic physical theory, and the one best understood by physicists. It describes accurately, as far as we know, the huge range of actual physical phenomema involving the materials encountered in daily life. It is also related to classical electrodynamics in a particularly beautiful and useful way. I take it as the basis of this work.

In this section I assume the reader to have some knowledge of the principles of quantum electrodynamics, and the notations used to describe it. I draw particularly on references [16] and [17], which describe in detail the natural connection between quantum electrodynamics and classical electrodynamics.

The brain is fairly transparent to electromagnetic radiation at frequencies that correspond to the first 10^3 standing wave modes, and the surrounding bone is a good insulator. This means that these modes should be quasi-stable. And these modes are fed by transitions between rotational states of proteins that should be activated by neuronal activity. Philip and Brian Stocklin have discussed these properties at length in connection with their own theory of consciousness.[18]

To represent the limited capacity of consciousness I assume, in this model, that the states of consciousness associated with a brain can be expressed in terms of this relatively small subset of the modes of the electromagnetic field in the brain cavity. (I also assume that events occurring outside the brain are keeping the state of the universe outside the brain cavity in a single state, so that the state of the brain can also be represented by a single state.) The brain is represented, in the method of Feynman, by a superposition of the collection of trajectories of the particles in it, with each element of the superposition accompanied by the state of the electromagnetic field that this collection of trajectories generates.

In the low-energy regime of interest here it should be sufficient to consider just the classical part of the photon interaction defined in [16]. Then the explicit expression for the unitary operator that describes the evolution from time t_1 to time t_2 of the quantum electromagnetic field in the presence of a set $L = \{L_i\}$ of specified classical charged-particle trajectories, with trajectory L_i specified by the function $x_i(t)$ and carrying charge e_i, is

$$U[L; t_2, t_1] = \exp < a^* \cdot J(L) > \exp < -J^*(L) \cdot a > \exp[-(J^*(L) \cdot J(L)/2)],$$

where, for any X and Y,

$$< X \cdot Y > \equiv \int d^4 k (2\pi)^{-4} 2\pi \delta^+(k^2) X(k) \cdot Y(k),$$

$$(X \cdot Y) \equiv \int d^4 k (2\pi)^{-4} i (k^2 + i\epsilon)^{-1} X(k) \cdot Y(k),$$

and $X \cdot Y = X_\mu Y^\mu = X^\mu Y_\mu$. Also,

$$J_\mu(L; k) \equiv \sum_i -i e_i \int_{L_i} dx_\mu \exp(ikx).$$

The integral along the trajectory L_i is

$$\int_{L_i} dx_\mu \exp(ikx) \equiv \int_{t_1}^{t_2} dt(dx_{i\mu}(t)/dt)\exp(ikx).$$

The $a^*(k)$ and $a(k)$ are the photon creation and annihilation operators:

$$[a(k), a^*(k')] = (2\pi)^3\delta^3(k-k')2k_0.$$

The operator $U[L; t_2, t_1]$ acting on the photon vacuum state creates the coherent photon state that is the quantum-theoretic analog of the classical electromagnetic field generated by classical point particles moving on the set of trajectories $L = \{L_i\}$ between times t_1 and t_2.

This $U[L; t_2, t_1]$ can be decomposed into commuting contributions from the various values of k. The general coherent state can be written

$$|q, p> \equiv \exp i(<q \cdot P> - <p \cdot Q>)|0>,$$

where $|0>$ is the photon vacuum state and

$$Q(k) = (a_k + a_k^*)/\sqrt{2}$$

and

$$P(k) = i(a_k - a_k^*)/\sqrt{2}.$$

The $q(k)$ and $p(k)$ are two functions defined (and square integrable) on the mass shell $k^2 = 0$, $k_0 \geq 0$, and the inner product of two coherent states is

$$<q, p|q', p'> = \exp -(<q-q'\cdot q-q'> + <p-p'\cdot p-p'> +2i<p-p'\cdot q+q'>)/4.$$

The coherent states $|q, p>$ can, for various mathematical and physical reasons, be regarded as the "most classical" of the possible states of the electromagnetic quantum field.[19] In the present model they will be the possible 'actual' states: i.e., the particular states into which the electromagnetic field inside the brain cavity can jump in a quantum actualization event. Thus the ith conscious event is represented by the transition

$$|\Psi_i(t_{i+1})> \longrightarrow |\Psi_{i+1}(t_{i+1})> = P_i|\Psi_i(t_{i+1})>,$$

where $P_i = |q, p; i><q, p; i|$ is a projection operator.

There is a decomposition of unity

$$I = \prod d^4 k (2\pi)^{-4} 2\pi \delta^+(k^2) \int dq_k dp_k / \pi$$

$$\times \exp(iq_k P_k - ip_k Q_k)|0_k> <0_k| \exp-(iq_k P_k - ip_k Q_k).$$

Here meaning can be given by quantizing in a box, so that that the variable k is discretized. Equivalently,

$$I = \int d\mu(q,p)|q,p> <q,p|,$$

where $\mu(q,p)$ is the appropriate measure on the functions q(k) and p(k). Then if the state $|\Psi> <\Psi|$ were to jump to $|q,p> <q,p|$ with probability density $<q,p|\Psi> <\Psi|q,p>$, the resulting mixture would be

$$\int d\mu(q,p)|q,p> <q,p|\Psi> <\Psi|q,p> <q,p|,$$

whose trace is

$$\int d\mu(q,p) <q,p|\Psi> <\Psi|q,p> = <\Psi|\Psi>.$$

Let the state of the electromagnetic field restricted to the modes that represent consciousness be called $|\Psi(t)>$. (Stricly speaking, one must use a density matrix formulation, but the generalization is trivial). Using the decomposition of unity one can write

$$|\Psi(t)> = \int d\mu(q,p)|q,p> <q,p|\Psi(t)>.$$

Hence the state at time t can be represented by the function $<q,p|\Psi(t)>$, which is a complex-valued function over the set of arguments $\{q_1, p_1, q_2, p_2, ..., q_n, p_n\}$, where n is the number of modes associated with $|\Psi>$. This formula expresses the state $|\Psi(t)>$, which in this model is the part of the state of the brain that corresponds to consciousness, as a superposition of states $|q,p>$. Each of these latter states $|q,p>$ is a maximally classical quantum state, and is supposed to a possible body-world schema; a representation of an experiencable picture of the body-world. Each state $|q,p>$ is like a musical one-thousand-note chord, and is one of the possible states into which $|\Psi(t)>$ can jump.

For each allowed value of k the pair of numbers (q_k, p_k) represents the state of motion of the kth mode of the electromagnetic field. Each of these modes is defined by a particular wave pattern that extends over the whole brain cavity. This pattern is an oscillating structure something like a sine wave or a cosine wave. Each mode is fed by the motions of all of the charged particles in the brain. Thus each mode is a representation of a certain integrated aspect of the activity of the brain, and the collection of values $q_1, p_1, ..., p_n$ is a compact representation of certain aspects the over-all electrical activity of the brain.

The state $|q, p>$ represents the conjunction, or collection over the set of all allowed values of k, of the various states $|q_k, p_k>$. The function

$$V(q, p, t) = <q, p|\Psi(t)><\Psi(t)|q, p>$$

satisfies $0 \leq V(q, p, t) \leq 1$, and it represents, according to orthodox thinking, the "probability" that a system that is represented by a general state $|\Psi(t)>$ just before the time t will be observed to be in the classically describable state $|q, p>$ if the observation occurs at time t.

In the absence of interactions, and under certain ideal conditions of confinement, the deterministic normal law of evolution entails that in each mode k there is an independent rotation in the (q_k, p_k) plane with a characteristic angular velocity $\omega_k = k_0$. Due to the effects of the motions of the particles there will be, added to this, a flow of probability that will tend to concentrate the probability in the neighborhoods of a certain set of "optimal" classical-type states $|q, p>$. The reason is that the function of brain dynamics is to produce some single template for action, and to be effective this template must be a "classical" state, because, according to orthodox ideas, only 'classical' states can be dynamically robust control states in the room-temperature brain [20].

According to a semi-classical description of the brain dynamics, only *one* of these classical-type states will be present, but according to quantum theory the state $|\Psi(t)>$ will usually be a superposition of many such states, unless collapses occurs at lower (i.e., microscopic) levels. The assumption here is that no collapses occur at the lower brain levels: there is absolutely no empirical evidence, or theoretical reason, for the occurrence of such lower-level brain events.

So in this model the probability will begin to concentrate around various locally optimal coherent states, and hence around the various (generally) isolated points (q, p) in the $2n-$dimensional space at which the quantity

$$V(q, p, t) = <q, p|\Psi_i(t)><\Psi_i(t)|q, p>$$

reaches a local maximum. Each of these points (q, p) represents a "locally-optimal solution" (at time t) to the search problem: as far as the myopic local mechanical process can see the state $|q, p>$ specifies an analog-computed best 'template for action' in the circumstances in which the organism finds itself.

According to an orthodox (Heisenberg-type) statistical idea there must eventually be a transition to a 'classically describable' state, and the selection of this state will be governed by 'pure chance', with the probability of the jump from the state $|\Psi(t)>$ to the possible classically describable state $|q, p>$ being given by the function $V(q, p, t) = <q, p|\Psi(t)><\Psi(t)|q, p>$. For reasons given earlier the aim here is construct, instead, a *causal* theory in which the selecting agent is not pure chance, but rather a conscious thought.

To create a causal quantum theory in which our conscious experiences control the brain activity we shall replace the above-mentioned statistical rule for the selection of the actual state by a causal dynamics in which the most 'optimal' (from a personal perspective) of these virtual experiences actualizes itself by actualizing the state $|q, p>$ that represents it.

There is in this model a natural mathematical definition of 'optimal', namely the function $V(q, p, t)$ defined above. The various conditions and constraints that characterize the overall situation in which the organism finds itself are dynamically expressed as the conditions that cause $V(q, p, t)$ to increase for arguments (q, p) that correspond to 'good solutions', as defined by the organism itself, to the problems that the organism faces, and to remain small for unsatisfactory solutions.

Note that the arguments (q, p) correspond to global features of the brain, and that each argument (q, p) specifies a possible experience. Thus a dynamics that is controlled by the function $V(q, p, t)$ is a nonlocal dynamics that is expressed in terms of the variables that identify certain indivisible experiential entities, our conscious thoughts.

To specify the dynamics let a certain time $t_{i+1} > t_i$ be defined by an (urgency) energy factor $E(t_{i+1}, t_i) = \hbar(t_{i+1} - t_i)^{-1}$. Let the value of (q, p) at the largest of the local-maxima of $V(q, p, t_{i+1})$ be called $(q(t_{i+1}), p(t_{i+1}))_{max}$. Then the simplest possible reasonable causal selection rule would be given by the formula

$$P_i = |(q(t_{i+1}), p(t_{i+1}))_{max} >< (q(t_{i+1}), p(t_{i+1}))_{max}|,$$

which entails that

$$|\Psi_{i+1} >< \Psi_{i+1}| / < \Psi_{i+1}|\Psi_{i+1} >= |(q(t_{i+1}), p(t_{i+1}))_{max} >< (q(t_{i+1}), p(t_{i+1}))_{max}|.$$

This dynamics could produce a tremendous speed up of the search process. Instead of waiting until all the probability gets concentrated in one state $|q, p >$, or into a set of isolated states $|q_i, p_i >$ [or choosing the state randomly, in accordance with the probability function $V(q, p, t_{i+1})$, which could often lead to a disastrous result], this simplest selection process would pick the state $|q, p >$ with the largest value of $V(q, p, t)$ at the time $t = t_{i+1}$. This process does not involve the complex notion of picking a random number, which is a physically impossible feat that is difficult even to define. It is a causal process, even though it is formulated in the space of variables (q, p) that label the person's possible conscious thoughts.

One important feature of this selection process is that it involves the state $\Psi(t)$ as a whole: the whole function $V(q, p, t_{i+1})$ must be known in order to determine where its maximum lies. This kind of selection process is not available in the semi-classical ontology, in which only one classically describable state exists at the macroscopic level. That is because this single classically describable macro-state state (e.g., some one actual state $|q, p, t_{i+1} >$) contains no information about what the $V(q, p, t_{i+1})$ associated with the other alternative possibilities would have been if the collapse to this state $|q, p, t_{i+1} >$ had not occurred.

There is no rational reason in quantum mechanics for such an earlier micro-level event to occur. Indeed, the only reason to postulate the occurrence of such premature reductions is to assuage the classical intuition that the action-potential pulse along each nerve "ought to be classically describable even when it is not observed", instead of being controlled when unobserved, as quantum theory normally requires, by the local deterministic equations of quantum field theory. The validity of this classical intuition is questionable if it severely curtails the ability of the brain to function optimally.

A second important feature of this selection process is that the actualized state Ψ_{i+1} is the state of the entire aspect of the brain that is connected to consciousness. So the feel of the conscious event will involve that aspect of the brain, taken as a whole. The "I" part of the state $\Psi(t)$ is its slowly changing part. This part is being continually re-actualized by the sequence of events, and hence specifies the slowly changing background part of the felt experience. It is this persisting stable background part of the sequence of templates for action that is providing the over-all guidance for the entire sequence of selection events that is controlling the on-going brain process itself.

The experiential aspect of the mind/brain dynamics is closely tied to the matter-like aspect represented by the wave function, in the sense that the various virtual experiences and their dynamical consequences are expressed in terms of the wave function and its evolution via the Schroedinger equation. But there is also the choice, or selecting, aspect of the thought, which is not reducible to the Schroedinger dynamics.

A somewhat more sophisticated search procedure would be to find the state $|(q,p)_{max}>$, as before, but to identify it as merely a candidate that is to be examined for its concordance with the objectives imbedded in the current template. This is what a good search procedure ought to do: first pick out the top candidate by means of a mechanical process, but then evaluate this candidate by a more refined procedure that could block its acceptance if it does not meet specified criteria.

This alternative proposed dynamics may seem overly sophisticated. But the generation of a truly random sequence is itself a very sophisticated (and indeed physically impossible) process, and all that the physical sciences have understood, so far, is merely the mechanical part of nature's two-part process. Here it is the not-well-understood selection process that is under consideration. I have imposed on this attempt to understand the selection process the naturalistic requirement that the whole process be expressible in natural terms, i.e., that the universal process be a causal self-controlling evolution of the Hilbert-space state vector in which all aspects of nature, including our conscious experiences, are efficacious.

No attempt is made here to show that the quantum statistical laws will hold for the aspects of the brain's internal dynamics controlled by conscious thoughts. No such result has been empirically verified. The validity of the statistical laws for events in the inanimate world is regarded as a consequence of our ignorance of the actual causes, and of certain a priori probability distributions. This is discussed in section 12.

10. Thoughts as Causes of Physical Effects.

It might be argued that even though the dynamics is expressed in terms of the variables (q, p) that specify our possible conscious thoughts, and although the dynamical process is what would follow from the idea that the 'optimal' virtual thought, acting as a unit, actualizes the corresponding quantum state, it would nevertheless also be possible to assert that the nonlocal causal process simply proceeds according to the nonlocal dynamical laws, with consciousness being merely an epiphenomenal consequence of the actualization process, instead of a bona fide cause of it.

In the classical-mechanics case the laws were formulated directly in terms of local properties, and all behaviour was therefore directly explainable, in principle, in terms of local entities. No dynamical property was given to any complex functional construct beyond what was explainable in principle by local physical properies acting in concert. Thus in the classical case one is able to 'see through' any complex physical entity down to the local causes, and recognize that all actions flow directly from the local properties, without a need to identify any basic causes other than these local properties: all complex structures become invisible, in classical mechanics, once one looks for the underlying or basic causes. Hence the neural correlates of thoughts, just like pistons and driveshafts, are seen as, fundamentally, nothing but high-level mechanical consequences of local properties.

But the in mind/brain dynamics described above consciousness is intrinsic: the model was constructed so as to complete quantum theory in a way that naturally brings conscious experiences, of the kind we actually experience, into the quantum theoretic description of the human mind/brain. In this model the functional constructs associated with conscious thoughts obey new laws that are not reducible to local laws, and these laws neither contravene, nor overide, or nor merely re-express, the local laws. Consequently, one can not see behind the quantum collapse any cause other than the conscious thought itself: if a 'causal' theory is demanded then one cannot, as in the classical case, 'see through' the functional entity, which in this case is an actualization event, and identify some other cause. Since conscious thoughts are known to exist, and, in any reasonable construal of the evidence, are the cause of the actions that the theory says they cause, the most parsimonious and reasonable ontological assumption is that they are in fact the causes of these events: there are no other natural candidates for the cause of these events, once pure chance is excluded.

11. Free Will?

How does this model cope with the problem of free will. The problem is that in a causal theory there can be no freedom, yet in a theory in which our choices are controlled by 'pure chance' the situation is even worse. For to be governed by mere whimsy, and complete lack of reason, is to be even less in personal control of our lives than in a world where consequence prevails.

The present model is causal: it does not involve any irreducible element of chance. On the other hand, the causality is not of the local-reductionistic type: actions are directly controlled by whole thoughts that are complex entities that are represented in the mathematical formalism by nonlocal (brain-wide) structures that contain, in a functionally effective form, the full content of the conscious thought. And these thoughts themselves are not determined by any lower-level realities or mechanisms, but they essentially extract themselves, in their wholeness, out of a quantum soup of virtual 'possibilities for what they might be'. And this extraction is effected by a nonlocal process that is controlled by the ongoing thought process itself, and that is honed, over the life of the organism, to elevate to actualness those thoughts that serve the needs that the organism itself defines with increasing precision over the course of years. This high-level sort of causality is very different from a local-reductionistic causality where everything is controlled fundamentally by microscopic events, and the high-level whole entities, our thoughts themselves, stand impotently outside the dynamical process.

12. Quantum Statistics.

If the process of selection and actualization of "the actual" in human brains is governed by a nonlocal causal process, rather than by pure chance, then one must naturally expect analogous causal processes to be occurring elsewhere in nature. If we assume that the selection process is in all cases controlled by a causal process then it must be explained why the statistical rules of quantum theory hold in those cases where they have been tested and validated.

An explanation can be constructed as follows. Consider an n-dimensional Hilbert space of points $(z_1, z_2, ..., z_n)$, where, each for each i,

$$z_i = x_i + iy_i = r_i \exp i\theta_i$$

is a complex number, and $r_i \geq 0$. This space can be imbedded in a 2n-dimensional real space of points $(x_1, y_1, x_2, y_2, ..., x_n, y_n)$, and each unitary transformation in the Hilbert space generates an orthogonal transformation in the real space. The volume in the real space defined by the intersection of the unit ball centered at the origin with the collection of rays from the origin that pass through a region R on the unit sphere is invariant under any orthogonal transformation, and hence also under the image in real space of any unitary transformation in the Hilbert space. Thus the volume (=surface area) of any region R of the unit sphere in the real space is invariant under the image of any unitary transformation in the Hilbert space.

Since dynamical evolution, and most symmetry operations in the the Hilbert space, are generated by unitary transformations, the a priori probability density of unit vectors in Hilbert space should be invariant under unitary transformations. Thus it is reasonable to assign to any region R on the surface of the real unit sphere **an a** priori probability equal to the volume (=surface area) of that region R.

This a priori probability rule can be used in the following way. Suppose that, as in our brain case, there is, for a given state Ψ_i, a rule that specifies a candidate projection operator P_i, and that if the passage from state Ψ_i to state $P_i\Psi_i$ is not "blocked" then the transition proceeds. If $P_i = I$, where I is the identity operator, then the passage is not blocked, since a change into itself is no change at all, and if $P_i = 0$ then the passage must be blocked, since a transition to the null state is not allowed.

But then what is the rule that determines whether the passage is blocked?

According to the idea behind the present theory everything that enters into the dynamics is represented in Hilbert space: nothing dynamically significant stands outside the Hilbert space of the universe! And the dynamics is to be specified in terms of the state of the universe, or perhaps in terms of the full history of states

$$(..., \Psi_{i-2}, \Psi_{i-1}, \Psi_i).$$

The simplest form for the "blocking rule" is that the states Ψ_i and $P_i\Psi_i$ determine a state Φ of unit norm that lies in the complex 2-dimensional subspace generated by Ψ_i and $P_i\Psi_i$, and that the transition from the state Ψ_i to the state $P_i\Psi_i$ proceeds unless for some representative of the state Ψ_i, which is defined only up to a phase factor, the direct path from Ψ_i to some representative of $P_i\Psi_i$ intersects the ray Φ

The geometric situation is this. The state Ψ_i can be represented in the 2-dimensional Hilbert space generated by Ψ_i and $P_i\Psi_i$ by the continuum of pairs of complex numbers

$$(z_1, z_2) = (\exp i\phi, 0); 0 \leq \phi \leq 1,$$

and the state $P_i\Psi_i$ can then be represented by the continuum of pairs

$$(\cos^2 \theta \exp i\phi, \sin \theta \cos \theta \exp i\phi \exp i\chi)$$

with $0 \leq \phi \leq 2\pi$ and $0 \leq \chi \leq 2\pi$. The overall phase factor $\exp i\phi$ drops out of all computations and can be set to unity. The phase factor χ reflects an arbitrary choice of the phase of the basis vector associated with the component z_2, and it is assumed that there is a representative of $P_i\Psi_i$ for each value of χ. The "direct path" from a representative of Ψ_i to a representative of $P_i\Psi_i$ can be traced out by allowing the value of θ to run from zero to its actual value. Allowing θ to run from zero to $\pi/2$ and χ to run from zero to 2π generates a 2-dimensional spherical surface $S_{1/2}$ of radius $1/2$ centered at $z_1 = 1/2$. The vectors Φ are defined as the set of unit-normed vectors from the origin $z_1 = z_2 = 0$, or as the equivalent parallel vectors of norm $1/2$ from the center of $S_{1/2}$. A uniform distribution of the unit-normed vectors Φ on the unit 2-sphere is equivalent to a uniform distribution of points on the spherical surface $S_{1/2}$. Notice that a point

$$(\cos^2 \theta', 0, \sin \theta' \cos \theta' \cos \chi', \sin \theta' \cos \theta' \sin \chi')$$

on $S_{1/2}$ blocks some direct path in $S_{1/2}$ from the representative $(1, 0, 0, 0)$ of Ψ_i to some representative of $P_i\Psi_i$ if and only if θ' satisfies $0 \leq \theta' \leq \theta$

In some situations, namely those in which the realities that are governing the second process are human conscious experiences, we have direct knowledge of what the governing realities are: they are exactly the conscious experiences

562

that are controlling the second process. But in cases where the collapse of the wave function is associated with, say, an event in a Geiger counter, we are not privy to the form of the controlling realities. So in these cases we must fall back to statistical considerations. According to the model described above, there is a vector Φ that determines whether or not the collapse will occur, but we are ignorant of what it is. But the a priori probability distribution for the location of the vector Φ corresponds to a uniform distribution over the spherical surface $S_{1/2}$. The probability that the transition from Ψ_i to $P_i\Psi_i$ will be blocked is then equal to the fraction of the surface area of $S_{1/2}$ that is covered as θ' runs from zero to θ. This probability is $1 - \cos^2\theta$. Hence the a priori probability that the transition will occur is $\cos^2\theta$. This is the same as $|P_i\Psi_i|^2/|\Psi_i|^2$, which is what quantum theory predicts. So in this model the statistical predictions of quantum theory would arise from a combination of our ignorance of the true causes, with an a priori uniform probability distribution over an appropriate 2-sphere of the real image of a Hilbert space vector Φ that determines whether the transition to a specified state occurs or not.

13. Two Final Remarks

1. Quantum brain theory has been characterized as "A solution in search of a problem". A first question, in this connection, is whether a semi-classical model of the brain—e.g., a model in which the action potential on every neuron is regarded as a well-defined classically describable electromagnetic pulse—is capable of generating solutions to search problems as quickly as the brain actually does it, or whether a quantum mechanism such as the hydrodynamic effect, or the picking of the 'optimal' solution discussed above, is needed. The way in which a classical brain could search for suitable templates for action (or recognize patterns) is not known at present in enough detail to make an estimate of the classically allowed rapidities possible . But it seems reasonable that nature would make use of the quantum possibilities for speeding up the search processes, and hence that the semi-classical model will, in the end, not be able to adequately explain the speed at which the brain works.

2. This question of speed is, however, not the only relevant consideration. Even if a semi-classical model were fast enough the question would arise why a dynamically inert psychical element is present at all in nature. Wigner emphasized that in the rest of physics every action of one thing upon another is

accompanied by a reaction of the second back on the first. A dynamically inert psychic reality could have no survival value, hence no physical reason to exist. Yet it seems absurd to think that something so different from its supposedly classical physical foundation could arise just by accident.

References

1. N. Bohr, See ref.15 p. 61/64.

2. A. Einstein, in A. Einstein: Philosopher-Scientist, ed. P.A. Schilpp, Tudor, New York, 1951. p.81.

3. W. Heisenberg, Physics and Philosophy, Harper Row, New York, 1958, Chapter III.

4. H. Everett III, Rev. Mod. Phys. 29, 463, (1957).

5. W.H. Zurek, in New Techniques and Ideas in Quantum Measurement Theory, ed, Daniel M. Greenberger, Annals of the New York Academy of Science **480** p.96

6. E. Joos, in New Techniques and Ideas in Quantum Measurement Theory, ed, Daniel M. Greenberger, Annals of the New York Academy of Science **480** p.12

7. M. Gell-Mann and James B. Hartle, Classical Equations for Quantum Systems, UCSBTH-91-15

8. R. Omnes, The Interpretation of Quantum Theory, Princeton University Press, Princeton NJ, 1994, p.348.

9. John Bell, in Sixty-Two Years of Uncertainty: Historical, Philosophical, and Physical Inquiries into the Foundations of Quantum Mechanics, ed. A.I. Miller, Plenum Press, New York and London.

10. A. Einstein, in A. Einstein: Philosopher-Scientist, ed. P.A. Schilpp, Tudor, New York, 1951. p.667-673.

11. Wm. James, The Principles of Psychology, Vol I, Dover, New York, 1950 (Reprinting of 1890 text) p.139

12. D. Mermin, Amer. J. Physics, **62**, 880, (1994); D. Bedford and H.P. Stapp, Synthese 102, 139-164, 1995; H.P. Stapp, Phys. Rev. A49, 4257, 1994; Ref.15, p.6.

13. N. Bohr, See ref. 15, Chapter 3

14. R.P. Feynman, The Feynman Lectures in Physics, R.P. Feynman, R.B. Leighton, and M.Sands, Addison-Wesley, (1965) Vol. III, Chapter 21.

15. H.P.Stapp, Mind, Matter, and Quantum Mechanics, Spinger-Verlag, Heidelberg, 1993. Chapter 6.

16. H.P.Stapp, Phys. Rev. D28, 1386 (1983)

17. T. Kawai and H.P. Stapp, Phys. Rev. D52, 2484-2532, (1995)

18. Philip L. Stocklin and Brian F. Stocklin T.I.T. J. of Life Sci., 1979, Vol 9, pp. 29-51; and Evidence for Endogenous Standing Microwaves as a Substrate for Consciousness. (Paper delivered at the conference "Toward a Scientific Basis for Consciousness, University of Arizona Tucson ,AR, 1994); Physical Basis for Pattern Processing in the Human Brain, 1992

19. R.J. Glauber, in Quantum Optics, S.M. Kay and A. Maitland, eds. Academic Press, London and New York, 1970; T. W. B. Kibble in ibid;
H.P. Stapp, in Quantum Implications: Essays in Honour of David Bohm, B.J. Hiley and F.David Peats eds., Routledge and Paul Kegan Ltd., London and New York, 1987.

20. H.P. Stapp, in Symposium on the Foundations of Modern Physics 1990, P.Lahti and P. Mittelstaedt eds., World Scientific, Singapore. Sec. 3.

23.
Mind and Matter: Aspects of the Implicate Order Described Through Algebra

Basil J. Hiley

Mind and Matter: Aspects of the Implicate Order Described Through Algebra

B. J. Hiley,

Physics Department, Birkbeck College, London, WC1E 7HX.

Abstract.

Bohm has argued that the fundamental problems in quantum mechanics arise because we insist on using the outmoded Cartesian order to describe quantum processes and has proposed that a more coherent account can be developed using new categories based on the implicate order. This requires that we take process as basic and develop an algebra of process. In this paper we discuss some aspects of the basis of this algebra and show how it accommodates some of the algebras fundamental to quantum mechanics. We also argue that this approach removes the sharp division between matter and mind and hence opens up new possibilities of exploring the relationship between mind and matter in new ways.

1. Introduction.

Over the years there has been considerable interest in the question as to whether quantum mechanics would be of any help in understanding the nature of mind. One of the earliest suggestions of a possible connection between mind and quantum mechanics was made by von Neumann (1955). In his approach to quantum mechanics there are distinct processes. One is the continuous evolution of the wave function in time determined by the Schrödinger equation. The other is the collapse of the wave function that occurs during measurement, a process that cannot be incorporated into the Schrödinger equation. Both von Neumann (1955) and Wigner (1968) argued that to complete quantum mechanics, the mind is needed to account for this second non-unitary process.

This approach has been dismissed by many physicists who have argued that we have given up too easily on the attempt to find a physical explanation for the collapse. Indeed a number of interesting possibilities along these lines are still under consideration. Notable among these are the proposals of Ghirardi, Rimini and Weber (1986) who suggest that there is a new physical process that spontaneously collapses the wave function, a process that becomes magnified in the presence of many particles. Thus in the presence of a measuring apparatus, a spontaneous collapse occurs rapidly.

There is also the more recent proposals of Penrose (1994) who argues that when gravitational effects are introduced, the non-linearity of general relativity takes over and, under appropriate conditions, a collapse of the wave function occurs. These processes still lack experimental verification but offer a way out that does not introduce mind.

A more subtle reason for a possible connection between quantum mechanics and mind starts with the recognition of the key role played by the notion of *indivisibility* or of the *wholeness* of quantum processes that was stressed again and again by Bohr (1961). Indeed the notion of wholeness forms an essential part of the Copenhagen interpretation. When quantum phenomena are considered in this approach, there appears to be a number of analogies with thought processes. Some of these have been discussed by Bohm (1951) and he has even suggested possible reasons for this connection. If these are correct, the connections become more than mere analogy. He suggested that although much of the brain may be acting in a classical way, there may be key points where the control mechanism is so sensitively balanced that they might be controlled by quantum processes.

Likewise Stapp (1993) in extending this idea goes even further. He argues that in Heisenberg's approach, the nature of atoms are not "actual" things. The "primal stuff" represented by the wave function has an "idealike" character rather than being simply "matterlike". To quote Stapp (1993, p.221):

> Indeed, quantum theory provides a detailed and explicit example of how an idealike primal stuff can be controlled in part by mathematical rules based in spacetime. The actual events in quantum theory are likewise idealike: each such happening is a *choice* that selects as the "actual", in a way not controlled by any known or purported mechanical law, one of the potentialities generated by the quantum-mechanical law of evolution.

For Stapp actual events occur in nature whether the observer is present or not. In other words the collapse process is a real actualisation, the nature of which is still unknown. When such an actualisation occurs in the brain, it becomes intimately related to thought and consciousness. Thus in the brain these events can be actualised within a large-scale metastable state of a collection of neurons and it is these actual events that correspond to a conscious experience.

What both Bohm and Stapp are implying is that there is no longer a sharp distinction between mind and matter in the quantum domain. However as far as I am aware neither exploit this connection in detail although through the notion of the implicate order Bohm was attempting to lay the foundation for exploring these relations further. The purpose of this paper is to bring out these connections in new ways.

2. Quantum Mechanics and Mind.

To begin the discussion of a possible relationship between quantum mechanics and the mind I would like to briefly review some of the important developments that have taken place. The thesis that mind has a crucial role to play in the final stage of any observation of a quantum system has been proposed by von Neumann (1955), Wigner (1968), Everett (1973), Lockwood (1989) and others. These authors do not relate the collapse to specific properties of the brain or of the mind. Rather they focus on the gap between what the formalism predicts and what is actually observed and suggest that it is the mind that, in some unspecified way, bridges the gap. It is not that the formalism is wrong, but rather it is not complete in the sense that the formalism contains all the possible outcomes of the experiment, whereas only one is actual. The actualisation is not accounted for within the formalism and it is this feature that has generated many debates and has produced many explanations.

For example, von Neumann (1955) who contributed so much to the mathematical foundations of quantum mechanics, argued quite clearly that the human observer must play an active and essential role in the final stage of any measurement process. For him this intercession cannot be further analysed and it is this feature that is the source of the indeterminism that is assumed to underlie all quantum processes.

Many physicists feel that the mind should not enter in this way and there are a number of alternative proposals to complete quantum mechanics. One such idea is that when the system is sufficiently "complex", it is possible to replace the human observer by a machine and then one can argue, as does Everett (1973), "the machine has perceived a particular result". Unfortunately, as far as I am aware, the question of how a machine can perceive is not discussed. In the hands of DeWitt (1973), Everett's original proposals were sanitised and became the "Many Worlds Interpretation". Rather than have a single machine actualise the result, the Universe itself splits into branches, each branch corresponding to one of the actualisations. In other words, all alternatives are actual, but exist in different universes. Apart from the extravagant multiplication of actual universes, there are other serious shortcomings, some of which have been discussed by Bohm and Hiley (1993).

The most recent development along these lines is what Murray Gell-Mann (1994) calls the "Modern Interpretation". This involves, among other things, introducing the notion of "information gathering and utilising systems" (IGUS) (See Gell-Mann and Hartle (1989)). Again I find their discussion is not sufficiently developed to provide a clear account of precisely what new principles are involved in these IGU systems. Rather it is assumed that they exist and the consequences are then further elaborated.

Not all physicists are satisfied with this type of solution and numerous attempts have been made to produce a collapse by some other more specific means. This usually involves introducing some form of non-linear term, either directly or indirectly, into the Schrödinger equation. Among the more notable attempts in this direction have been those of Bohm and Bub (1966), of Bialynicki-Birula and Mycielski (1976), of Zurek (1981, 1989), of Ghirardi, Rimini and Weber (1986), of Petrosky and Prigogine (1994), and of Penrose (1994). Unfortunately all of these present problems of one form or another.

3. Does Quantum Mechanics have any role to play in understanding Mind?

The previous paragraph shows the difficulty of resolving the problem of the collapse of the wave function, but it gives little indication of how quantum theory will help in understanding brain/mind function. Any positive evidence for this must come from another direction. In the introduction I pointed out that both Bohm (1951) and Stapp (1993) were trying to build on the analogies between quantum processes and thought processes. Central to both their approaches was the notion of "indivisibility" or "quantum wholeness". These notions are rather difficult to deal with in the way we currently think about physical processes. We insist on dividing processes into parts in an attempt to get a clear picture as to "what is going on". If we are told that two particles that are in different regions of space, are inseparable, we have great difficulty in thinking about this. To be spatially separate is to be "actually" separate. Is there any intermediate way of looking at this indivisibility that might help to bridge this gap? My work on the Bohm interpretation of quantum mechanics has convinced me that there is and when we go down this road, we find nature is not mechanistic but has new features that are in some sense "mind-like".

In Bohm and Hiley (1993), we tried to bring this point out although I fear we did not succeed! In the book we show that a re-formulation of quantum mechanics allows us to explore new notions that help to provide new insights into the way quantum processes are structured. I do not want to argue that these notions are *forced* on us by the quantum formalism, but by adopting them we are led to new ways of exploring quantum processes in a more coherent way. What I want to do here is to consider whether some of these new notions will be of any use in exploring the nature of thought and mind. To do this I will need to briefly highlight some of the relevant features of the Bohm interpretation.

In this approach, a detailed study of the properties of the quantum potential indicated that this potential is qualitatively different from all known classical potentials, a point that is often missed by some supporters of this approach. The reason for missing this point is because the potential appears alongside the classical potential in an equation that has the same form as Newton's equation of motion. Attempts to derive this potential from a deeper *mechanical* sub-structure have not been successful. My own feeling is that this attempt will always fail for reasons that I will not discuss here.

One feature that is of particular importance is that, unlike potentials derived from classical waves, the quantum potential is independent of the amplitude of the quantum wave. This means that a wave of very small amplitude can produce a large effect on the particle. Equally a wave of large amplitude can produce a small effect. In fact the force depends only on the form of the wave profile. To help understand this idea recall how an audio signal is carried by a radio wave. The audio signal modulates the profile of the high frequency carrier. Here the audio energy can be quite small, but its form can be amplified to produce a large effect in the radio itself. By analogy we have argued that the small energy in the quantum wave can be magnified by some as yet unknown internal process so as to produce a large effect on the particle. We have carried this idea further and proposed the quantum wave carries information about the environment to the particle.

In making this proposal it should be noted that we are using the word "information" in a different way from its usual use. It is not information in the sense of a passive list of instructions. Rather it is used in the sense of a dynamical process giving new form to the activity we call the particle. In Bohm and Hiley (1993) we have suggested that there are three types of information, active, passive and inactive which should be distinguished. With these new categories of information we have shown how all quantum processes can be accounted for.

All of this is very different from what we would expect from a physical theory. Instead of giving rise to a mechanical model, the properties of the quantum field suggest that its main role is to organise particles or groups of particles into co-ordinated movement which is shaped by the environment in which they find themselves. The appearance of nonlocality in such co-ordinated movements supports the idea that a mechanical interpretation of such movements is not adequate. In fact the structure that begins to emerge is very reminiscent of the proposals of Whitehead (1939) who suggested that matter should be regarded as "organic" rather than mechanistic.

Throughout my discussions with David Bohm on this interpretation, it seemed as if the differences between mind and matter were being eroded. Not only is thought organised and structured by active information, but particles and fields could be better understood if one regarded them as being influenced by active information. In the former case, it is information for us, while in the latter it is information for the particle or the field.

There is another factor that helps to make these ideas more plausible. The particles themselves emerge as structures in the field itself. This suggests that what we regard as the "solid" substance out of which all macroscopic matter emerges is in some sense "illusory". By this I do not want to suggest that in order to understand nature we have to resort to some sort of medieval mysticism, or ultimately attribute all existence to "pure spirit". This would mean a return to pre-science in which a kind of irrationality would dominate. Pauli (1984) has suggested such a position, but I reject it. I look at these new features as an extension of science into regions where the old categories are not adequate.

The real problem is that we do not have the correct categories with which to discuss quantum processes adequately. It is not clear as to precisely what these categories are at present and therefore it will be necessary to explore new categories which will be sensitive to the kind of changes that are needed to accommodate both matter and mind so that we do not have to resort to extreme mysticism or return to Cartesian dualism. Of course in embarking on such an undertaking it will be necessary to go beyond quantum mechanics and to explore more radical approaches that will ultimately enable us to deal with both mind and matter in one theory. I do not pretend this will be an easy matter, but I feel that if we are not radical we will remain engaged in a series of fruitless arguments.

4. New Concepts.

With this background in mind, I want to explore a different approach to physical processes. Let me motivate this by making the following comments. One of the basic assumptions that we have inherited from Decartes is the sharp distinction between matter, *res extensa*, and thought, *res cogitans*. Matter is defined as existing in space in the form of separated extended objects. It is further assumed that matter is rational, obeying immutable laws of physics. Newtonian physics assumes various local interactions between these objects which govern their movement. This movement is continuous and, more importantly, it is deterministic. For convenience, I call the categories necessary to carry through this programme "Cartesian categories".

Mind, on the other hand, does not exist in space. It does not, therefore have any notion of locality. Furthermore it can appear to be irrational, jumping all over the place! It certainly does not appear to be deterministic although it can become mechanical, going through routines that have been programmed in at some stage.

Let us now consider quantum mechanics from the standard view point. At the particle level it appears indeterministic. There are "quantum jumps" which cannot be accounted for in terms of a continuous process in space-time, and there is a kind of non-separablity or "wholeness". It is further assumed that it is impossible to analyse the behaviour of a particle in a way that is independent of the means of observation. But if we pay attention only to the wave function, then everything is local because all the wave functions are local. Continuity and determinism also hold in the sense that wave functions follow a well-defined differential equation, namely, the Schrödinger equation. In other words we have successfully embedded quantum mechanics into the Cartesian categories at the level of the wave function, but not at the level of the individual particle or field.

To carry this through successfully we must now place the emphasis solely on the mathematical equations, allowing these equations to take us into new areas where we are even further from finding any connection with the physics. Thus it is assumed that only through abstract equations can we begin to understand physics. I would like to claim that the implicit assumption that lies behind this approach is that science is only possible using Cartesian categories. It is this assumption that I would like to challenge.

Before making some specific proposals, I would like to comment on the Bohm interpretation. At first sight it seems that in this interpretation we can maintain the Cartesian categories at the level of the individual particle. We have trajectories along which the individual objects travel in a continuous and deterministic way, but we are forced to introduce nonlocality. Nonlocality is not a notion that can be maintained within the Cartesian categories. So even in this interpretation we are forced to abandon these categories. The main role of the Bohm interpretation is to show is that it is possible to provide an ontology for quantum phenomena, i.e., an objective structure lying behind the appearances that are revealed in measurement. It is this possibility that is denied by the standard interpretation and its many variants all of which are based on the Cartesian categories.

These remarks may be surprising to some. After all in my book with David Bohm, we showed that the Bohm interpretation is internally consistent and contains no contradictions with experiment. However the interpretation uses the differential manifold as a basic descriptive form. This assumes that space-time is continuous and local. In spite of this it is necessary to introduce nonlocality and it can be argued that this nonlocality introduces an incoherent feature into the description. It is as if the nonlocality is "plastered on" as an afterthought and does not arise naturally from within the basic structure.

Nonlocality in itself does not lead to any conflict with experiments in the non-relativistic domain, but will there be problems in the relativistic domain? In attempting to answer this question we were able to extended the Bohm interpretation to the Dirac equation and to a quantum field theory of boson fields. In the former, particle trajectories could still be maintained, while in the latter, particle trajectories have to be abandoned and their place taken by the field configurations which are treated as the "beables" in this case. These configurations have since been evaluated and they are found to evolve in a deterministic way, being controlled by a "super-quantum potential" (See Lamm and Dewdney (1994)).

In both cases nonlocality is still present in the generalised quantum potentials associated with each case. However the presence of this nonlocality did not lead to any conflict with experiment as it is not possible to use these effects to transfer energy across space-like connections. The remarkable feature of these theories was that while at the level of the beables, locality and Lorentz invariance are not maintained, at the level of observables, i.e., the statistical level, both locality and Lorentz invariance are maintained. What this seems to suggest is that the invariances that are assumed to be immutable laws of physics at the fundamental level, emerge as statistical features of a yet more fundamental domain of process. We are thus led to seriously consider whether space-time itself may be a statistical feature of this deeper underlying process.

One can begin to see the necessity of raising such a question by considering a different problem, namely, the problem of quantising gravity, a problem that presents many conceptual difficulties (See C. Isham (1987)). When a field is quantised (such as the electromagnetic field) it is subjected to fluctuations. If general relativity is a correct theory of gravity then we know that the metric of space-time plays the role of the gravitational potential. If the fields fluctuate, the metric must fluctuate. But the metric is intimately related to the geometry of space-time. It enables us to define angle, length, curvature, etc. In consequence if the metric fluctuates, the geometric property of space-time will also fluctuate. What does it mean to have a fluctuating space-time?

Let me propose that at a deeper level, let us say at the sub-quantum level, space-time has no meaning so that space-time itself is merely a statistical display at a higher level. Thus space-time with its local relations and Lorentz invariance are all statistical features and that underlying this is a structure that does not find a natural expression in the space-time continuum. The nonlocal features, which appear also in the standard approach to quantum theory, are then a macroscopic reflection of this deeper structure. In other words, this pre-space is not merely a curiosity manifesting itself at distances of the order of the Planck length ($\sim 10^{-33}$ cm). It has much more immediate consequences at the macroscopic level.

I am not alone in making a proposal to start from this position. Penrose (1972) writes:

> I wish merely to point out the lack of firm foundation for assigning any physical
> reality to the conventional continuum concept.......Space-time theory would be
> expected to arise out of some more primitive theory.

John Wheeler (1978) puts it much more dramatically.

It is NOT

| | Day One: | Geometry |
| | Day Two: | Quantum Physics. |

But

| | Day One: | The Quantum Principle |
| | Day Two: | Geometry. |

But if we deny that space-time is a fundamental descriptive feature, where do we start to develop a new theory?

5. Beyond Space-time.

Let us begin the discussion by asking what space-time enables us to do. Essentially it allows us to co-ordinate the positions of particles and fields, and to place them in a certain order, that order being mapped onto R^4. This order enables us to describe the way particles and fields evolve under the classical laws of physics. In doing this we have put the order of physical processes in one-to-one correspondence with a Cartesian co-ordinate grid. Our emphasis is then is on a linear, local order and we require forces to produce deviations from this linear order.

There is another approach that has been used to order physical structures and that is to use barycentric co-ordinates. Here it is not the linear order that is emphasised, but a more complex order. We order processes in terms of a set of elementary units called simplexes. The 0-simplex is a point, the 1-simplex is a section of a line, the 2-simplex is a triangle, the 3-simplex is a tetrahedron and so on. Using these elementary building blocks we construct a simplicial complex which we assume will enable us to order physical processes in a new way.

This mathematics enables us to describe physical structures that emphasise topological properties of the structure which involves questions such as what is on the boundary of a given sub-structure. Of course, all of this can also be done on the space-time manifold. Indeed if this is done then these structures give rise to the Grassmann algebra which finds considerable use in modern physics.

We want to use this algebra in a different way. In an earlier paper Bohm, Hiley and Stuart (1970) showed how all the basic equations of physics could be given very simple expression in terms of the vanishing of boundary operators of differential forms. John Wheeler's (1990) beautiful book "A Journey into Gravity and Spacetime" took these ideas much further in general relativity and showed that this theory could be summarised in the sentence "The boundary of a boundary: where the action is! "

Again all of this can still be embedded in a local continuous space-time manifold which may or may not contain topological features such as "wormholes" and "handles". However the point I want to make is that all of this can be made independent of any underlying continuous manifold. If we could give physical meaning to these simplexes in a way that is independent of space-time, then we would have begun to do what we set out to do, namely, order physical processes independently of space-time. This will not leave us too isolated from space-time because we always have the possibility of recovering the space-time structure by exploiting the isomorphism between the de Rahm cohomology and the abstract cohomology defined by these general simplexes. The de Rahm cohomology is the cohomology defined by the Grassmann algebra using differential forms. This isomorphism will enable us to make direct contact with space-time.

The key question that remains is "What physical processes could these simplicial complexes represent?" To answer this question, it is clear that we must give up the notion of the particle and the field as basic descriptive entities. They are relevant only when giving rise to the order of physical processes within the Cartesian framework.

To many this may be asking too much and in order to make this suggestion a little more palatable, let us ask the question: "Where is the 'substance' of matter?" Is it in the atom? The answer is clearly "no". The atoms are made of protons, neutrons and electrons. Is it then in the protons and neutrons? Again "no", because these particles are made of quarks and gluons. Is it in the quark? We can always hope it is, but my feeling is that these entities will be shown to be composed of "preons", a word that has already been used in this connection. But we need not go down that road to see that there is no ultimon. A quark and an antiquark can annihilate each other to produce photons (electromagnetic energy) and the photon is hardly what we need to explain the solidity of macroscopic matter such as a table. Thus we see the attempt to attribute the stability of the table to some ultimate "solid" entity is misguided. We have, in Whitehead's words, a "fallacy of misplaced concreteness".

I would like to suggest there is no ultimate "solid" material substance from which matter is constructed, there is only "energy" or perhaps we should use a more neutral term such as "activity" or "process" or even "flux". This is implicitly what most physicists assume when they use field theory. But field theory depends on continuity and local connection. As I have remarked above it is local continuum from which I want to breakaway.

I am suggesting that what underlies all material structure and form is the notion of activity, movement or process. I will use the term "process" as part of my minimum vocabulary to stand for pure activity or flux and regard all matter as being semi-autonomous, quasi-local invariant features in this back ground of continual change. Bohm preferred to call this fundamental form "movement" and he called the background from which all physical phenomena arose the "holomovement".(See Hiley (1991) for an extensive discussion of this notion in the present context). I have since learnt that the word "movement" invariably invokes the response "movement of what?" But in our terms, movement or process cannot be further analysed. It is simply a primitive descriptive form from which all else follows, but to avoid ambiguities we could refer to it as the "holoflux" Here process replaces the term "field" as a primitive descriptive form of present day physics. Thus in our approach, the continuity of substance, either particle or field, is being replaced by the continuity of form within process.

6. Grassmann's contribution.

While I was talking about the possibilities of developing a description of physical processes in terms of process and activity in the late seventies, my attention was drawn to Grassmann's (1894) own account of how he was lead to what we now call a Grassmann algebra. (See Lewis (1977) for a recent discussion of Grassmann's work). To begin with, Grassmann argued that mathematics was about thought not about material reality. Mathematics studies relationships in thought, not a relationship of content, but a relationship of forms within which the content of thought is carried. Mathematics is to do with ordering forms created in thought and is therefore of thought. Now thoughts are clearly not located in space-time. They cannot be co-ordinated within a Cartesian frame. They are "outside" of space.

Thought is about becoming, how one thought becomes another. It is not about being. Being is a relative invariant or stability in the overall process of becoming. What I would like to suggest is that there is a new general principle lying behind Grassmann's ideas, namely, "Being is the outward manifestation of becoming".

Bohm and I exploited this principle in "The Undivided Universe" (1993). There we argued that in the Bohm interpretation, the classical level is to be regarded as the relatively stable manifest level (literally that which can be held in the hand or thought), while the quantum level is the subtle level that is revealed in the manifest level. In the first part of the book we showed how these notions could be applied to matter.

We also showed how similar arguments go through for thought. Thought is always revealed in thought. One aspect of thought becomes manifest and stable through constant re-enforcement, either by repetition or learning. New and more subtle thoughts are then revealed in these older re-enforced thoughts. The newer subtler thoughts can, in turn, became stable and form the basis for revealing yet more subtle thoughts. In his way a hierarchy of complex thought structures can be built up into a multiplex of structure.

What I want to suggest in this paper is that both material process and thought can be treated by the same set of categories and hence by the same mathematics. They appear to be very different, thoughts being ephemeral, whereas matter is more permanent. For me it is a question of relative stability and that stability in the case of material process is compounded to produce the appearance of permanence to us. I want to suggest is that if we can find a common language for matter and thought, then it will be possible to remove the Cartesian barrier between them and we will have the possibility of a deeper investigation into the relation between matter and mind.

7. The Algebra of Process.

How are we to build up such a mathematical structure? I believe that Grassmann has already begun to show us how this might be achieved in terms of an algebra which we now find very useful for some purposes in physics, but the full possibilities have been lost because Grassmann's original motivation has been forgotten. With this loss the exploitation of its potentially rich structure has been stifled

To begin the discussion, let us ask how one thought becomes another? Is the new thought independent of the old or is there some essential dependence? The answer to the first part of the question is clearly "no" because the old thought contains the potentiality of the new thought, while the new thought contains a trace of the old.

Let us follow Grassmann and regard P_1 and P_2 as the opposite poles of an indivisible process of a thought. To emphasise the indivisibility of this process, Grassmann wrote the mathematical expression for this process between a pair of braces as $[P_1P_2]$ which we can represent in the form of a diagram

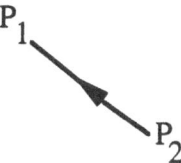

The braces and the arrow emphasise that P_1 and P_2 cannot be separated. It has the potentiality of being subdivided and if it is actually subdivided it becomes another process altogether. When applied to space, these braces were called extensives by Grassmann. For more complex structures, we can generalise these basic processes to:

$$[P_1P_2] \qquad [P_1P_2P_3] \qquad [P_1P_2P_3P_4]$$

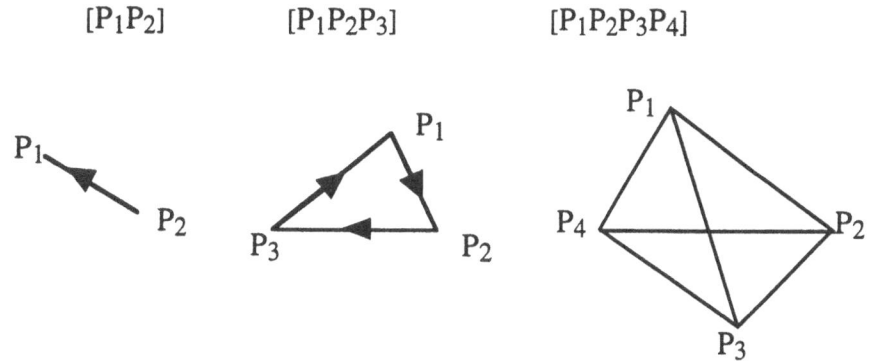

In this way we have a field of extensives from which we can construct a multiplex of relations of thought, process, activity or movement. The sum total of all such relations constitutes the holoflux.

For Grassmann then, space was a particular realisation of the general notion of process. Each point of space is a distinctive form in the continuous generation of distinctive forms, one following the other. It is not a sequence of independent points but each successive point is the opposite pole of its immediate predecessors so that points become essentially related in a dynamic way. In this view space cannot be a static receptacle for matter. It is a dynamic, flowing structure.

The structure of process can be regarded as an algebra over the real field in the following sense:

(i)   multiplication by a real scalar denotes the strength of the process.

(ii)   the process is assumed to be oriented. Thus $[P_1P_2] = -[P_2P_1]$.

(iii) the addition of two processes produces a new process. A mechanical analogy of this is the motion that arises when two harmonic oscillations at right angles are combined. It is well-known that these produce an elliptical motion when the phases are adjusted appropriately. This addition process can be regarded as an expression of the order of co-existence.

(iv) to complete the algebra there is a inner multiplication of processes defined by

$$[P_1P_2][P_2P_3] = [P_1P_3]$$

This can be regarded as the order of succession.

8. The Clifford Algebra of Process.

Let me now illustrate briefly how this algebra can carry the directional properties of space within it, without the need to introduce a co-ordinate system. I will only discuss the ideas that lead to what I call the directional calculus.

I start by assuming there are three basic movements, which corresponds to the fact that space has three dimensions. The process can be easily generalised to higher dimensional spaces with different metrics.

Let these three movements be $[P_0P_1]$, $[P_0P_2]$, $[P_0P_3]$. I then want to describe movements that take me from $[P_0P_1]$ to $[P_0P_2]$, from $[P_0P_1]$ to $[P_0P_3]$ and from $[P_0P_2]$ to $[P_0P_3]$. This means I need a set of movements $[P_0P_1P_0P_2]$, $[P_0P_1P_0P_3]$ and $[P_0P_2P_0P_3]$. At this stage the notation is looking a bit clumsy so it will be simplified by writing the six basic movements as [a], [b], [c], [ab], [ac], [bc].

We now use the order of succession to establish

$$[ab][bc] = [ac]$$
$$[ac][cb] = [ab]$$
$$[ba][ac] = [bc]$$

where the rule for the product (contracting) is self evident. There exist in the algebra, three two-sided units, [aa], [bb], and [cc]. For simplicity we will replace these elements by the unit element 1. This can be justified by the following results [aa][ab] = [ab] and [ba][aa] = [ba];   [bb][ba] = [ba], etc.
Now

$$[ab][ab] = -[ab][ba] = -[aa] = -1$$
$$[ac][ac] = -[ac][ca] = -[cc] = -1$$
$$[bc][bc] = -[bc][cb] = -[bb] = -1$$

There is the possibility of forming [abc]. This gives

$$[abc][abc] = -[abc][acb] = [abc][cab] = [ab][ab] = -1$$
$$[abc][cb] = [ab][b] = [a], \text{ etc.}$$

Thus the algebra closes on itself and it is straight forward to show that the algebra is isomorphic to the Clifford algebra C(2) which was called the Pauli-Clifford algebra in Frescura and Hiley (1980).

The significance of this algebra is that it carries the rotational symmetries and this is the reason I call it the directional calculus. The movements [ab], [ac] and [bc] generate the Lie algebra of SO(3), the group of ordinary rotations. For good measure this structure contains the spinor as a linear sub-space in the algebra, which shows that the spinors arises naturally from an algebra of movements. The background to all of this has been discussed in Frescura and Hiley (1984)

The generalisation to include translations has recently been carried out but the formulation of the problem is not as straight forward as we have to deal with an infinite dimensional algebra (see Frescura and Hiley (1984) and Hiley and Monk (1994)). It would be inappropriate to discuss this structure here. I will merely remark that our approach leads to the Heisenberg algebra, strongly suggesting that our overall approach is directly relevant to quantum theory.

9. The Multiplex and Neighbourhood.

Let us now return to consider structures that are not necessarily tied to space-time. For simplicity let us confine our attention to structures that can be built out of 0-dimension simplexes $\sigma_{(0)}$ and 1-dimension simplexes $\sigma_{(1)}$ only. Since we do not insist on continuity, any structure in the multiplex is constructed in terms of chains, i,e.,

$$C_{(0)} = \Sigma_i \, a_i \, \sigma_{(0)}{}^i \qquad \text{and} \qquad C_{(1)} = \Sigma_j \, b_j \sigma_{(1)}{}^j$$

where we will assume the coefficients a_i and b_j are taken over the reals or complex numbers

To those unfamiliar with the mathematics, it might be useful to have a simple example of a chain. Consider a newspaper photograph. Here the 0-dim simplexes are the spots of print, while the real weights a_i are the degree of blackness of the dots. Notice this description does not locate the positions of the dots. To do that we need to introduce the notion of a neighbourhood. This requires us to ask what is on the boundary of each simplex. The usual mathematical term that contains information about the neighbourhoods is called an incident matrix. This can be defined through the relations

$$\mathbf{B}\sigma_{(1)}{}^i = \Sigma_j \, {}_{(1)}\eta^i_j \sigma_{(0)}{}^j$$

Here **B** is the boundary operator and ${}_{(1)}\eta^i_j$ is the incident matrix.

Since we have no absolute neighbourhood relation, we can define different neighbourhood relations. Here again an illustration might help to understand the rich possibilities that this descriptive form may have. In one of his experiments on the development of concepts in young children, Piaget describes how when children are asked to draw a map of the local area around their home and are asked to position the playground, school, ice cream shop, dentist etc. The children place the ice cream shop and the playground close to home, but will place the dentist and the school far from home, regardless of their actual physical distance from their home. The children are using a neighbourhood relation which has to do with pleasure and not actual distance.

Thus by generalising the notion of neighbourhood, it is possible to have many different orders on the same set of points depending on what is taken to be the relevant criterion for the notion of neighbourhood in a particular context. Thus our description becomes context dependent and not absolute. This is very important both for thought and quantum theory.

In thought the importance of context is very obvious. How often do people complain that their meaning has been distorted by taking quotations out of context? The

579

importance of context in quantum theory has only recently begun to emerge, although it was always implicit in Bohr's notion of wholeness. However in the Bohm interpretation context dependence becomes crucial. Indeed the famous von Neumann "no hidden variable" theorem only goes through if it is assumed that physical processes are independent of context. Exploration using hidden variables was held up for a long time before the full significance of context was appreciated.

The new approach that I am suggesting has another interesting feature. It may be possible to have many different orders on the same set of points, or it may even be possible to define a different set of points, i.e., 0-dimensional simplexes, since the points themselves are to be regarded as particular movements i.e., of a movement into itself. Thus it may be possible to abstract many different orders from the same underlying process. To put it another way, the holomoflux contains many possible orders not all of which can be made explicit at the same time.

This general order has been called the "implicate order" by Bohm. The choice of one particular set of neighbourhood relations enables one to make one particular order manifest over some other. Any order that can be made manifest is called an "explicate order". So we have emerging from this approach a new set of ideas which fit the categories that Bohm was developing. It is in terms of these categories that we can give order to both physical and mental processes.

As we have already remarked one feature of this new description is that it is not always possible to make manifest all orders together at one time. This is an important new idea that takes us beyond the Cartesian order where it is assumed that it is always possible to account for all physical processes on one level, namely, in space-time. The new order removes the primacy of space-time allowing other important orders to be given equal importance. Thus in quantum mechanics the use of complementarity is now seen as a necessity arising out of the very nature of physical processes, rather than being a limitation on our ability to account for quantum processes.

Mathematically this new idea can be expressed through what we have called an "exploding" transformation. This is brought about by considering a structure built out of a set of basic simplexes $\sigma_{(0)}{}^i$ & $\sigma_{(1)}{}^j$ and then transforming the structure to one built out of a different set of basic simplexes $\Sigma_{(0)}{}^i$ & $\Sigma_{(1)}{}^j$. For simplicity we will call these basic simplexes "frames". Suppose these frames are related through the relations

$$\sigma_{(0)}{}^i = \Sigma_j \; a_j{}^i \; \Sigma_{(0)}{}^j \quad \text{and} \quad \sigma_{(1)}{}^i = \Sigma_j \; \beta_j{}^i \; \Sigma_{(1)}{}^j$$

We may now ask how the neighbourhood relations are related under such a transformation. Suppose we have

$$\mathbf{B}\sigma_{(1)}{}^i = \Sigma_j \; {}_{(1)}\eta^i{}_j \sigma_{(0)}{}^j \quad \text{and} \quad \mathbf{B}\Sigma_{(1)}{}^i = \Sigma_j \; {}_{(1)}\eta^i{}_j \; \Sigma_{(0)}{}^j$$

and then ask how the incidence matrices are related. We have

$$\mathbf{B}\sigma_{(1)}{}^i = \mathbf{B}(\Sigma_j \; \beta_j{}^i \; \Sigma_{(1)}{}^j) = \Sigma_{j,k} \; a_k{}^i \; {}_{(1)}\eta^k{}_j \; \Sigma_{(0)}{}^j = \Sigma_{\ell \,(1)} \eta^i{}_\ell \sigma_{(0)}{}^\ell$$

$$= \Sigma_\ell \; {}_{(1)}\eta^i{}_\ell \; a_n{}^\ell \; \Sigma_{(0)}{}^n$$

So that

580

$$_{(1)}\eta^i{}_1 = a_k{}^i \,_{(1)}\eta'^k{}_j \,(a_j^l)^{-1}$$

Or in matrix form

$$\eta = \alpha\,\eta'\,\alpha^{-1}$$

This is the exploding transformation so called because the original structure can look quite different after such a transformation. Furthermore what is local in one frame need not be not local in any other. Mathematically these transformations are similarity transforms or automorphisms. In the algebraic approach we have tried to exploit these automorphisms in our algebraic description of pre-space. (See Hiley and Monk (1994)).

10. The Unmixing Experiment.

A specific example of such a transformation was given by Bohm (1980). I would briefly like to recall this example to show how it fits into my argument.

Consider two concentric transparent cylinders that can rotate relative to each other. Between these cylinders there is some glycerine (See figure 1).

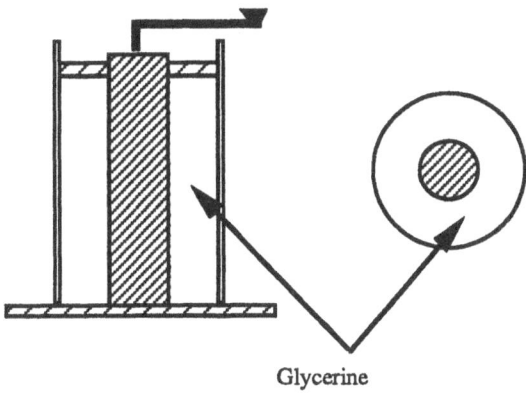

Glycerine

Figure 1 The unmixing experiment.

If a spot of dye is placed in the glycerine and the inner cylinder rotated, the spot of dye becomes smeared out and eventually disappears. There is nothing surprising about that, but what is surprising is that if we reverse the rotation, then the spot of dye reappears!

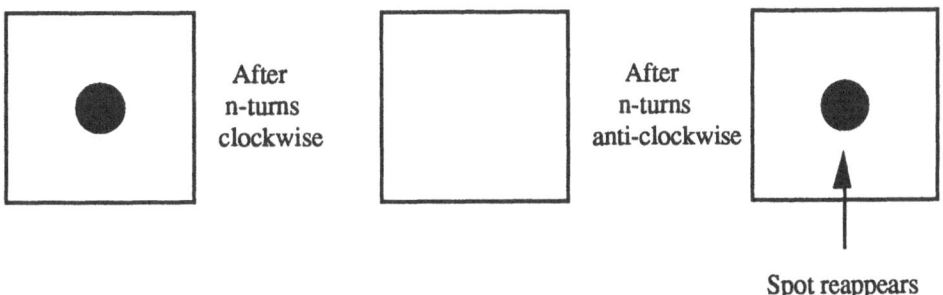

After
n-turns
clockwise

After
n-turns
anti-clockwise

Spot reappears

Figure 2 The reappearing spot.

This actually works in practice and is easily explained in terms of the laminar flow of the glycerine under slow rotation. What we want to illustrate here is that in the "mixed" state, there does not seem to be any distinctive order present. Yet the order is, as it were, *implicit* in the liquid and our activity of unmixing i.e., unwinding, makes manifest the order that is implicit in the glycerine.

To carry the idea further, we can arrange to put in a series of spots of dye, displaced from one another in the glycerine. Place one spot at x_1 and then rotate the inner cylinder n_1 times. Then place another spot at x_2 and rotate the inner cylinder again a further n_2 times and so on repeating N times in all. If we were then to unwind the cylinder, we would see a series of spots apparently moving through the glycerine. If the spots were very close together we would have the impression of the movement of some kind of 'object' starting from position x_1 and terminating at position x_N (see figure 2). But no object has actually moved anywhere! There is simply an unfolding and then enfolding movement which creates a series of distinguished forms that are made manifest in the glycerine. So what we have taken to be the continuous movement of substance is actually a continuous unfoldment of form.

Figure 3 The morphology of stuff.

Recalling my earlier remarks concerning the ephemeral nature of material processes, we can follow Whitehead (1957) and suggest that a quantum particle could be understood within the framework of these ideas. As Whitehead (1939) puts it "An actual entity is a process and is not describable in terms of the morphology of a stuff".

In this view, the cloud chamber photograph does not reveal a "solid" particle leaving a track. Rather it reveals the continual unfolding of process with droplets forming at the points where the process manifests itself. Since in this view the particle is no longer a point-like entity, the reason for quantum particle interference becomes easier to understand. When a particle encounters a pair of slits, the motion of the particle is conditioned by the slits even though they are separated by a distance that is greater than any size that could be given to the particle. The slits act as an obstruction to the unfolding process, thus generating a set of motions that gives rise to the interference pattern.

11. Evolution in the Implicate Order.

I would now like to show how we can arrive at an equation of unfoldment using the ideas of the last two sections. We assume that we have an explicate order symbolised by **e**. This could have considerable inner structure, but for simplicity we will simply use a single letter to describe it. We want to find an equation that will take us to a new explicate order **e'** as a result of the unfolding movement.

Let us again use the order of succession to argue that the explicate order is enfolded via the expression eM_1. Here M_1 is an element of the algebra that describes the enfolding process. The next unfolded explicate order will be obtained from the expression $M_2\mathbf{e'}$. Here M_2 is the process giving rise to the unfoldment. To express the continuity of form, we equate these two expressions to obtain

$$\mathbf{e}M_1 = M_2\mathbf{e'}.$$

or

$$\mathbf{e'} = M_2^{-1}\,\mathbf{e}\,M_1$$

Thus the movement is an algebraic automophism, analogous to the transformation that we called the exploding transformation.

Let us now assume for simplicity that $M_1 = M_2 = M$, where $M = \exp[iH\tau]$. Here H is some element of the algebra characterising the enfolding and τ is the enfolding parameter. For small τ we have

$$\mathbf{e'} = (1 - iH\tau)\,\mathbf{e}\,(1 + iH\tau)$$

so that

$$i\frac{(\mathbf{e'} - \mathbf{e})}{\tau} = H\mathbf{e} - \mathbf{e}H = [H,\mathbf{e}]$$

Therefore in the limit as $\tau \to \infty$, we obtain

$$i\frac{d\mathbf{e}}{d\tau} = [H,\mathbf{e}].$$

This equation has the same form as the Heisenberg equation of motion.

If we represent the explicate order $\mathbf{e}$ by a matrix and assume it is factorable i.e., $\mathbf{e} = \psi\phi$ we find

$$i\frac{d\psi}{d\tau}\phi + i\psi\frac{d\phi}{d\tau} = (H\psi)\phi - \psi(\phi H)$$

If we regard ψ and ϕ as independent, we can separate the equation into two

$$i\frac{d\psi}{d\tau} = H\psi \quad\text{and}\quad -i\frac{d\phi}{d\tau} = \phi H.$$

If H is identified with the Hamiltonian, then the first equation has the same form as the Schrödinger equation. If ϕ is regarded as $\psi^\dagger$ then the second equation has the same form as the complex conjugate of the Schrödinger equation. Thus the Schrödinger equation arises in a very simple from the unfolding process.

12. Conclusion.

In this paper I have tried to motivate a new way of looking at physical processes in which the sharp Cartesian division between mind and matter can be removed. We began by showing that it is possible to explore new ways of describing material process that does not begin with an *a priori* given space-time continuum. Instead by starting with the notion of activity or process which is taken as basic, we are able to link up with some of the mathematics used in algebraic geometry. In fact in the particular example we used, we were able to recover the Clifford algebra, implying that some aspects of the symmetries of space can be carried by the process itself, albeit in an implicit form. By extending these ideas to include Bohm's idea of the enfolding process, we were also able to construct an algebra similar to the Heisenberg algebra used in quantum theory.

The motivation for exploring this approach came from two different considerations. Firstly, it came from the problems of trying to understand what quantum mechanics seems to be saying about the nature of physical reality. Using the Cartesian framework we find that, rather than helping to clarify the physical order underlying quantum mechanics, we are led to the well-known paradoxes that make quantum theory so puzzling and often unacceptable to many. It seems to me that these difficulties will not be resolved by tinkering with the mathematics of present day quantum mechanics. What is called for is a radically new approach to quantum phenomena.

The second strand of my argument was inspired by the work of Grassmann who showed how by analysing thought, one could be led to new mathematical structures. In other words, by regarding thought as an algebraic process, Grassmann was led to a new algebra which we now call the Grassmann algebra. The scope of this algebra has become rather limited by being grounded in space-time as, unfortunately, the original motivations have been largely forgotten. By reviving these ideas, I have been exploring whether the similarities between thought and quantum processes that I have tried to bring out in this paper could possibly lead to new ways of thinking about nature.

What this means is that if we can give up the assumption that space-time is absolutely necessary for describing physical processes, then it is possible to bring the two apparently separate domains of *res extensa* and *res cogitans* into one common domain. What I have tried to suggest here is that by using the notion of process and its description by an algebraic structure, we have the beginnings of a descriptive form that will enable us to explore the relation between mind and matter in new ways.

In order to discuss these ideas further we must use the general framework of the implicate order introduced by Bohm (1980). An important feature within this order is that it is not possible to make everything explicit at same time. This feature is well illustrated in the unmixing experiment described above. Here when there are a series of spots folded into the glycerine, only one spot at a time can be made manifest. In order to make manifest another spot, the first spot must be enfolded back into the glycerine and so on. If we now generalise this idea and replace the spots by a series of complex structure-processes within the implicate order, then not all of these processes can be made manifest together. In other words, within the implicate order there exists the possibility of a whole series of non-compatible explicate orders, no one of them being more primary than any other.

This is to be contrasted with the Cartesian order where it is assumed that the whole of nature can be laid out in a unique space-time for our intellectual examination. Everything in the material world can be reduced to one level. Nothing more complicated is required. I feel this implicit dependence on the Cartesian order is the reason why it is such a shock when people first realise that quantum mechanics requires a principle of complimentarity. Here we are asked to look upon this as arising from the limitations of our human ability to construct a unique description, this ambiguity having its roots in the uncertainty principle. But it is not merely an uncertainty; it is a new ontological principle that arises from the fact that it is not possible to explore complementary aspects of physical processes together. Within the Cartesian order, complimentarity seems totally alien and mysterious. There exists no structural reason as to why these incompatibilities exist. Within notion of the implicate order, a structural reason emerges and provides a new way of looking for explanations.

Finally I would like to emphasise that it is not only material processes that require this mutually exclusivity. Such ideas are well known in other areas of human activity. There are many examples in philosophy and psychology. To illustrate what I mean here, I will give the example used by Richards (1974, 1976). He raises the question: "Are there ways of asking 'what does this mean?' which actually destroys the possibility of an answer?" In other words can a particular way of investigating some statement make it impossible for us to understand the statement? In general terms what this means is that we have to find the appropriate (explicate) order in which to understand the meaning of the statement. Context dependence is vital here as it is in quantum theory and this is ultimately a consequence of the holistic nature of all processes.

Such ideas cannot be accommodated within the Cartesian framework. If we embrace the notions of the implicate-explicate order proposed by Bohm, we have a new and more appropriate framework in which to describe and explore both material processes and mental processes

13. References.

I. Bialynicki-Birula and J. Mycielski, (1976) *Ann. Phys., (N.Y)*, **100**, 62-93.

D. Bohm, (1952) *Quantum Theory*, Prentice-Hall, Englewood Cliffs.

D. Bohm, (1952) *Phys. Rev.*, **85**, 66-179; **85**,180-193.

D. Bohm, (1980) *Wholeness and the Implicate Order*, Routledge, London.

D. Bohm and J. Bub, (1966) *Rev. Mod, Phys.*, **38**, 435-69.

D. Bohm and B. J. Hiley, (1993) *The Undivided Universe: an Ontological Interpretation of Quantum Theory*, Routledge, London.

D. Bohm, B. J. Hiley and A. E. G. Stuart, (1970) *Int. J. Theor. Phys.*, **3**, 171-183.

N. Bohr, (1961) *Atomic Physics and Human Knowledge*, Science Editions, New York.

H. Everett, (1973) *The Theory of the Universal Wave Function. In The Many-Worlds Interpretation of Quantum Mechanics*, ed., B. DeWitt and N. Graham, Princeton University Press, Princeton.

F. A. M. Frescura and B. J. Hiley, (1980) *Found. Phys.*, **10**, 7-31.

F. A. M. Frescura and B. J. Hiley, (1984) *Revista Brasilera de Fisica, Volume Especial, Os 70 anos de Mario Schonberg*, 49-86.

M. Gell-Mann, (1994) *The Jaguar and the Quark:Adventures in the Simple and Complex.* Little, Brown and Co., London.

M. Gell-Mann and J. B. Hartle, (1989) *Quantum Mechanics in the Light of Quantum Cosmology*, in Proc. 3rd Int. Symp. Found. of Quntum Mechanics, ed., S. Kobyashi, Physical Society of Japan, Tokyo.

G. C. Ghirardi, A. Rimini and T. Weber, (1986) *Phys. Rev.* **D34**, 470-491.

H. G. Grassmann, (1894) *Gesammeth Math. und Phyk. Werke*, Leipzig.

B. J. Hiley, (1991) *Vacuum or Holomovement*, in The Philosophy of Vacuum, ed., S. Saunders and H. R. Brown, Clarendon Press, Oxford.

B. J. Hiley and N. Monk, (1993) *Mod. Phys. Lett.*, **A8**, 3225-33.

C. J. Isham, (1987) *Quantum Gravity, Genreral Relativity and Gravitation*, Proc. 11th Int. Conf. on General Relativity and Gravitation (GR11), Stokholm, !986, Cambridge University Press, Cambridge.

M. M. Lamm and C. Dewdney, (1994) *Found. Phys.*, **24**, 3-60.

A. C. Lewis, (1977) *Ann. Sci. (N.Y)*, **34**, 104.

M. Lockwood, (1989) *Mind, Brain & the Quantum: the Compound 'I'*, Blackwell, Oxford.

J. von Neumann, (1955) *Mathematical Foundations of Quantum Mechanics*, Princeton University Press, Princeton.

R. Penrose, (1994) *Shadows of the Mind*, Oxford University Pess, Oxford.

T. Petrosky and I. Prigogine, (1994) *Chaos, Solitons and Fractals*, **4**, 311-359.

H. P. Stapp, (1993) *Mind, Matter and Quantum Mechanics*, Springer, Berlin.

W. Pauli, (1984) *Physik und Erkenntnistheorie*, ed. K von Meyenn Vieweg, Braunschweig.

R. Penrose, (1972) in *Magic without Magic: Essays in Honour of J. A. Wheeler,* ed. Klauder.

I. A. Richards, (1974) *Beyond,* Harbrace.

I. A. Richards, (1976) *Complementarities: Uncollected Essays,* Harvard University Press, Harvard.

J. A. Wheeler, (1990) *A Journey into Gravity and Spacetime,* Freeman, New York.

J. A. Wheeler, (1978) *Quantum Theory and Gravitation*, in Mathematical Foundations of Quantum Theory, ed. Marlow, Academic Press, New York.

A. N. Whitehead, (1957) *Process and Reality,* Harper & Row, New York.

A. N. Whitehead, (1939) *Science in the Modern World*, Penguin, London.

E. P. Wigner, (1986) in *The Scientist Speculates*, ed., I. H. Good, p. 232, SUNY Press, New York.

W. H. Zurek, (1981) *Phys. Rev.,* **24D**, 1516-25.

W. H. Zurek, (1989) *Nature* , **341**, 119-24.

Afterword

Karl H. Pribram and Joseph King

Appalachian IV: Afterword

by Joseph S. King and Karl H. Pribram

Introduction

The Fourth Appalachian Conference on Behavioral Neurodynamics centered on the theme: Learning as Self-Organization. When told of this theme, one colleague dismissed it as an appeal to a new fad called 'self-organization' despite the fact that the theme has been around for more than 30 years. In fact, the term is applied to two distinct processes: 1) autopoiesis, which is more aptly described as the self-maintenance of structure despite the replacement of its parts; and 2) its other formal meaning as described by one of its originators Ilya Prigogine, in the keynote address for Appalachian II. In this second sense, the use of the term 'self-organization' describes turbulence more often called chaos, the development of stabilities far from equilibrium.

In keeping with the interdisciplinary purposes of the conferences, the proceedings contain papers by noted neuroscientists, engineers, model builders and practitioners. As Pribram reminded us in his introduction to the conference, self-organization is the key to learning beyond the laboratory, learning as it occurs in a variety of educational settings from the classroom to facilities for special education. Learning in these situations serves the self-organizing needs of the individual, not simply the reduction of primary drives. To ignore or eliminate the existence the needs for self-organization views living organisms as nothing but collections of programs that are more reminiscent of the serial computer than of biology.

For a keynote to implement the theme of learning as self-organization, we chose to recall a paper presented by Konrad Lorenz at Harvard University as a part of the Lowell Lecture series on the biology of learning and remembering. The paper makes a classic contribution to the resolution of the age-old discussion concerning the roles of genetics and experience to memory storage and retrieval. In addition, this paper is important in that Lorenz conceptualizes learning as a self-organizational set of processes resulting in a new set of emergents operating at a new scale of organization. As always with biological systems, it is both the organism's environment and its physiology which constrain the organizational processes.

As the conference proceeded, it became obvious that self-organization is best approached in terms of the scale at which various processes operate. Failure to account for scale results in a category error and confusion: scientists talking past each other rather than with each other. Indeed, our primary accomplishment as a set of transdisciplinary conferences may turn out to be the description relations between processes occurring at different scales. The importance of scale was the theme of the Third Appalachian Conference, and continues to form the basis for the organization of these proceedings.

The Behavioral Level: Learning

To set the stage, a series of contributions inaugurated by Duane Rumbaugh indicate that, finally, experimental psychologists are addressing the issue of learning at a higher level than that encompassed by habituation, classical and operant conditioning. Hans-Lucas Teuber was fond of remarking that conditioned operants could be obtained, not only without the participation of the subject's cerebral cortex (true), but as well, without the participation of the experimenter's cerebral cortex (false). As can be seen from the contributions by McIlvane and Killeen, even if that had ever been the case, certainly operant conditioning as a technique has come of age. McIlvane provides developmental data to introduce his work on behavioral momentum, and Killeen uses operant data to describe a physics of learning with multidimensional forces which act through behavior to pattern an organism's actions. All of the papers in this section address learning as an emergent -- again a theme that echoes a previous (the second) conference.

The Network Level: Self-Organization

Part II addresses an issue left hanging to some extent in Part I (except for the paper by Lillian Greeley on the role of attractors as seen in discourse since the time of ancient Greece). In Part II, possible processes are discussed in terms of the how neural networks can learn. The importance of scale became apparent in the presentations of Michael Stadler, Stassinopoulos and Kak. These mathematicians and scientists are concerned with the more formal properties of self-organizing biological systems. Michael Stadler points out that the phase transitions involved in learning provide new challenges to the theory of self-organization. He notes that while the phase transitions involved in perception operate within the same degree of order, those involved in learning result in higher order states. These changes to higher ordered states are not continuous, as often assumed in biology, but are saltatory. Stadler then proceeds to delineate the procedures necessary for the identification of phase transitions in learning and their control parameters. His synergetic approach capitalizes on the dynamic and reciprocal relationships between macro and micro states as mutually organizing. He further suggests that the brief plateaus during learning often reflect the processes of reorganization involved in learning, outlines their behavior correlates, and distinguishes them from other factors that do not involve self-organization.

Stassinopoulos shows, in his contribution, that the formal description of self-organization itself portrays the relation between scales. Kak reminded us that self-organization, by definition, is the use of local interactions to affect global characteristics. To our mind, however, an important consideration regarding the properties of the reinforcement process is not addressed: How can the environmental contingencies (as contrasted to those inherent in the biology of the organism) be represented in brain tissue so as to influence the probability of the recurrence of the behavior being reinforced? Pribram (1995) has addressed this issue in a recent paper, "The Enigma of Reinforcement," by suggesting that storage in the spectral domain (in the frontal cortex) can accomplish the task.

Paul Werbos concludes this second section by defining a new field-neuroengineering, which uses mathematics as a bridge between brain processes and engineering. This approach sees the brain as operating on principles of approximate dynamic programming, rather than optimization principles which constitute the traditional engineering model. He then illustrates how these mathematical principles can be meaningfully applied to the understanding of brain function in a hierarchally arranged neural systems framework.

The Neural Systems Level: Process

Part III is composed of contributions that delineate the impact of experience on the neural substrate of learning. These contributions indicate where and how "it all happens". The expressed common goal was to relate neural representations to learning. Jason Brown opened the section with his discussion of the effects of brain injury as illustrative of a reorganization of function, rather than simply a dysfunction. His analysis results in the finding that such reorganizations represent earlier forms of cognitive function, and thus provide insights into the cognitive and neural reorganization that takes place during development. Hanlon continues the focus on development with her thorough analysis of developmental data from individuals aged .2 to 24 years. Her analysis indicates that brain reorganization, measured in terms of coherence patterns in EEGs, develops in systematic fashion and differs markedly in males and females.

Making use of tomographic imaging technology, Dan Schacter presented data implicating inferior temporal, hippocampal and lateralized frontal neural systems in different memory tasks. His explorations indicate that implicit, priming effects are carried out through sensory specific "association" (intrinsic) cortical areas, whereas hippocampal activation occurs when explicit memory requirements are added to a task, at least during recall. His newest data indicate that when effort is involved in recall during memory tasks, frontal lobe and anterior cingulate activation results, whereas the hippocampus is involved when encoding process are effortless. The term effort is used by Schacter in a somewhat different sense than it was used by Pribram and McGuinness (1975; 1992), who developed the theme that the hippocampus is involved in an effort-comfort dimension that leads to efficiencies in processing. According to these authors, the frontal lobes become involved when the need for effective processing has to override efficiency by flexibly reorganizing working memory. It is this flexible process that Schacter taps with his definitions of effort.

Ray Kesner used his considerable skill in interdigitizing animal and human neuropsychology to describe how different forms of memory aid in the representation of reinforcement and its context. His elegant neuropsychological testing of animals with controlled lesions of hippocampus and amygdala provide interesting and informative parallels in the comparative study of learning processes. According to his data, the amygdala is involved in "valuing," while the hippocampus acts to select information so that events can become separated in time and space. These results are consonant with those obtained by Pribram and his collaborators in monkeys (reviewed by Pribram 1991; 1995) in that "events" (from Latin ex-venire, out-comes) are seen as the reinforcing consequences of actions -- those consequences that are relevant (in location and duration) to the organism's processes.

Don Tucker illustrated the use of Electroencephalography (EEG) with new geodesic, dense array technology to provide the temporal as well as spatial resolution necessary to study many neuropsychological relationships. His paper views his data on the reprise of an occipital-frontal-occipital response evoked by visual stimulation and extends the interpretation of these results into a historical framework. Observations of these effects of brain lesions in patients, electrical recordings, tomographic displays and the results of controlled resections of specific brain systems have yielded a fascinating body of evidence. Our hope is that this evidence will again become a part of the mainstream of knowledge in the neurosciences which are currently dominated by the results of microelectrode and membrane chemical studies to the exclusion of observations made by other techniques.

The Social Level: The Organization of Self in Society

Parts IV and V need an explanation: What might social phenomena and the phenomena observed in quantum and classical physics have to do with behavioral neurodynamics? As can be seen from the contributions to these parts of the proceedings, a great deal. Much of what we learn is already organized for us in our environment. What the organism needs to do, in a manner of speaking, and, in actuality, is to resonate, through sensory transduction, to these external orderings. This was the theme propounded by James Gibson over the years, and it applies to perceiving the social environmental contingencies to be learned as much as it does to ongoing perception of physical objects and events.

Freeman emphasizes the biological roots of social behavior which can be ignored only by placing society in peril. Bradley's data, on the other hand, show that the organization of communal systems can be understood in terms of both biology and social control to provide an information flow as developed by Gabor. Both Freeman and Bradley add a highly creative dimension to the application of physical models to biological systems. Alan Gregory tops these endeavors in his analysis of language, specifically the use of vocabulary. Gregory's methods follow the traditions of Freud and Jung in guiding the individual's self-organizing propensities into useful social channels.

The Transcendental Level: Self-Organization on a Grand Scale

Strangely, in the "reductive" scheme of the sciences, those dealing with quantum microphysics have found it necessary to reduce it to include "observation;" that is, perception and therefore, psychology. Hiley traces for us the development of algebra from thought and the aspects of algebra that define the spectral domain (the implicate order) within which mind and matter become enfolded into one another. It is currently a "hot" theoretical issue to explore how some of the processes critical to learning are carried out in the brain at the quantum scale. The papers by Kak, Stapp and Hiley explore for us the contributions to self-organization which occur at this microlevel, contributions which herald the ancient wisdom "As above, so below".

Conclusion

Like the other Appalachian Conferences, this Fourth Conference provided the anticipated simulation and the concomitant challenge to integrate rather than isolate our knowledge in order to better understand principles through which organisms operate in and on their environments. Further, in this set of proceedings, we are reminded that this knowledge is not meant to sit in academe, but to be used to enhance and expand our competence in organizing such organism/environment transactions.

References

Pribram, K.H. (1995) The Enigma of Reinforcement. *Neurobehavioral Plasticity: Learning, Development and Response to Brain Insults.* Proceedings of the Bob Isaacson Symposium in Clearwater, FL. New Jersey: Lawrence Erlbaum Associates, pp. 381-403.

Pribram, K.H. & McGuinness, D. (1975) Arousal, activation, and effort in the control of attention. *Psychological Review, 82*, pp. 116-149.

Pribram, K.H. & McGuinness, D. (1992) Attention and para-attentional processing: Event-related brain potentials as tests of a model. In: D. Friedman & G. Bruder (Eds.), *Annals of the New York Academy of Sciences, 658* (pp. 65-92). New York: New York Academy of Sciences

List of Authors Cited

Abhinavagupta, R., Chap. 21
Abhyankar, K. V., Chap. 21
Abraham, F. D., Chap. 20
Adams, B. J., Chap. 2
Adler, M. J., Foreword
Adolphs, R., Chap. 16
Aggleton, J. P., Chap. 16
Albino, R. C., Chap. 16
Alcock, J., Chap. 1
Aldrich, H., Chap. 20
Alkon, D. L., Chaps. 10, 18
Allan, L. G., Chap. 3
Allison, J., Chap. 2
Alpert, N. M., Chap. 15
Alstrøm, P., Chap. 9
Amann, K., Chap. 16
Amari, S., Chap. 18
Amsel, A., Chap. 3
Anderson, A. K., Chap. 15
Anderson, C., Chap. 11
Anderson, J. A., Chaps. 10, 18
Anderson, M. L., Chap. 10
Andreasen, N. C., Chap. 15
Andrus, R. G., Chap. 16
Andy, O. J., Chap. 1
Apostol, T. M., Chap. 4
Appley, M. H., Chap. 3
Argiolas, A., Chap. 18
Arletti, R., Chap. 18
Arndt, S., Chap. 15
Ashby, W. R., Chap. 20
Ashe, J. H., Chap. 16
Atak, J. R., Chaps. 2, 3
Atkinson, J., Chap. 12
Altman, J., Chap. 13
Aurobindo, S., Chap. 21
Awh, E., Chap. 17
Baddeley, A. D., Chap. 17
Baer, L., Chap. 15
Baerends, G. P., Foreword
Bailey, J. T., Chap. 3
Bair, T. B., Chap. 13
Bajcsy, R., Chap. 7
Bak, P., Chap. 9
Ballesteros, S., Chap. 14
Bally, G., Foreword
Balsam, P. D., Chap. 3
Bank, B., Chap. 10
Banks, M., Chap. 12
Baras, J. S., Chap. 11
Bard, L., Chap. 12
Barkley, R. A., Chap. 2

Barlow. G. W., Chap. 3
Barnard, P., Chap. 17
Barnard, P. J., Chap. 17
Barron, A. R., Chap. 11
Bartlett, W., Chap. 11
Barto, A., Chap. 11
Barto, A. G., Chap. 9
Bauchot, R., Chap. 1
Baudry, M., Chaps. 10, 17
Baum, W. M., Chap. 2
Bayer, S. A., Chap. 13
Baylor, D. A., Chap. 7
Beale, R., Chap. 11
Bechara, A., Chap. 16
Becker, J. T., Chap. 16
Beis, J. S., Chap. 7
Beiser, D., Chap. 11
Bell, J., Chap. 22
Bell, J. S., Chap. 21
Bell, M. A., Chap. 13
Belrdichevsky, D., Chap. 15
Bender, M., Chap. 12
Benelli, A. A., Chap. 18
Bennett, J. G., Foreword
Berman, A. J., Foreword
Berman, R. F., Chap. 16
Bertalanffy, L. von, Foreword
Bertolini, A., Chap. 18
Best, C., Chap. 12
Best, P. J., Chap. 16
Bevins, R. A., Chap. 3
Bhalla, U. S., Chap. 10
Bishop, E., Chap. 12
Blackwell, K. T., Chap. 10
Blanchard, D. C., Chap. 3
Blanchard, R. J., Chap. 3
Blaney, P. H., Chap. 17
Blau, P. M., Chap. 20
Block, G. D., Chap. 21
Blodgett, H. C., Chap. 1
Blume, H., Chap. 17
Boakes, R. A., Chap. 3
Bohm, D., Chaps. 20, 21
Bohr, D., Chaps. 21, 22
Boles, Chap. 15
Bolland, B. L., Chap. 16
Bolles, R. C., Chap. 3
Bonner, J., Chap. 12
Booker, J., Chap. 15
Born, M., Foreword
Bougon, M., Chap. 20
Bowe, C. A., Chap. 3
Bower, G. H., Chap. 17

Bower, J., Chap. 10
Bowers, J., Chap. 15
Bradley, M. M., Chaps. 17, 20
Bradshaw, C. M., Chap. 2
Brakke, K. E., Chap. 1
Bratfisch, O., Chap. 3
Braver, T. S., Chap. 17
Breiter, H. C., Chap. 15
Breland, K., Chaps. 3, 4
Breland, M., Chaps. 3, 4
Brewer, W. F., Chap. 15
Brian, M. V., Chap. 7
Broadbent, D. E., Chap. 17
Brooks, W., Chap. 3
Brown, G. M., Chap. 15
Brown, J. S., Chap. 3
Brown, J. W., Chap. 12
Brown, T. H., Chap. 17
Browne, M. P., Chap. 3
Brunswik, E., Foreword
Bry, M., Chap. 8
Bryan, G. K., Chap. 1
Bryson, A., Chap. 11
Buckner, R., Chap. 14
Buckner, R. L., Chap. 15
Bunsey, M., Chap. 16
Burgos, J. E., Chap. 2
Burton, B., Chap. 16
Busemeyer, J. R., Chap. 3
Busse, T., Chap. 10
Butler, M. M., Chap. 3
Cahill, L., Chap. 16
Caldwell, J. D., Chap. 18
Callahan, T. D., Chap. 2
Campbell, D. T., Foreword
Cao, W., Chap. 7
Carey, S., Chap. 12
Carley, K., Chap. 20
Carlton-Ford, S., Chap. 20
Carpenter, G. A., Chap. 10
Carter, C. S., Chap. 18
Casey, B. J., Chap. 17
Catania, A. C., Chap. 3
Caviness, V., Chap. 13
Chabris, C. F., Chap. 15
Challis, B. H., Chap. 17
Changeux, J., Chap. 17
Changeux, J. P., Chaps. 10, 13
Chapin, J., Chap. 11
Chapman, P. F. Chap. 17
Chapuis, N., Chap. 3
Charnock, D. J., Chap. 3
Cheng, K., Chap. 3